Studies in Computational Intelligence 433

For further volumes:
http://www.springer.com/series/7092

Enrique Alba, Amir Nakib, and Patrick Siarry (Eds.)

Metaheuristics for Dynamic Optimization

Editors
Enrique Alba
E.T.S.I. Informática
University of Málaga
Málaga
Spain

Patrick Siarry
Laboratoire Images, Signaux et Systèmes
Intelligents (LISSI)
Université Paris-Est Créteil
Créteil
France

Amir Nakib
Laboratoire Images, Signaux et Systèmes
Intelligents (LISSI)
Université Paris-Est Créteil
Créteil
France

ISSN 1860-949X e-ISSN 1860-9503
ISBN 978-3-642-44370-1 ISBN 978-3-642-30665-5 (eBook)
DOI 10.1007/978-3-642-30665-5
Springer Heidelberg New York Dordrecht London

Softcover reprint of the hardcover 1st edition 2013

Printed on acid-free paper

Springer is part of Springer Science+Business Media (www.springer.com)

To my mother (E. Alba)
To my family (A. Nakib)
To my family (P. Siarry)

This work has been partially funded by the Spanish Ministry of Science and Innovation and FEDER under contracts TIN2008-06491-C04-01 (the MSTAR project) and TIN2011-28194 (the roadME project), and by the Andalusian Government under contract P07-TIC-03044 (the DIRICOM project).

Preface

This book is an updated effort in summarizing the trending topics and new hot research lines in solving dynamic problems using metaheuristics. An analysis of the present state in solving complex problems quickly draws a clear picture: problems that change in time, having noise and uncertainties in their definition are becoming very important. The tools to face these problems are still to be built, since existing techniques are either slow or inefficient in tracking the many global optima that those problems are presenting to the solver technique.

Thus, this book is devoted to include several of the most important advances in solving dynamic problems. Metaheuristics are the most popular tools to this end, and then we can find in the book how to best use genetic algorithms, particle swarms, ant colonies, immune systems, variable neighborhood search, and many other bio-inspired techniques. Also, neural network solutions are considered in this book. Both, theory and practice have been addressed in the chapters of the book. Mathematical background and methodological tools in solving this new class of problems and applications are included. From the applications point of view, not just academic benchmarks are dealt with, but also real world applications in logistics and bioinformatics are discussed here. The book then covers theory and practice, as well as discrete versus continuous dynamic optimization, in the aim of creating a fresh and comprehensive volume. This book is targeted to either beginners and experienced practitioners in dynamic optimization, since we took care of devising the chapters in a way that a wide audience could profit from its contents. We hope to offer a single source for up-to-date information in dynamic optimization, an inspiring and attractive new research domain that appeared in these last years and is here to stay.

The resulting work is in front of you, a book with 16 chapters covering the topic on various metaheuristics for dynamic optimization.

In the first chapter, Amir Nakib and Patrick Siarry present the different tools and benchmarks developed to evaluate the performances of competing algorithms aimed at solving Dynamic Optimization Problems (DOPs). They first cite various test problems currently exploited in the continuous case, and then focus on the two most used: the moving peaks benchmark, and the generalized approach to construct benchmark problems for dynamic optimization (also called GDBG).

Chapter 2, by Briseida Sarasola and Enrique Alba, is devoted to quantitative performance measures for DOPs. The chapter introduces two new performance tools to overcome difficulties that may appear while reporting results on DOPs. The first one is a measure based on linear regression to evaluate fitness degradation. The second measure is based on the area below the curve defined by some population attribute at each generation (e.g., the best-of-generation fitness)

In Chapter 3, Irene Moser and Raymond Chiong present the Moving Peaks Benchmark (MPB) that was devised to facilitate the comparison between competing approaches for solving DOPs. The chapter reviews all known optimization algorithms that have been tested on the dynamic MPB. The majority of these approaches are nature-inspired. The results of the best-performing solutions based on the MPB are directly compared and discussed.

Chapter 4, by Ignacio G. del Amo and David A. Pelta, is interested in a new technique for comparing multiple algorithms under several factors in DOPs. It allows to compact the results in a visual way, providing an easy way to detect algorithms' behavioral patterns. However, as every form of compression, it implies the loss of part of the information. The pros and cons of this technique are explained, with a special emphasis on some statistical issues that commonly arise when dealing with random-nature algorithms.

In Chapter 5, Philipp Rohlfshagen and Xin Yao point out that the role of representations and variation operators in evolutionary computation is relatively well understood for the case of static optimization problems, but not for the dynamic case. Yet they show that the fitness landscape analysis metaphor can be fruitfully used to review previous work on solving of evolutionary DOPs. This review highlights some of the properties unique to DOPs and paves the way for future research related to these important issues.

In Chapter 6, Kalyanmoy Deb presents two approaches for single as well as multi-objective dynamic optimization. Both methods are discussed and their working principles are illustrated by applying them to different practical optimization problems. Off-line optimization techniques can be exploited too, which strongly depend if the change in the problem is significant or not. Further applications of both proposed approaches remain as important future work in making on-line optimization task a reality in the coming years.

In Chapter 7, Mathys C. du Plessis and Andries P. Engelbrecht are interested in self-adaptive Differential Evolution (DE) for dynamic environments with fluctuating numbers of optima. Despite the fact that evolutionary algorithms often solve static problems successfully, DOPs tend to pose a challenge to evolutionary algorithms, and particularly to DE, due to lack of diversity. However, self-adaptive Dynamic Population DE developed by the authors is proved very efficient in case the number of optima fluctuates over time.

Mardé Helbig and Andries P. Engelbrecht propose in Chapter 8 a dynamic multi-objective optimization algorithm based on Particle Swarm Optimization (PSO). The chapter investigates the effect of various approaches to manage boundary constraint violations on the performance of that new algorithm when solving multi-objective DOPs. Furthermore, the performance of the authors' algorithm is compared against that of three other state-of-the-art dynamic multi-objective optimization algorithms. Ant Colony based algorithms for DOPs are investigated by Guillermo Leguizamon and Enrique Alba in Chapter 9. Ant algorithms use a set of agents which evolve in an environment to construct one solution The authors present a general overview of the most relevant works regarding the application of ant colony based algorithms for DOPs. They also highlight the mechanisms used in different implementations found in literature, and thus show the potential of this kind of algorithms for research in this area.

In Chapter 10, Julien Lepagnot, Amir Nakib, Hamouche Oulhadj and Patrick Siarry focus on elastic registration of brain cine-MRI sequences using a dynamic optimization algorithm. Indeed this registration process consists in optimizing an objective function that can be considered as dynamic. The obtained results are compared to those of several well-known static optimization algorithms. This comparison shows the relevance of using a dynamic optimization algorithm to solve this kind of problems.

An Artificial Immune System for solving constrained DOPs is presented in Chapter 11 by Victoria S. Aragon, Susana C. Esquivel, and Carlos A. Coello Coello. It is an adaptation of an existing algorithm, which was originally designed to solve static constrained problems. The proposed algorithm is validated with eleven dynamic constrained problems which involve the following scenarios: dynamic objective function with static constraints, static objective function with dynamic constraints and dynamic objective function with dynamic constraints.

Chapter 12, by Mostepha R. Khouadjia, Briseida Sarasola, Enrique Alba, El-Ghazali Talbi and Laetitia Jourdan, is devoted to metaheuristics for dynamic vehicle routing. The aim consists in designing the optimal set of routes for a fleet of vehicles in order to serve a given set of customers; routes must be reconfigured dynamically to take into account new customer orders arriving while the working day plan is in progress. A survey on solving methods, such as population-based metaheuristics and trajectory-based metaheuristics, is exposed.

Juan José Pantrigo and Abraham Duarte present in Chapter 13 a low-level hybridization of Scatter Search and Particle Filter to solve the dynamic Travelling Salesman Problem (TSP). To demonstrate the performance of that approach, they conducted experiments using two different benchmarks. Experimental results have shown that the new algorithm outperforms other population based metaheuristics, such as Evolutionary Algorithms or Scatter Search, by reducing the execution time without affecting the quality of the results.

In Chapter 14, Amir Hajjam, Jean-Charles Créput and Abderrafiaa Koukam address the standard dynamic and stochastic Vehicle Routing Problem (VRP). They propose a solving method which manipulates the self-organizing map (SOM) neural network similarly as a local search into a population based memetic algorithm. The goal is to simultaneously minimize the route lengths and the customer waiting time. The experiments show that the new approach outperforms the best competing operations research heuristics available in literature.

Chapter 15, by Pedro C. Pinto, Thomas A. Runkler and Joao M. C. Sousa, deals with the solving of dynamic MAX-SAT problems. The authors propose an ant colony optimization algorithm and a wasp swarm optimization algorithm, which are based on the real life behavior of ants and wasps, respectively. Both algorithms are applied to several sets of static and dynamic MAX-SAT instances and are shown to outperform the greedy hill climbing and simulated annealing algorithms used as benchmarks.

The last chapter of the book (Chapter 16), by Trung Thanh Nguyen and Xin Yao, is devoted to Dynamic Time-linkage optimization Problems (DTPs), which are special DOPs where the current solutions chosen by the solver can influence how the problems might change in the future. Although DTPs are very common in real-world applications, they have received very little attention from the evolutionary optimization community. This chapter attempts to fill this gap by addressing some characteristics that are not fully known about DTPs.

We do hope you will find the volume interesting and thought provoking. Enjoy!

April 2012

Málaga, Spain — Enrique Alba
Paris, France — Amir Nakib
Paris, France — Patrick Siarry

Contents

List of Contributors

Enrique Alba
E.T.S.I. Informática, Campus de Teatinos,
University of Málaga 29071 Málaga, Spain
e-mail: eat@lcc.uma.es

Ignacio G. del Amo
Models of Decision and Optimization Research Group (MODO),
Dept. of Computer Sciences and Artificial Intelligence, University of Granada.
I.C.T. Research Centre (CITIC-UGR), C/ Periodista Rafael Gómez, 2, E-18071,
Granada, Spain
e-mail: ngdelamo@ugr.es, dpelta@decsai.ugr.es

Victoria S. Aragón
Laboratorio de Investigación y Desarrollo en Inteligencia Computacional (LIDIC),
Universidad Nacional de San Luis - Ejército de los Andes 950 (5700)
San Luis, Argentina
e-mail: vsaragon@unsl.edu.ar

Raymond Chiong
Faculty of Higher Education, Swinburne University of Technology,
50 Melba Avenue, Lilydale, Victoria 3140, Australia
e-mail: rchiong@swin.edu.au

Carlos A. Coello Coello
CINVESTAV-IPN (Evolutionary Computation Group) - Computer Science
Department, Av. IPN No. 2508, Col. San Pedro Zacatenco,
México D.F. 07300, México
e-mail: ccoello@cs.cinvestav.mx

Jean-Charles Créput
Laboratoire Systèmes et Transports, U.T.B.M., 90010 Belfort, France
e-mail: jean-charles.creput@utbm.fr

Kalyanmoy Deb
Kanpur Genetic Algorithms Laboratory (KanGAL),
Department of Mechanical Engineering,
Indian Institute of Technology Kanpur, PIN 208016, India
e-mail: deb@iitk.ac.in

Abraham Duarte
Universidad Rey Juan Carlos, c/ Tulipán s/n Móstoles Madrid, Spain
e-mail: abraham.duarte@urjc.es

Andries P. Engelbrecht
Department of Computer Science, School of Information Technology,
University of Pretoria, Pretoria, 0002, South Africa
e-mail: engel@cs.up.ac.za

Susana C. Esquivel
Laboratorio de Investigación y Desarrollo en Inteligencia Computacional (LIDIC),
Universidad Nacional de San Luis - Ejército de los Andes 950 (5700)
San Luis, Argentina
e-mail: vsaragon@unsl.edu.ar

Amir Hajjam
Laboratoire Systèmes et Transports, U.T.B.M., 90010 Belfort, France
e-mail: amir.hajjam@utbm.fr

Mardé Helbig
CSIR Meraka Institute, Scientia, Meiring Naude Road, 0184,
Brummeria, South Africa; and Department of Computer Science,
University of Pretoria, 0002, Pretoria, South Africa
e-mail: mhelbig@csir.co.za

Laetitia Jourdan
INRIA Lille Nord-Europe, Parc scientifique de la Haute-Borne, Bâtiment A,
40 Avenue Halley, Park Plaza, 59650 Villeneuve d'Ascq, France
e-mail: laetitia.jourdan@inria.fr

Mostepha R. Khouadjia
INRIA Lille Nord-Europe, Parc scientifique de la Haute-Borne, Bâtiment A,
40 Avenue Halley, Park Plaza, 59650 Villeneuve d'Ascq, France
e-mail: mostepha-redouane.khouadjia@inria.fr

Abderrafiãa Koukam
Laboratoire Systèmes et Transports, U.T.B.M., 90010 Belfort, France
e-mail: abder.koukam@utbm.fr

Guillermo Leguizamón
Universidad Nacional de San Luis, Av. Ejército de Los Andes 950 (5700),
San Luis, Argentina
e-mail: legui@unsl.edu.ar

Julien Lepagnot
Université Paris-Est Créteil, Laboratoire Images,
Signaux et Systèmes Intelligents (LISSI, EA 3956)
61, avenue du Général de Gaulle 94010 Créteil (France)
e-mail: julien.lepagnot@u-pec.fr

Irene Moser
Faculty of Information & Communication Technologies,
Swinburne University of Technology, Victoria 3122, Australia
e-mail: imoser@swin.edu.au

Amir Nakib
Université Paris-Est Créteil, Laboratoire Images,
Signaux et Systèmes Intelligents (LISSI, EA 3956)
61, avenue du Général de Gaulle 94010 Créteil, France
e-mail: nakib@u-pec.fr

Trung Thanh Nguyen
School of Engineering, Technology and Maritime Operations,
Liverpool John Moores University, United Kingdom
e-mail: T.T.Nguyen@ljmu.ac.uk

Hamouche Oulhadj
Université Paris-Est Créteil, Laboratoire Images,
Signaux et Systèmes Intelligents (LISSI, EA 3956)
61, avenue du Général de Gaulle 94010 Créteil, France
e-mail: oulhadj@u-pec.fr

Juan José Pantrigo
Universidad Rey Juan Carlos, c/ Tulipán s/n Móstoles Madrid, Spain
e-mail: juanjose.pantrigo@urjc.es

David A. Pelta
Models of Decision and Optimization Research Group (MODO),
Dept. of Computer Sciences and Artificial Intelligence, University of Granada.
I.C.T. Research Centre (CITIC-UGR), C/ Periodista Rafael Gómez, 2, E-18071,
Granada, Spain
e-mail: ngdelamo@ugr.es, dpelta@decsai.ugr.es

Pedro C. Pinto
Bayern Chemie GmbH, MBDA Deutschland, Department T3R,
Liebigstr. 15-17 D-84544 Aschau am Inn - Germany
e-mail: pedro.caldas-pinto@mbda-systems.de

Mathys C. du Plessis
Department of Computing, Sciences, PO Box 77000,
Nelson Mandela Metropolitan University, Port Elizabeth, 6031, South Africa
e-mail: mc.duplessis@nmmu.ac.za

Philipp Rohlfshagen
School of Computer Science and Electrical Engineering, University of Essex, Colchester CO4 3SQ, United Kingdom
e-mail: prohlf@essex.ac.uk

Thomas A. Runkler
Siemens AG, Corporate Technology, Intelligent Systems and Control, CT T IAT ISC, Otto-Hahn-Ring 6, 81730 Munich, Germany
e-mail: thomas.runkler@siemens.com

Briseida Sarasola
E.T.S.I. Informática, Campus de Teatinos, University of Málaga 29071 Málaga, Spain
e-mail: briseida@lcc.uma.es

Patrick Siarry
Université Paris-Est Créteil, Laboratoire Images, Signaux et Systèmes Intelligents (LISSI, EA 3956)
61, avenue du Général de Gaulle 94010 Créteil, France
e-mail: siarry@u-pec.fr

João M. C. Sousa
Technical University of Lisbon,Instituto Superior Técnico, Dep. of Mechanical Engineering, IDMEC-LAETA, Avenida Rovisco Pais, 1049-001, Lisbon, Portugal
e-mail: jmsousa@ist.utl.pt

El-Ghazali Talbi
INRIA Lille Nord-Europe, Parc scientifique de la Haute-Borne, Bâtiment A, 40 Avenue Halley, Park Plaza, 59650 Villeneuve d'Ascq, France
e-mail: talbi@inria.fr

Xin Yao
School of Computer Science, University of Birmingham, Birmingham B15 2TT, United Kingdom,
e-mail: xin@cs.bham.ac.uk

List of Tables

List of Figures

Chapter 1
Performance Analysis of Dynamic Optimization Algorithms

Amir Nakib and Patrick Siarry

Abstract. In recent years dynamic optimization problems have attracted a growing interest from the community of stochastic optimization researchers with several approaches developed to address these problems. The goal of this chapter is to present the different tools and benchmarks to evaluate the performances of the proposed algorithms. Indeed, testing and comparing the performances of a new algorithm to the different competing algorithms is an important and hard step in the development process. The existence of benchmarks facilitates this step, however, the success of these benchmarks is conditioned by their use by the community. In this chapter, we cite many tested problems (we focused only on the continuous case), and we only present the most used: the moving peaks benchmark , and the last proposed: the generalized approach to construct benchmark problems for dynamic optimization (also called benchmark GDBG).

1.1 Introduction

The dynamic optimization problems (DOPs) can be met in many real-world cases. A dynamic optimization problem can be formulated as follows:

$$\begin{aligned} &\min\ f(\mathbf{x},t) \\ &\text{s.t. } \ h_j(\mathbf{x},t) = 0 \text{ for } j = 1,2,...,u \\ &\qquad g_k(\mathbf{x},t) \leq 0 \text{ for } k = 1,2,...,v, \end{aligned} \tag{1.1}$$

where $f(\mathbf{x},t)$ is the objective function of a minimization problem, $h_j(\mathbf{x},t)$ denotes the jth equality constraint, and $g_k(\mathbf{x},t)$ denotes the kth inequality constraint.

Amir Nakib · Patrick Siarry
Université Paris Est Créteil, Laboratoire Images,
Signaux et Systèmes Intelligents (LISSI, EA 3956)
61, avenue du Général de Gaulle 94010 Créteil, France
e-mail: nakib@u-pec.fr

E. Alba et al. (Eds.): Metaheuristics for Dynamic Optimization, SCI 433, pp. 1–16.
springerlink.com

The function f is deterministic at any point in time, but is dependent on time t. Consequently the position of the optimum changes over the time. Thus, the algorithm dealing with DOPs should be able to continuously track this optimum rather than requiring a repeated restart of the optimization process. The algorithms dedicated to solve this kind of problem take into account, during the optimization process, the information from previous environments to speed up optimization after a change.

The naive approach after a change of the environment consists in formulating each change as the arrival of a new optimization problem that has to be solved from scratch. Indeed, if there is enough time to solve the problem, this is an efficient way. However, the time for restarting the optimization process is, in most cases, short, and it is based on the assumption of the identification of changing events, which is not always the case.

To increase the convergence speed after a change, one direction consists in finding the best way to use the previous known information about the search space to speed up the search after a detected change. For example, if the new optimum is *close* to the old one, one can reduce the search space to the neighborhood of the previous optimum. Of course, taking into account the information from the past depends on the nature of the change. If the change is drastic, and the new problem has little similarity to the previous problem, restart may be the only viable option, and use of information from the past would be misguiding rather than helping the search.

In most real-world problems, however, the changes are smooth and using knowledge from the past would be a good way to speed up the optimization. The difficult question is: *what information should be kept, and how it is used to accelerate search after the environment has changed?*

However, the information transfer does not guarantee that the optimization algorithm is flexible enough to follow the optimum over the changes. Most Dynamic Optimization Algorithms (DOAs) converge during the run, at least when the environment has been static for some time, without losing their adaptability. Thus, besides transferring knowledge, a successful DOA for dynamic optimization problems has to maintain adaptability.

In order to evaluate the performances of the dynamic optimization algorithms, many researchers have applied several dynamic test problems to them. The most known are: the *moving peaks* benchmark (MPB) proposed by Branke [3], the DF1 generator introduced by Morrison and De Jong [17], the single- and multiobjective dynamic test problem generator by dynamically combining different objective functions of existing stationary multiobjective benchmark problems, suggested by Jin and Sendhoff [6], Yang and Yao's exclusive-or (XOR) operator [22][24][25], Kang's dynamic traveling salesman problem (DTSP) [10], and dynamic multi knapsack problem (DKP), etc.

The goal of this chapter is to present the different methods to evaluate the performances of a DOA. In the following section, the performance analysis of optimization algorithms methods is presented. In section 3, the Moving Peak benchmark is summarized. In section 4, we present the generalized benchmark. In section 5, the dynamic multiobjective optimization benchmark is presented. The conclusion is presented in section 6.

1.2 Performance Analysis Tools of Optimization Algorithms

In this section, we present some metrics and tools that can be used to analyze the performance of an optimization algorithm. Here, we do not focus only on dynamic optimization but we present a set of tools to evaluate any optimization algorithm.

1.2.1 Fitness Value

The fitness of a solution is a numerical value that provides an indication on how well the solution meets the objective(s) of the problem. The concept of fitness is central to nature inspired algorithms (metaheuristics). The concept of fitness is applied in most metaheuristics. In the case of benchmark problems such as the Sphere function, the mathematic formulation and the location of the global optimum are known. In such cases the fitness function corresponds to the distance to the global optimum. The best fitness is known and is often zero. A solution that is closer to the global optimum has a smaller error and a best fitness than a solution farther away.

Another kind of problem that can be met is when the global optimum is unknown. It may not even be known whether or not a global optimum exists, and, if it does, whether there are multiple global optima. Most examples of this type of problem are NP-hard and the fitness score is a function of the system output(s). Furthermore, the fitness score may be a weighted function of output parameters. An example is a pipe renewal problem in drinking water networks, where the numbers and types of pipes at hand, the provided pressure at demand points, and the cost may all be weighted and incorporated into fitness values [4].

It is known that the three spaces of adaptation of an algorithm are: decision variable space, system output space, and fitness space. System output space is the space defined by the dynamic range(s) of the output variable(s). The fitness space is the space used to define the *goodness* of the solutions in the output space. It is recommanded to scale the fitness to values between 0 and 1, where 0 or 1 is the optimal value, depending on the optimization problem (minimization or maximization). Thus, system output and fitness generally do not coincide.

Furthermore, the numerical value of the fitness rarely has a meaning. In most cases, we only use fitness values to rank solutions. A proposed solution with a fitness value of 0.950 is rarely exactly twice as good as a solution with a fitness value of 0.760. We simply have a rank-ordered list of how good a solution is relatively to other solutions.

It is a common practice to vary parameters of the algorithm such as population size and attempt to see what value produces a better cost. For example, run the metaheuristic fifty times with one population size and fifty times with another population size. Due to the stochastic nature of the algorithm, we may very well get a different fitness value each time.

How do we determine which solution is better? If all of the fitness values for one population size are better than those for another population size, the situation is clear: use the solution that consistently produces the best fitness values. However, the situation is not always so simple. Especially, when we are fine-tuning parameters to maximize system performance, we can meet situations that are difficult to analyze and interpret.

1.2.2 Computational Analysis

The computational analysis of an optimization algorithm can be provided using a theoretical analysis or an empirical one. In the first case, the worst-case complexity of the algorithm is computed. Usually, the asymptotic complexity is not enough to represent the computational performances of metaheuristics. If the probability distribution of the input instances is available, then average-case complexity is recommended and it is more practical.

To perform empirical analysis, different measures of the computation time of the metaheuristic used to solve a given instance are presented. The computation time corresponds to the CPU time or wall clock time, with or without input/output and preprocessing/postprocessing time.

The main drawback of computation time measure is its dependency on the used hardware (e.g., processor, GPUs, memories, ...), operating systems, language, and compilers on which the metaheuristic is executed. Some indicators that are independent of the computer system may also be used, such as the number of objective function evaluations. It is an acceptable measure for time-intensive and constant objective functions. Using this metric may be problematic for problems where the evaluation cost is low compared to the rest of the metaheuristics or is not time constant since it depends on the solution evaluated and time. This appears in some applications with variable length representations (genetic programming, robotics, etc.) and dynamic optimization problems.

Different stopping criteria may be used: time to obtain a given target solution, time to obtain a solution within a given percentage from a given solution (e.g., global optimal, lower bound, best known), and number of iterations [21].

1.2.3 Classical Metrics

Two classical metrics for the effectiveness of metaheuristics were described by De Jong (1975). These metrics, however, are appropriate for only the algorithms that evolve a population of solutions. De Jong named these metrics off-line performance and on-line performance.

When an optimization algorithm is being run off-line, many system configurations can be evaluated (the fitness calculated) and the best configuration selected.

For on-line work, however, configurations must be evaluated in real time, therefore the usual goal is to develop an acceptable solution as early as possible.

The on-line performance, which measures the ongoing performance of a system configuration, is defined in Equation 1.2, where $\bar{f}_c(g)$ is the average population fitness for a system configuration c during generation g and G is the number (index) of the latest generation:

$$S_c^{\text{Online}} = \frac{1}{G}\sum_{g=1}^{G} \bar{f}_c(g), \tag{1.2}$$

The off-line performance measures convergence of the algorithm and is defined in Equation 1.3, where $f^*_{c(g)}$ is the best fitness of any population member in generation g for system configuration c. Off-line (convergence) performance is thus the average of the best fitness values from each generation up to the present.

$$S_c^{\text{Offline}} = \frac{1}{G}\sum_{g=1}^{G} \bar{f}^*_{c(g)}, \tag{1.3}$$

1.2.4 Sensitivity Analysis

The definition of the sensitivity of the optimization algorithms, also called robustness, is not standardized inside the community. Different alternative definitions were proposed for sensitivity. In general, it corresponds to insensitivity against small deviations in the input instances (data) or the parameters of the algorithm. The lower the variability of the obtained solutions, the better the sensitivity. Sensitivity analysis that is related to the applications of optimization algorithms sometimes focuses on the problem and/or solution domain.

The parameters of the metaheuristics play an important role in their search capacity. Indeed, the sensitivity of a metaheuristic with respect to its parameters is critical to its performance and, therefore, its successful applications. Using this approach, we consider the parameters of an optimization algorithm as the input values to the sensitivity analysis, and its performance values as the output values. Many indicators can be used to evaluate the sensitivity; here we propose:

- *Parameter sensitivity*: it measures the effect of a given parameter on a given output when all other parameters are constrained to be constant.
- *Performance sensitivity*: it takes into account the fact that the effect of a given parameter on a given output differs with varying the data set values. This indicator measures the average effect of a given parameter on a given output over a set of data.

For metaheuristics algorithms, the output values can include parameters such as fitness value, convergence rate, and the maximum generation required to reach a good enough solution. The input values may be different for different algorithms. For example, for genetic algorithms, the input values can be mutation rate, crossover rate, population size, and so on. For particle swarm optimization algorithms, the input values can be inertia weight w, cognitive and social coefficients c_1 and c_2, and so on.

1.2.5 Statistical Analysis

Different statistical tests may be carried out to analyze and compare the metaheuristics. The statistical tests are performed to estimate the confidence of the results to be scientifically valid. The selection of a given statistical hypothesis testing tool is performed according to the characteristics of the data.

Under some assumptions (normal distributions), the most widely used test is the paired *t-test*. Otherwise, a nonparametric analysis may be realized, such as the Wilcoxon test and the permutation test. For a comparison of more than two algorithms, ANOVA models are well-established techniques to check the confidence of the results. Multivariate ANOVA models allow simultaneous analysis of various performance measures (e.g., both the quality of solutions and the computation time). Kolmogorov-Smirnov test can be performed to check whether the obtained results follow a normal (Gaussian) distribution. Moreover, the Levene test can be used to test the homogeneity of the variances for each pair of samples. The Mann-Whitney statistical test can be used to compare two optimization methods. According to a p-value and a metric under consideration, this statistical test reveals if the sample of approximation sets obtained by a search method S_1 is significantly better than the sample of approximation sets obtained by a search method S_2, or if there is no significant difference between both optimization methods.

These different statistical analysis procedures must be adapted for nondeterministic (or stochastic) algorithms. Indeed, most metaheuristics belong to this class of algorithms. Many trials (100 runs is the most used number) must be carried out to derive significant statistical results. From this set of trials, many measures may be computed: mean, median, minimum, maximum, standard deviation, the success rate that the reference solution (e.g., global optimum, best known, given goal) has been attained, and so on.

Below are some considerations about the use of statistical tools for analysing the performances of metaheuristics or stochastic-based optimization algorithms:

- *t-tests* require that certain assumptions be made regarding the format of the data. The one sample t-test requires that the data have an approximately normal distribution, whereas the paired t-test requires that the distribution of the differences is approximately normal. The unpaired t-test relies on the assumption that the data from the two samples are both normally distributed, and has the additional requirement that the standard deviations (SDs) from the two samples are approximately equal.

  Formal statistical tests are performed to examine whether a set of data are normal or whether two SDs (or, equivalently, two variances) are equal, although results from these should always be interpreted in the context of the sample size and associated statistical power in the usual way. However, the t-test is known to be robust to modest departures from these assumptions, and so a more informal investigation of the data may often be sufficient in practice.

If assumptions of normality are violated, then appropriate transformation of the data may be used before performing any calculations. Similarly, transformations may also be useful if the SDs are very different in the unpaired case. Finally, these methods are restricted to the case where comparison has to be made between one or two groups. This is probably the most common situation in practice but it is by no means uncommon to want to explore differences through three or more methods. This requires an alternative approach that is known as analysis of variance (ANOVA).

- *The nonparametric tests* require very few or very limited assumptions to be made about the format of the data, and can therefore be used in situations where classical methods, such as t-tests, may be inappropriate. They can be useful for dealing with unexpected, outlying observations that might be problematic with a parametric approach. Moreover, these methods are intuitive and are simple to carry out by hand, for small samples at least. Indeed, nonparametric methods are often useful in the analysis of ordered categorical data in which assignation of scores to individual categories may be inappropriate. In contrast, parametric methods require scores to be assigned to each category, with the implicit assumption that the effect of moving from one category to the next is fixed.

  However, nonparametric methods may lack power as compared with more traditional approaches. This is of particular concern if the sample size is small or if the assumptions for the corresponding parametric method (e.g. normality of the data) hold. Moreover, these methods are geared toward hypothesis testing rather than estimation of effects. It is often possible to obtain nonparametric estimates and associated confidence intervals, but this is not generally simple. In many cases an adjustment to the statistic test may be necessary.
- The Kruskal-Wallis, Jonckheere-Terpstra, and Friedman tests can be used to test for differences between more than two groups or treatments when the assumptions for analysis of variance are not held.
- *The P-value* is the probability that an observed effect is simply due to chance; it therefore provides a measure of the strength of an association. Moreover, it does not provide any measure of the size of an effect and cannot be used in isolation to inform about the best optimization algorithm. Indeed, P-values are affected both by the magnitude of the effect and by the size of the study from which they are derived, and should therefore be interpreted with caution. In particular, a large P-value does not always indicate that there is no difference and, similarly, a small P-value does not necessarily signify a high difference. The subdivision P-values into *significant* and *non-significant* are poor statistical practice and should be avoided. Finally, exact P-values should always be presented, along with estimates of effect and associated confidence intervals.

The success rate: The success rate (*SR*) represents the ratio between the number of successful runs and the number of trials:

Table 1.1 MPB parameters in scenario 2.

Parameter	Scenario 2
Number of peaks N_p	10
Dimension d	5
Peak heights	$[30,70]$
Peak widths	$[1,12]$
Change cycle α	5000
Change severity s	1
Height severity	7
Width severity	1
Correlation coefficient λ	0
Number of changes N_c	100

$$SR = \frac{\text{NbSuc}}{\text{NbR}}, \tag{1.4}$$

where *NbSuc* is the number of successful runs and *NbR* is the total number of runs.

The performance rate: The performance rate (PR) takes into account the computational effort to find the solution by considering the number of objective function evaluations:

$$PR = \frac{\text{NbSuc}}{\text{NbR} \times \text{Nbfeval}}, \tag{1.5}$$

where $Nbfeval$ is the total number of evaluations of the objective function.

1.3 The Moving Peaks Benchmark

The most commonly used benchmark for continuous dynamic optimization is the Moving Peaks Benchmark (MPB) [3].

MPB is a maximization problem that consists of a number of peaks that randomly vary their shape, position, and height upon time. At any time, one of the local optima can become the new global optimum. MPB generates DOPs consisting of a set of peaks that periodically move in a random direction, by a fixed amount s (the change severity). The movements are autocorrelated by a coefficient λ, $0 \leq \lambda \leq 1$, where 0 means uncorrelated and 1 means highly autocorrelated. The peaks change position every α evaluations, and α is called *time span*. The fitness function used for the landscape of MPB is formulated in Equation 1.6:

$$f(\mathbf{x},t) = max_{i=1,\ldots,N_p}\left(H_i(t) - W_i(t)\sqrt{\Sigma_{j=1}^{d}\left(x_j - X_{ij}(t)\right)^2}\right), \tag{1.6}$$

where N_p is the number of peaks, d is the number of dimensions, and $H_i(t)$, $W_i(t)$ and $\mathbf{X}_i(t)$ are the height, the width, and the position of the ith peak at the time t, respectively.

In order to evaluate the performance, the *offline error* is used. The offline error (oe) is defined in Equation 1.7:

$$oe = \frac{1}{N_c} \sum_{j=1}^{N_c} \left(\frac{1}{N_e(j)} \sum_{i=1}^{N_e(j)} (f_j^* - f_{ji}^*) \right) \tag{1.7}$$

where N_c is the total number of fitness landscape changes within a single experiment, $N_e(j)$ is the number of evaluations performed for the jth state of the landscape, f_j^* is the value of the optimal solution for the jth landscape, and f_{ji}^* is the current best fitness value found for the jth landscape. We can see that this measure has some weaknesses: it is sensitive to the overall height of the landscape, and to the number of peaks. It is important for an algorithm to find the global optimum quickly, thus minimizing the offline error. Hence, the most successful strategy is a multi-solution approach that keeps track of every local peak [19]. In [3], three sets of parameters, called scenarios, were proposed. It appears that the most commonly used set of parameters for MPB is scenario 2 (see Table 10.3).

Figure 10.1 illustrates an MPB landscape before and after a change (after one time span). More details about this benchmark will be given in the dedicated chapter.

Table 1.2 Comparison with competing algorithms on MPB using $s = 1, 2, ..., 6$.

Algorithm	Offline error					
	$s=1$	$s=2$	$s=3$	$s=4$	$s=5$	$s=6$
Moser and Chiong, 2010 [18]	0.25 ± 0.08	0.47 ± 0.12	0.49 ± 0.12	0.53 ± 0.13	0.65 ± 0.19	0.77 ± 0.24
Lepagnot *et al.* [9]	0.35 ± 0.06	0.60 ± 0.08	0.91 ± 0.10	1.23 ± 0.10	1.62 ± 0.13	2.00 ± 0.20
Lepagnot *et al.*, 2009 [7, 8]	0.59 ± 0.10	0.87 ± 0.12	1.18 ± 0.13	1.49 ± 0.13	1.86 ± 0.17	2.32 ± 0.18
Moser and Hendtlass, 2007 [18, 19]	0.66 ± 0.20	0.86 ± 0.21	0.94 ± 0.22	0.97 ± 0.21	1.05 ± 0.21	1.09 ± 0.22
Yang and Li, 2010 [23]	1.06 ± 0.24	1.17 ± 0.22	1.36 ± 0.28	1.38 ± 0.29	1.58 ± 0.32	1.53 ± 0.29
Liu *et al.*, 2010 [14]	1.31 ± 0.06	1.98 ± 0.06	2.21 ± 0.06	2.61 ± 0.11	3.20 ± 0.13	3.93 ± 0.14
Lung and Dumitrescu, 2007 [15]	1.38 ± 0.02	1.78 ± 0.02	2.03 ± 0.03	2.23 ± 0.05	2.52 ± 0.06	2.74 ± 0.10
Bird and Li, 2007 [1]	1.50 ± 0.08	1.87 ± 0.05	2.40 ± 0.08	2.90 ± 0.08	3.25 ± 0.09	3.86 ± 0.11
Lung and Dumitrescu, 2008 [16]	1.53 ± 0.01	1.57 ± 0.01	1.67 ± 0.01	1.72 ± 0.03	1.78 ± 0.06	1.79 ± 0.03
Blackwell and Branke, 2006 [2]	1.75 ± 0.06	2.40 ± 0.06	3.00 ± 0.06	3.59 ± 0.10	4.24 ± 0.10	4.79 ± 0.10
Li *et al.*, 2006 [13]	1.93 ± 0.08	2.25 ± 0.09	2.74 ± 0.09	3.05 ± 0.10	3.24 ± 0.11	4.95 ± 0.13
Parrott and Li, 2006 [20]	2.51 ± 0.09	3.78 ± 0.09	4.96 ± 0.12	5.56 ± 0.13	6.76 ± 0.15	7.68 ± 0.16

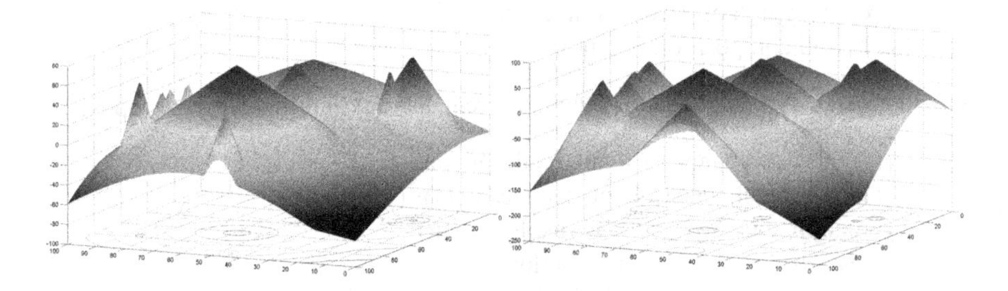

Fig. 1.1 An MPB landscape before and after a change.

Table 1.3 Static functions used to generate the GDBG problems.

Name	Function	Range
Sphere	$f(\mathbf{x}) = \Sigma_{i=1}^{d} x_i^2$	$[-100, 100]^d$
Rastrigin	$f(\mathbf{x}) = \Sigma_{i=1}^{d} (x_i^2 - 10\cos(2\pi x_i) + 10)$	$[-5, 5]^d$
Weierstrass	$f(\mathbf{x}) = \sum_{i=1}^{d} (\sum_{k=0}^{k_{max}} [a^k \cos(2\pi b^k (x_i + 0.5))]) - d \sum_{k=0}^{k_{max}} [a^k \cos(\pi b^k)]$ $a = 0.5, b = 3, k_{max} = 20$	$[-0.5, 0.5]^d$
Griewank	$f(\mathbf{x}) = \frac{1}{4000} \Sigma_{i=1}^{d} (x_i)^2 - \Pi_{i=1}^{d} \cos(\frac{x_i}{\sqrt{i}}) + 1$	$[-100, 100]^d$
Ackley	$f(\mathbf{x}) = -20\exp(-0.2\sqrt{\frac{1}{d}\sum_{i=1}^{d} x_i^2}) - \exp(\frac{1}{d}\sum_{i=1}^{d}\cos(2\pi x_i)) + 20 + \exp(1)$	$[-32, 32]^d$

Table 1.2 summarizes the last published scores of the dynamic optimization algorithms tested on MPB benchmark. Here, we will not comment on these results because it is out of the scope of this chapter.

1.4 The Generalized Dynamic Benchmark Generator

The Generalized Dynamic Benchmark Generator (GDBG) is the second benchmark described in this chapter, it was introduced in [11, 12]. It was provided for the CEC'2009 Special Session on Evolutionary Computation in Dynamic and Uncertain Environments. The functions used to create this benchmark are depicted in Table 10.2. These functions were rotated, composed, and combined to form six problems with different degrees of difficulty:

F_1: rotation peak function (with 10 and 50 peaks)
F_2: composition of Sphere's function
F_3: composition of Rastrigin's function
F_4: composition of Griewank's function
F_5: composition of Ackley's function
F_6: hybrid composition function

A total of seven dynamic scenarios with different degrees of difficulty was proposed:

T_1: small step change (a small displacement)
T_2: large step change (a large displacement)
T_3: random change (Gaussian displacement)
T_4: chaotic change (logistic function)
T_5: recurrent change (a periodic displacement)
T_6: recurrent with noise (the same as above, but the optimum never returns exactly to the same point)
T_7: changing the dimension of the problem

Table 1.4 GDBG parameters used during the CEC'2009 competition.

Parameter	Value
Dimension d (fixed)	10
Dimension d (changed)	$[5, 15]$
Change cycle α	$10000 \times d$
Number of changes N_c	60

The basic parameters of the benchmark are given in Table 1.4.

There are 49 test cases that correspond to the combinations of the six problems with the seven change scenarios (indeed, function F_1 is used twice, with 10 and 50 peaks, respectively). For each of them, the average best (Equation 1.8), average mean (Equation 1.9), average worst (Equation 1.10) values, and the standard deviation (Equation 1.11) of the absolute error are recorded:

$$Avg_{best} = \sum_{i=1}^{runs} \min_{j=1}^{N_c} \frac{E_{ij}}{runs} \tag{1.8}$$

$$Avg_{mean} = \sum_{i=1}^{runs} \sum_{j=1}^{N_c} \frac{E_{ij}}{runs \times N_c} \tag{1.9}$$

$$Avg_{worst} = \sum_{i=1}^{runs} \max_{j=1}^{N_c} \frac{E_{ij}}{runs} \tag{1.10}$$

$$STD = \sqrt{\frac{\sum_{i=1}^{runs} \sum_{j=1}^{N_c} \left(E_{ij} - Avg_{mean}\right)^2}{runs \times N_c}}, \tag{1.11}$$

where $E_{ij} = \left| f_j^* - \tilde{f}_{ji}^* \right|$, f_j^* is the value of the global optimum for the jth landscape, $\tilde{f}_{ji}^*$ is the value of the best solution found during the ith run of the tested algorithm, for the jth landscape, and *runs* is the number of runs of the tested algorithm on the benchmark, equal to 20 in our experiments.

The convergence graphs, showing the relative error $r_i(t)$ of the run with median performance for each problem, are also computed. For the maximization problem F_1, the formula used for $r_i(t)$ is defined in Equation 1.12, and for the minimization problems F_2 to F_6, it is defined in Equation 1.13:

$$r_i(t) = \frac{f_i(t)}{f_i^*(t)} \tag{1.12}$$

$$r_i(t) = \frac{f_i^*(t)}{f_i(t)}, \tag{1.13}$$

where $f_i(t)$ is the value of the best found solution at time t since the last occurrence of a change, during the ith run of the tested algorithm, and $f_i^*(t)$ is the value of the global optimum at time t.

A mark is calculated for each test case using the formula given in Equation 1.14:

$$mark_{pct} = mark_{max} \times \sum_{i=1}^{runs} \sum_{j=1}^{N_c} \frac{r_{ij}}{runs \times N_c}, \tag{1.14}$$

The sum of all marks $mark_{pct}$ gives a score, denoted by op, that corresponds to the overall performance of the tested algorithm. The maximum value of this score is 100. In Equation 1.14, r_{ij} is calculated using the formula defined in Equation 1.15, and $mark_{max}$ is a coefficient that defines the percentage of the mark of each test case in the final score. It is also the maximal mark that can be obtained by the tested algorithm on each test case.

$$r_{ij} = \frac{r_i(t_j + \alpha)}{1 + \sum_{s=1}^{\frac{\alpha}{s_f}} \frac{1 - r_i(t_j + s_f \times s)}{\frac{\alpha}{s_f}}}, \tag{1.15}$$

where t_j is the time at which the jth change occurred, s_f is the sampling frequency, here equal to 100, and α is the change cycle (as for MPB, it corresponds to the number of evaluations that makes a time span).

1.4.1 Illustration of Some Test Functions in GDBG

F_3: Composition of Rastrigin's function

Basic functions: $f_1 - f_{10}$ =Rastrigin's function

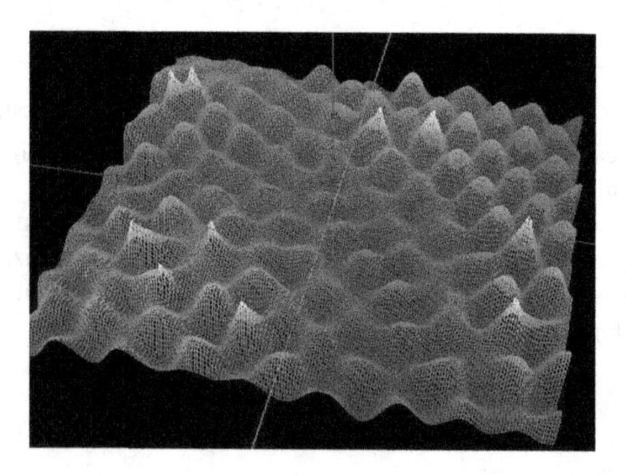

Fig. 1.2 3-D map for 2-D function of F_3.

Properties

Multi-modal
Scalable

Rotated
A huge number of local optima
$x \in [-5,5]^n$, Global optimum $x^*(t) = \mathbf{O_i}, F(x^*(t)) = H_i(t), H_i(t) = min_j^m H_j$

F_5:Composition of Ackley's function

Basic functions: $f_1 - f_{10}$ =Ackley's function

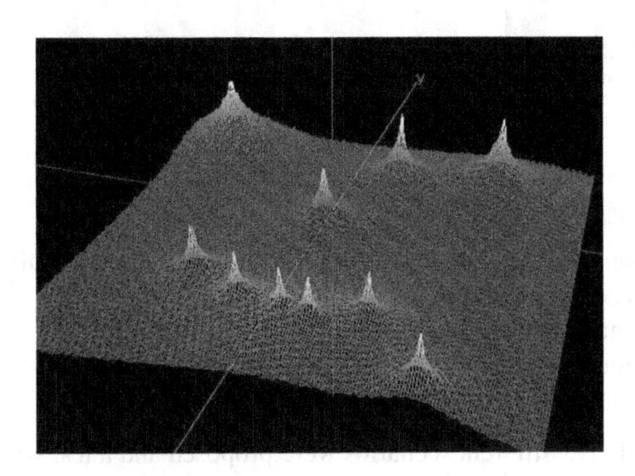

Fig. 1.3 3-D map for 2-D function of F_5.

Properties

Multi-modal
Scalable
Rotated
A huge number of local optima
$x \in [-5,5]^n$, Global optimum $x^*(t) = \mathbf{O_i}, F(x^*(t)) = H_i(t), H_i(t) = min_j^m H_j$

Figure 1.4 presents the scores of recent dynamic optimization algorithm on GDBG benchmark.

1.5 Dynamic Multiobjective Optimization Benchmark

In this section, the dynamic multiobjective optimization benchmark (DMOB) proposed by Farina *et al.* [5] is presented. Indeed, unlike in the single-objective dynamic optimization problems, where the ordering criterion in decision space is trivial, here, we are dealing with two distinct yet related spaces where an ordering criterion has to be considered: decision variable space and objective space. Such an increased complexity holds true for static problems and even more for dynamic problems, where there are four possible ways a problem can demonstrate a time-varying change:

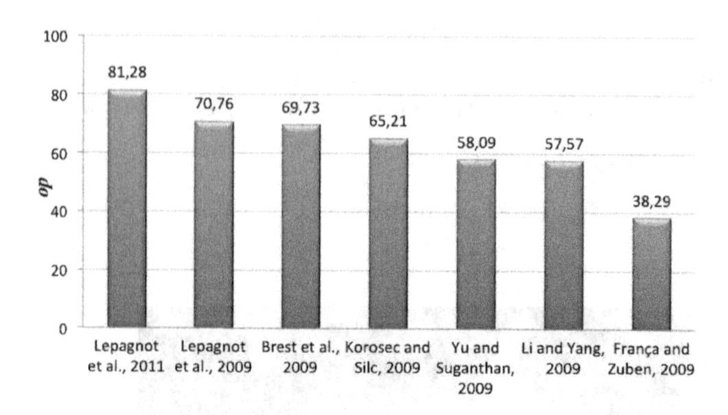

Fig. 1.4 Comparison with competing algorithms on GDBG.

1. The POS (optimal decision variables) changes, whereas the POF (optimal objective values) does not change.
2. Both POS and POF change.
3. POS does not change but POF varies.
4. The problem changes but POF and POS do not vary.

In this bechmark five different scenarios were proposed and a test suite for continuous problems was also proposed (FDA1 to FDA5). Based on the proposed scenarios, the five test problems involve ZDT and DTLZ test problems. However, these test problems can be used as a representative set of test problems in a study. The authors do not provide more functions but provide the procedure to build other more interesting problems.

1.6 Conclusions

In this chapter a suggested tool has been studied to analyse the performance of stochastic-based optimization algorithms. Moreover, a description of two continuous dynamic optimization benchmarks was presented. We also reported a review of scores of different algorithms on these benchmarks. We ended this chapter by a benchmark for dynamic multiobjective algorithms. We hope this chapter will help the researchers to speed up their development of new algorithms.

References

[1] Bird, S., Li, X.: Using regression to improve local convergence. In: Proc. Congr. Evol. Comput., Singapore, pp. 592–599. IEEE (2007)

[2] Blackwell, T., Branke, J.: Multi-swarms, exclusion and anti-convergence in dynamic environments. IEEE Transactions on Evolutionary Computation 10(4), 459–472 (2006)

[3] Branke, J.: The Moving Peaks Benchmark website (1999), http://www.aifb.unikarlsruhe.de/~jbr/MovPeaks

[4] Eberhart, R.C., Shi, Y.: Computational intelligence: concepts to implementation. Elsevier (2007)
[5] Farina, M., Deb, K., Amato, P.: Dynamic multiobjective optimization problems: test cases, approximations, and applications. IEEE Transactions on Evolutionary Computation 8(5), 425–442 (2004)
[6] Jin, Y., Sendhoff, B.: Constructing Dynamic Optimization Test Problems Using the Multi-objective Optimization Concept. In: Raidl, G.R., Cagnoni, S., Branke, J., Corne, D.W., Drechsler, R., Jin, Y., Johnson, C.G., Machado, P., Marchiori, E., Rothlauf, F., Smith, G.D., Squillero, G. (eds.) EvoWorkshops 2004. LNCS, vol. 3005, pp. 525–536. Springer, Heidelberg (2004)
[7] Lepagnot, J., et al.: Performance analysis of MADO dynamic optimization algorithm. In: Proc. IEEE Adaptive Computing in Design and Manufacturing. Int. Conf. on Intelligent Systems Design and Applications, Pisa, pp. 37–42. IEEE (2009)
[8] Lepagnot, J., Nakib, A., Oulhadj, H., Siarry, P.: A new multiagent algorithm for dynamic continuous optimization. International Journal of Applied Metaheuristic Computing 1(1), 16–38 (2010)
[9] Lepagnot, J., Nakib, A., Oulhadj, H., Siarry, P.: Brain cine-MRI registration using MLSDO dynamic optimization algorithm. In: IXth Metaheuristics International Conference, pp. S1–25–1–S1–25–9 (2011)
[10] Li, C., Yang, M., Kang, L.: A New Approach to Solving Dynamic Traveling Salesman Problems. In: Wang, T.-D., et al. (eds.) SEAL 2006. LNCS, vol. 4247, pp. 236–243. Springer, Heidelberg (2006)
[11] Li, C., Yang, S.: A Generalized Approach to Construct Benchmark Problems for Dynamic Optimization. In: Li, X., Kirley, M., Zhang, M., Green, D., Ciesielski, V., Abbass, H.A., Michalewicz, Z., Hendtlass, T., Deb, K., Tan, K.C., Branke, J., Shi, Y. (eds.) SEAL 2008. LNCS, vol. 5361, pp. 391–400. Springer, Heidelberg (2008)
[12] Li, C., Yang, S., Nguyen, T.T., Yu, E.L., Yao, X., Jin, Y., Beyer, H.-G., Suganthan, P.N.: Benchmark generator for CEC 2009 competition on dynamic optimization. Technical report, University of Leicester, University of Birmingham, Nanyang Technological University (2008)
[13] Li, X., Branke, J., Blackwell, T.: Particle swarm with speciation and adaptation in a dynamic environment. In: Proc. Genetic Evol. Comput. Conf., Seattle, Washington, USA, pp. 51–58. ACM (2006)
[14] Liu, L., Yang, S., Wang, D.: Particle swarm optimization with composite particles in dynamic environments. IEEE Trans. Syst. Man. Cybern. Part B 40(10), 1634–1648 (2010)
[15] Lung, R.I., Dumitrescu, D.: Collaborative evolutionary swarm optimization with a Gauss chaotic sequence generator. Innovations in Hybrid Intelligent Systems 44, 207–214 (2007)
[16] Lung, R.I., Dumitrescu, D.: ESCA: A new evolutionary-swarm cooperative algorithm. SCI, vol. 129, pp. 105–114 (2008)
[17] Morrison, R.W., De Jong, K.A.: A test problem generator for non-stationary environments. In: Proc. Congr. Evol. Comput., pp. 2047–2053 (1999)
[18] Moser, I., Chiong, R.: Dynamic function optimisation with hybridised extremal dynamics. Memetic Computing 2(2), 137–148 (2010)
[19] Moser, I., Hendtlass, T.: A simple and efficient multi-component algorithm for solving dynamic function optimisation problems. In: Proc. Congr. Evol. Comput., pp. 252–259. IEEE, Singapore (2007)

[20] Parrott, D., Li, X.: Locating and tracking multiple dynamic optima by a particle swarm model using speciation. IEEE Transactions on Evolutionary Computation 10(4), 440–458 (2006)
[21] Talbi, E.-G.: Metaheuristics: from design to implementation. John Wiley and Sons Inc. (2009)
[22] Yang, S.: Non-stationary problem optimization using the primal-dual genetic algorithm. In: Proc. Congr. Evol. Comput., pp. 2246–2253. IEEE, Canberra (2003)
[23] Yang, S., Li, C.: A clustering particle swarm optimizer for locating and tracking multiple optima in dynamic environments. IEEE Transactions on Evolutionary Computation (2010)
[24] Yang, S., Yao, X.: Experimental study on population-based incremental learning algorithms for dynamic optimization problems. Soft Computing – A Fusion of Foundations, Methodologies and Applications 9(11), 815–834 (2005)
[25] Yang, S., Yao, X.: Population-based incremental learning with associative memory for dynamic environments. IEEE Transactions on Evolutionary Computation 12(5), 542–562 (2008)

Chapter 2
Quantitative Performance Measures for Dynamic Optimization Problems

Briseida Sarasola and Enrique Alba

Abstract. Measuring the performance of algorithms over dynamic optimization problems (DOPs) presents some important differences when compared to static ones. One of the main problems is the loss of solution quality as the optimization process advances in time. The objective in DOPs is in tracking the optima as the landscape changes; however, it is possible that the algorithm gets progressively further from the optimum after some changes happened. The main goal of this chapter is to present some difficulties that may appear while reporting the results on DOPs, and introduce two new performance tools to overcome these problems. We propose a measure based on linear regression to measure fitness performance degradation, and analyze our results on the moving peaks problem, using several measures existing in the literature as well as our performance performance degradation measure. We also propose a second measure based on the area below the curve defined by some population attribute at each generation (e.g., the best-of-generation fitness), which is analyzed in order to see how it can help in understanding the algorithmic search behavior.

2.1 Introduction

The problem of finding good performance measures for dynamic optimization problems (DOPs) is not a trivial one. A good measure should at least describe what the researcher is actually perceiving. It should also have a lower set of restrictions (in order to be widely applicable) and allow a numerical (maybe statistical) treatment of its results. Some traditional measures from non-stationary problems, like offline performance [5] and accuracy [67], have been transferred to DOPs, although they often need to be modified and adapted to dynamic environments. Other measures have been specially designed for DOPs, like collective mean fitness [9]. Although

Briseida Sarasola · Enrique Alba
Universidad de Málaga
e-mail: {briseida,eat}@lcc.uma.es

E. Alba et al. (Eds.): Metaheuristics for Dynamic Optimization, SCI 433, pp. 17–33.
springerlink.com

there is a wide plethora of other measures available (window accuracy [67], best known peak error [4], and peak cover [5]), most current studies tend to use the three measures mentioned in the first place, as well as a visual analysis of the algorithm running performance.

However, at this moment there is no general consensus about what measure to use. While the great majority of studies use an average of the best current fitness, we think that the single value provided by best fitness averages is often not enough, since it is possible to obtain the same average value from very different sets of points and complete opposite search behaviors. In addition, few measures explore other aspects of the problem dynamics. Among these other measures we can cite the ones reporting the diversity (most typically entropy, inertia [10], and Hamming distance based measures [12–14]), as well as stability and ε-reactivity [67].

An important issue which is usually not taken into account in existing studies is the ability of a certain algorithm to obtain good solutions for a long time in a steady manner while the search landscape changes, i.e. to be able to track the moving optima for a big number of periods. A period is an interval of time without changes in the problem definition. Based on existing results and on our own experience, the performance of an algorithm over contiguous and continuous changes in the problem can degrade with time. Our first objective in this work is then to present a new metric to measure how an algorithm degrades as the search advances. For that purpose, we consider a scenario using the moving peaks benchmark and genetic algorithms. This is a fundamental issue since it characterizes the ability of the algorithm in the long term, giving the actual power of the algorithm for a real application or for its scalability in time. Our second objective is to study the curve defined by the algorithmic search behavior regarding some population attribute, and see how this tool can be used to learn more about algorithm performance .

The rest of the chapter is structured as follows: Section 2.2 explains the main performance measures in the literature; Section 2.3 presents a way of measuring fitness degradation ; Section 2.4 exposes the moving peaks problem, which will be used as case study in this paper; Section 2.5 analyzes how existing measures perform for a given case study; Section 2.6 introduces how the area below the fitness curve can be used to analyze the algorithmic search behavior; finally, some conclusions are drawn in Section 2.7.

2.2 Background and Motivation

The aim of using dynamic optimization is not only to find the global optima, but to be able to track the movement of these optima through the search time. Although it is still quite common in existing works to use line charts to visually compare the running fitness of algorithms, a number of numeric measures have been proposed. In this section we review and discuss the most widely used measures in the literature.

2.2.1 Existing Performance Measures for DOPs

In this section we describe the most popular measures in the literature.

Offline performance

Its usage for DOPs was proposed in [5]. It is calculated as the average of the best value found so far in the current period (see Eq. 2.1). For its computation it requires that changes in the landscape are known beforehand.

$$x^* = (1/N) \cdot \sum_{i=1}^{N} f(period\_best_i), \tag{2.1}$$

Collective mean fitness

It was introduced by Morrison in [9]. It is similar to offline performance , but considers the best value in the current generation, and thus does not require to know about changes in the search space (see Equation 2.2). It is also referred to in the literature as Best-of-Generation Average ($\overline{BOG}$).

$$F_C = (1/N) \cdot \sum_{i=1}^{N} f(generation\_best_i), \tag{2.2}$$

Accuracy, stability, and reactivity

This group of three measures was first proposed for static optimization problems, and explicitly adapted for dynamic environments in [67]. The accuracy measures how good the best solution in the current population is with respect to the best (Max_F) and worst (Min_F) known values in the search space. It ranges between 0 and 1, where a value closer to 1 means a higher accuracy level. The formula is often simplified with $Min_F = 0$. This measure is also known as the relative error.

$$accuracy_i = \frac{f(generation\_best_i) - Min_F}{Max_F - Min_F}, \tag{2.3}$$

Stability is also viewed as an important issue in DOPs. An algorithm is stable if the changes in the environment do not affect its accuracy severely (see Eq. 2.4).

$$stability_i = \max\{0, accuracy_i - accuracy_{i-1}\}, \tag{2.4}$$

where $stability \in [0,1]$. An algorithm is considered stable if stability is close to 0. Finally, another aspect to be taken into account is the ability of the algorithm to react quickly to changes. This is measured by the ε-reactivity, which ranges between 1 and the number of generations ($maxgen$) (a smaller value implies a higher reactivity):

$$reactivity_i = \min\{i' - i | i < i' \leq maxgen, i \in \mathbb{N}, \frac{accuracy_{i'}}{accuracy_i} \geq (1 - \varepsilon)\}, \tag{2.5}$$

2.2.2 Discussion

However, none of these measures reflects a very important aspect in DOPs. Algorithm performance can degrade after the landscape has changed several times, resulting in a loss of fitness quality in the following optimization stages. Fig. 2.1 shows an example for an algorithm over a general problem. On the left, the running best fitness and the known optimum are represented. It can be easily seen that the running best fitness is much closer to the known best fitness at first periods (d_i), while the distance between them becomes bigger at the last periods (d_j). The same situation is illustrated in the graph on the right, but this time we represent the accuracy for each period compared to the maximum possible accuracy value ($accuracy = 1.0$).

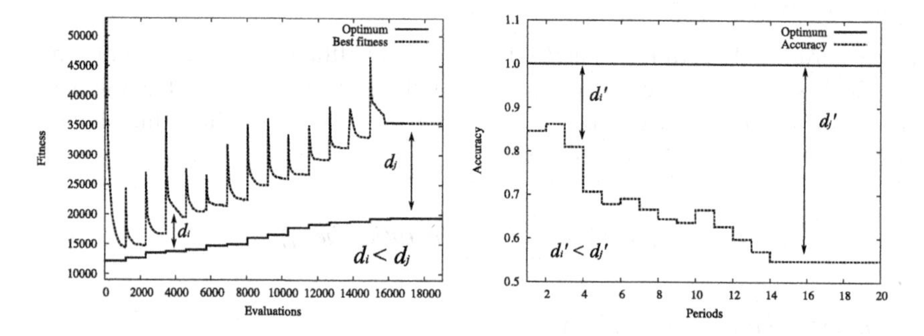

Fig. 2.1 An example to illustrate degradation in DOPs: Running best fitness compared to the best known fitness in each period (left) and the best accuracy obtained in each period compared to optimal accuracy (1.0) (right). Example taken from an experiment on the Dynamic Vehicle Routing Problem.

Besides, when using fitness averages, two or more algorithms can achieve the same average value with very different search behaviors: it is possible to arrive at the same average through diverging search dynamics. An example of this problem is shown in Figure 2.2: Algorithm 1 obtains better fitness values at the beginning, but then the average worsens; the performance of Algorithm 2 is the most stable one; finally, Algorithm 3 obtains the worst results in the first half of the simulation, but finally reaches the same $\overline{BOG}$. The standard experimental study would report that all three algorithms have the same $\overline{BOG} = 0.78$. However, the graph draws our attention toward the differences between them. This is a common problem in dynamic optimization, where several algorithms can provide statistically similar results according to their $\overline{BOG}$ values. Therefore, we consider that another approach is required to find out more information about the algorithmic performance.

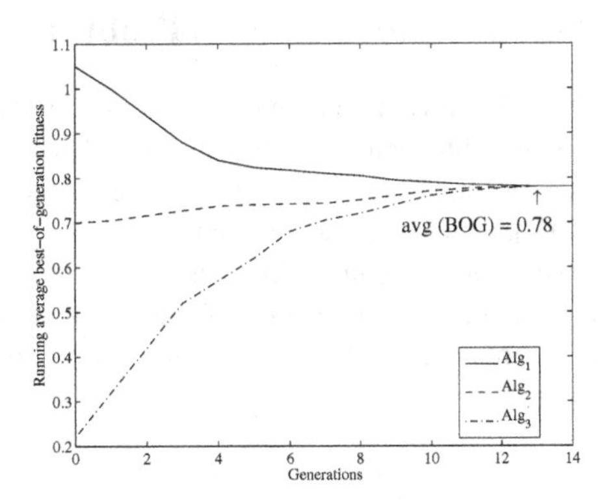

Fig. 2.2 An example showing the running $\overline{BOG}$ (top) for three algorithms. $\overline{BOG}$ by itself is not pointing out a clear performance ranking.

2.3 A Measure for Degradation: $\beta_{degradation}$

As already discussed, none of the previously mentioned measures accounts for the progressive degradation suffered by the algorithm. We define degradation as the loss of fitness quality which can affect the optimization process; this results in obtaining worse solutions as time advances. This loss of fitness quality is more obvious and serious when solving DOPs, because it is expected that the algorithm is able to achieve good solutions for a number of different landscapes which follow one another in time. This degradation is more evident as the execution runs for a longer time and affects any of the mentioned existing measures.

We propose then to measure degradation using linear regression over the consecutive accuracy values achieved at the end of each period [3]. We have selected the accuracy since it is consistently used as the most important error measure in most studies. For that purpose we use Eq. 2.6, where the variable y is an approximation to the overall accuracy, $\bar{\mathbf{x}}$ is a vector of size P, and $\beta_{degradation}$ is the slope of the regression line. P is the number of periods in the dynamic problem. Each x_i is the accuracy of the best solution found in period i averaged over all independent runs N (see Eq. 2.7, where $period\_best_{ji}$ denotes the best solution found in period i in the j-th run). A positive $\beta_{degradation}$ value indicates that the algorithm keeps a good improvement and still provides good solutions: the bigger the improvement, the higher the slope value will be. On the contrary, a negative value implies a degradation in the solution quality, where a smaller value implies a deeper loss of quality.

$$y = \beta_{degradation}\bar{\mathbf{x}} + \varepsilon, \tag{2.6}$$

$$x_i = \frac{1}{N}\sum_{j=1}^{N} f(period\_best_{ji}), \tag{2.7}$$

2.4 The Moving Peaks Problem

There are several bechmark problems for Dynamic Optimization (DO), such as the dynamic bit-matching, dynamic royal road, moving parabola, time-varying knapsack problem, etc. In this paper, we focus on one of the most widely used benchmark problems for DO: the moving peaks problem [5]. The moving peaks idea is to have an artificial multidimensional landscape consisting of several peaks where the height, width, and position of each peak are altered every time a change in the environment occurs. The cost function for N dimensions and m peaks has the following form:

$$F(\bar{\mathbf{x}},t) = \max\{B(\bar{\mathbf{x}}), \max_{i=1...m} P(\bar{\mathbf{x}}, h_i(t), w_i(t), \bar{\mathbf{p}}_i(t))\}, \tag{2.8}$$

where $B(\bar{\mathbf{x}})$ is a time-invariant "base" landscape, and P is the function defining a peak shape, where each of the m peaks has its own time-varying height (h), width (w), and location ($\bar{\mathbf{p}}$) parameters. Every certain number of evaluations, the height and width of every peak are changed by adding a random Gaussian variable. The location of every peak is changed by a vector v of fixed length s. A parameter λ determines if a peak change depends on the previous move or not. If $\lambda = 0$ then each change is completely random, while for $\lambda = 1$, the peak will always move in the direction of the former change.

2.5 Experimental Setting

This section is aimed at studying the behavior of the measures explained in Section 2.2. For that purpose, we consider the moving peaks problem, using a problem configuration which corresponds closely to the standard settings proposed by Branke [1]. We use a plane defined in $(0,100) \times (0,100)$ with 10 dimensions and 10 peaks. The peak heights are defined in the interval [30, 70] and the widths in [0.001, 0.2]. The height change severity is set to 7.0 and the width change severity to 0.01. Changes occur every 1000 evaluations and $\lambda = 0$.

We use an elitist generational genetic algorithm (genGA) as well as three well-known strategies which adapt metaheuristics to dynamic environments: hypermutation (hm) [6], memory (me) [8], and random immigrants (ri) [7]. This results in four algorithms: genGA, genGA+hm, genGA+me, genGA+ri. The population size is 100 individuals, parent selection is done by binary tournament, single point crossover probability p_c equals 1.0, and mutation probability for each solution is $p_m = 0.1$.

We also study the statistical significance of the results. First we check if the data follow a normal distribution using the Kolmogorov-Smirnov test. If the answer is positive, we perform a Levene's test to check the homogeneity of the variances. If the Levene test is positive, we do an ANOVA test to compare the means; otherwise

[1] `http://www.aifb.uni-karlsruhe.de/~jbr/MovPeaks/`

we perform a Welch's test. If the data do not follow a Gaussian distribution, we use a Kruskal-Wallis test. All these tests use a level of confidence of 95 %. Unless stated otherwise, the results are statistically significant.

We start by comparing the graphs for a mean execution of each algorithm, which is the most basic strategy (Section 2.5.1). Then we analyze the results with Morrison's collective mean fitness and Weicker's measures (see sections 2.5.2 and 2.5.3 respectively). The offline performance results are excluded since they are identical to those of collective mean fitness (since we are using an elitist genGA).

2.5.1 Studying the Graphical Representation of Mean Executions

The most immediate way of analyzing the performance of algorithms is comparing the graphs which show the running fitness. This is done in most existing works. The advantage is a fast and intuitive way of comparing performances; however, the results can often be difficult to interpret, they might not always fit in a figure of the desired size, and they can easily lead to confusion.

Figure 2.3 shows an example of this procedure, where the mean of all 100 runs for each algorithm is represented. The length of the run has been shortened to $40k$ evaluations for the sake of clarity. From this graph we conclude that genGA+ri gives usually better fitness values than the rest; genGA+me and canonical genGA obtain similar results in the first half of the simulation, although genGA+me comes closer to genGA+ri and outperforms genGA in the second half. However, does this analysis reflect the real behavior of the analyzed algorithms? We do not think so. The approach is insufficient, as it seems impossible to draw a satisfactory conclusion from these data.

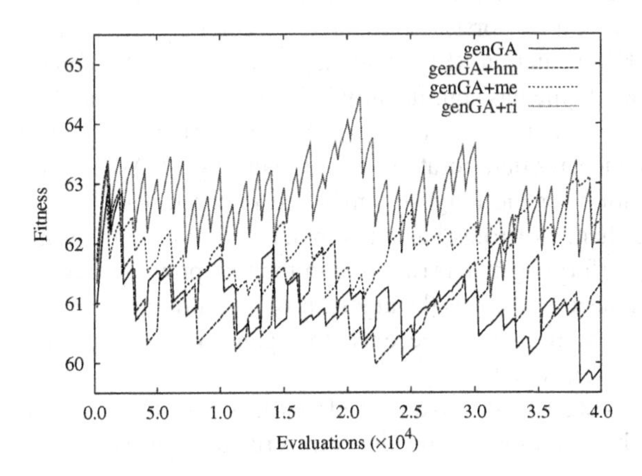

Fig. 2.3 Mean trace of four algorithms on the moving peaks problem.

2.5.2 Studying the Collective Mean Fitness

Collective mean fitness (F_C) is a metric with several advantages. First, it does not need to know when changes happen in the environment (as offline performance and error measures need). Second, the value is obtained from the fitness function, which is directly related to the problem. However, Morrison states that F_C should collect information over a representative sample of the fitness landscape dynamics, but there is no clue as what "a representative sample" means. To illustrate this, we will compare the F_C considering three different total run lengths: 40*k*, 100*k*, and 400*k* evaluations. Numerical results in Table 2.1 show how much the best F_C values depend on the maximum number of evaluations (stopping condition). We show with these results that even 100*k* evaluations are not useful for a representative sample of the landscape dynamics. However, the main issue here is how can we determine this number and whether a very large value (such as 400*k* evaluations) would provide a representative sample or not.

Table 2.1 F_C for each algorithm over three different `max_eval` values.

Evaluations	genGA	genGA+hm	genGA+me	genGA+ri
40,000	60.9735	61.0878	61.9364	**62.7696**
100,000	60.6000	60.9026	62.4002	**62.8713**
400,000	60.5353	60.8690	**63.2371**	62.8385

In order to visually understand this difficulty, we show in Fig. 2.4 the running F_C for the three considered window spans. Fig. 2.4(a) shows that genGA+ri clearly obtains the best results in the 40*k* evaluations case, while genGA+me, genGA, and genGA+hm (from best to worst) are not able to keep its pace. When we extend the simulation to 100*k* evaluations (Fig. 2.4(b)), genGA+ri is still pointed as the best algorithm, but genGA+me has significantly improved its performance during the last evaluations. Another change with respect to the shorter experiment is that canonical genGA performance has deteriorated with respect to genGA+hm. To conclude, we show the simulation for the longest considered interval, namely 400*k* evaluations (Fig. 2.4(c)). A change of roles has taken place here, as the trend now shows that genGA+me outperforms genGA+ri, while in the two previous cases it was exactly the contrary conclusion. It seems the algorithms have already been exposed to a significant part of the search landscape, except for genGA+me, whose F_C will improve if the simulation is extended.

In short, in this section we have proved that it is easy to get a given conclusion and its contrary by just allowing the algorithms run a bit further.

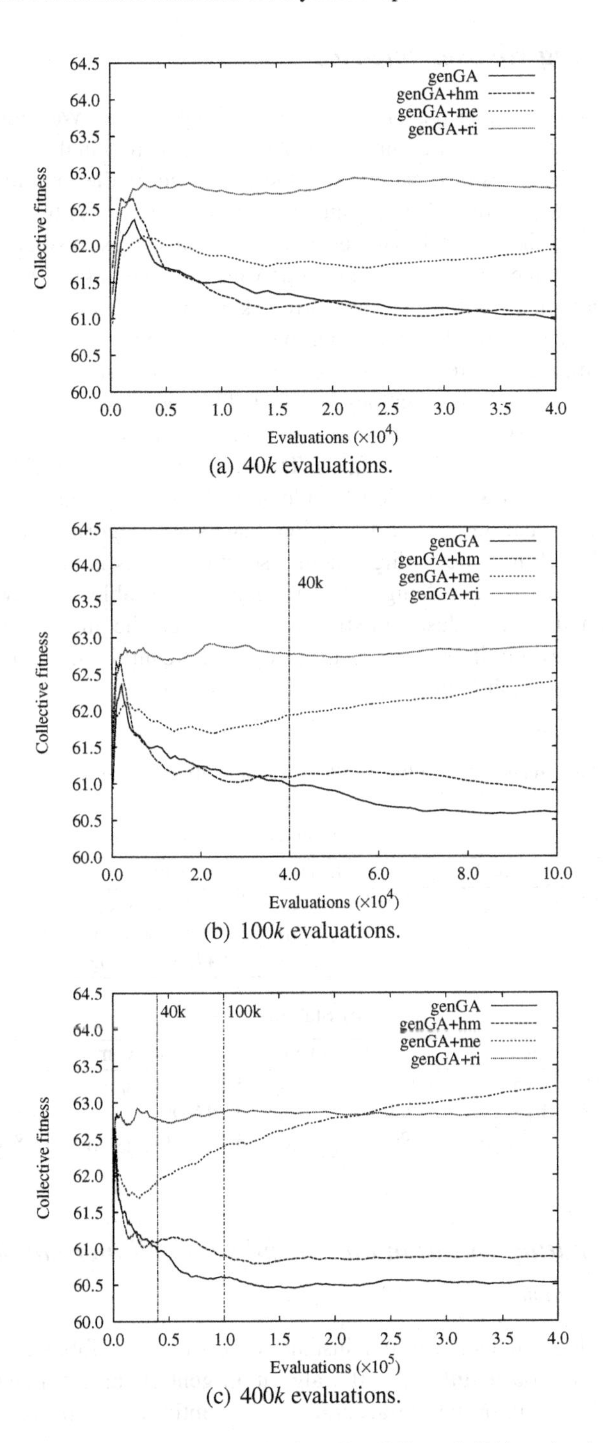

(a) 40k evaluations.

(b) 100k evaluations.

(c) 400k evaluations.

Fig. 2.4 Running F_C for three stopping criteria: 40k, 100k, and 400k evaluations.

2.5.3 Studying Weicker'smetrics

This section studies the usage of three metrics proposed by Weicker (accuracy, stability, and reactivity) to measure the performance of the analyzed algorithms. Among these three measures, accuracy is the main one, while the other two provide complementary results. The advantage of using accuracy is that it provides a bounded range of values in $[0,1]$. However, it is necessary to use a value as reference for the optimum in the current landscape; this value may be unknown in real-world problems, although we could use the current known optimum for such a search space, further research could find new optima and make our results deprecated.

We have compared the results achieved by the three measures and verified that several problems arise. First, accuracy is affected by the same problem as the previously described measures (Table 2(a)). Second, stability does not directly relate to the goodness of solutions (Table 2(b)). There, the canonical genGA is the most stable algorithm after 40*k* evaluations, while genGA+me is the most stable one for 100*k* and 400*k* evaluations. Interestingly, the least stable algorithm in all cases is genGA+ri, although it achieves high-quality solutions. This is due to high fitness drops when the environment changes. Even if genGA+ri is able to recover fast after changes happen, the severe descent in the accuracy affects the final stability. Finally, as stated in [1], ε-reactivity results are usually insignificant; in our experiments, all four algorithms obtain the same average ε-reactivity $= 1$.

Table 2.2 Weicker's measures results for 40*k*, 100*k*, and 400*k* evaluations.

(a) Accuracy

Evals	genGA	genGA+hm	genGA+me	genGA+ri
40,000	9.170e-01 $_{6.4e-03}$	9.171e-01 $_{8.7e-03}$	9.338e-01 $_{8.8e-03}$	**9.465e-01** $_{8.9e-03}$
100,000	9.133e-01 $_{9.0e-03}$	9.159e-01 $_{8.7e-03}$	9.396e-01 $_{9.3e-03}$	**9.486e-01** $_{8.5e-03}$
400,000	9.127e-01 $_{9.1e-03}$	9.167e-01 $_{9.1e-03}$	**9.544e-01** $_{1.3e-02}$	9.473e-01 $_{8.6e-03}$

(b) Stability

Evals	genGA	genGA+hm	genGA+me	genGA+ri
40,000	**3.588e-04** $_{1.3e-04}$	3.934e-04 $_{1.2e-04}$	3.901e-04 $_{2.5e-04}$	1.859e-03 $_{1.5e-03}$
100,000	3.852e-04 $_{1.1e-04}$	4.244e-04 $_{1.3e-04}$	**3.707e-04** $_{2.2e-04}$	1.810e-03 $_{1.3e-03}$
400,000	4.010e-04 $_{1.2e-04}$	4.290e-04 $_{1.1e-04}$	**3.234e-04** $_{2.4e-04}$	1.789e-03 $_{1.3e-03}$

2.5.4 Translating Intuition and Graphs into Numerical Values: $\beta_{degradation}$

The resulting slopes detected in our instances are shown in Table 2.3. Considering the longest period length $p = 400$, algorithm genGA+me degrades the least; in fact, it is able to improve the accuracy as the optimization process advances.

This trend is already detected with $p = 40$. Canonical genGA is the one most affected by degradation through all experiments. Fig. 2.5 shows the regression line obtained for the four algorithms. It is also visually evident that genGA+me obtains the ascending line with the steepest slope, which indicates the absence of degradation and higher improvement of solutions. Besides, genGA obtains the steepest descendant line, which indicates a faster degradation. We can remark that $\beta_{degradation}$ values are of the order of 10^{-4} to 10^{-6}. This is not only due to the high number of periods, but also to accuracy ranging in $[0,1]$.

Table 2.3 $\beta_{degradation}$ for each algorithm and number of periods.

Periods	genGA	genGA+hm	genGA+me	genGA+ri
40	-4.4835e-04	-1.9975e-04	**8.5789e-05**	-1.9254e-05
100	-2.8144e-05	-1.1124e-04	**1.5504e-04**	3.0762e-05
400	-3.4919e-06	1.5469e-06	**7.3754e-05**	-2.0230e-06

In summary, in this section we have shown that our degradation measure provides representative results for different period lengths, and more important, it is able to detect trends at early optimization stages which are later confirmed after longer simulation times.

2.6 A General Performance Measure for DOPs: *abc*

The objective in this section is introducing a more general tool to measure performance in DOPs. Basically, it consists in measuring the area below the curves obtained by the algorithm while exploring the search space [2].

2.6.1 Definitions

First we need to define the concept of population attribute. An attribute of a population is a value which can be measured at any time step. An attribute can depend uniquely on the current solutions in the population, but it can also partly depend on the previous population, or even on all successive generations which precede it. Examples of attributes are the best fitness in a population, the standard deviation, and the average best-of-generation fitness so far.

Next, we need to introduce the area below a curve (*abc*). It can be mathematically defined as follows. Let $p_A(x)$ be the function which determines a certain attribute value achieved by algorithm A in each generation. The area below $p_A(x)$ can be calculated as the definite integral of $p_A(x)$ over the interval $[1,N]$, where N is the total number of generations. We normalize the area by averaging it over the total number of generations (see Equation 2.9).

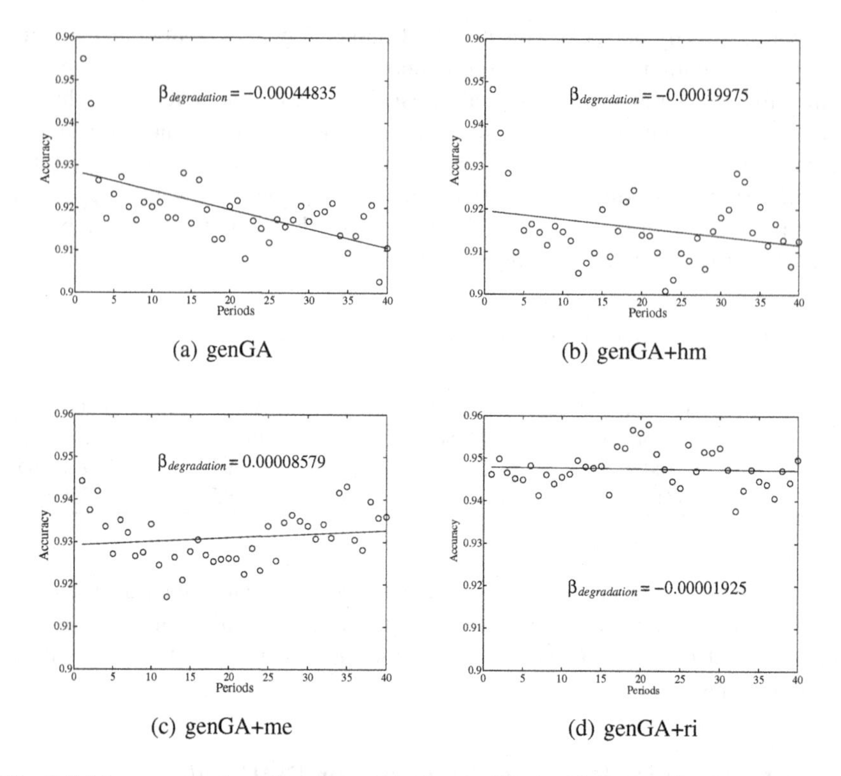

(a) genGA (b) genGA+hm

(c) genGA+me (d) genGA+ri

Fig. 2.5 Linear regression after 40 periods for genGA (top left), genGA+hm (top right), genGA+me (bottom left), and genGA+ri (bottom right).

$$abc_p^A = \frac{1}{N}\int_1^N p_A(x)\,dx, \tag{2.9}$$

Since $p_A(x)$ is defined as a discrete function, abc_A can be intuitively calculated as the sum of the differential areas for each pair of generations $[i, i+1]$. We can therefore calculate the area using the trapezoidal method with unit spacing (see Equation 2.10).

$$abc_p^A = \frac{1}{N}\sum_{i=1}^{N-1}\left(p_A(i) + \frac{|p_A(i+1) - p_A(i)|}{2}\right), \tag{2.10}$$

2.6.2 *Solving the Moving Peaks Problem*

In this case, we use a problem configuration which corresponds closely to the third standard scenario proposed by Branke. We use a plane defined in $(0,100) \times (0,100)$ with 5 dimensions and 50 peaks. The peak heights are defined in the interval [30.0,

70.0] and the widths in [1.0, 12.0]. The height change severity is set to 1.0 and the width change severity to 0.5. Changes occur every 1000 evaluations and $\lambda = 0.5$.

The results provided by the $\overline{BOG}$ can be categorized in two groups of algorithms with respect to performance . The first group (Group 1 from now on) is formed by genGA+me and genGA+ri, whose results are in $(34,36)$, and the second group (Group 2) is composed by genGA, genGA+re, and genGA+hm, with results ranging in $(27,30)$ (see Table 2.4, where ▲ denotes that the algorithm in the current row is significantly better than the algorithm in the column header, while - means no statistical difference could be asserted). Algorithms in the first group are not statistically different, and the same is true in the case of the second group (see Table 2.5). From these results it is concluded that memory-based and random immigrants algorithms are better than hypermutation, complete restart, and canonical genGA.

However, an essential part of the analysis is to remark that it is not possible to make any distinction between the behaviors of algorithms with no statistically significant differences between them. This lack of significant difference does not necessarily mean they behave similarly in the search landscape. In these cases, we can study the area below the $\overline{BOG}$ curve.

Table 2.4 $\overline{BOG}$ obtained for each algorithm on the moving peaks instance.

genGA	genGA+re	genGA+hm	genGA+me	genGA+ri
29.0590	28.1864	27.5579	34.2207	**35.0185**

Table 2.5 Statistical significance tests with respect to the $\overline{BOG}$ measure.

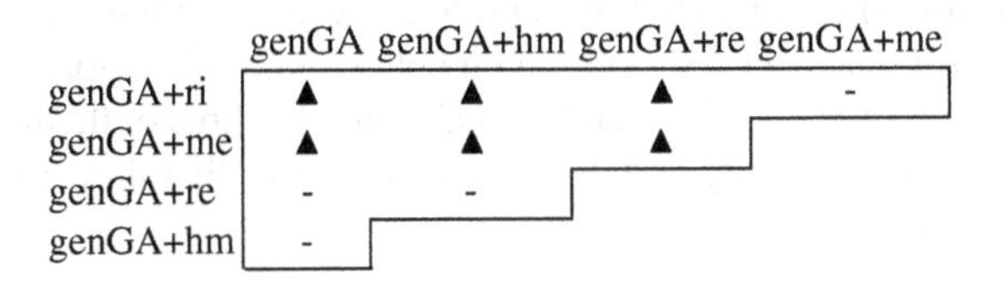

	genGA	genGA+hm	genGA+re	genGA+me
genGA+ri	▲	▲	▲	-
genGA+me	▲	▲	▲	
genGA+re	-	-		
genGA+hm	-			

Table 2.6 shows the results of *abc* with respect to the $\overline{BOG}$ curve. Now we are able to see the differences in the dynamics of the algorithms. In effect, algorithms genGA, genGA+hm, and genGA+re can be now distinguished: the first two of them obtain similar results, while genGA+re obtains a lower value. This means that the first two behave similarly, and we can expect they both obtain initially good solutions but lose quality with respect to the performance of genGA+re. On the contrary, genGA+re is probably more regular than genGA and genGA+hm. The results obtained by genGA+me and genGA+ri are also similar in this case, which suggests that the two algorithms have a similar behavior during the search.

It is important to notice that these values do not relate directly to the quality of the solutions. They should be used to study the dynamic behavior of the algorithms, and they are specially useful to compare statistically similar algorithms; in this case,

Table 2.6 Normalized $abc_{\overline{BOG}}$ obtained for each algorithm on the moving peaks instance.

genGA	genGA+re	genGA+hm	genGA+me	genGA+ri
31.9366	27.5357	31.3606	34.5355	**35.5767**

lower values are desirable, since they suggest a better adaptation to changing environments in the long term.

Figure 2.6 shows the evolution of the running fitness, the running best-of-generation average, and the running normalized area below the $\overline{BOG}$ curve. The three graphs illustrate the problem we have discussed in this chapter. Figure 2.6(a) shows that genGA+re is more regular in its behavior than genGA and genGA+hm, understanding regularity as being able to reach good solutions in most periods. On the contrary, genGA and genGA+hm find it more difficult to obtain good solutions after the environment changes, and we can expect this loss of solution quality continues. Although the average best-of-generation value is very similar in all these three algorithms, the search behavior is very different, as we can appreciate in Figure 2.6(b). In Figure 2.6(c) we can check that $abc_{\overline{BOG}}$ is able to detect this difference.

If we analyze the performance of the algorithms with respect to β-degradation (see Table 2.7), we can obtain a broader view on how algorithms behave. All the results in this table are statistically significant. Algorithm genGA+re achieves a greater value (the one closest to 0), which means it degrades the least: it is the steadiest algorithm regarding the quality of the found solutions. Algorithm genGA+ri obtains the second best β-degradation, that, together with the best $\overline{BOG}$, makes genGA+ri the best algorithm in this experimental setting. The worst results (lowest β-degradation values) are obtained by genGA and genGA+hm. In this way, we are able to tell the difference between the search behaviors of the algorithms, specially when studying genGA+re, genGA, and genGA+hm: the three of them obtain a similar final $\overline{BOG}$, however, using β-degradation we realize that they are behaving in quite a different way.

Table 2.7 β-degradation obtained by the algorithms on the studied moving peaks instance.

genGA	genGA+re	genGA+hm	genGA+me	genGA+ri
-0.02309	**-0.00363**	-0.02812	-0.01565	-0.00952

Since we cannot always rely on visually inspecting running graphs to extract conclusions of experiments, we consider that our new measures can help to understand the algorithmic behavior from a numerical point of view.

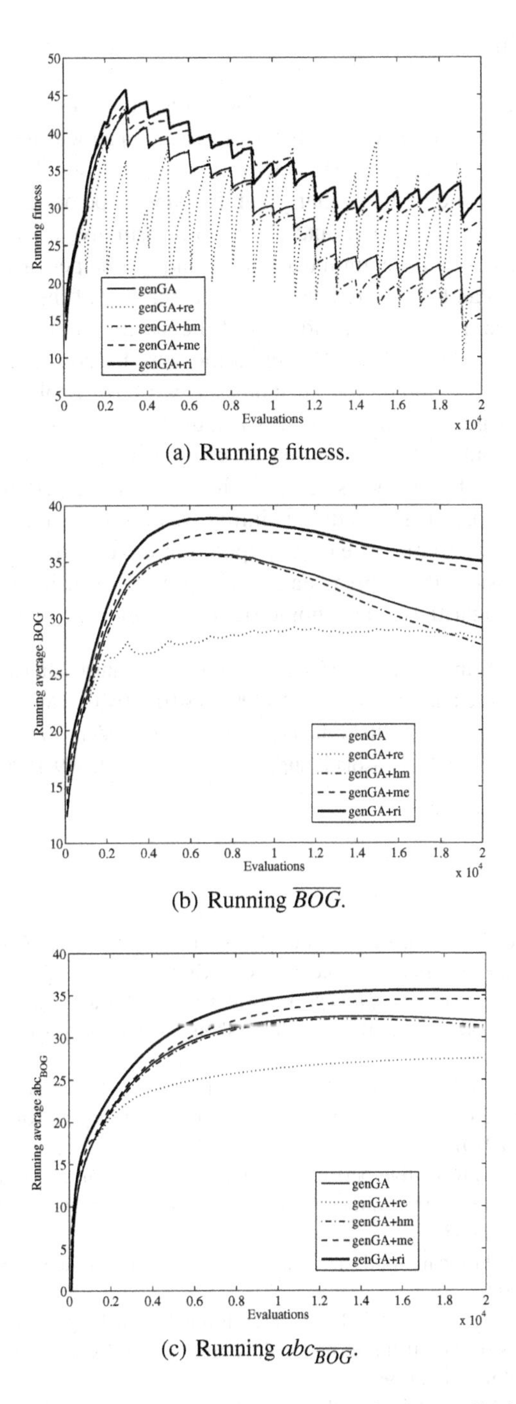

(a) Running fitness.

(b) Running $\overline{BOG}$.

(c) Running $abc_{\overline{BOG}}$.

Fig. 2.6 Running representation of (a) the running fitness, (b) the running best-of-generation, and (c) the running area below the $abc_{\overline{BOG}}$ curve for all five algorithms.

2.7 Conclusions

In this chapter, the most popular measures for DOPs have been studied and analyzed on the moving peaks benchmark . We have discussed on a measure for fitness degradation, which we call $\beta_{degradation}$. This measure provides a suitable way of quantifying the loss of solution quality as changes in the landscape occur. It is quantitative, not a mere inspection of a graph, and it is also informative about the behavior of the algorithm throughout time, unlike accuracy and offline performance . The trend in the performance can be detected earlier than using any of the other measures described in the literature and independently of the simulation time window defined by the researcher. We have also presented a measure based on the area below the curve defined by some population attribute and we have applied this measure to the average best-of generation fitness; this is a useful way to discriminate between algorithms which could not be distinguished using just standard fitness averages.

Future work includes further studies on how fitness degradation affects other DOPs, as well as an extensive study on similar ways to measure degradation . We also plan on using *abc* to measure other population features (e.g., diversity, robustness), and will consider the approximation of these curves using specific functions (e.g., polynomial, logarithmic, trigonometric functions).

Acknowledgments. Authors acknowledge funds from the Spanish Ministry of Sciences and Innovation and FEDER under contracts TIN2008-06491-C04-01 (M* project), TIN2011-28194 (roadME project), and CICE, Junta de Andalucía under contract P07-TIC-03044 (DIRICOM project). Briseida Sarasola is supported by grant AP2009-1680 from the Spanish government.

References

[1] Alba, E., Saucedo Badía, J.F., Luque, G.: A study of canonical GAs for NSOPs. In: Metaheuristics: Progress in Complex Systems Optimization, p. 245 (2007)
[2] Alba, E., Sarasola, B.: ABC, a new performance tool for algorithms solving dynamic optimization problems. In: IEEE Congress on Evolutionary Computation, pp. 1–7 (2010)
[3] Alba, E., Sarasola, B.: Measuring fitness degradation in dynamic optimization problems. In: Proceedings of the European Conference on the Applications of Evolutionary Computation (2010)
[4] Bird, S., Li, X.: Informative performance metrics for dynamic optimisation problems. In: Proceedings of the 9th Annual Conference on Genetic and Evolutionary Computation, pp. 18–25 (2007)
[5] Branke, J.: Evolutionary optimization in dynamic environments. Kluwer Academic Publishers (2002)
[6] Cobb, H.G., Grefenstette, J.J.: Genetic algorithms for tracking changing environments. In: Proceedings of the Fifth International Conference on Genetic Algorithms, pp. 523–530. Morgan Kaufmann (1993)
[7] Grefenstette, J.J.: Genetic algorithms for changing environments. In: Second International Conference on Parallel Problem Solving from Nature, pp. 137–144 (1992)

[8] Mori, N., Imanishi, S., Kita, H., Nishikawa, Y.: Adaptation to changing environments by means of the memory based thermodynamical genetic algorithm. In: Proceedings of the 7th International Conference on Genetic Algorithms, pp. 299–306 (1997)
[9] Morrison, R.: Performance measurement in dynamic environments. In: Branke, J. (ed.) GECCO Workshop on Evolutionary Algorithms for Dynamic Optimization Problems, pp. 5–8 (2003)
[10] Morrison, R.W., De Jong, K.A.: Measurement of Population Diversity. In: Collet, P., et al. (eds.) EA 2001. LNCS, vol. 2310, pp. 31–41. Springer, Heidelberg (2002)
[11] Weicker, K.: Performance Measures for Dynamic Environments. In: Guervós, J.J.M., et al. (eds.) PPSN 2002. LNCS, vol. 2439, pp. 64–73. Springer, Heidelberg (2002)
[12] Yang, S.: Genetic algorithms with memory- and elitism-based immigrants in dynamic environments. Evolutionary Computation 16(3), 385–416 (2008)
[13] Yang, S., Tinós, R.: A hybrid immigrants scheme for genetic algorithms in dynamic environments. International Journal of Automation and Computing, 243–254 (2007)
[14] Yang, S., Yao, X.: Population-based incremental learning with associative memory for dynamic environments. IEEE Trans. Evolutionary Computation 12(5), 542–561 (2008)

Chapter 3
Dynamic Function Optimization: The Moving Peaks Benchmark

Irene Moser and Raymond Chiong

Abstract. Many practical, real-world applications have dynamic features. If the changes in the fitness function of an optimization problem are moderate, a complete restart of the optimization algorithm may not be warranted. In those cases, it is meaningful to apply optimization algorithms that can accommodate change. In the recent past, many researchers have contributed algorithms suited for dynamic problems. To facilitate the comparison between different approaches, the Moving Peaks (MP) function was devised. This chapter reviews all known optimization algorithms that have been tested on the dynamic MP problem. The majority of these approaches are nature-inspired. The results of the best-performing solutions based on the MP benchmark are directly compared and discussed. In the concluding remarks, the main characteristics of good approaches for dynamic optimization are summarised.

3.1 Introduction

Dynamic function optimization is an active research area. Over the years, many different algorithms have been proposed to solve various types of dynamic optimization problems. The Moving Peaks (MP) benchmark, created by Branke [7] in 1999, has been used extensively by researchers in this domain to test their own approaches against others.

Irene Moser
Faculty of Information & Communication Technologies,
Swinburne University of Technology,
Victoria 3122, Australia
e-mail: imoser@swin.edu.au

Raymond Chiong
Faculty of Higher Education,
Swinburne University of Technology,
50 Melba Avenue, Lilydale, Victoria 3140, Australia
e-mail: rchiong@swin.edu.au

E. Alba et al. (Eds.): Metaheuristics for Dynamic Optimization, SCI 433, pp. 35–59.
springerlink.com

Designed to cover a large range of dynamically changing fitness landscapes, the MP problem consists of a multi-dimensional landscape with a definable number of peaks, where the height, width, and position of each peak are altered slightly every time a change in the environment occurs. Formally, it can be defined with the following function:

$$F(\mathbf{x},t) = max(B(\mathbf{x}), max_{1..m} P(\mathbf{x}, h_i(t), w_i(t), \mathbf{p}_i(t))), \tag{3.1}$$

where $B(\mathbf{x})$ is the base landscape on which the m peaks move, with each peak P having its height h, width w and location $\mathbf{p}$. It is necessary to note that the location, width, height, and movement of the peaks are all free parameters. For the purpose of performance comparison, three standardized sets of parameter settings, called *Scenario 1*, *Scenario 2* and *Scenario 3* respectively, were defined. With increasing numbers, the scenarios become more difficult to solve. Most of the benchmark results have been published predominantly for Scenario 2 with 10 peaks that move at an average distance of 1/10 of the search space in a random direction, mainly due to its appropriateness in terms of difficulty and solvability.

In this chapter, we review the existing approaches for solving the MP problem since its inception. These approaches range from evolutionary algorithms and swarm intelligence algorithms to hybrid algorithms. Both evolutionary and swarm intelligence algorithms are population-based stochastic search algorithms inspired by real dynamic processes such as natural evolution and swarm behaviour (see [13–15, 20]). Since they typically do not require extensive knowledge of the function to be optimised, this makes them highly suitable for solving dynamic optimization problems.

Before ending this section, it is worth noting that the focus of this review is solely on the MP problem created by Branke [7]. For reviews or surveys with wider scopes on dynamic optimization, interested readers are referred to the book by Branke [8] or the survey papers by Jin and Branke [27] and Cruz et al. [16].

The remainder of this chapter is organized as follows. In the next section (i.e., Section 8.2), we discuss the fundamental characteristics of the MP problem as well as the common performance measures used. Following this, we review various currently known approaches for solving the problem in Section 3.3. A discussion of the best solutions with comparable results is then given in Section 3.4, and concluding remarks are drawn in Section 3.5.

3.2 Background

3.2.1 Dynamic Optimization

As discussed before, the MP problem consists of a base landscape with a definable number of peaks created randomly within certain width and height limits. The solver's task is to follow the highest peak in a landscape of several mountains while the landscape is changing. There are many free parameters, hence many different scenarios can be created and different aspects of a given algorithm can be tested with this problem. To provide comparable configurations, three sets of standard parameter settings known as Scenario 1, Scenario 2 and Scenario 3 were published as standardized benchmark problems at the same time as the original problem definition.

The frequency of the landscape change Δe is stated as numbers of function evaluations available to the solver between changes. The movement of the peaks depends on a linear correlation between a completely random assignment of new coordinates and the direction of the previous move to the new coordinates. A parameter $0 \leq \lambda \leq 1$ is used to define the balance between random and predictable movements in the shift vector $\mathbf{v}_i$:

$$\mathbf{v}_i(t) = \frac{s}{|\,\mathbf{r} + \mathbf{v}_i(t-1)\,|}((1-\lambda)\mathbf{r} + \lambda \mathbf{v}_i(t-1)), \tag{3.2}$$

Here, $\mathbf{r}$ is the random shift vector and s is a definable parameter regulating the severity (length) of the movement. A λ of 0.0 results in a completely random direction of movement while, a value of 1.0 makes every move's direction depend entirely on the direction of the previous move. The effects of the parameter settings are explained in detail in [10].

When the landscape changes, the location of a peak is shifted by a random variable r drawn from a Gaussian distribution. The heights and widths of all peaks are also changed by adding Gaussian variables to the present values.

The standard scenarios, i.e. problem instances, most commonly use 5 dimensions. The majority of the existing approaches in the literature have used Scenario 2. The original listing for Scenario 2 leaves a choice with regard to a subset of the parameters, but typically the settings listed in Table 3.1 are used.

Table 3.1 The choice of parameter value where Scenario 2 admits a range of values – s is the severity of the change and correlation lambda describes the degree of correlation between the direction of the current and previous moves, where 0.0 stands for a completely random move (no correlation)

Parameters	Admissible Values	Most Common Choice of Value
number of peaks	$10-200$	10
s	$0.0-3.0$	1.0
correlation lambda	$0.0-1.0$	0.0
change every x evaluations	$1000-5000$	5000

3.2.2 *Performance Measures*

Measuring algorithm performance in a dynamic environment is not entirely straightforward. Early approaches used a specific measure, the "generation-best" for plotting the development of the population as it evolves over time (e.g. [1], [22] and

[56]). Ayvaz et al. [2] proposed measuring the run time until the best result is found. The drawback with this method is that the algorithm may have found the best solution long before the end of the run. To measure the time until a certain quality is found, the anticipated quality would have to be known in advance.

De Jong [29] proposed the use of online and offline performances, which differ in the fact that the online performance is based on every result the fitness function returns, whereas the offline performance records the fitness of the best solution known when a new solution is evaluated. They are useful in a context where the quality of the best possible solution (the highest peak) remains constant over time (even though its location may change). Grefenstette [22] observed that subtracting the current (online or offline) performance from the current best would compensate for this shortcoming and introduced the online and offline errors.

The offline error has since been used as the generally accepted performance measure in dynamic environments, although its shortcomings are well known: it favours algorithms that find relatively good solutions early (which is not necessarily a disadvantage); it skews the outcome over time due to the averaging factor; and it is sensitive to the average quality of all available solutions (the average height of the landscape). For example, when comparing performances on landscapes with varying numbers of peaks using the offline error, only comparisons on problem instances with an equal number of peaks are meaningful, and only if the peaks are of the same average width and height. Despite these disadvantages, the offline error has been used extensively as a standard performance measure by many researchers for dynamic problems.

3.3 Existing Approaches

3.3.1 Evolutionary Algorithms

Initially, most approaches applied to the MP problem were based on evolutionary algorithms. One of the pioneering solutions can be found in the work of the author of the MP benchmark. In his seminal work, Branke [7] presented a memory-based multi-population variant of a Genetic Algorithm (GA), which divides the real-value encoded population into two parts. One of the parts maintains a population with a good average fitness and stores fit individuals in memory. After a change, this population retrieves individuals from the memory. The second population also stores its best individuals in memory, but does not retrieve any individuals from it. After a change, the second part of the population is re-randomized.

In the same work, Branke [7] proposed a further alternative according to which the population is split into three islands. All three populations contribute to the memory while only one of the populations retrieves individuals from this memory. The individuals stored in the memory, although inactive, are counted as part of the population. Only the single best individual is added to the memory from time to time (every 10 generations in this case). For this method, several replacement strategies were explored. The two best-performing strategies were one which replaces the most similar individual and a second strategy based on replacing the most similar individual if it is of worse fitness. Individuals are retrieved from the memory after each change, detected through recalculation of the fitness of solutions in the memory.

Branke used a configuration with 5 peaks for the experiments on the MP benchmark. The performance of implementations with one, two and three populations, each combined with the memory facility, was compared with a single-population GA that does not use memory. The results show a superior performance of the implementation with two populations when the peaks change their heights but do not move, whereas the three-population variation performs best when the peaks change their locations as well as their heights.

While the memory-based GA shows interesting features, Branke conceded that the application of memory-based approaches is limited, since they are only truly useful in the case of an exact recurrence of a previous situation. This approach is less useful for more drastic changes to the landscape. With this in mind, he collaborated with other researchers to develop the self-organizing scouts approach [9]. Unlike earlier approaches, this algorithm does not attempt to detect the occurrence of a change in order to react to it.

The approach [9] is based on a GA with a forking mechanism. Initially, there is a single population. During the iterations, subpopulations are divided off, each with a designated search area and size. Each subpopulation's area size is allocated according to a predefined distance from the group's best individual, which defines the initial group at the time of the separation, then remains constant for the subpopulation's life cycle.

The number of individuals in the subpopulation varies according to a quality factor derived from the fitness of the group's best individual and the level of group dynamics, a measure of the group's improvement potential. The improvement potential is determined as the group's fitness increases during the latest generation. While the parent population is excluded from the areas of the child populations, the child populations may search each others' areas as long as the group's centre (the fittest individual) does not enter another child population's space. Whenever this happens, the subpopulation is eradicated.

Branke et al. [9] compared the algorithm's results from solving the MP problem to results produced by a standard GA on the same problem. The change in severity of the peak movement was set to 1, whereas the largest value used for the memory-based approach in [7] was 0.6. The landscape changes every 5000 evaluations. A crucial definable parameter of the self-organizing scouts approach is the range of admissible subpopulation sizes. Regardless of the choice of subpopulation sizes, this self-organizing scouts approach outperforms the standard GA in every instance. Better results were achieved using larger subpopulation sizes. Varying the overall population size was reported to have little effect on the outcome. The authors also concluded that increasing the change frequency of the landscape enhances the advantage of the proposed approach over the standard GA.

Nevertheless, the numerical results reported in both [7] and [9] used the offline performance as a measure, which is not universally comparable when the maximum height of the landscape is unknown. To accommodate comparisons of results from different publications, Branke and Schmeck [10] introduced the offline error as a new performance measure for the MP problem. Branke and Schmeck [10] also presented new empirical results obtained from the self-organizing scouts approach as well as from the memory-based "two subpopulations" GA proposed in [7], both trialled on an MP instance with 10 peaks. For the comparison, a standard GA and a memory-based GA were also implemented. The experiments recorded the offline error for different severities of change and the authors concluded that although the performance of the algorithms declines with increasing severity in all cases, the self-organizing scouts as well as the memory/search population approach from [7] perform far better than the standard versions, maintaining a far greater diversity over the landscape. Over a larger number of peaks, both the self-organizing scouts and the memory/search population with increased memory sizes were reportedly able to maintain a suitable diversity even with as many as 50 peaks. However, this diversity is only beneficial to the performance in the case of the self-organizing scouts.

Apart from the work of Branke et al., the MP benchmark was also used by Kramer and Gallagher [31] who devised different implementations of GA variations to find solutions to a hardware configuration problem. The hardware was intended as a controller for a device. Within the controller, the GA's task is to evolve a neural network with a suitable strategy to optimise a controlled device's performance based on sensor input. The GA receives scalar feedback on the controlled device's performance from a consumer. Initial comparisons between the performances of the GA implementations were made using the MP problem. Due to the specific needs of the controller project, the MP parameter settings used were somewhat original and did not compare to commonly used instances.

The GA approach [31] uses elitism, mutation and re-sampling. It evolves a real-value encoded probability vector which gets feedback from the binary encoded solutions. The re-sampling strategy re-evaluates the fitness of the current best solution to detect a change. This base implementation is then enhanced with hypermutation and random immigrants to form two different variations, which were tested against the performance of the base version.

As benchmark, an MP scenario with 25 narrow peaks was used with a single peak moving among a grid of smaller, equally sized and shaped peaks. Experimental results show that the algorithm variations using enhanced diversification outperform the basic approach as expected on both metrics used: the online error and the pre-shift error. While the online error is one of the optional metrics provided with the MP benchmark, the pre-shift error is a measure introduced by Kramer and Gallagher. It records the deviation from the current maximum immediately before the landscape is changed again.

The original memory-based multi-population GA by Branke [7] forms the basis of the algorithm developed by Zou et al. [59]. Instead of describing the partitioning as two populations and a memory, as in [7], Zou et al. defined them as three populations – memory, mutation and reinitialization. The memory is initialized at the time the initial population of all parts is created, and the "reinitialization" population performs the tasks of Branke's "search" population. This population only contributes to the memory (as was the case in Branke's approach), whereas the "mutation" population exploits the memory as well as contributes to it. The "reinitialization" population is recreated after a change has been detected (as was the case in Branke's "search" population). The only apparent difference between Zou et al.'s algorithm and Branke's lies in the use of a chaotic sequence for all random number generations in [59].

Zou et al. used two problems for their experimental validation, a parabolic function and the MP problem. They benchmarked their approach with an evolutionary algorithm introduced by Angeline [1] and reported a favourable comparison. One of the two problem instances used by Zou et al. seems identical with the one used by Branke in [7]. The offline performance given in both papers could therefore be expected to be comparable, indicating that the new algorithm described in [59] competes favourably with the earlier version. The authors, however, maintained that their performance evaluation was different from the method used in the original paper by Branke.

Moving away from memory-based approaches, Ronnewinkel and Martinetz [49] explored a speciation approach that applies the formation of colonies. Named the MPGA, it bears some resemblance to the self-organizing scouts approach in [9]. In Ronnewinkel and Martinetz's approach, colonies are formed around centroids defined as individuals with the best fitness in the region. The individuals are listed by distances. Slumps in the fitness scores of the individuals indicate borders between colonies. As with the parent population of the self-organizing scouts, the "search" population is allowed to roam the space between child populations (called colonies), but not to enter the space of a colony. New centroids are detected and colonies are divided off when two colonies split – it is the emerging fitness slumps between the members which give rise to a split – or when the "search" population finds a new individual whose fitness is competitive with the existing centroids. Colonies are merged when a valley between the fitnesses of their members disappears.

To further enhance diversity, Ronnewinkel and Martinetz kept a list of centroids which is longer than the actual number of individuals in the colony count. Potential candidates are stored at the end. When all centroids' fitnesses are re-evaluated, centroids may move up in the ranking list and start a new colony. To the same end, the general search population uses a diversity-preserving selection. One obvious advantage of their approach over the self-organizing scouts, mentioned also by the authors, is that no minimum or maximum diameter values have to be set beforehand.

For their experiments, Ronnewinkel and Martinetz used settings almost identical to Scenario 1 of the MP benchmark. The MPGA was compared with many GAs that maintain diversity by tag sharing, deterministic crowding, multi-niche crowding and restricted tournament selection. The results reported in terms of the offline error suggest that the MPGA outperforms all its competitors, although it is clearly superior only in the case of 5 peaks, less so when 50 peaks are present.

Even though the MP problem is single-objective in nature, Bui, Branke and Abbass [11] applied a multi-objective GA formulation to it by introducing a second objective that serves the purpose of maintaining diversity. To this end, the authors introduced six alternative "artificial" objectives: time-based, random, inversion of the primary objective function, distance to the closest neighbour, average distance to all individuals, and the distance to the best individual in the population. These variations were compared to a standard GA and a random immigrants version – both with elitism – on Scenario 2 of the MP problem, with a choice of 50 peaks. The average generation error was used as a performance measure. To obtain this metric, the height of each generation's best individual is subtracted from the maximum height in the landscape. The multi-objective variations that used the distance to all other individuals and the distance to the nearest individual as second objectives were reported to perform significantly better than all other GAs.

Bui, Branke and Abbass [12] later extended their work by applying the Non-dominated Sorting Genetic Algorithm version 2 (NSGA-II) [18]. The extension ranks the non-dominated solutions and maintains diversity using the crowding distance as a basis for the implementation of the six artificial objectives listed before. As in [11], a standard GA and a random immigrants version with varying crossover and mutation rates were used as benchmarks for the experimental study. Additionally, a benchmark algorithm called MutateHigh was used, implemented as a standard GA with a 50% mutation rate. Again, the average generation error was applied as a measure, but this time, the more generally used offline error was also given. In their experiments, the authors varied the extent of the changes in height and width. The variations of the multi-objective optimiser as described in [11] were applied and found to perform better than the standard GA on all 4 problem instances with different change severities. Each variant achieves its best results with different crossover and/or mutation rates.

Fentress [21] introduced a new concept – *exaptation*, defined as adaptation to a situation that may or may not occur in the future. He "preapted" solutions with potential to become global maxima using a GA with hill-climbing scouts. The algorithm performed well due to the smooth ascent provided by the instances of the MP problem used. So as not to exploit this trait, Fentress further developed his approach into a multi-population GA called mp-eGA with tribes centred around a randomly chosen point. A Gaussian distribution was used to produce the next generation in an attempt to keep the centre points of the tribes within a reasonable proximity. As the tribes evolve, the search space of the tribes is limited by the Gaussian distribution to prevent individuals from being located undesirably close to the centre of a different tribe. The population size is adapted dynamically depending on the quality of the tribe. The quality is measured as a function of the dynamism of the tribe (difference in fitnesses) and the aggregated fitness. This fitness measure was first introduced in [9].

The performance of this algorithm was compared to the author's implementation of a standard GA with the inclusion of the variations proposed by Branke [7] and Branke et al. [9]. The results, measured as the normalised offline performance, suggest that the mp-eGA outperforms the benchmarks used for comparison by

5%-10%. The question arises whether the employed method to enforce diversification in a multi-population GA deserves the term exaptation. This becomes particularly evident in the case of the hill-climbing GA: the populations simply maintain a position on the highest point of the existing peaks; evidence of the algorithm finding maxima before they emerge could not be found in the description.

Zeng et al. [58] proposed an orthogonal design based evolutionary algorithm, called the ODEA, where its population consists of "niches" instead of individuals. Each niche selects the best solution found so far as its representative. An orthogonal design method is then employed to work on these niches, guided by a mutation operator. Some niches – the "observer niches" – watch over known peaks, and others – the "explorer niches" – find new peaks. The former niches make full use of fitness similarities of the landscapes before and after a change, while the latter carry out a global search. The authors compared their algorithm to the self-organizing scouts introduced by Branke et al. [9]. Numerical results on different peak movements, number of peaks and dimensionalities show that the ODEA performs better than the self-organizing scouts.

Ayvaz et al. [2] presented a review of several evolutionary approaches with different characteristics. They compared the algorithms' performances on an undefined scenario, exploring the algorithms under different levels of severity of change. Their empirical studies indicated that the self-organizing scouts [9] might produce the best results among the compared evolutionary algorithms. The authors also proposed that the self-organizing scouts could perform even better when enhanced with a crossover-based local search introduced by Lozano et al. [38].

The efficacy of Branke et al.'s self-organizing scouts was further evident in the work of Woldesenbet and Yen [55], who proposed a new dynamic evolutionary algorithm that uses variable relocation for dynamic optimization problems. The algorithm of Woldesenbet and Yen is based on the idea that individuals in the population could be shifted if a change in the environment occurs to fit the resulting new locations of the optima better. For each individual, the fitness change is computed and compared to its parents as well as the movement vector relative to the centroid of its parents. A floating average is maintained on these values for each of the individuals. The information is apportioned to each dimension separately, averaged, and then used to compute a sensitivity value which relates the measured fitness change to changes in the decision variable which, again, is apportioned to each dimension. It is clear that a fitness change in an individual can arise from moving the individual in the decision space or from a change in the environment. If an environment change occurs, re-evaluating the population can provide an estimation of the sensitivity of the fitness towards the change. The change in fitness, be it positive or negative, is then used to compute a suitable relocation radius and direction for moving the individuals along the previously computed decision variable sensitivity vector. This movement is bounded by a minimum and maximum distance as well as the limits of the search space, and performed a specified number of times.

Woldesenbet and Yen tested their algorithm against several dynamic benchmark problems, one of which is the MP problem, and compared it to some state-of-the-art dynamic evolutionary approaches, including the self-organizing scouts. In their experiments, two variations of the proposed algorithm – one enhanced with memory, and another using several clusters to preserve diversity – were introduced. Numerical results based on the MP benchmark with varying numbers of peaks suggest that the variation enhanced with memory performs significantly better than other dynamic evolutionary algorithms, except the self-organizing scouts. For all but the single peak problem, the memory-enhanced variant provides either comparable or worse results than the self-organizing scouts. In their effort to further improve the algorithm's performance, Woldesenbet and Yen used a clustering technique for diversity preservation, and showed that this modification produces superior results to that of the self-organizing scouts in all comparisons.

3.3.2 Swarm Intelligence Algorithms

Particle Swarm optimization (PSO) is another popular method that has been used extensively in the dynamic optimization domain. Inspired by the flocking behaviour of swarms, each particle in PSO has a gradually diminishing velocity and is attracted to the swarm's best-known location. Some alternative implementations also use a personal best location as a secondary attractor. Typically, PSO's particles slow down to provide a more fine-grained exploration towards the end of the search process. This convergence behaviour, however, is less desirable in dynamic environments (such as the MP problem).

Blackwell [5], who introduced charged particles (hence CPSO) that repel each other and circle around neutral particles of the swarm for better convergence behaviour, was among the first to study PSO for the MP problem. Later, Blackwell and Branke [3] applied a multi-population version of the same approach as multi-CPSO to the same problem. They also introduced multi-Quantum Swarm optimization (multi-QSO), a variation whose charged particles move randomly within a cloud of a fixed radius centred around the swarm attractor. Both alternatives use an exclusion radius to stop swarms from exploring the same areas, reinitialising the worse-performing of two swarms when found in each others' exclusion zones.

Given a constant number of 100 particles, Blackwell and Branke demonstrated experimentally that the optimal number of swarms for Scenario 2 is around 10 (± 1), and that the usage of a single swarm leads to convergence to a local optimum for all algorithms (PSO, multi-CPSO, multi-QSO). Multi-QSO in particular has shown superior performance in almost all configurations and produced the best result in the test series.

Unlike the PSO approaches by Blackwell and Branke, Janson and Middendorf [26] proposed to respond explicitly to a change in the landscape after detection. Their hierarchical variation of the PSO (H-PSO), first explored in a static context (see [25]), was compared experimentally to variations of the same base algorithm. Instead of following a global attractor, each particle in H-PSO uses the best location found by the individual immediately above it in the tree structure in addition to its own best find. At each evaluation, the tree structure is adjusted root to leaf by swapping nodes if the child node has performed better. Using diverse attractors, this structure is likely to maintain diversity in a dynamic environment. Diversity maintenance is further enhanced by a feature that reinitialises a defined number of particles to a random position when a change has been detected.

A variation named Partitioned H-PSO (PH-PSO) was also introduced. It divides the tree structure into subtrees whose nodes are reinitialized if they are at a predefined level in the tree. This leaves the root node and a given level of children to preserve the memory of a previous location, while the lower branches have their new root nodes reinitialized. After a given period of time, the branches are rejoined to the tree according to the structure evaluation procedure.

A comparison between PSO, H-PSO and PH-PSO was reported on experiments with three different functions, the MP problem among them. It is not clear which scenario has been used, though change severities of 0.1 and 3.0 are stated. The offline errors are shown as a graph, not as a list of numerical values. It seems that H-PSO outperforms the other variations on instances with smaller severity, while PH-PSO is the most successful on more severe changes if all subtrees are reinitialized after a change.

In a concurrently developed study, Parrott and Li [46] adapted the speciation technique introduced for GA by Li et al. [33] to PSO. Given an area defined by a fixed radius around the attractor (the best particle in the species), the members of the species are defined by the area they are located in. In case of overlap, the member belongs to the species with superior attractor. When a maximum number of members to the species is exceeded, the worst members are reinitialized to random positions.

As a performance measure, the deviation of the best particle from the global optimum was used. Parrott and Li investigated different population sizes and speciation radii. They concluded that large radii and larger population sizes lead to a more even spread over the landscape[1] and therefore provide swifter reactions to change. Since the authors used the benchmark problem defined by De Jong and Morrison [28] instead of the MP problem, their results do not compare with other PSO approaches reviewed here. We included it because this approach forms the basis of another approach (see the later part of this section) to the MP benchmark.

Based on earlier work, Blackwell and Branke [4] added anti-convergence to the exclusion and quantum/charged particle features first conceived in [5] and [3]. Anti-convergence maintains diversity through a mechanism that reinitialises the worst-performing swarm as soon as all swarms have converged. As exclusion radii around the existing swarms are enforced, the new swarm is guaranteed to search for a new peak.

An analysis of the ideal parameter settings for swarm size and quantum cloud radius, which are derived from the peak shift distance, has also been included in [4]. Similarly, the number of swarms was set equal to the number of peaks. This assumes that both the distance and the peak count are known and do not vary greatly.

[1] See [54] for a detailed discussion of various issues in the fitness landscape.

Implementations based on quantum and charged particles were compared on Scenario 2 of the MP problem. The best-performing configuration is a quantum multi-swarm approach with 5 quantum and 5 neutral particles and as many swarms as there are peaks. The experimental results show that anti-convergence is beneficial when there are more swarms than peaks. Nevertheless, Blackwell and Branke pointed out that the performance deteriorates when there are more swarms than peaks due to the constant reinitialization of swarms that cannot find a peak (as all peaks are occupied). Consequently, this approach shows little promise for an environment where the number of peaks is unknown, which is likely to be the case in real-life optimization problems.

To overcome the limitation, Li, Branke and Blackwell [35] combined some aspects of the previous work of Blackwell and Branke [4] with the notion of speciation PSO (S-PSO) introduced in an earlier publication of Li [34]. The approach was designed to optimise problems with primarily unknown numbers of peaks. It tackles problem dynamics by detection and response. After each change, the species are reinitialized by sorting the particles according to the new fitness and adding them to an existing species if the landscape already has particles within the predefined species radius. If it did not have particles, the new particles are made new species centres. The algorithm observes a maximum member count of species, which was devised first by Parrot and Li [46]. Particles that exceed this threshold are initialized to a random location. Various anti-convergence techniques explored by Blackwell and Branke in [3] and [4] were added to different variations of the algorithm and the performances of all implementations were compared. As a new step towards diversity maintenance, a diversity measure is introduced for particles within the same species. Species that have converged past this threshold are reinitialized around the best-performing particle.

S-PSO was tested on Scenario 2 of the MP problem. The authors concluded that the best-performing variation seems to be a PSO with speciation and an initial population of neutral particles, in which the converged swarms are placed randomly within a radius of their previous attractor. A subswarm is considered as having converged when the distance between the central (attractor) particle and the farthest particle is less than 0.1. One-half of the particles are reinitialized as neutral, the other half as quantum particles as described by Blackwell and Branke [4]. The neutral particles are designed to perform the hill-climbing task.

Following the trend, Wang et al. [53] introduced Branke's technique of applying a multi-population approach, originally used with a GA [7], to PSO. The memory-based reinitialization of the population is triggered by the discovery of a peak. The intended diversity maintenance therefore responds not only to change but also to the challenge of a multi-modal landscape where several optima have to be discovered. Based on the MP benchmark, the authors carried out experiments to compare the performance of several PSO algorithms with that of the proposed triggered memory-based scheme. From their results, Wang et al. concluded that the memory mechanism can improve the performance of PSOs in dynamic environments. Moreover, the triggered memory method has shown to be more efficient than the traditional memory method in exploring the solution space.

Du and Li [19] proposed a new Multi-strategy Ensemble PSO (MEPSO) for dynamic optimization. For the purpose of achieving a good balance between exploration and exploitation, all particles in MEPSO are divided into two parts: the first part is designed for enhancing the convergence ability of the algorithm, and the other for extending the search area of the particle population to avoid being trapped in a local optimum. Two new strategies, Gaussian local search and differential mutation, were introduced to develop the individuals of each of these parts separately. Comparing MEPSO to other PSOs – including multi-QSO, PH-PSO and a standard PSO with reinitialization – on the MP problem, Du and Li demonstrated that their algorithm can outperform the rest in all tested conditions (varying numbers of peaks, varying degrees of severity and different dimensionalities) when the dynamic environment is unimodal. Although multi-QSO tends to perform better than MEPSO in multimodal environments (especially when the number of swarms is larger than the number of peaks), it has been observed that multi-QSO is more sensitive to parameter settings than MEPSO.

Inspired by Branke et al.'s self-organizing scouts [9], Li and Yang presented a Fast Multi-Swarm optimization (FMSO) algorithm [32] to locate and track multiple optima in dynamic environments. In FMSO, two types of swarms are used: a parent swarm for maintaining diversity and detecting the most promising area in the search space when the environment changes, and a group of child swarms for exploring the local areas in their own subspaces. Each child swarm has a search radius, and there is no overlap among all child swarms as the radii act as exclusion zones. If the distance between two child swarms is less than their radius, the whole swarm of the worse one is removed. This guarantees that no more than one child swarm covers a single peak.

Li and Yang compared the performance of FMSO with that of the OMEA [58], an algorithm that was shown to outperform the self-organizing scouts (see Section 3.3.1). Numerical results based on the MP problem indicate that FMSO performs better than the OMEA. In addition, the experiments also demonstrated that FMSO can find not only the global or near-global optimum, but also track the moving best solution in a dynamic environment. As pointed out by the authors, however, the performance of FMSO is quite sensitive to the radius of child swarms.

A new kind of particles, called "compound particles", was introduced by Liu et al. [36] to form part of the PSO population. The compound PSO has both independent and compound particles. Independent particles behave according to the PSO paradigm of shifting particles in the direction of both a local and global best position. Compound particles, on the other hand, consist of three individual particles. When the swarm has been randomly initialized, some of the swarm members are chosen to become compound particles. Each of these particles is then combined with two other particles, which are chosen such that the connecting edges between them are of length L, and the three particles form a triangle in a two-dimensional space.

When a compound particle moves, the worst of the three member particles, denoted by W, is chosen as a base. A point between the other particles, A and B, is created and denoted by C. W and C are subsequently used to form a triangle with a new point R, which is created with the help of a reflection vector. If R has a better fitness than W, a new point E is created along the extension of the line between W and R. The new compound particle then consists of the point E and the original points A and B.

Using the MP problem as a benchmark, Liu et al. compared their compound PSO to a basic PSO as well as Parrot and Li's [46] S-PSO. Unfortunately, none of the 9 scenarios used coincides with Branke's scenarios. Based on 10 peaks, the scenarios differ in the severity of change as well as the change interval, which is not defined in terms of function evaluations but numbers of generations. As we do not know their population sizes and cannot infer the number of evaluations used, we have to rely on the authors' comparison with Parrot and Li's implementation, which is reported to outperform the compound PSO only sporadically. This new PSO approach was further investigated in [37].

For the purpose of preserving diversity, Hashemi and Meybodi [23] proposed a cellular PSO algorithm and tested it on the MP problem. In this algorithm, a cellular automaton partitions the search space into cells. At any time, in some cells of the cellular automaton a group of particles will search for a local optimum using their best personal experience and the best solution found in their neighbouring cells. To prevent diversity loss, a limit is imposed on the number of particles in each cell. Experimental studies, which compare this approach to the well-known PSO approach by Blackwell and Branke in [4], suggest that the cellular PSO method can be highly competitive. In addition, the authors claim that cellular PSO requires less computational effort since it does not need to compute the distances between every pair of particles in the swarm on every iteration.

Subsequently, Kamosi et al. [30] presented a multi-swarm algorithm with very similar ideas to those behind the FMSO algorithm proposed by Li and Yang [32]. As in FMSO, two types of swarms are used: a parent swarm for exploring the search space, and several non-overlapping child swarms for exploiting a promising area found by the parent swarm. To prevent redundant search around the same area, two procedures were adopted. First, a parent particle will be reinitialized if it collides with a child swarm. Second, if two child swarms collide, the one with the least fitness will be removed. In addition, to track the local optima after detecting a change in the environment, particles in each child swarm change their behaviour to quantum particles temporarily and perform a random search around the child swarm's attractor. Experimental results on the MP benchmark suggest that the proposed algorithm outperforms other PSO algorithms in comparison, including FMSO and the cellular PSO method [23] previously introduced by the same group of authors, on all tested configurations.

While multi-swarm approaches have proved useful in dynamic optimization, there are several important issues to consider (e.g., how to guide particles to different promising subregions, how to determine the proper number of subswarms, how to calculate the search area of each subswarm, how to create subswarms, etc.) for this kind of approaches to work well. To address these issues, Yang and Li [57] investigated a clustering PSO where a hierarchical clustering method is used to locate and track multiple peaks. With this clustering method, the appropriate number of subswarms is automatically determined and the search area of each subswarm is also automatically calculated. A local search method is introduced to search optimal solutions in a promising subregion found by the clustering method.

Yang and Li conducted numerous experiments to test the performance of their clustering PSO, and compared it to several state-of-the-art multi-swarm algorithms from the literature. Experimental results suggest that their clustering PSO outperforms all others on the MP problem with different shift severities. The performance of the clustering PSO also seems to scale well with different numbers of peaks. When the number of peaks is less than 20, the clustering PSO performs better than all the other peer algorithms. When optimising problem instances with more than 20 peaks, the clustering PSO's results are still superior to those produced by other approaches with the exception of a hybrid algorithm by Lung and Dumitrescu [39] (which will be discussed in Section 3.3.3).

3.3.3 Hybrid Approaches

Besides evolutionary algorithms and PSO, there also exist several highly effective hybrid/cooperative approaches[2] for solving the MP problem. One such example can be found in the work of Mendes and Mohais [41] who experimented with a multi-population approach of Differential Evolution (DE) and explored seven different schemes in combination with three ways of developing elite individuals, one of them based on the idea of quantum particles introduced by Blackwell and Branke [3]. Their experiments used Scenario 2 whose peak count of 10 is assumed to be a known constant and subsequently used as the fixed number of populations and in determining the diameter of the exclusion zones. In case of an overlap, the population with an inferior best solution is reinitialized.

Mendes and Mohais reported having achieved their best results using a scheme that involves the best individual and the difference between four random population-best individuals. This best-performing implementation uses 40% elite individuals created from the population-best individual using a Gaussian distribution. Experimental results show that these settings equal the performance observed by Blackwell and Branke [3]. The authors also observed that smaller populations achieve better results and they attributed this to the frequency of change in Scenario 2, which is once in 5000 iterations.

The good performances of Blackwell and Branke's PSO and Mendes and Mohais' DE encouraged Lung and Dumitrescu [39] to develop a hybrid algorithm that combines PSO and a Crowding DE, called Collaborative Evolutionary-Swarm optimization (CESO), in which equal populations of both methods collaborate. The Crowding DE maintains diversity by replacing the closest individual if it is fitter. The PSO's task is then to converge to the global optimum. Whenever a change is

[2] It is never easy to clearly distinguish/categorize different types of methods, as some of the previously discussed algorithms in Sections 3.3.1 and 3.3.2 could be seen as hybrid/-cooperative approaches too. Our categorization here is partly dictated by the need of our presentation as well as the sequence of relevant publications.

detected, the PSO swarm is reinitialized to the Crowding DE population. Their results on Scenario 2 with 10 peaks surpass those of Blackwell and Branke's as well as Mendes and Mohais'.

Afterwards, Lung and Dumitrescu extended their work with a new collaborative model called Evolutionary Swarm Cooperative Algorithm (ESCA) [40]. ESCA uses three equal-sized populations of individuals: a main population evolving by the rules of Crowding DE, a particle swarm population, and another Crowding DE population acting as a memory for the main population. Three types of collaboration mechanisms are used to transmit information between populations, and they differ in the mode and quantity of information transmitted. The performance of ESCA was evaluated using numerical experiments on the MP benchmark, and reported an outstanding result on Scenario 2 with the standard settings. Even though the offline error obtained is not as good as CESO, ESCA is shown to be able to cope better with severe changes in the fitness landscape than CESO.

Despite these impressive results, the best solution in the literature comes in a very simple algorithm first presented by Moser and Hendtlass [44], called Multiphase Multi-individual Extremal optimization (MMEO). As suggested by the name, MMEO is a multi-phase, multi-individual version of the Extremal optimization (EO) approach originally designed by Boettcher and Percus [6]. Based on a very simple principle of mutating a single solution according to a power-law distribution, EO attempts to exclude bad solutions rather than to find good solutions. In other words, EO was not designed to show any convergence behaviour. This characteristic makes it a very promising choice for dynamic problems where the time of change is not known.

Devised especially for the MP problem, however, MMEO is better suited to the tracking task of following the peaks and scores high on the offline error metric. It consists of separating global and local search phases with deterministic sampling strategies and devices to save function evaluations. Consequently, it outperforms all the available approaches mentioned above based on the offline error on Scenario 2.

Although the results were exceptional, the local search component of MMEO still carries out redundant steps which cause unnecessary function evaluations. Furthermore, the step lengths used in [44] were chosen without careful consideration. As such, the use of Hooke-Jeeves (HJ) pattern search to further enhance the performance of MMEO has been examined by Moser and Chiong [43]. Proposed by Hooke and Jeeves [24] almost 50 years ago, the HJ pattern search is still among the first choices for researchers in need of a deterministic local search. It starts with a base point followed by exploratory moves that change the values of the dimensional variables one at a time. A combined pattern move then repeats all changes that are found to be successful in the exploratory move and uses a single function evaluation to evaluate the effect of the combined change. This procedure goes on until no improving change can be made in any dimension.

The hybrid EO + HJ approach presented by Moser and Chiong [43] differs from the original MMEO only in the local search part. That is, this hybrid EO algorithm uses the HJ pattern search with different step length sequences. In addition, the HJ-based local search also records directions of successful steps of the preceding exploratory move. Experimental results on different scenarios, varying numbers of peaks and varying dimensionalities show that the hybrid EO + HJ approach significantly improves the results of MMEO.

In a concurrent attempt to improve the local search phase of MMEO, Moser and Chiong [45] presented a new MMEO algorithm by systematically exploring appropriate step length sequences and removing unnecessary redundant steps in its local search procedure. During the development of the original MMEO, it became clear that the step lengths used for the local search were crucial for its performance. The original local search in [44] did not made use of the information as to which direction the previous successful move had taken. In most cases, moving in the same direction as the last successful step will further improve the fitness. In some cases, the last move, although successful, has "overshot the goal" and needs retracting. It is therefore useful to attempt a move in the opposite direction, but only when following the previous move's direction has failed. The new local search in [45] addressed these issues. Extensive numerical experiments on different configurations of the MP problem confirmed that the improvement is significant, in which the new MMEO outperforms all other available solutions in the literature.

3.3.4 Other Approaches

Other types of solutions available for the MP problem include the Stochastic Diffusion Search (SDS) [42], the B-Cell Algorithm (BCA) [50] and the agent-based cooperative strategies approach [47, 48].

The SDS is a metaheuristic inspired by neural networks. Meyer et al. [42] offered a detailed description of the approach, which involves subdividing the objective function and assigning different parts of it to agents who evaluate it for their adopted solution. Initially, solutions are assigned randomly. The success of the partial solution evaluation by the agent decides whether the agent is active. Inactive agents randomly choose another agent and adopt its solution if the chosen agent is active. If not, a random solution is created. Although enhanced with some factors aimed at maintaining diversity, this approach does not compare favourably to local search and PSO on a pattern matching problem. The authors provided neither a detailed description of the adaptation of SDS to the MP problem nor numerical results; the performance was simply described as "following the peak almost perfectly".

The BCA belongs to the class of immune-based algorithms (see [17]). Like evolutionary algorithms, it is a population-based search which evolves a set of elements called B-cells or antibodies to cope with antigens representing locations of unknown optima of a given function. Trojanowski [50] applied the algorithm to the MP benchmark, and used Scenario 2 in the experimental study to investigate various properties of the BCA as well as the dynamic environment. The results show that the algorithm is a viable means for solving the MP problem, and that it copes well with the increasing number of moving peaks. However, no direct comparison to other approaches in the literature has been given.

Subsequently, Trojanowski and Wierzchoń [52] presented a comprehensive analysis of five different immune-based algorithms (one of which is the BCA) based on the MP problem. Scenarios 1 and 2 with 5 and 50 peaks, respectively, were used in their experiments. Different types of mutations were studied, and the authors showed that the BCA and another algorithm known as the Artificial Immune Iterated Algorithm [51] produce the best results among the compared cases.

The cooperative strategies system proposed by Pelta et al. [47, 48] is a decentralized multi-agent approach, where a population of cooperative agents move over a grid containing a set of solutions to the problem at hand. The grid of solutions is used as an implicit diversity mechanism. The authors also implemented a simple strategy to maintain diversity explicitly in two stages: perturbation of a certain percentage of solutions and a random re-positioning of the agents. In [47], these diversity-maintaining mechanisms as well as several communication schemes of the approach were analysed using four different scenarios of the MP problem, but without any comparison to other solutions in the literature. The results indicate that the implicit mechanism based on a grid of solutions can be an effective way of preventing diversity loss. However, the explicit mechanism proposed has not performed as well as expected.

In [48], two kinds of control rules to update the solutions set were studied: one using a simple frequency based resampling (probabilistic) rule and another one a fuzzy-set based rule. Apart from studying the behaviour of these two control rules, the authors also compared their results to Blackwell and Branke's multi-CPSO and multi-QSO on Scenario 2 of the MP problem. Numerical results show that the offline errors obtained by the cooperative strategies are either similar or lower than multi-CPSO and multi-QSO. Nevertheless, no statistical tests were given to ascertain the significance of the results. In terms of the two control rules, the fuzzy one has obtained better results than the probabilistic counterpart.

3.4 Comparison and Discussion

The results of the approaches described in Section 3.3 are listed here if they were stated in a comparable form. The scenarios with the given parameters do not represent instances of the problem, as the resulting instances depend on the random number generator used. This limitation, however, is reduced in proportion to the number of trials performed. Therefore, the number of trials is listed with the results in Table 3.2.

As can be seen from the table, Moser and Chiong's new MMEO [45] and EO + HJ [43] have achieved hitherto unsurpassed results on Scenario 2 of the MP problem. Among the PSO-based solutions with comparable results, Yang and Li's clustering PSO [57] obtained the best offline error of 1.06 followed by Blackwell and Branke's PSO with anti-convergence [4] on an offline error of 1.80 from solving Scenario 2. Li et al.'s S-PSO [35] has an offline error of 1.93, while Blackwell and Branke's earlier version PSO in [3] achieved an offline error of 2.16.

Table 3.2 Comparison of the results based on Scenario 2; all authors used the common choice of value listed in Table 3.1

Publications	Base algorithm	No. of trials for average	Offline errors
Blackwell & Branke [3]	PSO	50	2.16 ± 0.06
Li et al. [35]	PSO	50	1.93 ± 0.06
Blackwell & Branke [4]	PSO	50	1.80 ± 0.06
Mendes & Mohais [41]	DE	50	1.75 ± 0.03
Lung & Dumitrescu [40]	PSO + DE	50	1.53 ± 0.01
Lung & Dumitrescu [39]	PSO + DE	50	1.38 ± 0.02
Yang & Li [57]	PSO	50	1.06 ± 0.24
Moser & Hendtlass [44]	EO	100	0.66 ± 0.20
Moser & Chiong [43]	EO	50	0.25 ± 0.10
Moser & Chiong [45]	EO	50	0.25 ± 0.08

Unfortunately, the majority of the solutions based on evolutionary algorithms did not provide detailed numerical results. Nevertheless, the graphical plots from several publications as well as some numerical results given in [55] suggest that their results are no competition for the approaches listed here. Other approaches were not listed either because no specific scenario was used, information on the settings used was incomplete or the numerical results used an incomparable unit of measurement.

Table 3.3 presents the experimental results (offline error and standard error) of seven state-of-the-art algorithms for solving Scenario 2, where the results are taken directly from the corresponding papers. The information presented in Table 3.3 partly coincides with that presented in Table 3.2, which lists all approaches that provide numeric results for Scenario 2 with 10 peaks. From the table we see that the clustering PSO by Yang and Li [57] outperforms all others on one peak, and the algorithm's performance is not significantly affected when the number of peaks increases. Generally speaking, increasing the number of peaks makes it harder for algorithms to track the optima. An interesting observation, however, is that the offline error of Yang and Li's clustering PSO actually decreases when the number of peaks is larger than 20. Similar trends can also be observed on some of the other algorithms. According to Yang and Li, the reason behind this is that when the number of peaks increases, there will be more local optima that have a similar height as the global optimum, hence there will be a higher probability for algorithms to find good but local optima.

The EO-based approaches [43–45] perform poorly on single-peak landscapes. According to the authors, this is mainly because their local search has long distances to cross (incurring many function evaluations while the best-known solution is still poor) using steps that have been calibrated for smaller "mountains". After a change, the peak has to be found again from a single stored solution, the previous peak. While the hybrid EO algorithms have not performed well with one peak, they outperform all other algorithms on all other instances (see [45] for details).

Table 3.3 Comparison of the results (offline error and standard error) for varying numbers of peaks based on Scenario 2

Peaks	[4]	[57]	[39]	[40]	[44]	[43]	[45]
1	5.07	**0.14**	1.04	0.98	11.3	7.08	7.47
	±0.17	±0.11	±0.00	±0.00	±3.56	±1.99	±1.98
10	1.80	1.06	1.38	1.53	0.66	0.25	**0.25**
	±0.06	±0.24	±0.02	±0.01	±0.20	±0.10	±0.08
20	2.42	1.59	1.72	1.89	0.90	**0.39**	0.40
	±0.07	±0.22	±0.02	±0.04	±0.16	±0.10	±0.11
30	2.48	1.58	1.24	1.52	1.06	**0.49**	0.49
	±0.07	±0.17	±0.01	±0.02	±0.14	±0.09	±0.10
40	2.55	1.51	1.30	1.61	1.18	**0.56**	**0.56**
	±0.07	±0.12	±0.02	±0.02	±0.16	±0.09	±0.09
50	2.50	1.54	1.45	1.67	1.23	**0.58**	0.59
	±0.06	±0.12	±0.01	±0.02	±0.11	±0.09	±0.10
100	2.36	1.41	1.28	1.61	1.38	**0.66**	**0.66**
	±0.04	±0.08	±0.02	±0.03	±0.09	±0.07	±0.07

Table 3.4 Comparison of the results (offline error and standard error) for varying shift severity based on Scenario 2

Severity	[4]	[57]	[39]	[40]	[44]	[43]	[45]
0	1.18	0.80	0.85	1.72	0.38	**0.23**	0.25
	±0.07	±0.21	±0.02	±0.03	±0.19	±0.10	±0.14
1	1.75	1.06	1.38	1.53	0.66	0.25	**0.25**
	±0.06	±0.24	±0.02	±0.01	±0.20	±0.10	±0.08
2	2.40	1.17	1.78	1.57	0.86	0.52	**0.47**
	±0.06	±0.22	±0.02	±0.01	±0.21	±0.14	±0.12
3	3.00	1.36	2.03	1.67	0.94	0.56	**0.49**
	±0.06	±0.28	±0.03	±0.01	±0.22	±0.14	±0.12
4	3.59	1.38	2.23	1.72	0.97	0.64	**0.53**
	±0.10	±0.29	±0.05	±0.03	±0.21	±0.16	±0.13
5	4.24	1.58	2.52	1.78	1.05	0.71	**0.65**
	±0.10	±0.32	±0.06	±0.06	±0.21	±0.17	±0.19
6	4.79	1.53	2.74	1.79	1.09	0.90	**0.77**
	±0.10	±0.29	±0.10	±0.03	±0.22	±0.17	±0.24

The experimental results presented in Tables 3.2 and 3.3 were obtained based on the shift severity of 1. This shift severity value controls the severity of the change: the higher the values, the more severe the changes and hence the more difficult the problem becomes. Taken from the corresponding papers, numerical results of the seven same algorithms on shift severity values from 0 to 6 are compared in Table 3.4.

From the table, we observe that the values reported by the EO variants are again superior than other algorithms. This is largely expected since EO is known to be resilient of severity of change, mainly due to its deliberate lack of convergence behaviour. Among the EO algorithms, the results of improved MMEOs [43, 45] are expectedly better than the original MMEO [44].

While it is clear that the EO algorithms are coping better with larger shifts, they do not cope well with the variations of shift severity as compared to Lung and Dumitrescu's ESCA [40]. Taking performance decay into consideration, ESCA seems to have adapted much better than the EOs.

3.5 Conclusions

In this chapter, we have reviewed all of the existing dynamic optimization approaches we are aware of whose performances have been tested using the MP test function. The majority of these approaches fall into the class of nature-inspired algorithms (see [13, 14]). This type of algorithms shares two common characteristics: (1) they are typically population-based, thus allowing the transfer of past information which is often helpful in dynamic optimization; and (2) they are adaptive. From the papers analysed, however, we see that the algorithms in their conventional form do not perform well in dynamic environments. Our review here shows that good approaches to dynamic function optimization must be capable of tracking the optima over time. They must also be able to maintain diversity in the population so as to avoid population convergence. Many of the good solutions store information from the past explicitly (i.e., the use of memory) in order to improve performance after future changes.

While carrying out the review, we notice that most of the studies reported only means and standard deviations in their comparisons. Although this could give us an indication of how competitive a particular algorithm is, the lack of statistical testing in the results renders the findings inconclusive. In view of this, there is a need to promote the use of proper statistical tests (e.g., the Wilcoxon rank-sum test or Mann-Whitney U test, t-test, Kruskal-Wallis test, etc.) in order to make the claims/results more solid. To make this possible, however, the authors would have to make either their implementation or their experimental data publicly available.

Acknowledgements. We would like to thank the editors for inviting us to contribute to this volume, and one of the anonymous referees for his/her useful comments.

References

[1] Angeline, P.J.: Tracking extrema in dynamic environments. In: Angeline, P.J., McDonnell, J.R., Reynolds, R.G., Eberhart, R. (eds.) EP 1997. LNCS, vol. 1213, pp. 335–345. Springer, Heidelberg (1997)

[2] Ayvaz, D., Topcuoglu, H., Gurgen, F.: A comparative study of evolutionary optimisation techniques in dynamic environments. In: Proceedings of the Genetic and Evolutionary Computation Conference (GECCO 2006), Seattle, WA, USA, pp. 1397–1398 (2006)

[3] Blackwell, T., Branke, J.: Multi-swarm Optimization in Dynamic Environments. In: Raidl, G.R., Cagnoni, S., Branke, J., Corne, D.W., Drechsler, R., Jin, Y., Johnson, C.G., Machado, P., Marchiori, E., Rothlauf, F., Smith, G.D., Squillero, G. (eds.) EvoWorkshops 2004. LNCS, vol. 3005, pp. 489–500. Springer, Heidelberg (2004)

[4] Blackwell, T., Branke, J.: Multi-swarms, exclusion and anti-convergence in dynamic environments. IEEE Transactions on Evolutionary Computation 10(4), 51–58 (2006)

[5] Blackwell, T.M.: Swarms in dynamic environments. In: Proceedings of the Genetic and Evolutionary Computation Conference (GECCO 2003), Chicago, IL, USA, pp. 1–12 (2003)

[6] Boettcher, S., Percus, A.G.: Extremal optimization: Methods derived from co-evolution. In: Proceedings of the Genetic and Evolutionary Computation Conference (GECCO 1999), Orlando, FL, USA, pp. 825–832 (1999)

[7] Branke, J.: Memory enhanced evolutionary algorithms for changing optimization problems. In: Proceedings of the IEEE Congress on Evolutionary Computation (CEC 1999), Washington, DC, USA, pp. 1875–1882 (1999)

[8] Branke, J.: Evolutionary Optimization in Dynamic Environments. Springer (2001)

[9] Branke, J., Kaußler, T., Schmidt, C., Schmeck, H.: A multi-population approach to dynamic optimization problems. In: Parmee, I.C. (ed.) Adaptive Computing in Design and Manufacturing (ACDM 2000), pp. 299–308. Springer, Berlin (2000)

[10] Branke, J., Schmeck, H.: Designing evolutionary algorithms for dynamic optimization problems. In: Tsutsui, S., Ghosh, A. (eds.) Theory and Application of Evolutionary Computation: Recent Trends, pp. 239–362. Springer, Berlin (2002)

[11] Bui, L.T., Branke, J., Abbass, H.A.: Diversity as a selection pressure in dynamic environments. In: Proceedings of the Genetic and Evolutionary Computation Conference (GECCO 2005), Washington, DC, USA, pp. 1557–1558 (2005)

[12] Bui, L.T., Branke, J., Abbass, H.A.: Multiobjective optimization for dynamic environments. In: Proceedings of the IEEE Congress on Evolutionary Computation (CEC 2005), Edinburgh, UK, pp. 2349–2356 (2005)

[13] Chiong, R., Neri, F., McKay, R.I.: Nature that breeds solutions. In: Chiong, R. (ed.) Nature-Inspired Informatics for Intelligent Applications and Knowledge Discovery: Implications in Business, Science and Engineering, ch. 1, pp. 1–24. IGI Global, Hershey (2009)

[14] Chiong, R. (ed.): Nature-Inspired Algorithms for Optimisation. Springer (2009)

[15] Clerc, M.: Particle Swarm Optimization. John Wiley and Sons (2006)

[16] Cruz, C., González, J.R., Pelta, D.A.: Optimization in dynamic environments: a survey on problems, methods and measures. In: Soft Computing – A Fusion of Foundations, Methodologies and Applications (2010), doi:10.1007/s00500-010-0681-0:(online first)

[17] de Castro, L.N., Timmis, J.: Artificial Immune Systems: A New Computational Approach. Springer (2002)

[18] Deb, K., Agrawal, S., Pratap, A., Meyarivan, T.: A Fast Elitist Non-Dominated Sorting Genetic Algorithm for Multi-objective Optimization: NSGA-II. In: Deb, K., Rudolph, G., Lutton, E., Merelo, J.J., Schoenauer, M., Schwefel, H.-P., Yao, X. (eds.) PPSN 2000. LNCS, vol. 1917, pp. 849–858. Springer, Heidelberg (2000)

[19] Du, W., Li, B.: Multi-strategy ensemble particle swarm optimization for dynamic optimization. Information Sciences 178, 3096–3109 (2008)

[20] Eiben, A.E., Smith, J.E.: Introduction to Evolutionary Computation. Springer (2003)

[21] Fentress, S.W.: Exaptation as a means of evolving complex solutions. Master's thesis, University of Edinburgh, UK (2005)
[22] Grefenstette, J.J.: Evolvability in dynamic fitness landscapes: a genetic algorithm approach. In: Proceedings of the IEEE Congress on Evolutionary Computation (CEC 1999), Washington, DC, USA (1999)
[23] Hashemi, A.B., Meybodi, M.R.: Cellular PSO: A PSO for Dynamic Environments. In: Cai, Z., Li, Z., Kang, Z., Liu, Y. (eds.) ISICA 2009. LNCS, vol. 5821, pp. 422–433. Springer, Heidelberg (2009)
[24] Hooke, R., Jeeves, T.: Direct search solutions of numerical and statistical problems. Journal of the Association for Computing Machinery 8, 212–229 (1961)
[25] Janson, S., Middendorf, M.: A hierachical particle swarm optimizer. In: Proceedings of the IEEE Congress on Evolutionary Computation (CEC 2003), Canberra, Australia, pp. 770–776 (2003)
[26] Janson, S., Middendorf, M.: A Hierarchical Particle Swarm Optimizer for Dynamic Optimization Problems. In: Raidl, G.R., Cagnoni, S., Branke, J., Corne, D.W., Drechsler, R., Jin, Y., Johnson, C.G., Machado, P., Marchiori, E., Rothlauf, F., Smith, G.D., Squillero, G. (eds.) EvoWorkshops 2004. LNCS, vol. 3005, pp. 513–524. Springer, Heidelberg (2004)
[27] Jin, Y., Branke, J.: Evolutionary optimization in uncertain environments: a survey. IEEE Transactions on Evolutionary Computation 9(3), 303–317 (2005)
[28] De Jong, K.A., Morrison, R.W.: A test problem generator for non-stationary environments. In: Proceedings of the IEEE Congress on Evolutionary Computation (CEC 1999), Washington, DC, USA, pp. 2047–2053 (1999)
[29] De Jong, K.A.: An analysis of the behavior of a class of genetic adaptive systems. PhD thesis, University of Michigan (1975)
[30] Kamosi, M., Hashemi, A.B., Meybodi, M.R.: A New Particle Swarm Optimization Algorithm for Dynamic Environments. In: Panigrahi, B.K., Das, S., Suganthan, P.N., Dash, S.S. (eds.) SEMCCO 2010. LNCS, vol. 6466, pp. 129–138. Springer, Heidelberg (2010)
[31] Kramer, G.R., Gallagher, J.C.: Improvements to the *CGA enabling online intrinsic. In: NASA/DoD Conference on Evolvable Hardware, pp. 225–231 (2003)
[32] Li, C., Yang, S.: Fast multi-swarm optimization for dynamic optimization problems. In: Proceedings of the 4th International Conference on Natural Computation, Jinan, Shandong, China, pp. 624–628 (2008)
[33] Li, J., Balazs, M.E., Parks, G.T., Clarkson, P.J.: A species conserving genetic algorithm for multimodal function optimization. Evolutionary Computation 10(3), 207–234 (2002)
[34] Li, X.: Adaptively Choosing Neighbourhood Bests Using Species in a Particle Swarm Optimizer for Multimodal Function Optimization. In: Deb, K., et al. (eds.) GECCO 2004. LNCS, vol. 3102, pp. 105–116. Springer, Heidelberg (2004)
[35] Li, X., Branke, J., Blackwell, T.: Particle swarm with speciation and adaptation in a dynamic environment. In: Proceedings of the Genetic and Evolutionary Computation Conference (GECCO 2006), Seattle, WA, USA, pp. 51–58 (2006)
[36] Liu, L., Wang, D.-W., Yang, S.: Compound Particle Swarm Optimization in Dynamic Environments. In: Giacobini, M., Brabazon, A., Cagnoni, S., Di Caro, G.A., Drechsler, R., Ekárt, A., Esparcia-Alcázar, A.I., Farooq, M., Fink, A., McCormack, J., O'Neill, M., Romero, J., Rothlauf, F., Squillero, G., Uyar, A.Ş., Yang, S. (eds.) EvoWorkshops 2008. LNCS, vol. 4974, pp. 616–625. Springer, Heidelberg (2008)

[37] Liu, L., Yang, S., Wang, D.: Particle swarm optimization with composite particles in dynamic environments. IEEE Transactions on Systems, Man, and Cybernetics—Part B: Cybernetics 40(6), 1634–1648 (2010)
[38] Lozano, M., Herrera, F., Krasnogor, N., Molina, D.: Real-coded memetic algorithms with crossover hill-climbing. Evolutionary Computation 12, 273–302 (2004)
[39] Lung, R.I., Dumitrescu, D.: A collaborative model for tracking optima in dynamic environments. In: Proceedings of the IEEE Congress on Evolutionary Computation (CEC 2007), Singapore, pp. 564–567 (2007)
[40] Lung, R.I., Dumitrescu, D.: Evolutionary swarm cooperative optimization in dynamic environments. Natural Computing 9(1), 83–94 (2010)
[41] Mendes, R., Mohais, A.: Dynde: A differential evolution for dynamic optimization problems. In: Proceedings of the IEEE Congress on Evolutionary Computation (CEC 2005), Edinburgh, UK, pp. 2808–2815 (2005)
[42] Meyer, K.D., Nasut, S.J., Bishop, M.: Stochastic diffusion search: Partial function evaluation in swarm intelligence dynamic optimization. In: Abraham, A., Grosan, C., Ramos, V. (eds.) Stigmergic Optimization. SCI, vol. 31, pp. 185–207. Springer, Berlin (2006)
[43] Moser, I., Chiong, R.: A Hooke-Jeeves Based Memetic Algorithm for Solving Dynamic Optimisation Problems. In: Corchado, E., et al. (eds.) HAIS 2009. LNCS (LNAI), vol. 5572, pp. 301–309. Springer, Heidelberg (2009)
[44] Moser, I., Hendtlass, T.: A simple and efficient multi-component algorithm for solving dynamic function optimisation problems. In: Proceedings of the IEEE Congress on Evolutionary Computation (CEC 2007), Singapore, pp. 252–259 (2007)
[45] Moser, I., Chiong, R.: Dynamic function optimisation with hybridised extremal dynamics. Memetic Computing 2(2), 137–148 (2010)
[46] Parrott, D., Li, X.: A particle swarm model for tracking multiple peaks in a dynamic environment using speciation. In: Proceedings of the IEEE Congress on Evolutionary Computation (CEC 2004), Portland, OR, USA, pp. 98–103 (2004)
[47] Pelta, D., Cruz, C., González, J.R.: A study on diversity and cooperation in a multiagent strategy for dynamic optimization problems. International Journal of Intelligent Systems 24(7), 844–861 (2009)
[48] Pelta, D., Cruz, C., Verdegay, J.L.: Simple control rules in a cooperative system for dynamic optimisation problems. International Journal of General Systems 38(7), 701–717 (2009)
[49] Ronnewinkel, C., Martinetz, T.: Explicit speciation with few a priori parameters for dynamic optimization problems. In: GECCO Workshop on Evolutionary Algorithms for Dynamic Optimization Problems, San Francisco, CA, USA, pp. 31–34 (2001)
[50] Trojanowski, K.: B-cell algorithm as a parallel approach to optimization of moving peaks benchmark tasks. In: Proceedings of the 6th International Conference on Computer Information Systems and Industrial Management Applications, Elk, Poland, pp. 143–148 (2007)
[51] Trojanowski, K., Wierzchoń, S.T.: Studying properties of multipopulation heuristic approach to non-stationary optimisation tasks. In: Klopotek, M.A., Wierzchoń, S.T., Trojanowski, K. (eds.) Proceedings of the International Conference on Intelligent Information Processing and Web Mining (IIPWM 2003). Advances in Soft Computing, vol. 22, pp. 23–32. Springer, Berlin (2003)
[52] Trojanowski, K., Wierzchoń, S.T.: Immune-based algorithms for dynamic optimization. Information Sciences 179, 1495–1515 (2009)

[53] Wang, H., Wang, D.-W., Yang, S.: Triggered Memory-Based Swarm Optimization in Dynamic Environments. In: Giacobini, M. (ed.) EvoWorkshops 2007. LNCS, vol. 4448, pp. 637–646. Springer, Heidelberg (2007)
[54] Weise, T., Zapf, M., Chiong, R., Nebro, A.J.: Why Is Optimization Difficult? In: Chiong, R. (ed.) Nature-Inspired Algorithms for Optimisation. SCI, vol. 193, pp. 1–50. Springer, Heidelberg (2009)
[55] Woldesenbet, Y.G., Yen, G.G.: Dynamic evolutionary algorithm with variable relocation. IEEE Transactions on Evolutionary Computation 13(3), 500–513 (2009)
[56] Yang, S.: Memory-based immigrants for genetic algorithms in dynamic environments. In: Proceedings of the Genetic and Evolutionary Computation Conference (GECCO 2005), Washington, DC, USA, pp. 1115–1122 (2005)
[57] Yang, S., Li, C.: A clustering particle swarm optimizer for locating and tracking multiple optima in dynamic environments. IEEE Transactions on Evolutionary Computation 14(6), 959–974 (2010)
[58] Zeng, S., de Garis, H., He, J., Kang, L.: A novel evolutionary algorithm based on an orthogonal design for dynamic optimization problems. In: Proceedings of the IEEE Congress on Evolutionary Computation (CEC 2005), Edinburgh, UK, pp. 1188–1195 (2005)
[59] Zou, X., Wang, M., Zhou, A., Mckay, B.: Evolutionary optimization based on chaotic sequence in dynamic environments. In: Proceedings of the IEEE International Conference on Networking, Sensing and Control, Taipei, Taiwan, pp. 1364–1369 (2004)

Chapter 4
SRCS: A Technique for Comparing Multiple Algorithms under Several Factors in Dynamic Optimization Problems

Ignacio G. del Amo and David A. Pelta

Abstract. Performance comparison among several algorithms is an essential task. This is already a difficult process when dealing with stationary problems where the researcher usually tests many algorithms, with several parameters, under different problems. The situation is even more complex when dynamic optimization problems are considered, since additional dynamism-specific configurations should also be analyzed (e.g. severity, frequency and type of the changes, etc). In this work, we present a technique to compact those results in a visual way, improving their understanding and providing an easy way to detect algorithms' behavioral patterns. However, as every form of compression, it implies the loss of part of the information. The pros and cons of this technique are explained, with a special emphasis on some statistical issues that commonly arise when dealing with random-nature algorithms.

4.1 Introduction

An essential task in the optimization area is to evaluate an algorithm against variations of different factors, either to determine the best combination of parameters for it (step size, population, etc) or to verify its robustness over several settings of a problem (number of local optima, dimensionality, etc). When dealing with Dynamic Optimization Problems (DOPs) [6], this situation is even harder, since these problems have some extra features that need to be analyzed (frequency of changes in the environment, severity of the change, etc). Moreover, it is also usual to compare

Ignacio G. del Amo · David A. Pelta
Models of Decision and Optimization Research Group (MODO),
Dept. of Computer Sciences and Artificial Intelligence, University of Granada,
I.C.T. Research Centre (CITIC-UGR), C/ Periodista Rafael Gómez, 2,
E-18071, Granada, Spain
e-mail: ngdelamo@ugr.es, dpelta@decsai.ugr.es
http://modo.ugr.es

E. Alba et al. (Eds.): Metaheuristics for Dynamic Optimization, SCI 433, pp. 61–77.
springerlink.com

multiple algorithms at the same time, for example to verify if a new proposal outperforms previous state-of-the-art techniques (1-vs-all), or to determine the best one of a set of methods in a competition when no prior knowledge about their performance exists (all-vs-all).

Apart from that, if metaheuristics and non-exact algorithms are used to solve these problems, their random nature will make it necessary to perform several independent repetitions of each experiment in order to obtain a set of result samples representative enough of its underlying distribution. There are several ways of presenting these samples, ranging from giving the mean or the median, to include statistics like the standard deviation, the first and third quartile, etc. In general, the more the information provided of the sample, the better the analysis that can be performed on the results. On the other hand, if there is too much information, it will be more complex to present it to the reader, and also, it will be increasingly difficult to manage it and grasp its meaning. Furthermore, this set of result samples is a random variable, and it is no longer a question of comparing two single values for deciding which algorithm is better; it is necessary to use statistical tools. One of the most used techniques is hypothesis testing, where a null hypothesis is stated (for example, that the underlying distribution of the results of two algorithms is the same) against an alternative hypothesis (that the distributions are not the same), and it is checked if the data from the samples support the null hypothesis at a certain significance level. If the data are too unlikely at that level, the null hypothesis is rejected.

Depending on the approach used for the design of the experiments (fractional, 2^k, full-factorial, etc. [1, 2]), the amount of obtained results can vary greatly. There are several techniques commonly used when presenting these data, ranging from the traditional numerical tables to specifically designed graphs (line charts, barplots, boxplots, qqplots, etc). When there are few results, the use of numerical tables is probably one of the best options, since they provide a complete and precise description of the data. However, if the amount of results increases, tables become rapidly intractable, due to their extension, along with the difficulty of comprehending the meaning of so much numerical data. Graphs allow to alleviate this situation, summarizing the data in a visual way. But again, if several factors are analyzed at the same time in a single experiment, it may be necessary to further compress the information, since the number of plots may grow to unsustainable levels (an example of this will be shown in Sect. 4.2). Several special-purpose graphs can be used in this situation to cope with high amounts of data (dendograms, combinations of pie-charts and scatter plots at the same time), although the type of graph that better suits each case tends to be dependent on the specific problem at hand. For an extensive showcase of visualization techniques, the interested reader is referred to [5].

The goal of this chapter is to introduce a technique named SRCS (Statistical Ranking Color Scheme) specifically designed to analyze the performance of multiple algorithms in DOPs over variations of several factors. This technique is based on the creation of a ranking of the results for each algorithm using statistical tests, and then presents this ranking using a color scheme that allows to compress it. It should be noted that other authors have already addressed this topic in related areas with similar approaches. For example, in the machine learning field, Demšar [8] uses statistical tests for comparing several classifiers over multiple data sets, proposing also a graphical way for presenting statistically non-differentiable classifiers by means of connected lines. And in the multi-objective optimization area, Fonseca et al. [9] devise a technique based on statistical tests to determine the level of attainment of each objective by several algorithms, while López-Ibáñez et al. [17] use this technique along with a graphical color-plotting scheme to visually remark the differences of the algorithms along their obtained Pareto-front.

Before giving the details of the SRCS technique, we will illustrate the application scenario with a practical example.

4.2 Typical Research Case: Comparing Multiple Algorithms over Several Configurations of a Problem

Let us suppose that we want to compare the performance of a set of metaheuristic algorithms over a DOP, for example, the Moving Peaks Benchmark (MPB).

The MPB is a test benchmark for DOP's originally proposed in [3]. It is a maximization problem consisting in the superposition of **m** peaks, each one characterized by its own height (**h**), width (**w**), and location of its centre (**p**). The fitness function of the MPB is defined as follows: Moving Peaks Benchmark

$$\mathbf{MPB}(\mathbf{x}) = \max_j \left(h^j - w^j \sqrt{\sum_{i=1}^{n} (x_i - p_i^j)^2} \right), \; j = 1, \ldots, m, \tag{4.1}$$

where n is the dimensionality of the problem. The highest point of each peak corresponds to its centre, and therefore, the global optimum is the centre of the peak with the highest parameter **h**.

Dynamism is introduced in the MPB by periodically changing the parameters of each peak j after a certain number of function evaluations (ω):

$$\mathbf{h}_j(t+1) = \mathbf{h}_j(t) + \mathbf{h_s} \cdot \mathbf{N}(0,1) \tag{4.2}$$

$$\mathbf{w}_j(t+1) = \mathbf{w}_j(t) + \mathbf{w_s} \cdot \mathbf{N}(0,1) \tag{4.3}$$

$$\mathbf{p}_j(t+1) = \mathbf{p}_j(t) + \mathbf{v}_j(t+1) \tag{4.4}$$

$$\mathbf{v}_j(t+1) = \frac{\mathbf{s}}{|\mathbf{r} + \mathbf{v}_j(t)|}((1-\lambda)\mathbf{r} + \lambda \mathbf{v}_j(t)). \tag{4.5}$$

Changes to both width and height parameters depend on a given severity for each of them ($\mathbf{w_s}$ and $\mathbf{h_s}$). Changes to the centre position depend on a shift vector $\mathbf{v}_j(t+1)$, which is a linear combination of a random vector $\mathbf{r}$ and the previous shift vector $\mathbf{v}_j(t)$ for the peak, normalized to length **s** (position severity, shift distance, or simply *severity*). Finally, parameter λ indicates the linear correlation with respect to the previous shift, where a value of 1 indicates "total correlation" and a value of 0 "pure randomness".

The MPB has been widely used as a test suite for different algorithms in the presence of dynamism. One of the most used configurations for this purpose is Scenario 2, which consists of the set of parameters indicated in Table 4.1.

Table 4.1 Standard settings for Scenario 2 of the Moving Peaks Benchmark

Parameter	Value
Number of peaks (m)	$\in [10, 200]$
Number of dimensions (d)	5
Peaks heights (h_i)	$\in [30, 70]$
Peaks widths (w_i)	$\in [1, 12]$
Change frequency (ω)	5000
Height severity (h_s)	7.0
Width severity (w_s)	1.0
Shift distance (s)	$\in [0.0, 3.0]$
Correlation coefficient (λ)	$\in [0.0, 1.0]$

Once the problem has been defined, we need to decide which performance measures are we going to use for evaluating and comparing the algorithms. There are different options, like using directly the fitness of the algorithm at every moment in time, or the absolute error in case the optimum is known. However, these measures have the problem that they are expressed in absolute units, and they do not give an idea of how close was an algorithm of reaching the optimum, nor allow us to easily compare the results between changes in the environment. For example, if at a given instant in time the fitness of an algorithm can be in the interval $[0, 10]$, an absolute error of 9 units is a very bad result, while if, in another instant, the fitness can be in the interval $[0, 1000]$, the same absolute error of 9 units is a remarkably good result. In order to allow an easier comparison of the results, a *relative* performance measure would be desirable.

Therefore, for the examples of this chapter, we will assume that the optimum is known, and we will use Weicker's definition [21] of the *accuracy* performance measures!accuracy of an algorithm A over a function F at a given instant in time t, as the basic performance measure:

$$accuracy_{F,A}^{(t)} = \frac{F(sol_A^{(t)}) - Min_F^{(t)}}{Max_F^{(t)} - Min_F^{(t)}}, \tag{4.6}$$

where $sol_A^{(t)}$ is the solution generated by the algorithm A at the instant t, and $Min_F^{(t)}$ and $Max_F^{(t)}$ are, respectively, the minimum and maximum values of the function F at the instant t. This measure has the advantage of always being bounded between 0 and 1, 0 being the worst possible value, and 1 the best (note that this is true independently of whether the problem is of the maximization or minimization type). In Weicker's original definition, $best_A^{(t)}$ is used instead of $sol_A^{(t)}$, referring to the best value of the algorithm at time t. This definition assumes that the algorithm is of an evolutionary type, where a population of solutions is evaluated at once, and $best_A^{(t)}$ refers to the best individual in that population. However, this cannot be always assumed, since the algorithm may not be population-based, or the problem may not allow that type of concurrent evaluation, thus forcing us to evaluate each solution sequentially. With the aim of not restricting the study to any given implementation of the algorithm nor the problem, we will use $sol_A^{(t)}$ without loss of generality over the *accuracy*. In return, we will define, independently of the algorithm, the *bestAccuracy* measure performance measures!best accuracy as:

$$bestAccuracy(t,t_0) = \begin{cases} accuracy(t) & \text{if } t = t_0 \\ \max\{accuracy(t), bestAccuracy(t-1)\} & \text{if } t > t_0 \end{cases} \quad (4.7)$$

where $t = t_0$ indicates the instant of time immediately after a change in the environment (variable t is "reset" in every change), such that the *bestAccuracy* refers only to the time elapsed since the last change.

We will now extend the *accuracy* to its *offline* and *average offline* performance measures!offline accuracy versions for several consecutive performance measures!average offline accuracy changes in the environment, using De Jong [7] and Branke [4] definitions:

$$offlineAccuracy(t_0,T) = \frac{1}{T-t_0}\sum_{t=t_0}^{T} bestAccuracy(t,t_0) \quad (4.8)$$

$$avgOfflineAccuracy(N_c) = \frac{1}{N_c}\sum_{n=0}^{N_c} offlineAccuracy(\tau_0(n), \tau_T(n)), \quad (4.9)$$

where N_c is the total number of changes considered, and $\tau_0(n)$ and $\tau_T(n)$ are functions that return, respectively, the first and last instant of time t of the stationary period n. A graphical explanation of these measures can be seen in Figs. 4.1 and 4.2.

At this point we can summarize an execution or *run* of an algorithm with the *avg. offline accuracy*. However, as it has been mentioned in the previous section, when dealing with stochastic algorithms it is necessary to perform a series of independent repetitions of the experiments in order to obtain a representative sample of its performance. Therefore, we will execute N_r runs of the algorithm, thus obtaining N_r measurements of the *avg. offline accuracy*.

Now that we have already defined how are we going to measure the performance, let us suppose that we want to compare 4 hypothetical algorithms for a given configuration of the MPB (for example, the widely used Scenario 2). In Sect. 4.4 we will analyze in more detail the influence of N_r in the results of the statistical tests, but for the moment, let us just assume that we perform a fixed amount of independent repetitions, say $N_r = 30$, for each algorithm. An example of the results that could be obtained is presented in Table 4.2 and Fig. 4.3.

In order to determine the existence of statistically significant differences in the results, we need to perform a series of hypothesis tests. Several authors have already

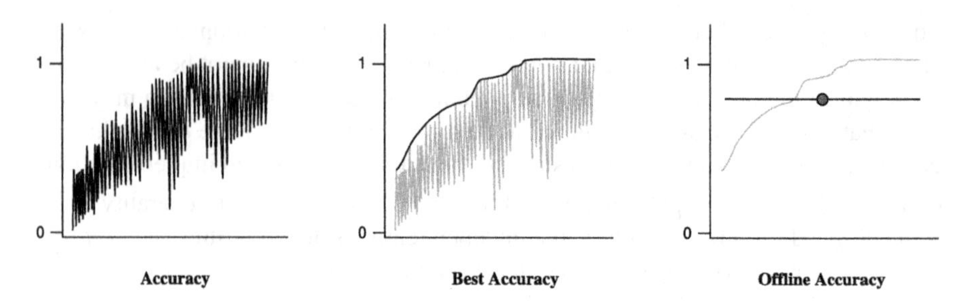

Fig. 4.1 Performance measure of an algorithm using different versions of *accuracy*

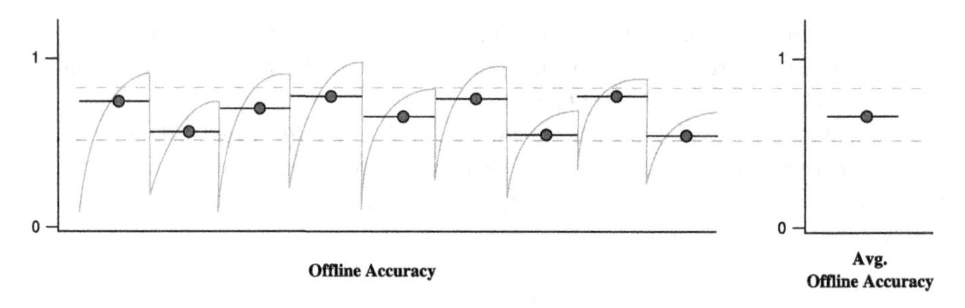

Fig. 4.2 Performance measure of an algorithm over several consecutive changes in the environment using the *offline accuracy* (the *best accuracy* is displayed in the background). The sudden drops of the *best accuracy* values indicate a change in the environment. The average of all *offline accuracy* measures is displayed in the right

pointed out that these results, in general, do not follow a normal distribution [11], therefore recommending the use of non-parametric tests for their analysis [13, 19]. We will use a significance level $\alpha = 0.05$, meaning that we are willing to assume a probability of mistakenly rejecting the null hypothesis of, at most, 0.05. The first issue that needs to be addressed is the fact that we are comparing multiple algorithms at the same time. Therefore, we need to use a test that allows to compare more than 2 groups simultaneously. For this example, we will perform a Kruskal-Wallis (KW) test [15], among all the samples of the 4 algorithms to check if there are global differences at the 0.05 significance level. If the KW test concludes that there are statistically significant differences, we will then perform a series of pair-wise tests between each pair of algorithms, to see if we can determine which are the ones that are causing those differences. In this case, we will use the Mann-Whitney-Wilcoxon test Mann-Whitney-Wilcoxon (MWW) [18, 22] test to compare each pair of algorithms. The combination of these tests is suitable, since the KW test can be considered as the natural extension of the MWW test to multiple groups. It is important to note that in order to guarantee the α-level achieved by the KW test (global), we need to adjust the α-level of each MWW test (pair-wise) to a certain value, usually much smaller than the first one. For this purpose, we will use Holm'shypothesis testing!Holm correction correction [14], although other techniques are also available (for example, Hochberg's, Hommel's, etc; for an in-depth comparison on the use of these techniques, the interested reader is referred to [8, 10, 11]). The results of the tests are shown in Table 4.3, where individual comparisons between each pair of algorithms can be seen, along with the sign of the comparison.

Table 4.2 Performance results of several algorithms on a single problem configuration (mean and standard deviation values of the *avg. offline accuracy*)

	Algorithm 1	**Algorithm 2**	**Algorithm 3**	**Algorithm 4**
Avg. Offline Accuracy	0.78 ±0.05	0.84 ±0.02	0.95 ±0.01	0.89 ±0.03

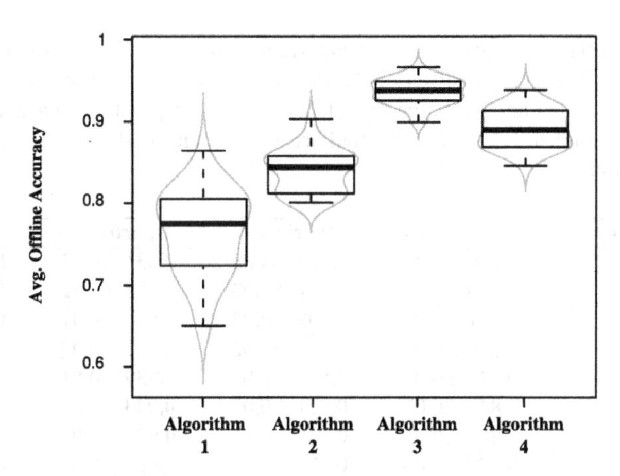

Fig. 4.3 Graphical representation of the results in Table 4.2. The distributions are displayed using a boxplot (dark lines) in combination with a kernel estimation of the distribution density (light lines)

Until now, the way of presenting the results (numerical data tables, boxplot graphs, and statistical tests tables) has been appropriated, and the data are comprehensible. We now contemplate extending the experimental framework. We want to know if the conclusions obtained for the algorithms follow any kind of pattern related to some characteristic of the problem (e.g., whether algorithm 3 is good only for Scenario 2 of the MPB, or if this is a general behaviour linked to, for example, low change frequencies). In order to answer this question, we perform more experiments, keeping all problem parameters constant, except for the change frequency, which we vary progressively.

We can see now that the number of results increases, and its presentation begins to be a problem, both at a table level, because of its extension and difficulty to comprehend the data (Table 4.4), and at a graphical level, because of its complexity (Fig. 4.4). However, it is still feasible to show the results this way, since, although data are

Table 4.3 Pairwise statistical differences among the *avg. offline accuracy* distribution of the algorithms. A '+' sign indicates that there are statistically significant differences between the algorithm in the row and the algorithm in the column, and that the sign of the comparison favors the algorithm in the row (i.e., is "better"). A '-' sign indicates the opposite, that the algorithm in the row is "worse". Finally, the word 'no' indicates no statistically significant differences

	Algorithm 1	**Algorithm 2**	**Algorithm 3**	**Algorithm 4**
Algorithm 1	no	-	-	-
Algorithm 2	+	no	-	-
Algorithm 3	+	+	no	+
Algorithm 4	+	+	-	no

Table 4.4 Performance results of several algorithms on multiple problem configurations (mean and standard deviation values of the *avg. offline accuracy*). The different configurations are based on systematic variations of one factor, the problem's change frequency, expressed in the number of evaluations. Boldface values indicate the best algorithm for the given configuration

Change Frequency	**Algorithm 1**	**Algorithm 2**	**Algorithm 3**	**Algorithm 4**
200	**0.630 ±0.03**	**0.632 ±0.03**	**0.631 ±0.03**	**0.630 ±0.03**
500	0.750 ±0.02	0.751 ±0.02	**0.783 ±0.02**	**0.781 ±0.02**
1000	0.811 ±0.02	0.798 ±0.02	**0.886 ±0.02**	**0.886 ±0.02**
1500	0.825 ±0.02	0.854 ±0.02	**0.922 ±0.02**	0.871 ±0.02
2000	0.840 ±0.02	0.859 ±0.01	**0.939 ±0.01**	0.862 ±0.02
2500	0.843 ±0.01	0.871 ±0.01	**0.943 ±0.01**	0.889 ±0.01
3000	0.852 ±0.01	0.880 ±0.01	**0.950 ±0.01**	0.913 ±0.01
3500	0.871 ±0.01	0.901 ±0.01	**0.959 ±0.01**	0.921 ±0.01
4000	0.860 ±0.01	0.906 ±0.01	**0.964 ±0.01**	0.932 ±0.01
4500	0.863 ±0.01	0.910 ±0.01	**0.968 ±0.01**	0.939 ±0.01
5000	0.869 ±0.01	0.911 ±0.01	**0.970 ±0.01**	0.941 ±0.01

now more difficult to grasp and manage, it is nevertheless still understandable (in Fig. 4.4 it is reasonably easy to see which algorithm is the best, and this can also be accomplished in Table 4.4 by enhancing the best algorithm's result using a boldface type). Anyway, it is worth noting that individual differences between each pair of algorithms in the statistical tests are now too lengthy to be shown, since they imply a comparison of the type *all against all* for each problem configuration, which, in general, is not practical for a publication (we are talking of 11 tables like Table 4.3).

Finally, when we consider to simultaneously analyze several factors (e.g., change frequency and severity of change), data grows exponentially, and the presentation in the form of tables and figures becomes intractable. In the literature, some examples of works can be found, where the magnitude of the study and the amount of obtained results force the authors to use such a high number of tables and graphs that the comprehension of the data gets obscured:

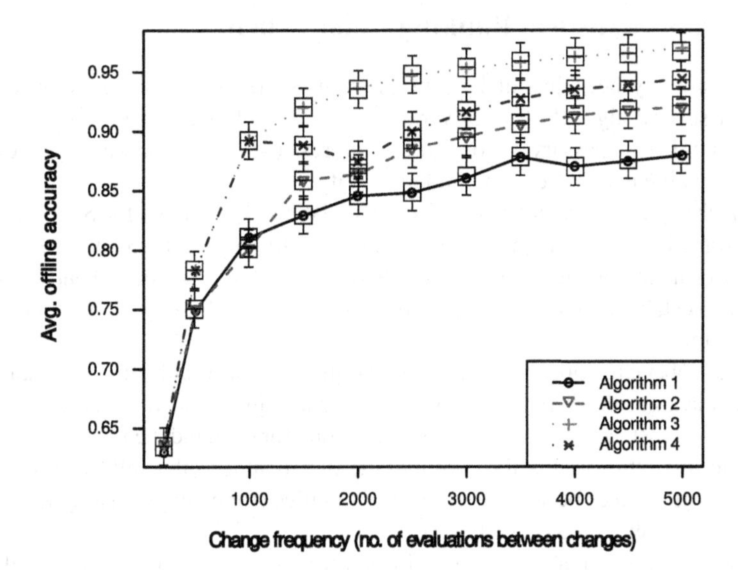

Fig. 4.4 Graphical representation of the results in Table 4.4, where each point corresponds to the results of an algorithm on a configuration of the problem. The results for each configuration are shown using a boxplot of the distribution

- In [23], the author uses 48 graphs and 45 tables of statistical comparisons to analyze the behaviour of 8 different versions of algorithms.
- In [12] a huge number of results are presented in the form of boxplot graphs for several algorithms on a single problem configuration, using 42 graphs for that purpose, although without statistical comparisons tables.
- In [20], the authors compare up to 19 different techniques in 32 tables full of numerical results, using for it an ad-hoc solution based in vectors that compress the information, since they explicitly admit the difficulty in performing so many comparisons.

These are only a small sample of the difficulties that a researcher may find when presenting the results of comparing multiple algorithms, multiple versions of them, multiple problem configurations, or combinations of all the previous. It should be pointed out, however, that each particular case is different from the rest, and that not always such a high number of tables and graphs must imply a bigger difficulty in the understanding of the data. In many cases, the skill of the author for grouping and presenting data is crucial to facilitate its comprehension. However, in general, it is easier for the reader to understand some concise results than some extensive ones.

In order to better confront this situation, we are going to introduce a proposal for compressing the information to be presented using color schemes obtained from the results of the statistical tests.

4.3 SRCS: Statistical Ranking Color Scheme

As it has already been justified in the previous sections, the presentation of the results of several algorithms over variations of multiple factors in a DOP can be problematic. It is necessary to somehow compress the information in order for the reader to be able to capture it and understand it.

The technique we present here, SRCS (Statistical Ranking Color Scheme), has been designed for those situations in which the main interest is to analyze the *relative* performance of the algorithms, rather than the *absolute* one. That is, when we want to establish an ordering or ranking of the algorithms for a given problem configuration.

The first obstacle appears precisely at the moment of establishing that ranking. Ordering the algorithms would be easy if we had a single value to measure their performance, but instead, we have a set of values (one for each independent execution). In order to solve this, we use the output of the statistical tests that tells us if there are significant differences between each pair of samples (for example, using the KW + MWW tests combination of Sect. 4.2).

The way of doing this will be as follows: for a given DOP configuration, all the algorithms begin with an initial ranking of 0. We first compare the results of all the algorithms using a multiple comparison test (e.g., the KW test) in order to determine if there are global differences. In case there are no differences among all, that would be the end of the process, and the algorithms would finish with their initial 0 rank. If, however, significant differences were found, an adjusted pair-wise test (e.g., MWW + Holm) would be performed between each pair of algorithms, in order to assess individual differences. If the pair-wise test says there are significant differences for a given pair of algorithms, the one with the best performance value (the median of the sample) adds $+1$ to its ranking, and the one with the worst value, -1. If there were no differences according to the pair-wise test (a tie), neither algorithm adds anything, but both maintain their previous ranking. At the end, every algorithm will have an associated ranking value, ranging in the interval $[-(N_a-1),+(N_a-1)]$, where N_a is the number of algorithms to be compared. A ranking value of $+r$ for a given algorithm indicates that its performance is significantly better than r algorithms, and a value of $-r$, that it is significantly worse than r algorithms (at the end of this chapter, in the Appendix, we provide an implementation of this ranking calculation using the R programming language).

However, until now we have only shifted the problem, since we have a ranking, but it is still numerical, and therefore, difficult to fully understand when presented in the form of tables if there are too many data. The solution to this comes from human's ability to better manage images and colors than numbers. Starting off from this ranking, we associate a color (for example white) to the maximum ranking value that can be obtained, $+(N_a-1)$, and another very different color (a dark one preferably) to the minimum ranking value that can be obtained, $-(N_a-1)$. All the intermediate ranking values are associated to an interpolated color between the two

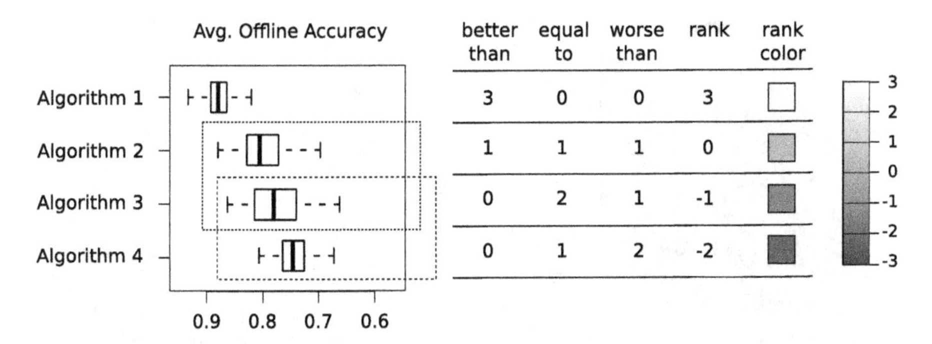

Fig. 4.5 Rank explanation. The boxplot shows the distribution of the performance measures of every algorithm, ordered by its median value. Dotted rectangles indicate those algorithms for which no statistical differences were found at the specified significance level (algorithms 2-3 and 3-4). The table in the right displays, for every algorithm, how many times it shows a significantly better performance ("better than"), no significant differences ("equal to") or significantly worse performance ("worse than") when it is compared with respect to the other 3 algorithms, and its final rank with the correspondent color key.

previous ones. Figure 4.5 explains the calculation of the ranking and the color association of the 4 algorithms we have been using previously, for a given problem configuration.

Color codes obtained from the ranking can now be used to represent the *relative* performance of each algorithm with respect to the others in a graphical way. This representation allows us to visualize the results of many configurations at once, giving the researcher the possibility to identify behavioural patterns of the algorithms more easily.

For example, let us suppose that we have the 4 algorithms of the previous examples, and we want to extend the study of their performance in the MPB with different variations of two factors: *severity*, and *change frequency*. As it has already been justified, presenting the results of these experiments in the form of tables may not be feasible. However, using the SRCS technique, we can arrange the rank colors of each configuration to create the images shown in Fig. 4.6. In this figure, the same color scheme as the one appearing in the explanation in Fig. 4.5 has been used, where a darker color indicates a worse performance, and a lighter one a better. Taking a quick glance at Fig. 4.6, and without having to examine any type of numerical data, we can obtain valuable overall information, like:

- in general, *algorithm 1* is the worst in almost all configurations
- *algorithm 3* has, in almost all configurations, a good or very good performance
- for higher change frequencies (higher number of evaluations between changes), *algorithm 3* is the best
- for lower change frequencies, *algorithm 4* is the best
- variations of the severity have, in general, less influence in the performance of the algorithms than variations of the change frequency

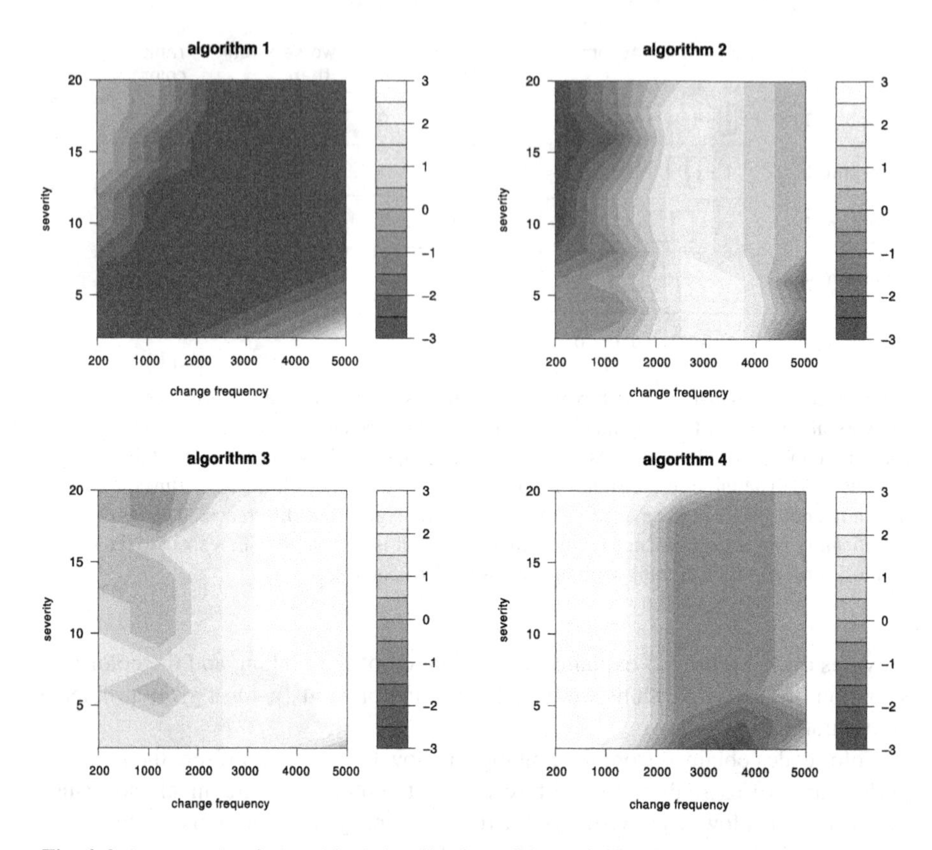

Fig. 4.6 An example of a graphical visualization of the rank-based color scheme for 4 hypothetical algorithms. The visualization shows a comparison of the results of 4 algorithms for different configurations of the factors *severity* and *change frequency* for a problem

Also, figures created using SRCS can be arranged to visualize variations of more than 2 factors (see Fig. 4.7), depending on the practitioner's creativity. These figures can help us to further detect behavioral patterns of the algorithms, and increase our understanding of them.

Finally, although the examples in this chapter used the *avg. offline accuracy* as performance measure, and the KW + MWW combination as statistical tests, the SRCS technique is not restricted to these methods. Other performance measures (avg. offline error, reactivity, etc.) and statistical tests (Friedman, Iman-Davenport, etc) are also valid, as long as their usage is appropriated. In-depth examples of the use of non-parametric statistical tests for comparing optimization algorithms can be found in [8, 10, 11].

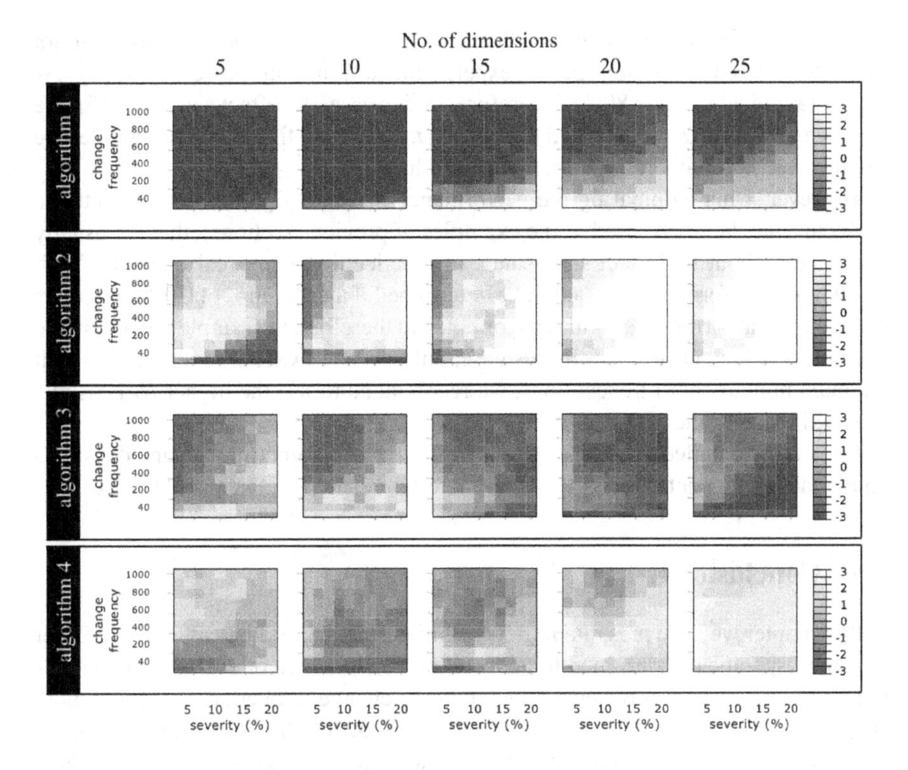

Fig. 4.7 An arrangement of the graphical color scheme of the rankings for visualizing variations of 3 different factors: severity, change frequency, and dimensionality.

4.4 Some Considerations on Statistical Tests

In a statistical hypothesis test, some of the main parameters that determine its outcome are:

- n, the sample size used.
- α, the significance level, or the probability of making a Type I error (*false positive*), i.e., rejecting the null hypothesis H_0 when it is true.
- θ, the effect size, or the minimum difference that can be detected by the test in order to be considered significant, in absolute units.
- π, the power of the test or the sensitivity level, equal to $1-\beta$, where β is the probability of making a Type II error (*false negative*), i.e., accepting the null hypothesis H_0 when it is false.

These parameters are interrelated, and the values of some of them are usually determined from those of the rest, which may be fixed as a result of the experiment's requirements. For example, in a clinical essay for a drug, it could be determined that a minimum increase in blood pressure of 15 mm Hg must be observed in order to consider its effect of practical significance, with a 99% confidence. Although not explicitely stated, in these types of experiments a minimum power is usually expected (a typical value is 80%). Therefore, in this case, the effect size ($\theta = 15$), the significance level ($\alpha = 0.01$), and the power ($\pi = 0.8$) are fixed, and the sample size should be adjusted in order to obtain those values.

However, when comparing the performance of some algorithms in a synthetic problem (like the MPB used in the examples of previous sections), there is usually no concern about the effect size, since, in practice, it has no real meaning. With no additional information, it cannot be determined if a difference of 0.1 fitness units between two algorithms is significant or not, and therefore, the sample size (N_r in the examples) is not constrained. In this case, unless some external requirements limit the maximun amount of executions, we recommend to use the higher N_r possible, as it will increase the power of the test.

For a more detailed introduction to the use of non-parametric tests and questions on the factors that determine them, the interested reader is referred to [13, 16, 19].

4.5 Conclusions

In this chapter we have presented a new technique, SRCS (Statistical Ranking Color Scheme), specifically designed to analyze the performance of multiple algorithms in DOPs over variations of several factors (e.g., change frequency, severity, dimensionality, etc). This technique is especially well-suited when we want to compare algorithms in a all-vs-all manner, for example, when we want to determine which are the best performing ones in a wide range of scenarios.

SRCS uses statistical tests to compare the performance of the algorithms for a given problem configuration, producing a ranking. Since the results of metaheuristics and non-exact algorithms do not generally follow a normal distribution, non-parametric tests are usually preferred. As a practical guideline, a multiple-comparison test must be performed first, like the Kruskal-Wallis test, in order to determine if there are global differences in the performance of the algorithms. Then, a pair-wise test is used, in order to assess individual differences between algorithm pairs, like the Mann-Whitney-Wilcoxon test. This pair-wise test must be adjusted in order to compensate for the family-wise error derived from the performance of multiple comparisons, using, for example, Holm's method. However, these tests are only suggestions that do not affect the way in which SRCS works, and other options can be used (Friedman's test, Iman-Davenport, etc).

The ranking produced is later used to associate color codes to each algorithm result, such that the *relative* performance of each algorithm with respect to the others can be represented in a graphical way. This representation allows us to visualize the results of many algorithms on many configurations in a much more compact way by enhancing differences between the results, and giving thus the researcher the possibility of identifying behavioural patterns more easily.

Like any information compressing technique, SRCS left out part of the information, so its use, either isolated or as a complement to other traditional ways for displaying results (tables and plots), should be evaluated in each case. With SRCS, using rankings for stressing out the differences among algorithms implies not displaying absolute performance values.

Acknowledgements. This work has been partially funded by the project TIN2008-01948 from the Spanish Ministry of Science and Innovation, and P07-TIC-02970 from the Andalusian Government.

Appendix

In this Appendix we provide an implementation of the ranking method of Sect. 4.3, using the R programming language [1].

```
#-----------------------------------------------------------------------------------------------------
# The function for calculating the rank of a set of algorithms using their performance results on a
# *single* problem configuration. Returns a list indexed by algorithm, with the corresponding rank value
# for each of them.
# Parameters:
# - data: a vector with the results of all the algorithms, that is, the 'n' independent repetitions of
#     the performance measure, for each algorithm.
# - group: a vector of factors for the data, indicating, for each corresponding entry of the data vector,
#     the algorithm it belongs to.
# - alpha: the minimum p-value for the tests to assess a significant difference. Defaults to 0.05
# - max: TRUE or FALSE. TRUE means the higher the performance measure, the better. FALSE means the
#     opposite. For example, if the performance measure is the error, set max = FALSE; if the performance
#     measure is the accuracy, set max = TRUE. Defaults to TRUE.
# Example input:
# - data <- c( 2.5, 2.3, ..., 1.2, 0.7, ..., 3.5, 4.1 )
# - group <- factor( "alg1", "alg1", ..., "alg2", "alg2", ..., "alg3", "alg3" )
# Example output:
# - rankList : [ ["alg1"][0], ["alg2"][2], ["alg3"][-1] ]
#-----------------------------------------------------------------------------------------------------
rank <- function( data, group, alpha=0.05, max=TRUE ) {
    # initialize the ranks to 0
    algorithms <- unique( group )
    rankList <- list()
    for( algorithm in algorithms ) {
        rankList[[algorithm]] <- 0
    }

    # calculate the vector of medians for all the algorithms' measures
    medians <- tapply( data, group, median )

    # perform a Kruskal-Wallis test to assess if there are differences among all the results
    dataframe <- data.frame( group, data )
    kruskal <- kruskal.test( data ~ group, data=dataframe )
    if( !is.na( kruskal$p.value ) && kruskal$p.value < alpha ) {

        # post-hoc test: perform a pairwise Mann-Whitney-Wilcoxon (MWW) rank sum test
        # with Holm correction to assess individual differences
        wilcoxon <- pairwise.wilcox.test( data, group, p.adj="holm", exact=FALSE )

        for( algorithm1 in rownames( wilcoxon$p.value ) ) {
            for( algorithm2 in colnames( wilcoxon$p.value ) ) {
                if( !is.na( wilcoxon$p.value[algorithm1,algorithm2] ) &&
                    wilcoxon$p.value[algorithm1,algorithm2] < alpha ) {

                    # there is a significant difference between algorithm1 and algorithm2;
                    # we need to identify which one is the best and which one the worst,
                    # we'll use the median for that purpose, since it is coherent with the
                    # use of the MWW method, which also uses medians
                    if( medians[algorithm1] > medians[algorithm2] ) {
                        best <- algorithm1
                        worst <- algorithm2
                    } else {
                        best <- algorithm2
```

[1] http://www.r-project.org/

```
                    worst <- algorithm1
                }

                # if max==FALSE, swap best and worst
                if( !max ) {
                    tmp <- best
                    best <- worst
                    worst <- tmp
                }

                # update ranks
                rankList[[best]] <- rankList[[best]] + 1
                rankList[[worst]] <- rankList[[worst]] - 1
            }
        }
    }
    }

    return( rankList )
}
```

References

[1] Bartz-Beielstein, T.: Experimental Research in Evolutionary Computation: The New Experimentalism. Natural Computing Series. Springer, Heidelberg (2006), doi:10.1007/3-540-32027-X

[2] Bartz-Beielstein, T., Chiarandini, M., Paquete, L., Preuss, M. (eds.): Experimental Methods for the Analysis of Optimization Algorithms. Springer, Heidelberg (2010), doi:10.1007/978-3-642-02538-9

[3] Branke, J.: Memory enhanced evolutionary algorithms for changing optimization problems. In: Proceedings of the 1999 IEEE Congress on Evolutionary Computation (CEC 1999), vol. 3, pp. 1875–1882. IEEE (1999), doi:10.1109/CEC.1999.785502

[4] Branke, J.: Evolutionary Optimization in Dynamic Environments. Genetic algorithms and evolutionary computation, vol. 3. Kluwer Academic Publishers, Massachusetts (2001)

[5] Chen, C.-H., Härdle, W., Unwin, A., Friendly, M.: Handbook of Data Visualization. Springer Handbooks of Computational Statistics. Springer, Heidelberg (2008), doi:10.1007/978-3-540-33037-0

[6] Cruz, C., González, J., Pelta, D.: Optimization in dynamic environments: a survey on problems, methods and measures. In: Soft Computing, pp. 1–22 (2010), doi:10.1007/s00500-010-0681-0

[7] De Jong, K.: An analysis of the behavior of a class of genetic adaptive systems. PhD thesis, University of Michigan, Ann Arbor, MI, USA (1975)

[8] Demšar, J.: Statistical comparisons of classifiers over multiple data sets. Journal of Machine Learning Research 7(1) (2006)

[9] Fonseca, V.G., Fonseca, C.M.: The attainment-function approach to stochastic multiobjective optimizer assessment and comparison. In: Bartz-Beielstein, T., Chiarandini, M., Paquete, L., Preuss, M. (eds.) Experimental Methods for the Analysis of Optimization Algorithms, pp. 103–130. Springer, Heidelberg (2010), doi:10.1007/978-3-642-02538-9_5

[10] García, S., Herrera, F.: An extension on "statistical comparisons of classifiers over multiple data sets" for all pairwise comparisons. Journal of Machine Learning Research 9, 2677–2694 (2008)

[11] García, S., Molina, D., Lozano, M., Herrera, F.: A study on the use of non-parametric tests for analyzing the evolutionary algorithms' behaviour: a case study on the cec'2005 special session on real parameter optimization. Journal of Heuristics 15(6), 617–644 (2009), doi:10.1007/s10732-008-9080-4

[12] Gräning, L., Jin, Y., Sendhoff, B.: Individual-based management of meta-models for evolutionary optimization with application to three-dimensional blade optimization. In: Yang, S., Ong, Y.-S., Jin, Y. (eds.) Evolutionary Computation in Dynamic and Uncertain Environments. SCI, vol. 51, pp. 225–250. Springer, Heidelberg (2007), doi:10.1007/978-3-540-49774-5_10

[13] Hollander, M., Wolfe, D.: Nonparametric Statistical Methods, 2nd edn. John Wiley & Sons, Inc. (1999)

[14] Holm, S.: A simple sequentially rejective multiple test procedure. Scandinavian Journal of Statistics 6(2), 65–70 (1979)

[15] Kruskal, W.H., Allen Wallis, W.: Use of ranks in one-criterion variance analysis. Journal of the American Statistical Association 47(260), 583–621 (1952)

[16] Russell, V.: Lenth. Some practical guidelines for effective sample size determination. The American Statistician 55(3), 187–193 (2001), doi:10.1198/000313001317098149

[17] López-Ibáñez, M., Paquete, L., Stützle, T.: Exploratory analysis of stochastic local search algorithms in biobjective optimization. In: Bartz-Beielstein, T., Chiarandini, M., Paquete, L., Preuss, M. (eds.) Experimental Methods for the Analysis of Optimization Algorithms, pp. 209–222. Springer, Heidelberg (2010), doi:10.1007/978-3-642-02538-9_9

[18] Mann, H.B., Whitney, D.R.: On a test of whether one of two random variables is stochastically larger than the other. The Annals of Mathematical Statistics 18(1), 50–60 (1947), doi:10.1214/aoms/1177730491

[19] Randles, R.H., Wolfe, D.: Introduction to the Theory of Nonparametric Statistics. John Wiley & Sons, Inc. (1979)

[20] Reyes-Sierra, M., Coello, C.: A study of techniques to improve the efficiency of a multi-objective particle swarm optimizer. In: Yang, S., Ong, Y.-S., Jin, Y. (eds.) Evolutionary Computation in Dynamic and Uncertain Environments. SCI, vol. 51, pp. 269–296. Springer, Heidelberg (2007), doi:10.1007/978-3-540-49774-5_12

[21] Weicker, K.: Performance Measures for Dynamic Environments. In: Guervós, J.J.M., Adamidis, P.A., Beyer, H.-G., Fernández-Villacañas, J.-L., Schwefel, H.-P. (eds.) PPSN 2002. LNCS, vol. 2439, pp. 64–73. Springer, Heidelberg (2002), doi:10.1007/3-540-45712-7_7

[22] Wilcoxon, F.: Individual comparisons by ranking methods. Biometrics Bulletin 1(6), 80–83 (1945), doi:10.2307/3001968

[23] Yang, S.: Explicit memory schemes for evolutionary algorithms in dynamic environments. In: Yang, S., Ong, Y.-S., Jin, Y. (eds.) Evolutionary Computation in Dynamic and Uncertain Environments. SCI, vol. 51, pp. 3–28. Springer, Heidelberg (2007), doi:10.1007/978-3-540-49774-5_1

Chapter 5
Dynamic Combinatorial Optimization Problems: A Fitness Landscape Analysis

Philipp Rohlfshagen and Xin Yao

Abstract. The role of representations and variation operators in evolutionary computation is relatively well understood for the case of static optimization problems thanks to a variety of empirical studies as well as some theoretical results. In the field of evolutionary dynamic optimization very few studies exist to date that explicitly analyse the impact of these elements on the algorithm's performance. In this chapter we utilise the fitness landscape metaphor to review previous work on evolutionary dynamic combinatorial optimization. This review highlights some of the properties unique to dynamic combinatorial optimization problems and paves the way for future research related to these important issues.

5.1 Introduction

The field of evolutionary dynamic optimization (see [7, 20, 27, 44]) is concerned with the application of evolutionary algorithms (EAs) to the class of dynamic optimization problems (DOPs). Unlike static optimization problems, the specifications of DOPs are time-variant and potentially affect the problem's fitness landscape (see section 5.3) structurally over time. This often necessitates the adaptation of solutions found so far to maintain satisfactory quality and feasibility, particularly in the case of online optimization where solutions need to be implemented continuously.

Philipp Rohlfshagen
School of Computer Science and Electrical Engineering,
University of Essex, Colchester CO4 3SQ,
United Kingdom
e-mail: prohlf@essex.ac.uk

Xin Yao
School of Computer Science,
University of Birmingham, Birmingham B15 2TT,
United Kingdom
e-mail: xin@cs.bham.ac.uk

E. Alba et al. (Eds.): Metaheuristics for Dynamic Optimization, SCI 433, pp. 79–97.
springerlink.com

Considerable effort has thus been devoted in recent years to develop new techniques that efficiently *track* high quality solutions as closely as possible over time.

The tracking of high quality solutions requires the dynamics of a problem to have some exploitable structure (similar to the requirement that static optimization problems must have an exploitable structure) and most practitioners assume that the global optima of successive problem instances encountered by the algorithm are correlated to some extent. However, such correlations or indeed any structural properties of the problem's fitness landscape depend strongly on the algorithm used to solve the problem: whereas the actual dynamics of the problem are beyond the control of the algorithm[1], the dynamics *observed* by the algorithm are determined by the chosen representation and variation [9, 33]. In other words, "in a dynamic environment, in addition to the (static) characteristics of the fitness landscape, the representation influences the characteristics of the fitness landscape dynamics..."[8, p 765]. It is thus surprising to note that "... the role of representations in dynamic environments has been largely neglected so far" [8, p 764]. This observation is particularly relevant to the combinatorial domain where algorithm-independent fitness landscapes may be difficult to construct.[2] This has had a fundamental impact on evolutionary dynamic optimization: the probably two most significant consequences concern the construction of general benchmark problems and the identification and classification of problem dynamics. These aspects form the core of this chapter.

We first introduce the field of evolutionary dynamic optimization with special emphasis on combinatorial optimization problems and their dynamics in section 5.2. In section 5.3 we formally introduce the notion of a problem's fitness landscape and extend the framework to the dynamic domain. We next review some novel representations and variation operators proposed for the dynamic domain and assess the most commonly used benchmark generators in section 5.4. We subsequently review a selection of studies concerned with the dynamics observable in dynamic combinatorial optimization problems (section 5.5) and conclude the chapter in section 5.6 where we highlight some important directions for future research in evolutionary dynamic optimization.

5.2 Evolutionary Dynamic Optimization

The field of evolutionary computation provides a variety of nature-inspired metaheuristics that have been utilised successfully to obtain high quality solutions to NP-hard optimization problems. Evolutionary algorithms (EAs) are population-based

[1] This is not true for the case of *time-linkage* where the solution quality obtained by an algorithm at any moment in time has an impact on the dynamics of the problem in the future (see [4, 5]); although we do not consider such cases in this chapter, it should be noted that the dynamics observed by the algorithm are still determined by the algorithm's representation and variation operators.

[2] The natural order of the numerical domain entails an implicit specification of algorithm-independent fitness landscapes (e.g., surface of a real-valued function).

global search algorithms inspired loosely by the general principles of evolutionary systems and attempt to obtain solutions of increasing quality by means of *selection*, *crossover* and *mutation*: selection favours those individuals in the algorithm's population (the multiset P) that represent solutions of higher quality (exploitation) whereas crossover and mutation, the algorithm's *variation operators*, generate offspring from those individuals to advance the search (exploration); mutation operators usually perturb individuals by a small degree whereas crossover operators combine two or more individuals to yield offspring that inherit genetic material from all parents.

EAs are usually understood to be *black-box algorithms* (see [13]) as they assume no knowledge regarding the probability distribution of problem instances encountered. EAs are thus often applied to difficult problems about which little or no information exists. An optimization problem $f : X \times \Delta \rightarrow \mathbb{R}$ (also known as the *objective function*) is a mapping from a search space $X = \{x_1, x_2, \ldots\}$ to the domain of reals, where $\Delta = \{\delta_1, \delta_2, \ldots\}$ are the parameters of the function; the value $f(x_i, \delta_j) \in \mathbb{R}$ indicates the quality of x_i with respect to problem instance δ_j (see [15]); the space of all f-values is denoted $\hat{f}$. The goal is usually to find the *global optimum* $x^\star \in X$ such that $f(x^\star, \delta) \succ f(x, \delta),\ \forall x \in X$, where $\succ \in \{\geq, \leq\}$.[3] Finally, the majority of real-world problems have inequality and/or equality constraints a solution must satisfy, specified by the functions $g : X \times \Delta \rightarrow \mathbb{R}^m$ and $h : X \times \Delta \rightarrow \mathbb{R}^p$, respectively; a solution is considered feasible if $g_i(x, \delta) \leq 0,\ i = 1, \ldots, m$ and $h_j(x, \delta) = 0,\ j = 1, \ldots, p$ (see [29]).

The class of dynamic optimization problems (DOPs) is more difficult to define as, in principle, any component of f may change over time (we assume that the definition of f entails the specification of the search space X) and the Handbook of Approximation Algorithms and Metaheuristics [25] states that a general definition of DOPs does not exist (as of 2007 and we are not aware of any widely accepted definition since). In this chapter we concentrate exclusively on those problems where the dynamics affect the constraints or parameters of the objective function as this seems to correspond to the most commonly considered scenario in the literature. The dynamic equivalent of the static problem $f(x, \delta)$ is $f(x, \delta(t))$; the dependency of δ on time $t \geq 0$, equivalent to the problem's dynamics, describes a trajectory through the space of all problem instances:

$$\delta(T) \longrightarrow \delta(T+1) \longrightarrow \delta(T+2) \longrightarrow \ldots$$

where $T\tau \leq t < (T+1)\tau$ and τ, a constant, is the period (duration) of change; we assume that time advances with every call to the objective function. The transitions are governed by a mapping $\mathscr{T} : \Delta \rightarrow \Delta$ that maps from one problem instance to another such that $\delta(T+1) = \mathscr{T}(\delta(T))$.[4]

[3] It should be noted that $\max_x\{f(x)\} \equiv \min_x\{-f(x)\}$; the terms *fitness* and *fitness landscape* are usually associated with maximisation whereas the terms *cost* and *cost surface* are the preferred choice in the case of minimisation.

[4] A more general definition allows for both $\mathscr{T}$ and τ to be time-variant. However, these cases are rarely considered in the literature.

5.3 Dynamic Fitness Landscapes

EAs have become a popular choice of search algorithm to tackle difficult optimization problems, particularly when traditional approaches are not applicable. In this section we review the role played by the algorithm's representation and variation operators. This is often assessed by means of the fitness landscape metaphor which is reviewed also. The remainder of this section discusses how the fitness landscape metaphor may be extended from static functions to those that are dynamic.

5.3.1 Representations and Variation Operators

The choice of representation and variation operators is crucial to the success of the EA in obtaining high quality solution to the problem of interest: "... the representation of an individual in the population and the set of operators used to alter its genetic code constitute probably the two most important components of the system, and often determine the system's success or failure." [2, C3.1:1]. This observation is particularly relevant to the field of combinatorial optimization which entails the set of all functions that have discrete search spaces.[5] In order to efficiently traverse a space X, it is often necessary to transform X into a domain that is suitable for the algorithm. This transformation is achieved using *representations*: a representation is a mapping $f_g : X_g \to X_p$ that transforms an element $x_g \in X_g$ to an element $x_p \in X_p$; the former is commonly referred to as a *genotype* and the latter as a *phenotype*. The objective function then corresponds to the composite mapping $f = f_p \circ f_g = f_p(f_g(x_g), \delta)$, where $f_p : X_p \times \Delta \to \mathbb{R}$ (see [35]). If no representation is used, $x_{gi} = x_{pi}$, $\forall x_{gi} \in X_g$. The space in which the algorithm search is generally referred to as the *search space* and the elements of x_g correspond to the problem's *decision variables*.

A representation may be injective (approximate), bijective (1:1) or surjective (redundant) and the exact choice of representation often depends both on f and the algorithm to be used: "... search algorithms are usually efficient in using a particular representation and not so efficient in using other types of representation." [2, C1.1:1]. It is generally recommended to use the most natural representation for the problem in question. In other words, the characteristics of the problem should determine the representation and not the other way around and if a representation is used at all (i.e., if it is not possible to search the space X_p directly), it should be "kept as simple as possible and obey some structure preserving conditions ..." [2, C.1.2:3].

Different types of representations have often led to different classes of algorithms: genetic algorithms, for instance, are often associated with binary representations, whereas evolution strategies are used almost exclusively in the continuous domain. One reason for this divergence of techniques is the need for appropriate variation operators given the choice of representation: whereas the representation determines the nature of the elements in X_g, the topology of the space is determined

[5] X here refers to the space of *actual* solutions, not their representation, as discussed in the remainder of this section.

by the algorithm's neighbourhood structure: the EA traverses the search space by means of variation operators that determine the points in X_g to be *considered* next (selection subsequently determines which points will be *utilised* by the algorithm), usually based on the points already sampled by the algorithm.

The points considered by the algorithm form the *neighbourhood* $N(x)$ of a point $x \in X$: $N(x) = \{y \in X \mid p(y = \xi(x)) > 0\}$ (see [23]), where ξ are the algorithm's variation operators. In the case of a simple pseudo-Boolean hill climber, the neighbourhood may be defined as $N(x) = \{y \in X \mid d_g(x,y) = c\}$, where $d_g : X_g \times X_g \to \mathbb{R}$ is a distance metric that describes how similar two points x_i and x_j are. A typical distance measure for pseudo-Boolean spaces is the *Hamming* distance d_H:

$$d_H(x,y) = \sum_{i=1}^{n} |x_i - y_i|, \; x,y \in \mathbb{B}^n, \tag{5.1}$$

If $c = 1$, for instance, the point $x = 000$ has a neighbourhood $N(x) = \{001, 010, 100\}$, where each element is a Hamming distance of 1 away from x.

EAs typically use stochastic variation operators and traditionally employ both *mutation* and *crossover* operators. Most mutation operators are *global* search operators as there is a non-zero probability any element in X_g may be sampled. Bit-wise mutation, for instance, inverts any element in x with probability p_m (mutation rate). The neighbourhood of any point $x \in X_g$ under mutation should thus be viewed as a probability distribution (possibly defined over an appropriate distance measure such as the Hamming distance). Crossover operators, on the other hand, combine two or more individuals with certain probability (crossover rate p_c) to yield an offspring that inherits genetic material from all parents. Given a particular algorithm, any two points a and b may then be considered neighbours if b can be reached via a single application of ξ to a. In the case of stochastic operators, it is the likelihood that $\xi(a) = b$ that determines the proximity in the space. It follows that the distance between non-neighbours is the length of the shortest path between them (if one exists) [23].

The following example illustrates the role played by the representation and variation operators. We assume some function $f : \mathbb{N} \times \Delta \to \mathbb{R}$ and consider the probably simplest EA, the $(1+1)$–EA, which manipulates a single solution x at any moment in time. A common choice to represent integers is binary where an integer y is given by some binary vector x such that $y = \sum_{i=1}^{n} x_i 2^{i-1}$. A binary representation allows one to use well-established mutation operators: an offspring is created using mutation only with bit-wise probability $p_m = 1/n$, such that $x_i = 1 - x_i$ if $r < p_m$ for $0 \leq i \leq n$ where r is a uniform random number in $[0,1]$. However, it is immediately obvious that the transformation from binary to integer is non-linear: the neighbouring integers 7 and 8, for instance, are represented by 0111 and 1000, respectively, which are a maximum Hamming distance apart. *Gray codes* were proposed to address the issue of *Hamming cliffs* [36]: Gray codes transform binary strings such that any neighbouring integers are represented by neighbouring bit-strings. This transformation gives rise to a different neighbourhood structure (assuming that the mutation operator remains the same) and hence may alter the trajectory of the search.

Finally, one can search the space of integers directly, eliminating the need to transform genotypes to phenotypes in the first place. This, however, requires alternative variation operators and these may be more difficult to define.

A tool commonly used to analyse the impact of genotype-phenotype transformations and the role played by the variation operators is the fitness landscape metaphor; this is discussed next.

5.3.2 *Fitness Landscapes*

A significant number of different representations have been proposed for a variety of domains and their impact on the algorithm's behaviour is often estimated by means of the *fitness landscape* metaphor. The concept of a fitness landscape has its origin in evolutionary biology (Sewall Wright; 1920s) and corresponds to an abstract visualisation of the problem's topological properties in terms of a landscape that consists of peaks, valleys and ridges. It was popularised in the realm of optimization by Jones [21]. A fitness landscape may be specified by the tuple:[6]

$$\mathscr{L} = (X_g, f_{p \circ g}, d_g), \tag{5.2}$$

The distance metric d_g should ensure non-negativity, symmetry and the triangle-inequality (see [30, p 454]) and may often be formulated independently of the algorithm (e.g., Hamming distance). However, the true distances between elements in the search space are determined by the algorithm and may differ from those defined by an auxiliary distance metric.[7] The distances between elements in X_p are also determined by the algorithm (through the combination of variation operators and representation) and again, these distances may deviate from those determined by an auxiliary and algorithm-independent measure. Finally, a distance d_f may be defined for $\hat{f}$ which most commonly corresponds to $d_{\hat{f}}(f(x), f(y)) = |f(x) - f(y)|$.

The algorithm's neighbourhood structure influences the topological features of the space X_g, including the presence of *local optima*: a local optimum is a locally optimal point x^* such that $f(x^*) \succ f(x), \forall x \in N(x^*)$. As mentioned previously, most EAs employ global stochastic variation operators and hence the neighbourhood corresponds to a probability distribution over all elements in X_g; nevertheless, local optima still exist as some points are statistically unlikely to be sampled within the time available.

It is clear from the definition of local optima that neither the chosen representation nor the algorithm's variation operators may be assessed in isolation as it is their combined effect that determines the actual neighbourhood structures utilised

[6] It should be noted that the mapping f_p may be composite itself and may thus further influence the structural properties of the fitness landscape: *a priori* heuristic knowledge, for instance, may be used to scale f-values to account for additional aspects such as *effective distance* [3].

[7] It is common to approximate the distances imposed by the algorithm using a simpler, non-stochastic, measure. For instance, fitness distance correlations [22] are often determined using the Hamming distance as an approximation of bit-wise mutations with $m_p = 1/n$.

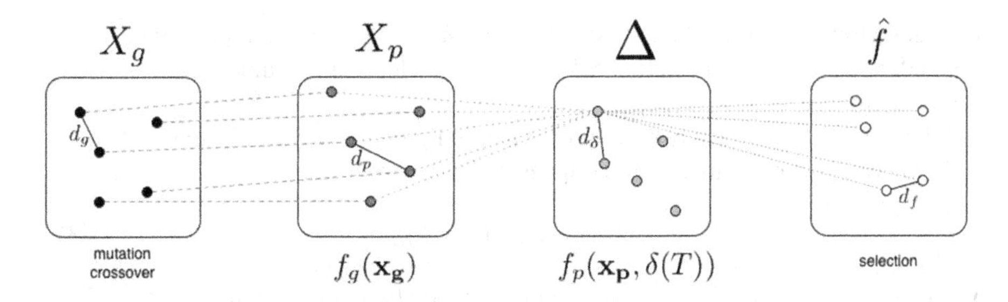

Fig. 5.1 Four different spaces that determine a dynamic fitness landscape, alongside the metrics used to define distances between the members of each space: the algorithm traverses the genotype space, X_g, by means of crossover and mutation and the mapping f_g maps genotypes to phenotypes. The latter corresponds to solutions to the problem of interest that may be mapped onto $\mathbb{R}$ depending on the current problem instance $\delta(T)$.

by the algorithm: whereas crossover and mutation act on elements in X_g, selection acts on elements in X_p and the relationship between these spaces is determined by the representation.[8]

5.3.3 Dynamic Fitness Landscapes

The notion of a fitness landscape, as defined in section 5.3.2, still applies in the dynamic case. However, the fitness landscape may now change over time, depending on $\mathscr{T}$ and τ and hence the space Δ needs to be considered in addition to X_g, X_p and $\hat{f}$. We assume the space Δ to be metric and define a distance measure d_Δ that indicates the degree of similarity between any two instances δ_i and δ_j. This is commonly known as the *magnitude of change*. However, it is immediately obvious that the distance between two instances is just one aspect of the transition. Other aspects include, for instance, the number of parameters affected and the degree to which each individual parameter has been changed; the *direction of change* is equally important (see [41]). The transformations involved in dynamic fitness landscapes in the combinatorial domain are depicted in Figure 5.1.

It follows that the choice of representation and variation operators not only determines the structure of the static fitness landscape at time T but also the transition from one fitness landscape to another. Numerous (high-level) classifications were proposed to describe different types of dynamics: in [9], Branke considers frequency of change, severity of change, predictability of change and cycle length/cycle accuracy. De Jong [12] (cited in [9]) distinguishes between alternating problems, problems with changing morphology, drifting landscapes, and abrupt and discontinuous problems. Younes et al. [52] further differentiates between dimensionality and non-dimensionality changes as well as dynamically significant and insignificant

[8] In many cases it is possible to achieve similar effects by *either* changing the representation or the variation operators. This is known as *isomorphism of the fitness landscape*; see [31] cited in [35, p 82]

changes. However, none of these classifications takes the aforementioned algorithm-dependency into account; section 5.5.2 reviews a selection of studies that explicitly looked at this issue.

The most extensive study regarding dynamic fitness landscapes is due to Richter [32] where a dynamic fitness landscape is defined as follows:

$$\mathscr{L}_D = (X, N, \Gamma, F, \phi), \tag{5.3}$$

where X, N are the search space and neighbourhood function as defined previously.[9] Γ is a time set that defines the temporal aspects of the dynamics and F is the set of functions encountered by the algorithm, such that for every $f \in F$, $f : X \times \Gamma \to \mathbb{R}$. Finally, the mapping ϕ is used to describe the trajectory of functions encountered: $\phi : F \times X \times \Gamma \to F$. Richter utilises this definition of a dynamic fitness landscape to propose a hierarchy of fitness landscapes, both for discrete and continuous time and analyses their topological properties in terms of modality, ruggedness, information content as well as dynamic severity; the focus is on dynamic fitness landscapes that exhibit spatio-temporal chaotic behaviour. The remainder of this chapter approaches the concept from the algorithm-design point of view, looking at different representations and variation operators that have been proposed to improve an algorithm's performance in the dynamic domain.

5.4 Dynamic Fitness Landscapes in Practice

This section first reviews some of the representations and variation operators developed specifically for the dynamic domain, highlighting the potential benefit of novel neighbourhood structures given the dynamics of the problem. This is followed by a review and assessment of popular benchmark problems used to evaluate the performance of such new algorithms.

5.4.1 Neighbourhood Structures for the Dynamic Domain

Although EAs are generally considered promising candidates for the class of DOPs, numerous issues were identified that may limit an algorithm's ability to efficiently track the moving optimum. In particular, convergence of the algorithm's population has been identified as a primary concern and numerous techniques were subsequently proposed to address this and other issues. These techniques, which are usually modifications of existing EAs, may broadly be classified as follows: diversity-preserving techniques, memory, representations, variation operators, memetic algorithms, speciation and multi-populations and anticipation and prediction. Below we review some techniques that propose novel representations or variation operators for dynamic combinatorial optimization problems.

[9] Richter includes the neighbourhood function rather than a distance measure in the definition of the fitness landscape; since the distance measure follows directly from the neighbourhood function, these definitions are considered synonymous.

Novel representations are amongst the earliest approaches in evolutionary dynamic optimization. The design of representations has focussed on two primary aspects: an implicit memory that learns short-term dynamics and flexible genotypes that may map to a variety of distinct phenotypes, subject to very minor modifications. Diploid (e.g., [17]) and polyploid (e.g., [18]) representations were suggested as a form of implicit memory: each individual essentially encodes multiple distinct solutions to the problem of interest, only one of which is expressed at any one time. Such approaches are also thought to generate additional diversity. Gaspar and Collard [16] propose the Folding GA where the representation includes meta-genes that control a transcription step to modify the expression (i.e., the mapping from X_g to X_p) of the individual. The transcription makes use of mappings based on duality and partial mirroring operations. A similar concept has been explored by Yang [48], who proposes a duality scheme for dynamic functions that is enforced externally. Each individual consists of a primal chromosome; a dual chromosome is defined as the chromosome that is maximum distance from the primal in some Euclidean space. The individual expresses whichever chromosome is of higher fitness, allowing individuals to respond to significant changes in the environment; subsequent work has proposed an adaptive variant [42]. Dasgupta and McGregor [11] explore similar principles, resulting in the Structured GA, which uses redundant genetic material and a gene activation mechanism that exploits a multi-layered chromosome structure.

Similar to the case of representations, one of the earliest approaches to evolutionary dynamic optimization corresponds to a novel variation operator: hypermutations [10] attempt to maintain high levels of diversity in an evolving population by temporarily increasing the algorithm's mutation rate following a change (also known as *triggered hypermutation*); an appropriately chosen mutation rate should allow the algorithm to search in the vicinity of previously found high-quality solutions. Similarly, many additional mutation operators have been proposed that allow for enlarged neighbourhoods, either constantly or reactively (i.e., following a change). However, almost all techniques focus on the continuous domain and very few apply in the combinatorial case. An example of the latter is due to Woldesenbet and Yen [45] who propose a variable (self-adaptive) relocation operator that moves the entire population that is (partially) adapted to the current state of the problem: using the individuals' history of f-values, a new population is generated that is better suited for the new environment encountered. The authors mention that this mechanism ensures the algorithm reuses as much information as possible from the previous evolutionary history.

This brief review highlights three important issues: first, the majority of novel variation operators have been proposed for the continuous domain where the function's fitness landscape is defined unambiguously by the function definition itself. This allows the specification of general (algorithm-independent) dynamic benchmark functions; as the next section will show, creating general benchmark problems for the combinatorial domain has proven difficult. Second, almost all techniques aim to widen the neighbourhood of points considered by the algorithm, either consistently or reactively (i.e., when a change takes place); the new neighbourhood is sampled with a considerable degree of stochasticity, expressing the uncertainty about the future structure of the fitness landscape. In other words, a large number of points in the vicinity of previously high-quality solutions are considered equiprobable in their utility following a change. Although this may allow the algorithm to sample points that are further away more easily, it also removes local structural properties that may be exploited otherwise. Third, the comparison of these techniques to their unmodified counterparts (i.e., the algorithms developed for static functions) often demonstrates an increase in performance yet the modifications are not commonly employed in the static case. This implies that the modifications shown beneficial in the dynamic case do not actually improve the algorithm's performance on the individual problem instances encountered; limiting the negative impact of these modifications on the individual problem instances may thus improve the algorithm's overall performance.

The systematic evaluation and comparison of different algorithms, such as those considered above, necessitates the use of unified benchmark problems these algorithms may be tested on (see [52]). In order to account for a variety of different dynamics, numerous benchmark problems have been proposed in the past; these are discussed next.

5.4.2 Dynamic Problem Benchmark Generators

The concept of fitness landscapes is often applied retrospectively to a specific optimization problem in order to deduce particular attributes that may determine whether an algorithm will perform well. In particular, numerous measures for problem difficulty have been developed that attempt to predict whether a problem is difficult or not by looking at the structural properties of the problem's fitness landscape. A well-known example is the *fitness distance correlation* (FDC; [22]): problems that have a FDC close to -1 are considered easy as points of higher fitness are, on average, closer to the global optimum. Conversely, if the FDC is close to 0, the f-value of a point does not imply anything about its genotypic distance to the global optimum. However, it is relatively easy to show counter-examples that have FDCs close to 1 (i.e., almost all search points lead away from the global optimum) yet are easy for an EA (e.g., short path problems; also see [1, 19]). Numerous other measures have been proposed also, including ruggedness (epistasis), neutrality, modality and information content. It is important to bear in mind that none of these measures may predict perfectly the difficulty of a complex problem [19].

The fitness landscape metaphor has also been used prospectively to design new problems: the NK fitness landscape [24], for instance, provides a general mechanism to generate problems with a tuneable degree of epistasis; the interdependencies of the problem's decision variables subsequently materialise as different degrees of ruggedness (lack of local correlation) in the fitness landscape. The increase in ruggedness is due to a higher information content of the underlying function, incrementally removing exploitable structural properties and thus often making the problem harder to solve. Similarly, some of the most popular benchmarks in dynamic evolutionary computation were modelled after the fitness landscape metaphor as reviewed next.

The three most widely used benchmarks are due to Branke (MOVING PEAKS; [7]), Morrison (DF1; [27]) and Yang (XOR; [49]). The former two are based on the continuous domain and model the search space as a "field of cones" [28], where each cone may be controlled individually to model different ranges of dynamics. In the two-dimensional case of DF1, for instance, the base function is given by

$$f(x,y) = \max_{i=1,\dots,N} [h_i - r_i \cdot \sqrt{(x-x_i)^2 + (y-y_i)^2}], \tag{5.4}$$

where N is the number of cones, each at location (x_i, y_i), and with height h_i and slope r_i. The initial morphology is randomly generated within the bounds specified by the user and the dynamics are modelled using a logistic function $y(t) = \alpha y(t-1)(1-y(t-1))$, where $\alpha \in [1,4]$ is a constant. The logistic function may be used to generate different trajectories (depending on α) ranging from static to recurrent and chaotic.

These continuous benchmarks are a literal realisation of the fitness landscape metaphor where the structure (and dynamics) of the fitness landscape may be controlled precisely by a small set of parameters. This is made possible by the natural order of the continuous domain[10] and a similar degree of control is significantly more difficult to achieve in the combinatorial domain as Younes et al. [52] point out: "... in discrete optimization, we cannot define an algorithm-independent landscape that can be made time-dependent to simulate dynamic environments." [52, p 27].[11] The only widely accepted benchmark for the combinatorial domain sidesteps this issue and is exclusively defined for problems where $x_p \in \mathbb{B}^n$.

XOR generates a dynamic version of any static binary problem: given a static fitness function $f(x)$, where $x \in \{0,1\}^n$, its dynamic equivalence is simply

$$f(x(t) \oplus m(T)) \tag{5.5}$$

where $\oplus$ is the bit-wise exclusive-or operator [49]. The period index $T = \lceil t/\tau \rceil$ is determined by the update period τ (i.e., $1/\tau$ is the frequency of change). The vector $m(T) \in \{0,1\}^n$, initially $m(0) = 0$, is a binary mask for period k, generated as follows: $m(T) = m(T-1) \oplus p(T)$, where $p(T) \in \{0,1\}^n$ is a randomly created

[10] It should be noted that although the continuous domain entails an algorithm-independent way to specify fitness landscapes, the true fitness landscape traversed by the algorithm still depends on the specifications of the algorithm.

[11] A fitness landscape may be viewed as a *fixed* (given some algorithm-independent metric d_p) topological structure that the algorithm traverses: the analysis of the landscape may hint at specific properties an algorithm should possess in order to efficiently locate regions of higher fitness. For instance, a specific mutation operator may allow the algorithm to "jump" off a local optimum to reach another basin of attraction. However, the ability of an algorithm to do so in effect transforms the fitness landscape and the relevance of a chosen variation operator should be evaluated by the *resulting* fitness landscape, rather than the algorithm's behaviour in a particular predetermined fitness landscape.

template for period T that contains exactly $\lfloor \rho n \rfloor$ ones. The value of $\rho \in [0,1]$ thus controls the magnitude of change which is specified as the Hamming distance between two binary points. It follows that the algorithm used to optimise the function is required to invert ρn bits to return to its previous position. XOR has also been extended to generate cyclical and noisy cyclical environments (see [50, 51]).

An initial analysis of XOR highlights that the problem's search space is fully preserved by the xor operation: as all search points are rotated to the same degree, their positions relative to one another are preserved [39] and an extended analysis using dynamical systems shows that the rotations are equivalent to an additional mutational step that may be taken whenever a change is to take place [40]. There is thus a need for dynamic combinatorial benchmarks that are structurally time-variant and numerous studies have made use of dynamic variants of the single and the (multi-dimensional) knapsack problem has been used on occasions to test and validate different EAs (e.g., [8, 10]). However, as Branke et al. [9, p 1433] point out: "little has been done to characterize and understand the nature of change in a real-world problems". This makes it difficult to assess how general individual scenarios are and hence whether the EA would behave similarly on different problem. Uyar and Uyar [41] raise a similar concern: the authors comment how the magnitude of change by itself is insufficient to characterise the dynamics of the problem as other attributes such as the direction of change (especially constraints) play an important role in the fitness landscape transformation.

5.5 Understanding Combinatorial Problem Dynamics

The majority of work in the field of evolutionary dynamic optimization attempts to improve an algorithm's performance by *transferring knowledge* from one moment in time to another. According to a recent review of the field, "a natural attempt to speed up optimization after a change would be to somehow use knowledge about the previous search space to advance the search after a change. If, for example, it can be assumed that the new optimum is "close" to the old one, it would certainly be beneficial to restrict the search to the vicinity of the previous optimum." [20, p 311]. However, as the previous sections have highlighted, the notion of distance (and hence what constitutes "small") is determined by the algorithm's implemented neighbourhood structure. In this section we first evaluate why (genotypic) distances between successive global optima have played such an important role in most empirical studies and subsequently discuss some of the issues that arise from the algorithm-dependent structure of a problem's fitness landscape.

5.5.1 Distance to the Optimum

In order to locate a function's global optimum $x^{\star}$, algorithms need to traverse the search space X, usually starting from a uniformly random point, using as guidance only the values in $\hat{f}$ that correspond to the points sampled by the algorithm (*black-box* scenario; see [13]). In order to solve a class of problem instances, one usually expects some structural properties and indeed, it has been shown that over all functions $f : A \rightarrow B$ (A and B are finite sets, B totally ordered [14]) closed under permutation, any two algorithms will perform equal given the lack of such exploitable structures (No Free Lunch; see [14, 46]). It follows that practitioners are not generally interested in the set of all functions but instead assume the functions to be "reasonable" [38] in that they are "simple" and "natural" [14]; an algorithm may subsequently be superior than another depending on how well *aligned* it is with the underlying probability distribution of functions encountered [46, p67].

One of the most common assumptions is that small changes to a point $x \in X_g$ will, at least on average, result in small changes in $f_p(f_g(x_g))$ (i.e., $d_g \propto d_{\hat{f}}$).[12] Such assumptions are reflected in an algorithm's design and Droste et al. [14] consider as *reasonable* algorithms "all search heuristics which have no a priori preference of search regions, which prefer to base their search more on evaluated search points with a high f-value (fitness-based selection), and which prefer to look at nearer (Hamming) neighbors." [14, p139]. The latter assertion (i.e., that algorithms should prefer to look at nearer Hamming neighbours) is strongly dependent on the *locality* of the algorithm's neighbourhood structures: if $d_p \propto f_{\hat{f}}$ holds, then $d_g \propto d_p$ ensures that these structural properties are exploitable. In other words, locality describes how well the algorithm *preserves* the natural structure of the problem; if locality is absent, any two neighbouring points in X_g may correspond to an arbitrary pair of points in X_p.

The majority of studies in dynamic evolutionary computation assume that successive problem instances encountered are correlated to one another and in particular that genotypic distances between successive global optima are small. In the case of pseudo-Boolean optimization, for instance, a randomly chosen initial search point x_0 is, in expectation, a Hamming distance of $n/2$ from $x^\star$ and has a distance of at least $n/3$ to the global optimum with overwhelming probability (see [43]). Thus, if the displacement of the global optimum is less than the expected distance to the global optimum following a random restart, re-locating the global optimum may be more efficient: as the majority of EAs search in the vicinity of already sampled points, using as guidance the f-values of new samples, the distance to the global optimum is crucial to the time it may take the algorithm to locate $x^\star$.

This assumption that the displacement of the global optimum is relatively small may be labelled the *distance-based assumption*: the likelihood that a particular point $x \in X_g$ may correspond to the next global optimum $x^\star(T+1)$ depending on the distance $d_g(x, x^\star(T+1))$. The distance-based assumption is probably the most widely considered one in the literature as it allows practitioners to define the *magnitude of change* in similar terms as the *frequency of change*; in fact, XOR implements this concept precisely (equation 5.5). However, there are at least two drawbacks of the distance-based assumptions: first, the genotypic displacement of the global

[12] It should be noted that this assumption is about *local* structural properties and not the *global* structure of a problem's fitness landscape (i.e., $d_g \propto f_{\hat{f}} \not\Rightarrow d_{\hat{f}} \propto d_g$).

optimum is algorithm-dependent and second, the distance-based assumption ignores all other elements in X_g. The next section proposes an alternative that, at least partly, addresses these issues.

5.5.2 Properties of Dynamic Fitness Landscapes

In comparison to the significant number of empirical studies, some of which were reviewed in the previous section, the number of studies that explicitly attempt to gain a better understanding of a problem's fitness landscape dynamics is relatively small. Intuition suggests that changes in the parameter space may have an arbitrary impact on the fitness landscape, a phenomenon that has been noted frequently in the literature. Branke, for example, notes that in the classical dynamic $n = 17$ knapsack problem [17] with varying capacity $c(T)$, optimal solutions may become infeasible if $c(T+1) < c(T)$ and hence it seems unlikely that information from the previous time steps may be reused [6].[13] Similarly, Yamasaki et al. [47] note that in binary optimization problems, small changes in the parameters may result in (disproportionally) large changes in the objective function. This property generally contradicts the notion of small scale changes from one environment to the next and may be responsible for the predominant focus on the continuous domain.

As mentioned in section 5.3.3, numerous classifications were proposed to characterise different dynamics. However, the utility of such classifications is somewhat limited as the attributes considered tend to be very abstract and none of the measures further quantifies the impact of these attributes on the actual dynamic fitness landscape. Weicker [44], on the other hand, proposes a lower level analysis of the decomposable fitness landscape generated by DF1: Weicker uses the concepts of coordinate transformations, fitness rescalings and stretching factors. Nevertheless, Weiker comments that these properties cannot easily be applied outside the artificial domain (also see [9]). This problem is a common one across most studies that attempt to improve our understanding of the properties of DOPs: the lack of a general framework makes it difficult to generalise findings across a wider class of problems and this is partly due to the algorithm-dependency of the combinatorial domain as demonstrated throughout this chapter.

A more concrete analysis of fitness landscape dynamics in the combinatorial domain is due to Branke et al. [9]: the authors analyse the impact of three different representations on the performance of a simple EA on a dynamic variant of the multi-dimensional knapsack problem.[14] The analysis of landscape changes focusses primarily on change severity and fitness correlation amongst successive problem instances, measured by taking samples of the search space before and after the

---

[13] This, however, depends crucially on the constraint-handling method employed; if chosen appropriately, even points that become infeasible following a change may still be valuable (see [8, 9]).

[14] The constraints of the MKP may be dealt with in numerous ways (e.g., penalty functions, decoders, repair algorithms, constraint-preserving operators, multi-objective problem representation) adding a further degree of complexity to the role played by the representation.

change. These measures include change severity, fitness correlation and value of past optima. As the authors note, for most of these measurements, knowledge regarding the problem's global and local optima is required. The authors compare two different representations: real-valued vectors with weight coding and a binary representation with penalty function and show how the measured outcomes vary considerably across the different representations. This study is extended in [8] where additional attributes are examined, including a comparison to random restarts and hypermutations; a permutation representation is considered also. The authors found that the representation that performed best on the static variant of the multi-dimensional knapsack problem, a real-valued vector with weight coding, also performed best on the dynamic variant of the problem.

A different study looked at the relationship between the fitness distance correlation of a problem instance and the impact of different types of dynamics on the transitions from one problem instance to another [33, 34]: the analysis of the dynamic subset sum problem has shown how the attributes of the problem instance significantly affect the distances between successive global optima. Small problem instances were evolved with different fitness distance correlations, ranging from 0 (random) to -1 (fully correlated).[15] The problem was altered by small modifications to the problem's parameters and constraints (the dynamics considered altered either the parameters or constraint, or both) and the subsequent analysis showed how the displacement of the global optimum was determined entirely by the FDC of the instance and not the magnitude of change. In other words, the structure of the fitness landscape determined entirely how much the global optimum was displaced by.

The transformation from one fitness landscape in time to another (assuming constant dimensionality) may be viewed as the scaling of the values in $\hat{f}$. Changes in the overall structure of the fitness landscape are thus determined by the arrangement of elements that correspond to the f-values. This topology is determined by the algorithm's representation and variation. It is possible to remove the algorithm dependency from the distance measure used to correlated different fitness landscapes by looking at the correlations between $\hat{f}(T)$ and $\hat{f}(T+1)$ directly. This *fitness-based assumption* may be seen as a generalisation of the distance-based assumption and defines the likelihood that some point $x \in X_g$ will be the next global optimum as $p(x^\star(T+1) = x) := z(f(x))$, where $z : \mathbb{R} \to \mathbb{R}$ is some (domain-specific) function that reflects the relevance in a point's f-value following a change. This concept may be extended to define the *rank-based difference* between two fitness landscapes $\mathscr{L}(T)$ and $\mathscr{L}(T+1)$: the degree to which the ranking of search points has been affected by the transition from one fitness landscape in time to another. The notion of rank-based difference is particularly suitable for EAs as the majority of EAs perform their search based on the relative f-values encountered, not the absolute f-values.

[15] The authors considered a binary representation of the subsets. The algorithm considered was a simple $(1+1)-$EA with bit-wise mutation probability $1/n$. The metric d_g was subsequently approximated using Hamming distances, with a fixed neighbourhood distance of $c = 1$.

5.6 Conclusions

The majority of algorithms developed for dynamic combinatorial optimization problems assume that successive problem instances are correlated and in most cases, this assumption is limited to the genotypic distances between successive global optima. However, the structural properties of the problem's fitness landscape are determined not only by the objective function but equally by the representation and variation operators of the algorithm: the dynamics describe a trajectory through the space Δ, causing the values $\hat{f}$ to change over time. The spatial arrangement of the elements these f-values correspond to is determined by the algorithm. In particular, the variation operators define a neighbourhood over X_g while the representation maps this neighbourhood to elements in X_p. In the case of dynamic optimization problems, the chosen neighbourhood structure of an algorithm not only influences the individual fitness landscapes encountered but also their transitions.

One of the fundamental goals in evolutionary dynamic optimization is the transfer of knowledge from one problem instance in time to another: "a natural attempt to speed up optimization after a change would be to somehow use knowledge about the previous search space to advance the search after a change." [20, p 311]. However, the identification of the kind of knowledge that may be used for this purpose in non-trivial. In the majority of cases, such knowledge is restricted to previous search points sampled by the algorithm. This chapter has shown that the utility of such points depends on the chosen representation and variation operators and is thus algorithm dependent. It is thus vital to identify some general properties that may be used to design more efficient algorithms for the dynamic domain, including the choice of representation, which has a fundamental impact on the difficulty of the problem. One fundamental issue in evolutionary dynamic optimization is the requirement for the algorithm to perform well on the individual problem instances as well as their transitions. It is often the case that modifications suggested specifically for the dynamic domain hinder the algorithm's performance on the individual problem instances (hence they are not used in static optimization). It is thus paramount to gain a better understanding of how the performance of an algorithm may be improved on *both* the individual problem instances encountered as well as their transition; this is likely to include some form of learning and adaptation

Liepins and Vose [26] (cited in [35, p 74]) demonstrate that a fully deceptive problem $f(x) = f_p(f_g(x))$ may be transformed to an easy problem by a transformation T: $g(x) = f[T(x)]$. However, as Rothlauf [35] argues, the choice of representation in the absence of domain specific information (black-box scenario) is non-trivial and universal mappings from difficult to easy problems do not exist (NFL; [46]). Subsequently, a representation should at the very least ensure it does not make problems of bounded difficulty more difficult. In the dynamic case, however, domain-specific knowledge may be accumulated over time and hence the choice of representation and variation operators may be improved (i.e., the requirement for locality may be relaxed); this may be achieved, for instance, using adaptive representations [37]. In the end, as an algorithm is applied to a DOP for an increasing amount of time, the algorithm's knowledge about the problem should monotonically increase (up to a limit); this necessarily reduces the black box uncertainty about future problem instances and should allow the algorithm to usefully increase the *heuristic bias* [30] of its representation and variation operators.

Acknowledgements. This work was partially supported by an EPSRC grant (No. EP/E058884/1) on "Evolutionary Algorithms for Dynamic Optimisation Problems: Design, Analysis and Applications."

References

[1] Altenberg, L.: Fitness distance correlation analysis: An instructive counter-example. In: ICGA, pp. 57–64 (1997)

[2] Bäck, T., Fogel, D.B., Michalewicz, Z. (eds.): Handbook of Evolutionary Computation. CRC Press (1997)

[3] Borenstein, Y., Poli, R.: Information landscapes. In: GECCO 2005: Proceedings of the 2005 Conference on Genetic and Evolutionary Computation, pp. 1515–1522. ACM, New York (2005)

[4] Bosman, P.A.N.: Learning, anticipation and time-deception in evolutionary online dynamic optimization. In: Proceedings of the 2005 Workshop on Genetic and Evolutionary Computation, pp. 39–47 (2005)

[5] Bosman, P.A.N., Poutrè, H.L.: Learning and anticipation in online dynamic optimization with evolutionary algorithms: the stochastic case. In: Proceedings of the 2007 Genetic and Evolutionary Computation Conference, pp. 1165–1172 (2007)

[6] Branke, J.: Memory enhanced evolutionary algorithms for changing optimization problems. In: Proceedings of the 1999 IEEE Congress on Evolutionary Computation, vol. 3, pp. 1875–1882. IEEE (1999)

[7] Branke, J.: Evolutionary Optimization in Dynamic Environments. Kluwer (2002)

[8] Branke, J., Orbayı, M., Uyar, Ş.: The Role of Representations in Dynamic Knapsack Problems. In: Rothlauf, F., Branke, J., Cagnoni, S., Costa, E., Cotta, C., Drechsler, R., Lutton, E., Machado, P., Moore, J.H., Romero, J., Smith, G.D., Squillero, G., Takagi, H. (eds.) EvoWorkshops 2006. LNCS, vol. 3907, pp. 764–775. Springer, Heidelberg (2006)

[9] Branke, J., Salihoglu, E., Uyar, S.: Towards an analysis of dynamic environments. In: Beyer, H.-G., et al. (eds.) Genetic and Evolutionary Computation Conference, pp. 1433–1439. ACM (2005)

[10] Cobb, H.G.: An investigation into the use of hypermutation as an adaptive operator in genetic algorithms having continuous, time-dependant nonstationary environments. Technical report, Naval Research Laboratory, Washington, USA (1990)

[11] Dasgupta, D., McGregor, D.R.: Nonstationary function optimization using the structured genetic algorithm. In: Männer, R., Manderick, B. (eds.) Parallel Problem Solving from Nature, vol. 2, pp. 145–154. Elsevier, Amsterdam (1992)

[12] De Jong, K.: Evolving in a Changing World. In: Raś, Z.W., Skowron, A. (eds.) ISMIS 1999. LNCS, vol. 1609, pp. 512–519. Springer, Heidelberg (1999)

[13] Droste, S., Jansen, T., Tinnefeld, K., Wegener, I.: A new framework for the valuation of algorithms for black-box optimization. In: Proceedings of the Seventh Foundations of Genetic Algorithms Workshop (FOGA), pp. 197–214 (2002)

[14] Droste, S., Jansen, T., Wegener, I.: Optimization with randomized search heuristics – the (a)nfl theorem, realistic scenarios, and difficult functions. Theoretical Computer Science 287 (2002)

[15] Garey, M.R., Johnson, D.S.: Computers and Intractability: A Guide to the Theory of NP-Completeness. W. H. Freeman and Company (1979)
[16] Gaspar, A., Collard, P.: From GAs to artificial immune systems: Improving adaptation in time dependent optimization. In: Proceedings of the IEEE International Congress on Evolutionary Computation, pp. 1867–1874 (1999)
[17] Goldberg, D.E., Smith, R.E.: Nonstationary function optimization using genetic algorithms with dominance and diploidy. In: Grefenstette, J.J. (ed.) Second International Conference on Genetic Algorithms, pp. 59–68. Lawrence Erlbaum Associates (1987)
[18] Hadad, B.S., Eick, C.F.: Supporting Polyploidy in Genetic Algorithms Using Dominance Vectors. In: Angeline, P.J., McDonnell, J.R., Reynolds, R.G., Eberhart, R. (eds.) EP 1997. LNCS, vol. 1213, pp. 223–234. Springer, Heidelberg (1997)
[19] He, J., Reeves, C., Witt, C., Yao, X.: A note on problem difficulty measures in black-box optimization: Classification, realizations and predictability. Evolutionary Computation 15(4), 435–443 (2007)
[20] Jin, Y., Branke, J.: Evolutionary optimization in uncertain environment - a survey. IEEE Transactions on Evolutionary Computation 9(3), 303–317 (2005)
[21] Jones, T.: Evolutionary algorithms, fitness landscapes and search. PhD thesis, Citeseer (1995)
[22] Jones, T., Forrest, S.: Fitness distance correlation as a measure of problem difficulty for genetic algorithms. In: Proceedings of the Sixth International Conference on Genetic Algorithms, pp. 184–192 (1995)
[23] Kallel, L., Naudts, B., Reeves, C.R.: Properties of fitness functions and search landscapes. In: Theoretical Aspects of Evolutionary Computing, pp. 175–206 (2001)
[24] Kauffman, S.A.: The Origins of Order. Oxford University Press (1993)
[25] Leguizamon, G., Blum, C., Alba, E.: Evolutionary Computation. In: Handbook of Approximation Algorithms and Metaheuristics, pp. 24.1–24.X. CRC Press (2007)
[26] Liepins, G.E., Vose, M.D.: Representational issues in genetic optimization. Journal of Experimental & Theoretical Artificial Intelligence 2(2), 101–115 (1990)
[27] Morrison, R.W.: Designing Evolutionary Algorithms for Dynamic Environments. Springer, Berlin (2004)
[28] Morrison, R.W., De Jong, K.A.: A test problem generator for non-stationary environments. In: Congress on Evolutionary Computation, vol. 3, pp. 2047–2053. IEEE (1999)
[29] Papadimitriou, C.H., Steiglitz, K.: Combinatorial Optimization: Algorithms and Complexity. Dover (1998)
[30] Raidl, G.R., Gottlieb, J.: Empirical analysis of locality, heritability and heuristic bias in evolutionary algorithms: A case study for the multidimensional knapsack problem. Evolutionary Computation 13(4), 441–475 (2005)
[31] Reeves, C.R.: Landscapes, operators and heuristic search. Annals of Operations Research 86, 473–490 (1999)
[32] Richter, H.: Evolutionary Optimization and Dynamic Fitness Landscapes. In: Zelinka, I., Celikovsky, S., Richter, H., Chen, G. (eds.) Evolutionary Algorithms and Chaotic Systems. SCI, vol. 267, pp. 409–446. Springer, Heidelberg (2010)
[33] Rohlfshagen, P., Yao, X.: Dynamic combinatorial optimization problems: An analysis of the subset sum problem. To appear in Soft Computing
[34] Rohlfshagen, P., Yao, X.: Attributes of Dynamic Combinatorial Optimisation. In: Li, X., Kirley, M., Zhang, M., Green, D., Ciesielski, V., Abbass, H.A., Michalewicz, Z., Hendtlass, T., Deb, K., Tan, K.C., Branke, J., Shi, Y. (eds.) SEAL 2008. LNCS, vol. 5361, pp. 442–451. Springer, Heidelberg (2008)
[35] Rothlauf, F.: Representations for Genetic and Evolutionary Algorithms. Springer (2002)

[36] Schaffer, J.D., Caruana, R.A., Eshelman, L.J., Das, R.: A study of control parameters affecting online performance of genetic algorithms for function optimization. In: Proceedings of the Third International Conference on Genetic Algorithms, pp. 51–60. Morgan Kaufmann Publishers Inc. (1989)

[37] Schnier, T., Yao, X.: Using multiple representations in evolutionary algorithms. In: Proceedings of the 2000 Congress on Evolutionary Computation, pp. 479–486. IEEE Press (2000)

[38] Thompson, R.K., Wright, A.H.: Additively decomposable fitness functions. Technical report, University of Montana, Computer Science Department (1996)

[39] Tinos, R., Yang, S.: Continuous dynamic problem generators for evolutionary algorithms. In: Proceedings of the 2007 IEEE Congress on Evolutionary Computation, pp. 236–243 (2007)

[40] Tinós, R., Yang, S.: An Analysis of the XOR Dynamic Problem Generator Based on the Dynamical System. In: Schaefer, R., Cotta, C., Kołodziej, J., Rudolph, G. (eds.) PPSN XI. LNCS, vol. 6238, pp. 274–283. Springer, Heidelberg (2010)

[41] Uyar, Ş., Uyar, H.: A Critical Look at Dynamic Multi-dimensional Knapsack Problem Generation. In: Applications of Evolutionary Computing, pp. 762–767 (2009)

[42] Wang, H., Yang, S., Ip, W.H., Wang, D.: Adaptive primal-dual genetic algorithms in dynamic environments. IEEE Transactions on Systems, Man, and Cybernetics Part B: Cybernetics 39(6), 1348–1361 (2009)

[43] Wegener, I.: Methods for the analysis of evolutionary algorithms on pseudo-boolean functions. In: Evolutionary Optimization, pp. 349–369 (2002)

[44] Weicker, K.: Evolutionary Algorithms and Dynamic Optimization Problems. Der Andere Verlag (2003)

[45] Woldesenbet, Y.G., Yen, G.G.: Dynamic evolutionary algorithm with variable relocation. IEEE Transactions on Evolutionary Computation 13(3), 500–513 (2009)

[46] Wolpert, D.H., MacReady, W.G.: No free lunch theorems for optimization. IEEE Transactions on Evolutionary Computation 1(1), 67–82 (1997)

[47] Yamasaki, K., Kitakaze, K., Sekiguchi, M.: Dynamic optimization by evolutionary algorithms applied to financial time series. In: Proceedings of the 2002 Congress on Evolutionary Computation (2002)

[48] Yang, S.: PDGA: the primal-dual genetic algorithm. In: Design and Application of Hybrid Intelligent Systems, pp. 214–223. IOS Press (2003)

[49] Yang, S.: Non-stationary problem optimization using the primal-dual genetic algorithms. In: Sarker, R., Reynolds, R., Abbass, H., Tan, K.-C., McKay, R., Essam, D., Gedeon, T. (eds.) Proceedings of the 2003 IEEE Congress on Evolutionary Computation, vol. 3, pp. 2246–2253 (2003)

[50] Yang, S.: Memory-enhanced univariate marginal distribution algorithms for dynamic optimization problems. In: Proceedings of the 2005 IEEE Congress on Evolutionary Computation, vol. 3, pp. 2560–2567 (2005)

[51] Yang, S., Yao, X.: Population-based incremental learning with associative memory for dynamic environments. IEEE Transactions on Evolutionary Computation 12(5), 542–561 (2008)

[52] Younes, A., Calamai, P., Basir, O.: Generalized benchmark generation for dynamic combinatorial problems. In: Proceedings of the 2005 Workshop on Genetic and Evolutionary Computation, pp. 25–31 (2005)

Chapter 6
Two Approaches for Single and Multi-Objective Dynamic Optimization

Kalyanmoy Deb*

Abstract. Many real-world optimization problems involve objectives, constraints, and parameters which constantly change with time. However, to avoid complications, such problems are usually treated as static optimization problems demanding the knowledge of the pattern of change a priori. If the problem is optimized in its totality for the entire duration of application, the procedure can be computationally expensive, involving a large number of variables. Despite some studies on the use of evolutionary algorithms in solving single-objective dynamic optimization problems, there has been a lukewarm interest in solving dynamic multi-objective optimization problems. In this paper, we discuss two different approaches to dynamic optimization for single as well as multi-objective problems. Both methods are discussed and their working principles are illustrated by applying them to different practical optimization problems. The off-line optimization approach in arriving at a knowledge base which can then be used for on-line applications is applicable when the change in the problem is significant. On the other hand, an off-line approach to arrive at a minimal time window for treating the problem in a static manner is more appropriate for problems having a slow change. Further approaches and applications of these two techniques remain as important future work in making on-line optimization task a reality in the coming years.

Kalyanmoy Deb
Kanpur Genetic Algorithms Laboratory (KanGAL)
Department of Mechanical Engineering
Indian Institute of Technology Kanpur,
PIN 208016, India
e-mail: deb@iitk.ac.in

* Kalyanmoy Deb is a Professor at IIT Kanpur, India and is also an Adjunct Professor at Aalto University School of Economics, Helsinki, Finland and a Visiting Professor at University of Skövde, Sweden.

E. Alba et al. (Eds.): Metaheuristics for Dynamic Optimization, SCI 433, pp. 99–116.
springerlink.com

6.1 Introduction

A dynamic optimization problem involves objective functions, dynamic optimization constraint functions, and problem parameters which can change with time. Such problems often arise in real-world problem solving, particularly in optimal control problems or problems requiring an on-line optimization. There are two computational procedures usually followed. In one approach, optimal control laws or rules (more generally a knowledge base) are evolved by solving an off-line optimization problem formed by evaluating a solution on a number of real scenarios of the dynamic problem [10, 11]. This approach is useful in solving problems which change frequently and are also computationally expensive for any optimization algorithm to be applied on-line. The other approach is a direct optimization procedure on-line in which an off-line study is suggested for finding a minimal time window within which the problem will be treated as unchanged. In the latter case, the problem is considered stationary for some time period and an optimization algorithm can be allowed to find optimal or near-optimal solution(s) within the time span in which the problem remains stationary. Thereafter, a new problem is constructed based on the current problem scenario and a new optimization is performed for the new time period. Although this procedure is approximate due to the static consideration of the problem during the time for optimization, efforts are made to develop efficient optimization algorithms which can track the optimal solution(s) within a small number of iterations so that the required time period for fixing the problem is small and the approximation error is reduced.

Both approaches are applicable for single as well as multi-objective optimization problems. In the case of single-objective dynamic optimization problems, the optimal solution changes during the optimization procedure and the task of an efficient optimization algorithm would be to track the optimum solution as closely as possible with a minimal computational effort. Although single-objective dynamic optimization has received some attention in the past [2], the dynamic multi-objective optimization is yet to receive significant attention. When a multi-objective optimization problem changes with time, the task of a dynamic evolutionary multi-objective optimization (EMO) procedure is to find or track the Pareto-optimal front as and when there is a change. Since a front of trade-off solutions changes with time, dynamic multi-objective optimization is expected to be harder than dynamic single-objective optimization. A previous study [6] illustrated different possibilities of a change in the optimal front. But since this study, there has been a lukewarm interest on this topic [7, 8].

In the remainder of this paper, we discuss in details the philosophies of both approaches in Section 6.2. The first approach in which an off-line optimization study is needed to obtain an optimal knowledge base for on-line optimization is described next in Section 6.3. This section also shows how the procedure can be applied to a dynamic robot navigation problem for a single-objective function of minimizing the overall time of travel and satisfying constraints related to avoidance of collision with moving obstacles. Due to the uncertain and imprecision nature of the associated variables, a fuzzy knowledge base is developed by an off-line application of an evolutionary algorithm. Later, the obtained fuzzy rule base is used to navigate on-line in unseen test scenarios. Section 6.4 then describes the second approach, in which an idea of the minimal time window for considering the problem as a static problem is determined based on an off-line study. The approach is applied to a hydro-thermal power dispatch problem in which power demand is considered as a changing parameter with time. The study shows how the obtained minimal time window allows the approach to be used on-line in a multi-objective version of the problem. The issue of automated decision-making required in the case of dynamic multi-objective optimization is discussed in Section 6.4.4. Conclusions are drawn in Section 6.5.

6.2 Solving Dynamic Optimization Problems

Many search and optimization problems in practice change with time and therefore must be treated as on-line optimization problems. The change in the problem with time t can be either in its objective functions or in its constraint functions or in its variable boundaries or in any problem parameters or in any combination of above. Such an optimization problem ideally must be solved instantly at every time instant t or whenever there is a change in any of the above functions with t. However, practically speaking, an optimization task requires a finite amount of computational time (τ_{opt}) to arrive at a solution reasonably close to the true optimum. In such problems, there are two time frames which are intertwined: (i) computational time in arriving at a solution (denoted as τ) and (ii) real time in which the problem undergoes a change (denoted as t). While an optimization run is underway in the time frame of τ, the problem gets also changed in the time frame of t. Here, we shall assume equivalence of both time frames and any time spent in one frame affects the same amount in the other time frame.

It now becomes obvious that in an on-line optimization task, as an optimization task is performed (taking a finite time) the optimization problem gets changed and the optimization task is not solving the same problem with which it started. The relevance and accuracy of the obtained optimum in the current context largely depends on the rate at which the problem changes with time. If the rate of change in the problem is fast compared to the time taken by the optimization algorithm in arriving at the optimal solution, the relevance of the optimal solution of earlier problem to the current context may be questionable. In such a situation, performing an optimization task on-line may not make much sense. On the other hand, if the rate in change is slow, an optimization task can be performed and the obtained optimal solution can still be meaningful. Based on these two scenarios, we suggest two different techniques for a possible on-line optimization task:

1. Develop an optimal rule base off-line and use it for on-line application, and
2. Develop an on-line optimization procedure by considering the problem to be static for a minimal time window.

We discuss each of these techniques in the following sections.

6.3 Approach 1: Off-Line Development of an Optimal Rule Base

This approach is more suitable to problems which change quickly with time or which require a computationally expensive evaluation procedure. In this approach, a number of instantiations of the dynamically changed problem are first collected. An off-line optimization task is then used to find a set of optimal rules (other classifier based approaches can also be adopted here) that would correctly work in the chosen instantiations. It is then believed that since the obtained optimal rule base worked on a number of cases in solving the task optimally, it would also work on new cases on-line. Thus, the obtained optimal rule base can be used to quickly find a reasonable solution to the changing problem. Since new instantiations can be somewhat different from the earlier chosen instantiations, the optimization task of finding the optimal rule base can be repeated in a regular interval during the on-line application process. For this purpose, the new and structurally different instantiations can be stored in an archive. The optimization task can continue in the background without disturbing the on-line application process. We describe one such application through an optimization based approach applied to dynamic robot navigation problem [4].

6.3.1 Off-Line Optimization Approach Applied to a Robot Navigation Problem

Figure 6.1 shows the suggested off-line optimization based approach. On a set of instantiations, an optimization algorithm is applied to find a knowledge base using rules or by other means. The optimization task would find a set of rules or classifiers which will determine the nature of the outcome based on the variable values at any time instant. In the following, we describe the procedure in the context of an on-line robot navigation problem.

The purpose of the dynamic motion planning (DMP) problem of a robot is to find an obstacle-free path which takes a robot from a point A to a point B with minimum time. There are essentially two parts of the problem:

1. Learn to find *any* obstacle-free path from point A to B, and
2. Learn to choose that obstacle-free path which takes the robot in a minimum possible time.

Both these problems are somewhat similar to the learning phases a child would go through while solving a similar obstacle-avoidance problem. If a child is kept in a similar (albeit hypothetical) situation (that is, a child has to go from one corner of a room to another by avoiding a few moving objects), the child learns to avoid incoming obstacle by taking detour from his/her path. It is interesting that while taking the detour he/she never calculates the precise angle of deviation. This process of avoiding an object can be thought as if the child is using a rule of the following sort:

If an object is very near and is approaching, then turn right to the original path.

Because of the imprecise definition of the deviation in this problem, it seems natural to use a fuzzy logic technique here.

The second task of finding an optimal obstacle-free path arises from a simile of solving the same problem by an experienced versus an inexperienced child. An inexperienced child may take avoidance of each obstacle too seriously and deviate by a large angle each time he/she faces an obstacle. This way, this child may lead away from the target and take a long winding distance to reach the target. Whereas, an experienced child may deviate barely from each obstacle, thereby taking the quickest route. If we think of how the experienced child has learned this trick, the answer is through experience of solving many such problems in the past. Previous efforts helped us to find a set of good rules to do the task efficiently. This is precisely the task of an optimizer which needs to discover the optimal set of rules needed to avoid obstacles and reach the target point in a minimum possible time. This is where the genetic algorithm (GA) is a natural choice.

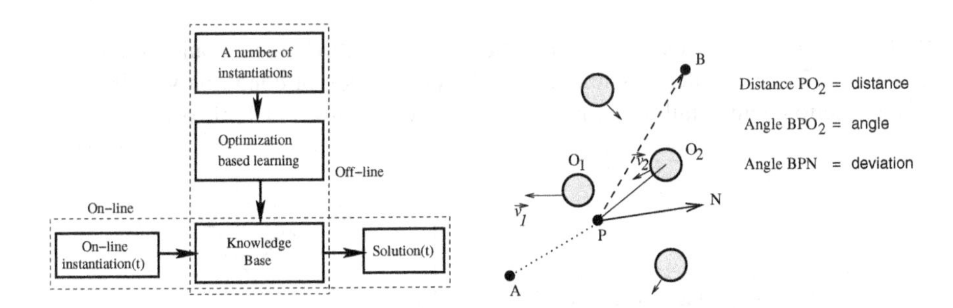

Fig. 6.1 Approach 1 is illustrated.

Fig. 6.2 A schematic showing condition and action variables for the robot navigation problem.

In the proposed genetic-fuzzy approach, a GA is used to create the knowledge base comprising of fuzzy rules for navigating a robot off-line. For on-line application, the robot uses its optimal fuzzy rule base to find an obstacle-free path for a given input of parameters depicting the state of moving obstacles and the state of the robot.

6.3.2 Representation of a Solution in a GA

A solution to the DMP problem is represented by a set of rules which a robot will use to navigate from point A to point B (Fig. 6.2). Each rule has three conditions: distance, angle, and relative velocity. The distance is the distance of the nearest obstacle forward from the robot. Four fuzzy values of distance are chosen: very near (VN), near (N), far (F), and very far (VF). The angle is the relative angle between the path joining the robot and the target point and the path to the nearest obstacle forward. The corresponding fuzzy values are left (L), ahead left (AL), ahead (A), ahead right (AR), and right (R). The relative velocity is the relative velocity vector of the nearest obstacle forward with respect to the robot. In our approach, we eliminate this variable by using a practical incremental procedure. Since a robot can sense the position and velocity of each obstacle at any instant of time, the critical obstacle ahead of the robot can always be identified. In such a case (Fig. 6.2), even if an obstacle O_1 is nearer compared to another obstacle O_2, and the relative velocity $\mathbf{v}_1$ of O_1 directs away from robot's path toward the target point B, whereas the relative velocity $\mathbf{v}_2$ of O_2 directs toward the robot (Position P), the obstacle O_2 is assumed to be the critical obstacle forward.

The action variable is deviation of the robot from its path toward the target (Fig. 6.2). This variable is considered to have five fuzzy values: L, AL, A, AR, and R. Triangular membership functions are considered for each membership function (Fig. 6.3). Using this rule base, a typical rule will look like the following:

If distance is VN and angle is A, then deviation is AL.

With four choices for distance and five choices for angle, there could be a total of 4 × 5 or 20 valid rules possible. For each combination of condition variables, a suitable action value (author-defined) is associated, as shown in Table 6.1.

Table 6.1 All possible rules are shown.

distance \ angle	L	AL	A	AR	R
VN	A	AR	AL	AL	A
N	A	A	AL	A	A
F	A	A	AR	A	A
VF	A	A	A	A	A

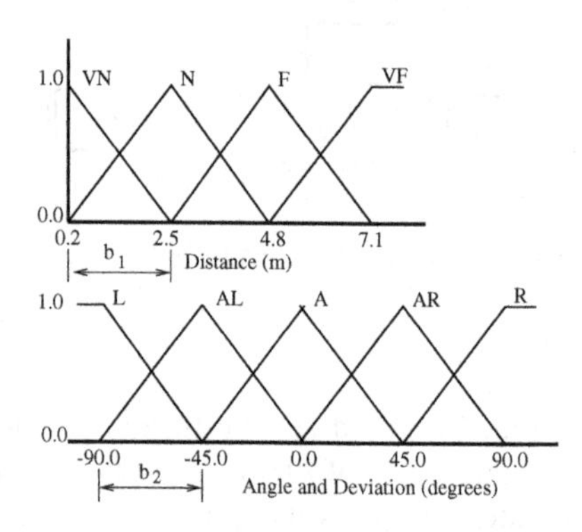

Fig. 6.3 Author-defined membership functions.

The task of GA is to find which rules (out of 20) should be present in the optimal rule base. We represent the presence of a rule by a `1` and the absence by a `0`. Thus, a complete solution will have a 20-bit length string of `1` and `0`. The value of i-th position along the string marks the presence or absence of the i-th rule in the rule base.

6.3.3 Evaluating a Solution in the GA

A rule base (represented by a 20-bit binary string) is evaluated by simulating a robot's performance on a number of scenarios and keeping track of the travel time, T. Since a robot may not reach the destination using a lethal rule base, the robot is allowed a maximum travel time. An average of travel times in all scenarios is used as the *fitness* of the solution.

Now, we shall discuss some details which will be necessary to calculate the actual travel time T. As mentioned earlier, the robot's total path is a collection of a number of small straight line paths traveled for a constant time ΔT in each step. To make the matter as practical as possible, we have assumed that the robot starts from zero velocity and accelerates during the first quarter of the time ΔT and then maintains a constant velocity for the next one-half of ΔT and decelerates to zero velocity during the remaining quarter of the total time ΔT. For constant acceleration and deceleration rates (a), the total distance covered during the small time step ΔT is $3a\Delta T^2/16$. At the end of the constant velocity travel, the robot senses the position and velocity of each obstacle and decides whether to continue moving in the same direction or to deviate from its path. This is achieved by first determining the predicted position of each obstacle, as follows:

$$P_{\text{predicted}} = P_{\text{present}} + (P_{\text{present}} - P_{\text{previous}}). \tag{6.1}$$

The predicted position is the linearly extrapolated position of an obstacle from its current position P_{present} along the path formed by joining the previous P_{previous} and present positions. Thereafter, the nearest obstacle forward is determined based on $P_{\text{predicted}}$ values of all obstacles and the fuzzy logic technique is applied to find the obstacle-free direction using the rule base dictated by the corresponding 20-bit string. If the robot has to change its path, its velocity is reduced to zero at the end of the time step; otherwise the robot does not decelerate and continues in the same direction with the same velocity $a\Delta T/4$. It is interesting to note that when the latter case happens (the robot does not change its course) in two consecutive time steps, there is a saving of $\Delta T/4$ second in travel time per such occasion. Overall time of travel (T) is then calculated by summing all intermediate time steps needed for the robot to reach its destination. This approach of robot navigation can be easily incorporated in a real-world scenario[1].

6.3.4 Results on Robot Navigation Problem

We consider five different techniques:

Technique 1: Author-defined fuzzy-logic controller. In this approach, a fixed set of 20 rules and author-defined membership functions are used. No optimization

[1] In all simulations here, $\Delta T = 4$ sec and $a = 1$ m/s^2 are chosen. These values make the velocity of the robot in the middle portion of each time step equal to 1 m/sec.

method is used to find optimal rule base or to find the optimal membership function distributions.

Technique 2: Optimizing membership functions alone. A set of all 20 author-defined rule base is assumed and the membership function distributions of condition and action variables are optimized. The shape of the membership functions is assumed to be triangular. The bases of the membership functions are considered as variables. The bases b_1 and b_2 (refer Fig. 6.3) are coded in 10 bit substrings each, thereby making a GA string equal to 20 bits. The base b_1 is decoded in the range (1.0, 4.0) cm and the base b_2 is decoded in the range (25.0, 60.0) degrees. Symmetry is maintained in constructing other membership function distributions. In all simulations here, the membership function distribution for deviation is kept the same as that in angle.

Technique 3: Optimizing rule base alone. The rule base is optimized in this study, while using an author-defined membership function. Here, the GA string is a 20-bit string (of `1` and `0` denoting the presence or absence of rules).

Technique 4: Optimizing membership functions and rule base. In this study, both optimization of finding optimized membership functions and finding an optimized rule base are achieved simultaneously. Here, a GA string is a 40-bit string with first 20 bits denoting the presence or absence of 20 possible rules, next 10 bits are used to represent the base b_1, and the final 10 bits are used to represent the base b_2.

In all runs, we use binary tournament selection (with replacement), the single-point crossover operator with a probability p_c of 0.98 and the bit-wise mutation operator with a probability p_m of 0.02. A maximum number of generations equal to 40 are used. In every case, a population size of 100 is used. In all cases, 10 different author-defined scenarios are used to evaluate a solution.

We now apply all five techniques to eight-obstacle problems (in a grid of 20×24 m^2). The optimized travel distance and time for Techniques 1 to 4 are presented in Table 6.2. Ideally multiple applications of a GA from different initial populations must be used to make a comprehensive evaluation, but here, we show results from a single simulation in each case. However, a visual inspection of multiple runs has shown similar results. The first three rows in the table show the performance of all approaches on scenarios that were used during the optimization process and the last three rows show their performance on new test (unseen) scenarios. The table shows that in all cases, Techniques 2, 3, and 4 have performed better than Technique 1 (no optimization). Paths obtained using all four approaches for scenario 4 (unseen) are shown in Fig. 6.4. It is clear that the paths obtained by Techniques 3 and 4 are shorter and quicker than those obtained by Techniques 1 and 2. The optimized rule bases obtained using Techniques 3 and 4 are shown in Tables 6.3 and 6.4.

The optimized membership functions obtained using Techniques 2 and 4 are shown in Figs. 6.5 and 6.6, respectively.

Here, Technique 4 (simultaneous optimization of rules and membership functions) has elongated the membership function distribution, so that classification of

Table 6.2 Travel distance D (in meter) and time T (in sec) obtained by five approaches for the eight-obstacle problem.

	Scenario	Technique 1		Technique 2		Technique 3		Technique 4	
		D	T	D	T	D	T	D	T
Training	1	27.203	28.901	26.077	27.769	26.154	27.872	26.154	27.872
	2	26.957	28.943	25.966	27.622	26.026	26.564	26.026	26.546
	3	29.848	36.798	28.623	35.164	26.660	34.547	27.139	35.000
Testing	4	33.465	43.364	26.396	27.907	26.243	27.512	26.243	29.512
	5	32.836	41.781	27.129	33.000	26.543	32.390	27.041	33.000
	6	33.464	43.363	28.001	31.335	27.164	31.000	27.164	31.000

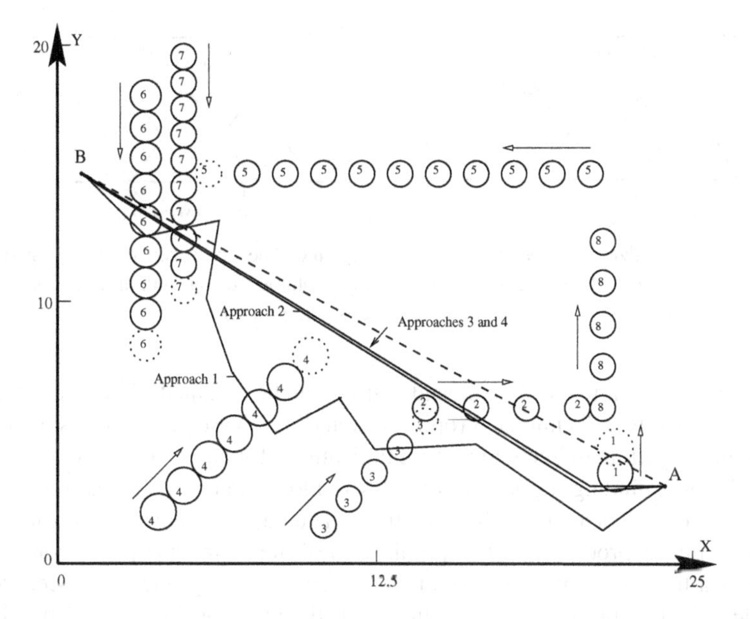

Fig. 6.4 Optimized paths found by all four approaches for the eight-obstacle problem are shown. There are seven obstacles and their movements are shown by an arrow. The location of the critical obstacle (that is *closest* to the robot and is considered for the fuzzy logic analysis at each time step) is shown by a dashed circle. In each case, the robot is clear from the critical obstacle.

relative angle is uniform in the range of $(-90, 90)$ degrees. Because only 10 scenarios are considered during the optimization process, it could have been that in most cases the critical obstacles come in the left of the robot, thereby causing more rules specifying L or AL to appear in the optimized rule base. By considering more scenarios during the optimization process, such bias can be avoided and equal number

Table 6.3 Optimized rule base (having nine rules only) obtained using Technique 3 for eight-obstacle problem.

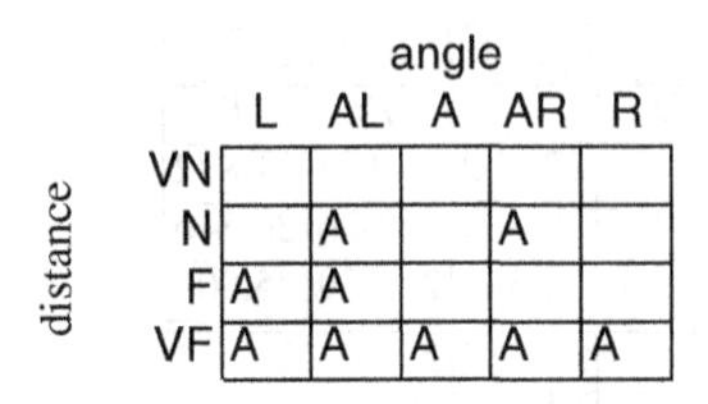

distance \ angle	L	AL	A	AR	R
VN					
N		A		A	
F	A	A			
VF	A	A	A	A	A

Table 6.4 Optimized rule base (having five rules only) obtained using Technique 4 for eight-obstacle problem.

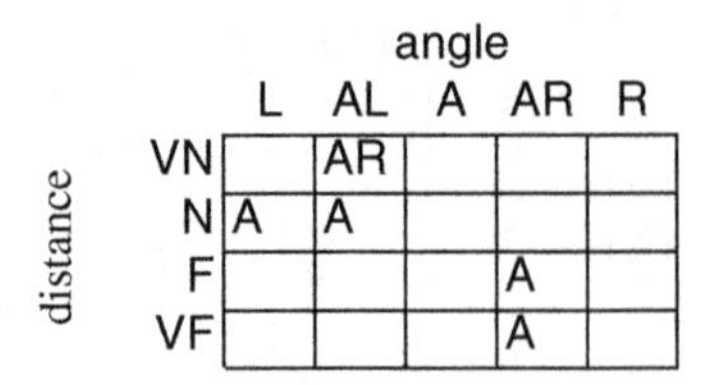

distance \ angle	L	AL	A	AR	R
VN		AR			
N	A	A			
F				A	
VF				A	

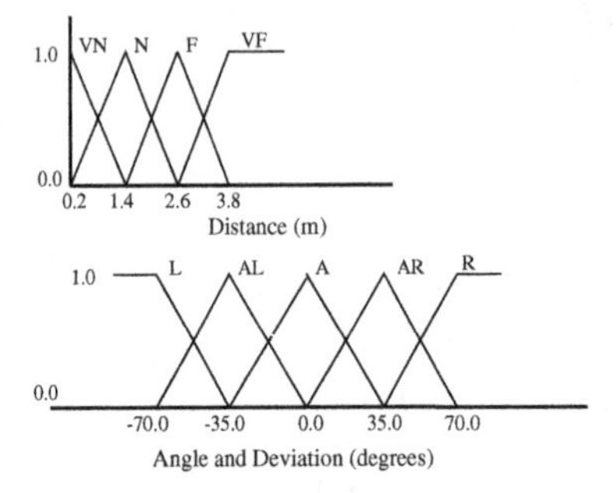

Fig. 6.5 The optimized membership function obtained using Technique 2 for eight-obstacle problem.

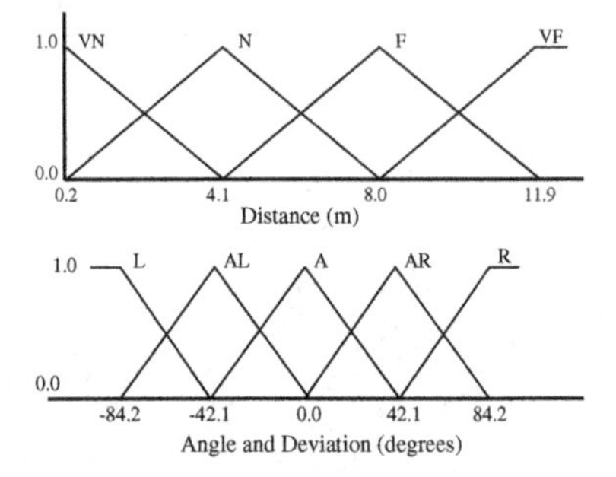

Fig. 6.6 The optimized membership function obtained using Technique 4 for eight-obstacle problem.

of rules specifying left and right considerations can be obtained. From Table 6.2, it can be observed that Technique 3 (optimization of rule base only) has resulted in a much quicker path than Technique 2 (optimization of membership function only). This is because finding a good set of rules is more important for the robot than finding a good set of membership functions. Thus, the optimization of rule base is a rough-tuning process and the optimization of the membership function distributions is a fine-tuning process. Among both tables, in only one case (Scenario 6 in Table 6.2) the optimization of membership function for a optimized rule base has improved the solution slightly (Technique 4). In all other cases, the optimized solutions are already obtained during the optimization of rule-base only and optimization of membership function did not improve the solution any further.

Although the performance of Techniques 3 and 4 is more-or-less similar, we would like to highlight that Technique 4 is a more flexible and practical approach. The similarity in the performances of Techniques 3 and 4 reveals that optimizing rule base has a significant effect and the optimization of the membership functions is only a secondary matter. Since the membership functions used in Technique 3 are developed by the authors and are reasonably good in these two problems, the performance of Technique 3 turns out to be good. However, for more complicated problems, we recommend using Technique 4, since it optimizes both the rule base and membership functions needed in a problem.

6.4 Approach 2: On-Line Optimization with a Minimal Time Window

For a steady change in a problem (which is most usual in practice), we suggest an on-line optimization technique which we discuss next.

Let us assume that the change in the optimization problem is gradual in t. Let us also assume that each optimization iteration requires a finite time G and that τ_T iterations are needed to track the optimal frontier within an allowable performance level. An assumption we make here is that the problem does not change (or is assumed to be constant) within a time interval t_T, such that $G\tau_T < t_T$. Thus, an initial $G\tau_T$ time is taken up by the optimization algorithm to track the new trade-off frontier and to make a decision for implementing a particular solution from the frontier. We expect that only a fraction of overall time is taken by the optimization algorithm, that is, $\alpha = G\tau_T/t_T$ is expected to be a small value (say 0.25). After the optimal frontier is tracked, $(1-\alpha)t_T$ time is spent on using the optimized solution for the rest of the time period. Fig. 6.7 illustrates this dynamic procedure. The objective function $f(x)$, hence also the optimum of $f(\mathbf{x})$, changes with time.

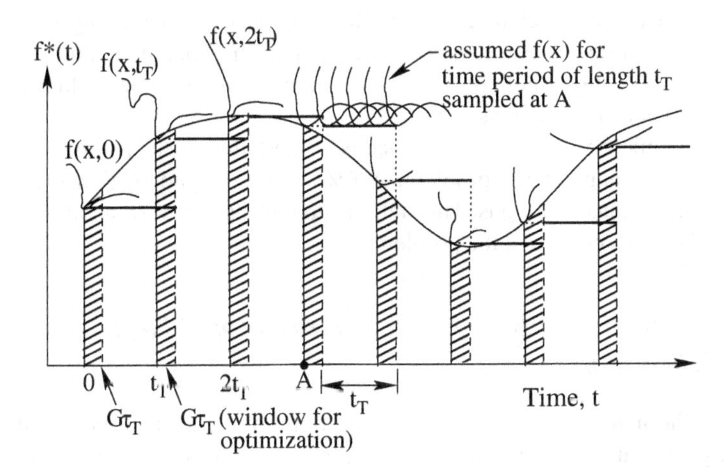

Fig. 6.7 The on-line optimization procedure adopted in this study. For simplicity, only one objective is shown.

The choice of the time window is a crucial matter. If we allow a large value of t_T (allowing a proportionately large number of optimization iterations τ_T), a large change in the problem is expected, but the change occurs only after a large number of iterations of the optimization algorithm. Thus, despite the large change in the problem, the optimization algorithm may have enough iterations to track the trade-off optimal solutions. On the other hand, if we choose a small τ_T, the change in the problem is frequent (which approximates the real scenario more closely), but a lesser number of iterations are allowed to track new optimal solutions for a problem which has also undergone a small change. Obviously, there lies a lower limit to τ_T below which, albeit a small change in the problem, the number of iterations is not enough for an optimization algorithm to track the new optimal solutions adequately. Such a limiting τ_T will depend on the nature of the dynamic problem and the chosen algorithm, but importantly allows the best scenario (and closest approximation to the original problem) which an algorithm can achieve. We suggest using an off-line study to find the limiting time window for an on-line optimization problem.

6.4.1 Dynamic NSGA-II for Handling Dynamic Multi-objective Optimization Problems

Here, we illustrate the working of the above on-line optimization approach on a dynamic multi-objective optimization problem. For this purpose, we suggested a modified NSGA-II procedure in an earlier study [5].

First, we introduce a test to identify whether there is a change in the problem at every generation. For this purpose, we randomly pick a few solutions from the parent population (10% population members used here) and re-evaluate them. If there is a substantial change in any of the objectives and constraint function values, we establish that there is a change in the problem. In the event of a change, all parent solutions are re-evaluated before merging parent and child population into a bigger pool. This process allows both offspring and parent solutions to be evaluated using the changed objectives and constraints.

In the dynamic NSGA-II, we introduce new randomly created solutions whenever there is a change in the problem. A ζ% of the new population is replaced with randomly created solutions. This helps us to introduce new (random) solutions whenever there is a change in the problem.

6.4.2 Application to Bi-objective Hydro-thermal Power Scheduling

In a hydro-thermal power generation system, both the hydro-thermal and thermal generating units are utilized to meet the total power demand. The optimum power scheduling problem involves the allocation of power to all concerned units, so that the total fuel cost of thermal generation and emission properties are minimized, while satisfying all constraints in the hydraulic and power system networks [12]. The problem is dynamic due to the changing nature of power demand with time. Thus, ideally the optimal power scheduling problem is truly an on-line dynamic optimization problem in which solutions must be found as and when there is a change in the power demand. In such situations, what can be expected of an optimization algorithm is that it tracks the new optimal solutions as quickly as possible, whenever there is a change.

The original formulation of the problem was given in Basu [1]. Let us also assume that the system consists of N_h number of hydro-thermal (P_{ht}) and N_s number of thermal (P_{st}) generating units sharing the total power demand, such that $\mathbf{x} = (P_{ht}, P_{st})$. The bi-objective optimization problem is given as follows:

$$
\begin{aligned}
&\text{Minimize } f_1(\mathbf{x}) = \textstyle\sum_{t=1}^{M}\sum_{s=1}^{N_s} t_T\,[a_s + b_s P_{st} + c_s P_{st}^2 + |d_s \sin\{e_s(P_s^{min} - P_{st})\}|],\\
&\text{Minimize } f_2(\mathbf{x}) = \sum_{t=1}^{M}\sum_{s=1}^{N_s} t_T\,[\alpha_s + \beta_s P_{st} + \gamma_s P_{st}^2 + \eta_s \exp(\delta_s P_{st})],\\
&\text{subject to } \textstyle\sum_{s=1}^{N_s} P_{st} + \sum_{h=1}^{N_h} P_{ht} - P_{Dt} - P_{Lt} = 0, \quad t = 1,2\ldots,M,\\
&\qquad \textstyle\sum_{t=1}^{M} t_T(a_{0h} + a_{1h}P_{ht} + a_{2h}P_{ht}^2) - W_h = 0, \quad h = 1,2,\ldots,N_h,\\
&\qquad P_s^{min} \le P_{st} \le P_s^{max}, \quad s = 1,2,\ldots,N_s, t = 1,2,\ldots,M,\\
&\qquad P_h^{min} \le P_{ht} \le P_h^{max}, \quad h = 1,2,\ldots,N_h, t = 1,2,\ldots,M.
\end{aligned}
\tag{6.2}
$$

The transmission loss P_{Lt} term at the t-th interval is given as follows:

$$
P_{Lt} = \sum_{i=1}^{N_h+N_s}\sum_{j=1}^{N_h+N_s} P_{it}B_{ij}P_{jt}. \tag{6.3}
$$

This constraint involves both thermal and hydro-thermal power generation units. Due to its quadratic nature, it is handled directly to repair solution [5]. The problem is dynamic due to changing nature of demand P_{Dt}. To make the demand varying in a continuous manner, we make a piece-wise linear interpolation of power demand values with the following (t, P_{dm}) values: (0, 1300), (12, 900), (24, 1100), (36, 1000), and (48, 1300) in (Hrs, MW). We keep the overall time window of $T = 48$ hours, but increase the frequency of changes (that is, increase M from four to 192, so that the time window t_T for each demand level varies from 12 hours to 48/192 hours or 15 minutes. It will then be an interesting task to find the smallest time window for keeping the problem static, which our dynamic NSGA-II can handle adequately. We run the dynamic NSGA-II procedure for $960/M$ (M is the number of changes in the problem) generations for each change in the problem, so as to have the overall number of function evaluations identical.

6.4.3 Results on Hydro-thermal Power Dispatch Problem

We apply a dynamic NSGA-II procedure – an elitist non-dominated sorting genetic algorithm [3] – discussed above to solve the dynamic optimization problem. In this case, we have considered $\alpha = 1$, that is, the time window is equal to the time required for completion of the NSGA-II optimization run. To evaluate the performance of the dynamic NSGA-II procedure at the end of each time window, we initially treat each problem as a static optimization problem and apply the original NSGA-II procedure [3] for a large number (500) of generations so that no further improvement is likely. We call these fronts as ideal fronts and compute the hyper-volume measure using a reference point which is the nadir point of the ideal front. Thereafter, we apply our dynamic NSGA-II procedure and find an optimized non-dominated front after each time window. Then for each front, we compute the hyper-volume using the same reference point and then compute the ratio of this hyper-volume value with that of the ideal front. This way, the maximum value of the ratio of hyper-volume for an algorithm is one and as the ratio becomes smaller than one, the performance of the algorithm gets poorer. In all runs here, a single NSGA-II run from an initial random population is used in order to simulate an actual application.

Figures 6.8 to 6.11 show the hyper-volume ratio for different number of changes ($\tau_T = 4$ to 192) in the problem with different proportion of addition of random solutions, ζ. The figures also mark the 50th, 90th, 95th, and 99th percentile of hyper-volume ratio, meaning the cut-off hyper-volume ratio which is obtained by the best 50, 90, 95, and 99 percent of M frontiers in a problem with M changes in power demand. Figures reveal that as M increases, the performance of the algorithm gets poorer due to the fact that a smaller number of generations ($\tau_T = 960/M$) were allowed to meet the time constraint. If a 90% hyper-volume ratio is assumed to be the minimum required hyper-volume ratio for a reasonable performance of the dynamic NSGA-II and if we base our confidence on the

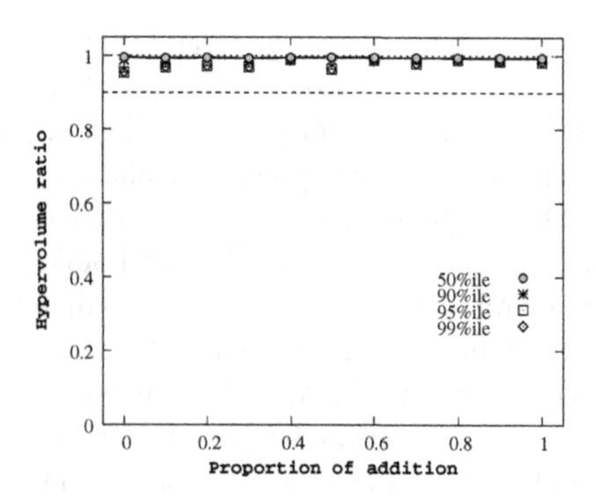

Fig. 6.8 3-hourly ($M = 16$) change with dynamic NSGA-II (Acceptable).

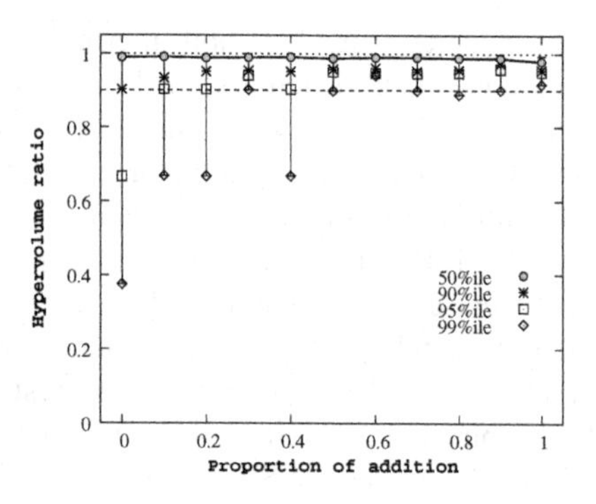

Fig. 6.9 1-hourly ($M = 48$) change with dynamic NSGA-II (Acceptable).

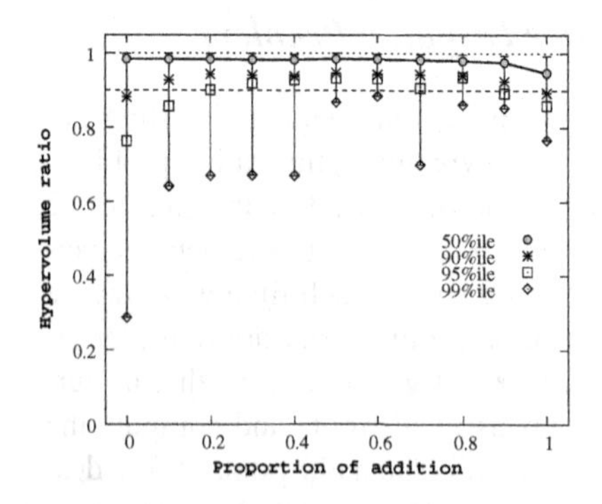

Fig. 6.10 30-min. ($M = 96$) change with dynamic NSGA-II (Acceptable and minimal).

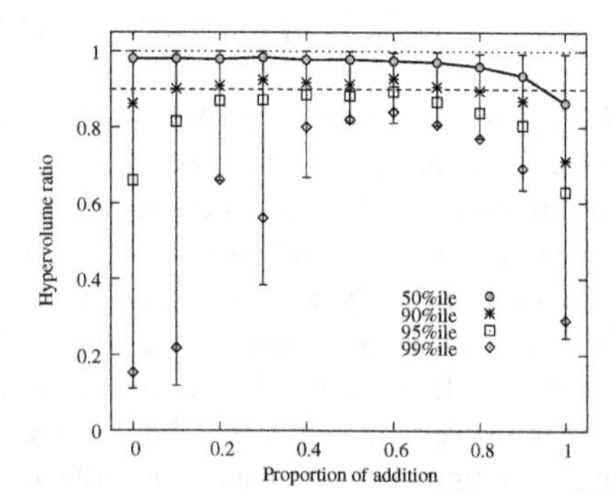

Fig. 6.11 15-min. ($M = 192$) change with dynamic NSGA-II (Not acceptable).

95-th percentile performance, the figures indicate that we can allow a maximum of $M = 96$ changes (with a time window of 30 min.) in the problem. For this case, about 20 to 70% random solutions can be added whenever there is a change in the problem to start the next optimization and an acceptable performance of the dynamic NSGA-II can be obtained. Too low addition of random points does not introduce much diversity to start the new problem and too large addition of random solutions destroys the population structure which would have helped for the new problem. The wide range of addition for a successful run suggests the robustness of the dynamic NSGA-II procedure for this problem.

6.4.4 Automated Decision Making in a Dynamic Multi-objective Optimization

A decision-making task is essential in a multi-objective optimization task to choose a single preferred solution from the obtained trade-off solution set. In a dynamic multi-objective optimization problem, there is an additional problem with the decision making task. A solution is to be chosen and implemented as quickly as the trade-off frontier is found, and before the next change in the problem has taken place. This definitely calls for an automatic procedure for decision-making with some pre-specified utility function or some other procedure. Automated decision-making is not available even in the multi-criteria decision making (MCDM) literature and is certainly a matter of future research, particularly if dynamic EMO is to be implemented in practice.

Here, we suggest a utility function based approach which works by providing different importances to different objectives. First, we consider a case in which equal importances to both cost and emission are given. As soon as a frontier is found for the forthcoming time period, we compute the pseudo-weight w_1 (for cost objective) for every solution $\mathbf{x}$ using the following term:

$$w_1(\mathbf{x}) = \frac{(f_1^{\max} - f_1(\mathbf{x}))/(f_1^{\max} - f_1^{\min})}{(f_1^{\max} - f_1(\mathbf{x}))/(f_1^{\max} - f_1^{\min}) + (f_2^{\max} - f_2(\mathbf{x}))/(f_2^{\max} - f_2^{\min})}. \tag{6.4}$$

Thereafter, we choose the solution with $w_1(\mathbf{x})$ closest to 0.5. A little thought will reveal that this task is different from performing a weighted-sum approach with equal weights for each objective. The task here is to choose the middle point in the trade-off frontier providing a solution equi-distant from individual optimal solutions (irrespective of whether the frontier is convex or non-convex). Since the Pareto-optimal frontier is not known a priori, getting the frontier first and then choosing the desired solution is the only viable approach for achieving the task and such information cannot be utilized a priori.

To demonstrate the utility of this automated decision-making procedure, we consider the hydro-thermal problem with 48 time periods (meaning an hourly change in the problem). Fig. 6.12 shows the obtained frontiers in solid lines and the

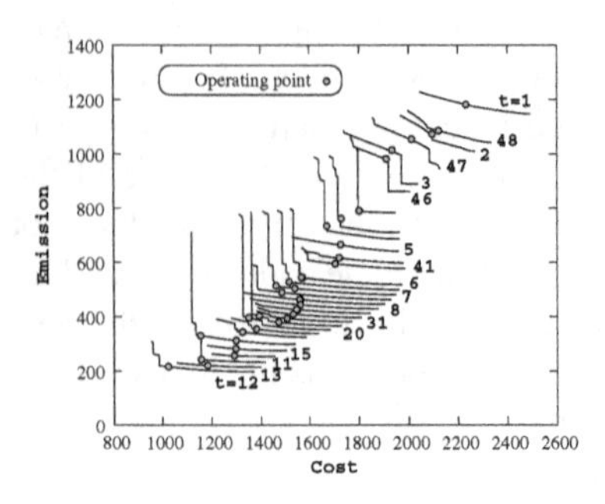

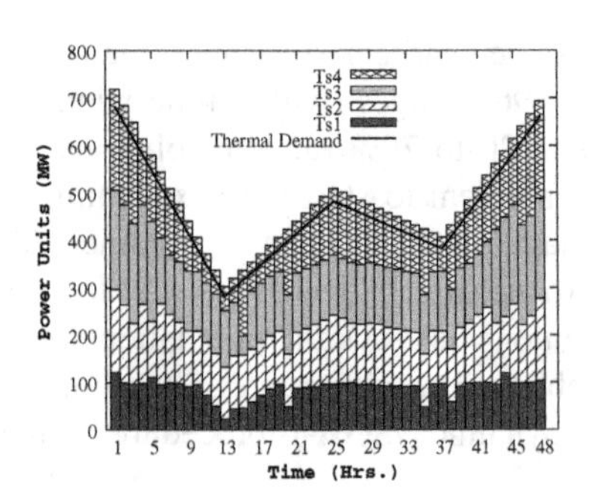

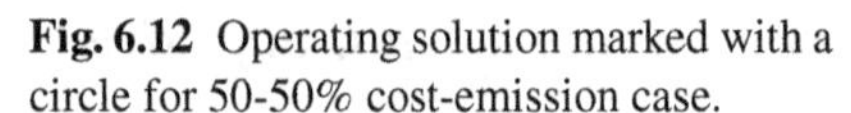

Fig. 6.12 Operating solution marked with a circle for 50-50% cost-emission case.

Fig. 6.13 Variation of thermal power production for 50-50% cost-emission case.

Table 6.5 Different emphasis between cost and emission used in the automated decision-making process produces equivalent power generation schedules.

Case	Cost	Emission
50-50%	74239.07	25314.44
100-0%	69354.73	27689.08
0-100%	87196.50	23916.09

corresponding preferred (operating) solution with a circle. It can be observed that due to the preferred importance of 50-50% to cost and emission, the solution comes nearly in the middle of each frontier. To meet the water availability constraint, the hydro-thermal units of $T_{h1} = 219.76$ MW and $T_{h2} = 398.11$ MW are computed and kept constant over time. However, four thermal power units must produce power to meet the remaining demand and these values for all 48 time periods are shown in Fig. 6.13. The changing pattern in overall computation of thermal power varies similar to that in the remaining demand in power. The figure also shows a slight over-generation of power to meet the loss term P_{Lt} given in equation 6.3.

Next, we compare the above operating schedule of power generation with two other extreme cases: (i) 100-0% importance to cost and emission and (ii) 0-100% importance to cost and emission. Table 6.5 shows the trade-off between cost and emission values for the three cases. Although this is one approach for executing an automated decision-making task, further ideas of choosing a preferred solution from an obtained set of trade-off solutions instantly using some previously defined multi-criteria decision making principles [9] must be worked on.

6.5 Conclusions

In this paper, we have dealt with solving on-line optimization problems using evolutionary optimization (EO) algorithms. Specifically, we have suggested and demonstrated the working of two approaches: (i) an off-line optimization task in which a set of optimal knowledge base is developed as guiding rules for handling changing problems on-line, and (ii) an on-line optimization approach in which the problem is considered static for a minimal amount of time windows. It has been argued that for a problem having a rapid change in any of its parameters, objectives, and constraints, the first approach may be more suitable. For a slow changing problem, the latter approach is more appropriate. Due to the flexibility in their operators and population approach, EO methods are potential for dynamic optimization.

The working of the first approach is demonstrated by solving a robot navigation problem in which the obstacles move while the optimization is underway. A fuzzy rule base system is used to store the knowledge of an optimal action based on two given input parameters. The off-line optimization is able to find an optimal rule-base for achieving the task adequately on on-line scenarios. Since it is an on-line approach, the use of an EO with its flexibility and population approach can be fully exploited.

The working of the second approach is demonstrated by solving a power dispatch problem that changes due to the ever-changing need of demand of power with time. To illustrate the generic nature of the approach, a two-objective version of the problem (minimizing cost and minimizing emission) has been considered. Based on a permissible performance limit, a dynamic NSGA-II approach has been able to identify a minimum time window (of 30 min.) for which the power demand can be considered static. Any faster consideration in the change of the problem was found to be too fast for the chosen algorithm to track the Pareto-optimal solutions of the problem. An offline estimation of an optimal time window exploits an EO's population approach to handle multiple conflicting objectives in their true sense without using any a priori and fixed decision-making principle.

Dynamic multi-objective optimization raises an important issue: an *automated decision making* task that must be performed as soon as the optimized trade-off front is found, as otherwise a delay in decision making may cause the problem to change significantly before the next round of optimization is performed. This chapter has demonstrated one such automated decision-making approach, but this remains to be an open area of further research.

The approaches of this chapter clearly demonstrate the potential for applying single and multi-objective evolutionary (or other) optimization techniques to on-line optimization tasks. A true implementation of these ideas in practice and further research for more sophisticated approaches would be the next step and way forward.

Acknowledgments. The results and some texts of this chapter are borrowed from author's earlier publications (references [4] and [5]). For details, readers are encouraged to refer these papers.

References

[1] Basu, M.: A simulated annealing-based goal-attainment method for economic emission load dispatch of fixed head hydrothermal power systems. Electric Power and Energy Systems 27(2), 147–153 (2005)

[2] Branke, J.: Evolutionary Optimization in Dynamic Environments. Springer, Heidelberg (2001)
[3] Deb, K., Agrawal, S., Pratap, A., Meyarivan, T.: A fast and elitist multi-objective genetic algorithm: NSGA-II. IEEE Transactions on Evolutionary Computation 6(2), 182–197 (2002)
[4] Deb, K., Pratihar, D.K., Ghosh, A.: Learning to Avoid Moving Obstacles Optimally for Mobile Robots Using a Genetic-Fuzzy Approach. In: Eiben, A.E., Bäck, T., Schoenauer, M., Schwefel, H.-P. (eds.) PPSN 1998. LNCS, vol. 1498, pp. 583–592. Springer, Heidelberg (1998)
[5] Deb, K., Udaya Bhaskara Rao, N., Karthik, S.: Dynamic Multi-objective Optimization and Decision-Making Using Modified NSGA-II: A Case Study on Hydro-thermal Power Scheduling. In: Obayashi, S., Deb, K., Poloni, C., Hiroyasu, T., Murata, T. (eds.) EMO 2007. LNCS, vol. 4403, pp. 803–817. Springer, Heidelberg (2007)
[6] Farina, M., Deb, K., Amato, P.: Dynamic multiobjective optimization problems: Test cases, approximations, and applications. IEEE Transactions on Evolutionary Computation 8(5), 425–442 (2000)
[7] Hatzakis, I., Wallace, D.: Dynamic multi-objective optimization with evolutionary algorithms: A forward-looking approach. In: Proceedings of the 8th Annual Conference on Genetic and Evolutionary Computation, pp. 1201–1208 (2006)
[8] Jin, Y., Sendhoff, B.: Constructing Dynamic Optimization Test Problems Using the Multi-objective Optimization Concept. In: Raidl, G.R., Cagnoni, S., Branke, J., Corne, D.W., Drechsler, R., Jin, Y., Johnson, C.G., Machado, P., Marchiori, E., Rothlauf, F., Smith, G.D., Squillero, G. (eds.) EvoWorkshops 2004. LNCS, vol. 3005, pp. 525–536. Springer, Heidelberg (2004)
[9] Miettinen, K.: Nonlinear Multiobjective Optimization. Kluwer, Boston (1999)
[10] Pratihar, D., Deb, K., Ghosh, A.: Fuzzy-genetic algorithms and time-optimal obstacle-free path generation for mobile robots. Engineering Optimization 32, 117–142 (1999)
[11] Pratihar, D.K., Deb, K., Ghosh, A.: Optimal path and gait generations simultaneously of a six-legged robot using a ga-fuzzy approach. Robotics and Autonomous Systems 41, 1–21 (2002)
[12] Wood, A.J., Woolenberg, B.F.: Power Generation, Operation and Control. John-Wiley & Sons (1986)

Chapter 7
Self-Adaptive Differential Evolution for Dynamic Environments with Fluctuating Numbers of Optima

Mathys C. du Plessis and Andries P. Engelbrecht

Abstract. In this chapter, we introduce the algorithm called: SADynPopDE, a self-adaptive multi-population DE-based optimization algorithm, aimed at dynamic optimization problems in which the number of optima in the environment fluctuates over time. We compare the performance of SADynPopDE to those of two algorithms upon which it is based: DynDE and DynPopDE. DynDE extends DE for dynamic environments by utilizing multiple sub-populations which are encouraged to converge to distinct optima by means of exclusion. DynPopDE extends DynDE by: using competitive population evaluation to selectively evolve sub-populations, using a midpoint check during exclusion to determine whether sub-populations are indeed converging to the same optimum, dynamically spawning and removing sub-populations, and using a penalty factor to aid the stagnation detection process. The use of self-adaptive control parameters into DynPopDE, allows a more effective algorithm, and to remove the need to fine-tune the DE crossover and scale factors.

7.1 Introduction

Despite the fact that evolutionary algorithms often solve static problems successfully, dynamic optimization problems tend to pose a challenge to evolutionary algorithms [25]. Differential evolution (DE) is one of the evolutionary algorithms

Mathys C. du Plessis
Department of Computing Sciences, P.O. Box 77000,
Nelson Mandela Metropolitan University,
Port Elizabeth, 6031, South Africa
e-mail: mc.duplessis@nmmu.ac.za

Andries P. Engelbrecht
Department of Computer Science,
School of Information Technology,
University of Pretoria, Pretoria, 0002, South Africa
e-mail: engel@cs.up.ac.za

E. Alba et al. (Eds.): Metaheuristics for Dynamic Optimization, SCI 433, pp. 117–145.
springerlink.com

that does not scale well to dynamic environments due to lack of diversity [35]. A significant body of work exists on algorithms for optimizing dynamic problems (see Section 7.3). Recently, several algorithms based on DE have been proposed [10, 13–15, 23].

Benchmarks used to evaluate algorithms aimed at dynamic optimization (like the moving peaks benchmark [6] and the generalized benchmark generator [20, 21]), typically focus on problems where a constant number of optima move around a multi-dimensional search space. While some of these optima may be obscured by others, these benchmarks do not simulate problems where new optima are introduced, or current optima are removed from the search space. Dynamic Population DE (DynPopDE) [14] is a DE-based algorithm aimed at dynamic optimization problems where the number of optima fluctuates over time. This chapter describes the subcomponents of DynPopDE and then investigates the effect of hybridizing DynPopDE with the self-adaptive component of *jDE* [10] to form a new algorithm, Self-Adaptive DynPopDE (SADynPopDE).

The following sections describe dynamic environments and the benchmark function used in this study. Related work by other researchers is presented in Section 7.3. Differential evolution is described in Section 7.4. The components of DynPopDE, the base algorithm used in this study, are described and motivated in Section 7.5. The incorporation of self-adaptive control parameters into DynPopDE to form SA-DynPopDE and the experimental comparison of these algorithms are described in Section 7.6. The main conclusions of this study are summarized in Section 7.7.

7.2 Dynamic Environments

It is not uncommon for the solution to real-world optimization problems to vary over time. In order to evaluate and compare algorithms aimed at dynamic optimization problems, researchers created benchmark functions. The benchmark used in this study is a variation of the moving peaks benchmark.

7.2.1 Moving Peaks Benchmark

Branke [6] created the moving peaks benchmark (MPB) to address the need for a single, adaptable benchmark that can be used to compare the performance of algorithms aimed at dynamic optimization problems. The multi-dimensional problem space of the moving peaks function contains several peaks, or optima, of variable height, width, and shape. These peaks move around with height and width changing periodically. The MPB allows the following parameters to be set:

- Number of peaks
- Number of dimensions
- Maximum and minimum peak width and height
- Change period (the number of function evaluations between successive changes in the environment)

- Change severity (how much the locations of peaks are moved within the search space)
- Height and width severity (standard deviations of changes made to the height and width of each peak)
- Peak function
- Correlation (between successive movements of a peak)

Three scenarios of settings of MPB parameters are suggested by Branke [6]; however, the majority of researchers using the MPB employ only variations of Scenario 2 settings. The Scenario 2 settings are listed in Table 7.1.

Table 7.1 MPB Scenario 2 settings

Setting	Value
Number of Dimensions	5
Number of Peaks	10
Max and Min Peak height	[30,70]
Max and Min Peak width	[1.0,12.0]
Change period	5000
Change severity	1.0
Height severity	7.0
Width severity	1.0
Peak function	Cone
Correlation	[0.0,1.0]

The performance measure suggested by Branke and Schmeck [7] is the *offline error*. The offline error is the running average of the lowest-so-far error found since the last change in the environment:

$$Offline\ Error = \frac{\sum_{t=1}^{T} E(\mathbf{x}_{best}(t),t)}{T}, \tag{7.1}$$

where T is the total number of function evaluations and $E(\mathbf{x}_{best}(t),t)$ is the error of the best individual found since the last change in the environment (referred to as the *current error*).

7.2.2 Extensions to the Moving Peaks Benchmark

This study investigates dynamic environments in which the number of peaks fluctuates over time. The MPB was therefore adapted by the current authors to allow the number of peaks to change when a change occurs in the environment. For the adapted MPB, the number of peaks, $m(t)$, is calculated as (Equation 7.2):

$$m(t) = \begin{cases} \max\{1, m(t-1) - M \cdot U(0,1) \cdot \mathscr{C}\} \\ \quad \text{if } U(0,1) < 0.5 \\ \min\{M, m(t-1) + M \cdot U(0,1) \cdot \mathscr{C}\} \\ \quad \text{otherwise} \end{cases}, \tag{7.2}$$

where M is the maximum number of peaks and $\mathscr{C}$ is the maximum fraction of M that can be added or removed from the population after a change in the environment. $\mathscr{C}$ thus controls the severity of the change in the number of peaks. For example, $\mathscr{C} = 1$ will result in up to M peaks being added or removed, while $\mathscr{C} = 0.1$ will result in a change of up to 10% of M. M and $\mathscr{C}$ are included as parameters to the benchmark function.

7.3 Related Work

A considerable body of work exists on optimization for dynamic environments. Algorithms directly related to this study are briefly described in this section.

A change in the environment may cause a local optimum to become the global optimum. Information regarding the position of local optima can thus be useful to find the global optimum after a change in the environment. This information is generally gathered by using multiple, independent populations to locate various optima. A key feature of these approaches is that independent populations are allowed to search for optima in parallel. Three of the seminal GA algorithms employing this strategy are self-organizing scouts (SOS) [1], [5], [11], shifting balance GA (SBGA) [28], and multinational GA (MGA) [34]. All three of these approaches make use of different techniques to distribute individuals among the populations intelligently.

Parrott and Li [29] suggested a multiple swarm PSO approach to optimizing dynamic problems, called speciation. Multiple swarms correspond to the idea of multiple populations in GAs, and have the same benefits: increased diversity and parallel discovery of optima. When using speciation, the social component of PSO provides an simple method to divide the swarm into sub-swarms. A particle is allocated to a sub-swarm if the Euclidean distance between the position of the particle and the best particle in the sub-swarm is below a certain threshold value. The global best value of each particle within a sub-swarm is set to the personal best value of the best particle. The sizes of sub-swarms are dynamic. Particles can migrate to another population by moving too far away from their current sub-swarm's best particle or by moving closer to another sub-swarm's best particle. It is allowable for a sub-swarm to contain only one particle. As a mechanism to prevent sub-swarms becoming too large, a maximum sub-swarm size is defined. When a sub-swarm's size exceeds this limit, the particles with the lowest fitness are randomly reinitialized and allocated to an appropriate sub-swarm.

Blackwell and Branke [4] introduced a multiple population PSO based algorithm that contains three components: exclusion, anti-convergence and quantum particles.

In contrast to earlier algorithms, in their approach, all sub-swarms contain the same number of particles. The aim of having multiple sub-swarms is that each sub-swarm should be positioned on its own, promising optimum in the environment. Unfortunately, sub-swarms often converge to the same optimum, hence decreasing diversity. Exclusion [3] is a technique meant to prevent sub-swarms from clustering around the same optimum by means of reinitializing sub-swarms that stray within a threshold Euclidean distance from better performing sub-swarms. This threshold distance is called the exclusion radius. Anti-convergence is meant to prevent stagnation of the particles in the search space. Consequently, if it is found that all sub-swarms have converged to their respective optima, the weakest sub-swarm is randomly reinitialized. Convergence of a sub-swarm is detected if all particles of the sub-swarm fall within a threshold Euclidean distance of each other. This threshold is called the convergence radius. *Quantum particles* is a means of increasing diversity by reinitializing a portion of the swarm of particles within a hyper-sphere, centered around the best particle in the swarm. Blackwell [2] further adapted the PSO-based algorithm of Blackwell and Branke [4] by self-adapting the number of swarms in the search space. This algorithm is aimed at situations where the number of optima in the dynamic environment is unknown. Swarms are generated when the number of free swarms that have not converged to a optimum (M_{free}) has dropped to zero. Conversely, swarms are removed if M_{free} is higher than n_{excess}, a parameter of the algorithm. The algorithms of Blackwell and Branke [4] and Blackwell [2] have the disadvantage that the severity of changes in the environment is given as a parameter to the optimization algorithms. While this information is readily available when employing a benchmark function, it is unlikely that this information will be known to the algorithms in real-world dynamic optimization problems.

Li *et al.* [22] improved the speciation algorithm of Parrott and Li [29], by introducing ideas from Blackwell and Branke [3], namely quantum particles to increase diversity, and anti-convergence to detect stagnation and subsequently reinitialize the worst performing populations. This algorithm is called Speciation-based PSO (SPSO).

An approach similar to quantum individuals, called Brownian individuals, is utilized by DynDE, a DE based algorithm for dynamic environments [23]. Use of Brownian individuals involves the creation of individuals close to the best individual by adding a small random value, sampled from a normal distribution, to each component of the best individual. Mendes and Mohais [23] adapted the ideas from Blackwell and Branke [4] to create their multi-population algorithm, DynDE, which uses exclusion to prevent populations from converging to the same peak. Mendes and Mohais [23] showed that DynDE was at least as effective as its PSO based counterparts. DynDE will be discussed in detail in Section 7.5.

Recently, Brest *et al.* [10] proposed a self-adaptive multi-population DE algorithm for optimizing dynamic environments, called *jDE*. This work focused on adapting the DE scale factor and crossover probability and is based on a previous algorithm for static optimization [9]. *jDE* also contains several components that are similar to other dynamic optimization algorithms. A technique similar to exclusion is used to prevent sub-populations from converging to the same optimum. An aging metaphor is used to reinitialize sub-populations that have stagnated on a local optimum. Each individual's age is incremented every generation. Offspring inherit the age of parents, but this age may be reduced if the offspring performs significantly better than the parent. Sub-populations of which the best individual is too old, are reinitialized. Within sub-populations a further mechanism is used to prevent convergence. An individual is reinitialized if the Euclidean distance between the individual and the best individual in the population is too small. The algorithm also utilizes a form of memory called an archive. The best individual is added to the archive every time a change in the environment occurs. One of the sub-populations is always created by randomly selecting an individual from the archive and adding small random numbers to each of the individual's components.

For a survey of algorithms for dynamic optimization see [19].

7.4 Differential Evolution

The purpose of this section is to describe the Differential Evolution (DE) algorithm and to discuss variations of the basic algorithm. The original DE algorithm is discussed in Section 7.4.1, followed by a discussion on DE schemes in Section 7.4.2 and DE control parameters in Section 7.4.3. Section 7.4.4 describes the performance of DE in dynamic environments and Section 7.4.5 discusses methods of detecting changes in dynamic environments.

7.4.1 Basic Differential Evolution

Differential Evolution (DE) is an optimization algorithm based on Darwinian evolution [12], created by Storn and Price [30], [33]. Several variants to the DE algorithm have been suggested, but the original algorithm is given in Algorithm 7.1.

7.4.2 Differential Evolution Schemes

Most variations of DE (called schemes) are based on different approaches to creating each of the temporary individuals, $\mathbf{v}_i$ (see equation (12.1)), and different approaches to the method of creating offspring [31]. One of two crossover schemes is typically used to create offspring. The first, binary crossover, is used in Equation (12.2). The second common approach is called exponential crossover.

By convention, schemes are labelled in the form DE/*a*/*b*/*c*, where *a* is the method used to select the base vector, *b* is the number of difference vectors, and *c* is the method used to create offspring. The scheme used in Algorithm 7.1 is referred to as DE/rand/1/bin.

Algorithm 7.1. Basic Differential Evolution

Generate a population, P, of I individuals by creating vectors of random candidate solutions, $\mathbf{x}_i$, $i = 1, \cdots, I$ and $|\mathbf{x}_i| = J$;
Evaluate the fitness, $F(\mathbf{x}_i)$, of all individuals.;
while *termination criteria not met* **do**
  foreach $i \in \{1, \ldots, I\}$; `// Create I individuals for a trial population`
  do
    Select three individuals, $\mathbf{x}_1$, $\mathbf{x}_2$ and $\mathbf{x}_3$, at random from the current population such that $\mathbf{x}_1 \neq \mathbf{x}_2 \neq \mathbf{x}_3$;
    Create a new trial vector $\mathbf{v}_i$ using:

$$\mathbf{v}_i = \mathbf{x}_1 + \mathscr{F} \cdot (\mathbf{x}_2 - \mathbf{x}_3) \tag{7.3}$$

    where $\mathscr{F} \in (0, \infty)$ is known as the scale factor and $\mathbf{x}_1$ is referred to as the base vector;
    Add $\mathbf{v}_i$ to the trial population;
  end
  foreach $\mathbf{x}_i$ *in the current population (referred to as the target vector), select the corresponding trial vector* $\mathbf{v}_i$ *from the trial population;* `// Perform crossover`
  do
    Create offspring $\mathbf{u}_i$ as follows:

$$u_{i,j} = \begin{cases} v_{i,j} \text{ if } (U(0,1) \leq Cr \text{ or } j == j_{rand}) \\ x_{i,j} \text{ otherwise} \end{cases} \tag{7.4}$$

    where $Cr \in (0,1)$ is the crossover probability and j_{rand} is a randomly selected index, i.e. $j_{rand} \sim U(1, J+1)$;
    Evaluate the fitness of $\mathbf{u}_i$;
    If $\mathbf{u}_i$ has a better fitness value than $\mathbf{x}_i$ then replace $\mathbf{x}_i$ with $\mathbf{u}_i$;
  end
end

Several methods of selecting the base vector have been developed and can be used with either of the crossover methods. Popular base vector selection methods include [31][32][24] (in each case the selected vectors are assumed to be unique):

DE/rand/2: Two pairs of difference vectors are used:

$$\mathbf{v}_i = \mathbf{x}_1 + \mathscr{F} \cdot (\mathbf{x}_2 + \mathbf{x}_3 - \mathbf{x}_4 - \mathbf{x}_5) \tag{7.5}$$

DE/best/1: The best individual in the population is selected as the base vector:

$$\mathbf{v}_i = \mathbf{x}_{best} + \mathscr{F} \cdot (\mathbf{x}_1 - \mathbf{x}_2) \tag{7.6}$$

7.4.3 Differential Evolution Control Parameters

The Differential Evolution algorithm has several control parameters that can be set. Ignoring extra parameters introduced by some DE schemes, the main DE control parameters are population size, scale factor ($\mathscr{F}$), and crossover factor (Cr).

The scale factor ($\mathscr{F}$) controls the magnitude of the difference vector and consequently the amount by which the base vector is perturbed. Large values of $\mathscr{F}$ encourage large scale exploration of the search space but could lead to premature convergence, while small values result in a more detailed exploration of the local search space while increasing convergence time.

The crossover factor (Cr) controls the diversity of the population, since a large value of Cr will result in a higher probability that new genetic material will be incorporated into the population. Large values of Cr result in fast convergence while smaller values improve robustness [17].

General guidelines for the values of parameters that work reasonably well on a wide range of problems are known; however, the best results in terms of accuracy, robustness, and speed are found if the parameters are tuned for each problem individually [17].

7.4.4 Differential Evolution in Dynamic Environments

Like most evolutionary algorithms, the initial population of DE is formed from individuals that are randomly dispersed in the search space. This ensures that the algorithm explores a large area of the search space during the first generations. Eventually, however, the algorithm converges to an optimum, with all the individuals clustered around a single point.

The amount by which the locations of individuals in the population differ is referred to as diversity. If individuals are uniformly distributed in the search space then their diversity is high. Alternatively, when individuals are clustered around a single point, the diversity is low. The diversity measure used in this work calculates the diversity, D, of a population of size I in J dimensions as

$$D = \sum_{i=1}^{I} \|\mathbf{d} - \mathbf{x}_i\|_2, \tag{7.7}$$

where $\mathbf{d}$ is the average location of all individuals, calculated as

$$d_j = \frac{\sum_{i=1}^{I} x_{i,j}}{I}, \ \forall\, j \in J, \tag{7.8}$$

In static environments this loss of diversity is not a bona fide problem and is in most cases desirable, since having all individuals clustered around the optimum assists with the fine-grained optimization at the end of the search process.

In dynamic environments, however, loss of diversity is a major cause of evolutionary algorithms being ineffective. When changes occur in the environment, the population lacks the diversity necessary to locate the position of the new global optimum [35]. Consider Figure 7.1 which depicts the offline error, current error and diversity averaged over 30 runs on the MPB of the basic DE algorithm described in Section 7.4. Ten changes in the environment are illustrated. Changes occur once every 5 000 function evaluations.

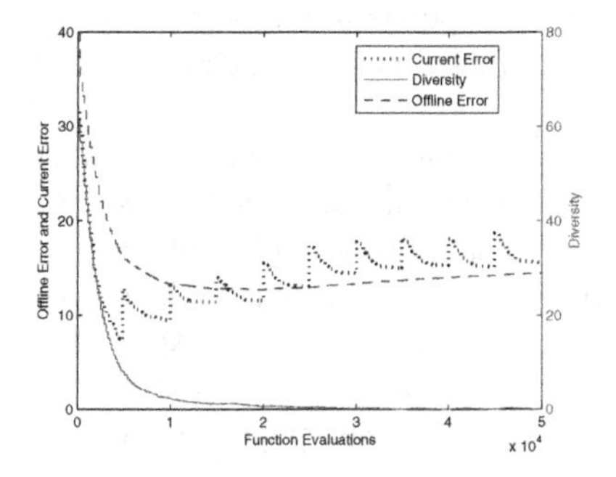

Fig. 7.1 Diversity, Current error and Offline error of DE with re-evaluation after changes

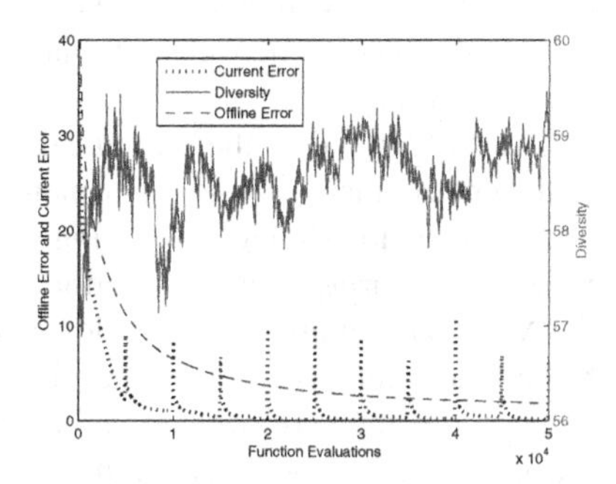

Fig. 7.2 Diversity, Current error and Offline error of DynDE on the MPB

During the period between the commencement of the algorithm and the first change in the environment, the current error and the offline error drops sharply due to the DE algorithm converging to one or more optima in the environment. The fact that individuals are clustered increasingly closely together is proven by the sharp drop in diversity. Directly after the first change in the environment, the current error increases, since the optima have shifted. Although the current error does improve after the first and subsequent changes, it never reaches the low value that was found before the first change occurred. The diversity continues to drop until it reaches a value close to zero shortly after the second change in the environment. This means that all individuals are clustered around a single optimum and that the rest of the search space is left completely unexplored. In DE mutations are performed based on the spacial differences between individuals. If individuals are spatially close then very small mutations are applied which hampers further exploration of the search space. Over 30 runs the average diversity per generation was found to be 0.422. This value is extremely low and explains why normal DE is not effective in dynamic environments.

7.4.5 Detecting Changes in the Environment

Most evolutionary dynamic optimization algorithms respond to changes in the environment. When the period between changes is known beforehand, it is necessary for the algorithm to detect changes automatically. Detecting changes is not trivial, since changes could be localized in small areas of the search space, or could involve the introduction of an optimum into an area of the search space where no individuals are located. It is thus necessary to have an appropriate strategy to detect changes. Standard approaches include periodically re-evaluating the best particle in the swarm [18], stationary sentinels [25], or random points in the problem space [11]. Discrepancies between current and previous fitness values indicate a change in the environment.

The MPB requires the optimization algorithm to automatically detect changes in the environment. It was experimentally determined by the authors that a simple change detection strategy, involving only the re-evaluation of a single sentinel individual once during every generation, is sufficient to detect all changes when using the MPB. This approach was followed throughout this study, but is not expected to be sufficient for all dynamic optimization problems. However, a more robust approach, for example re-evaluating the best individual in each sub-population, can be introduced relatively easily should it be required.

7.5 Dynamic Population Differential Evolution

Dynamic Population Differential Evolution (DynPopDE) [14], an extension of DynDE and CDE [15], is aimed at dynamic optimization problems where the number of optima in the search space is unknown or fluctuates over time.

DynDE is a differential evolution algorithm developed by Mendes and Mohais [23] to solve dynamic optimization problems. Sections 7.5.1 to 7.5.5 provides an overview of the components inherited by DynPopDE from DynDE. The two algorithmic components that make up CDE are described and motivated in Sections 7.5.6 and 7.5.7. DynPopDE is an extension of CDE consisting of three components, namely: spawning new populations (Section 7.5.8), removing populations (Section 7.5.9), and the introduction of a penalty factor on the performance value used to determine which sub-population is to evolve at a given time (Section 7.5.10).

7.5.1 Multiple Populations

Typically, a static problem space may contain several local optima. These optima typically move around in a dynamic environment and also change in height and shape. This implies that an entirely different optimum may become the global optimum once a change in the environment occurs. Consequently, not only the movement of the global optimum in the problem space must be tracked, but also the local optima. An effective method of tracking all optima is to maintain several independent sub-populations of DE individuals, one sub-population on each optimum. In their most successful experiments Mendes and Mohais [23] used 10 sub-populations, each containing 6 individuals.

7.5.2 *Exclusion*

In order to track all optima, it is necessary to ensure that all sub-populations converge to different optima. If all populations converged to the global optimum, it would defeat the purpose of having multiple populations. Mendes and Mohais [23] used exclusion to prevent sub-populations from converging to the same optimum. Exclusion compares the locations of the best individuals from each sub-population. If the spatial difference between any two of these individuals becomes too small, the entire sub-population of the inferior individual is randomly reinitialized. A threshold is used to determine if two individuals are too close. The threshold, or exclusion radius, is calculated as

$$r_{excl} = \frac{X}{2p^{\frac{1}{d}}}, \tag{7.9}$$

where X is the range of the d dimensions (assuming equal ranges for all dimensions), and p is the number of peaks. Equation (7.9) shows that the exclusion threshold increases with an increase in the number of dimensions and decreases if the number of peaks is increased. Because knowledge of the number of peaks is generally not available, it was proposed that the exclusion threshold be calculated using the number of sub-populations as follows [14]:

$$r_{excl} = \frac{X}{2K^{\frac{1}{d}}}, \tag{7.10}$$

where K is the number of populations, thus making the threshold dependent on the number of available populations. The same equation was used by Blackwell [2] in the self-adapting multi-swarms algorithm.

7.5.3 *Brownian Individuals*

In cases where a change in the environment results in the positional movement of some of the optima, it is unlikely that all of the sub-populations will still be clustered around the optimal point of their respective optima, even if the change is small. In order for individuals in the sub-populations to track the moving optima more effectively, the diversity of each population should be increased. Mendes and Mohais [23] successfully used Brownian individuals for this purpose. In every generation a predefined number of the weakest individuals are flagged as Brownian. These individuals are then replaced by new individuals created by adding a small random number, sampled from a zero centered Gaussian distribution, to each component of the best individual in the sub-population. A Brownian individual, $\mathbf{x}_{brown}$, is thus created from the best individual $\mathbf{x}_{best}$ using

$$\mathbf{x}_{brown} = \mathbf{x}_{best} + \mathcal{N}(0, v), \tag{7.11}$$

where v is the standard deviation of the Gaussian distribution. Mendes and Mohais [23] showed that a suitable value of v to use is 0.2.

7.5.4 DE Scheme

Mendes and Mohais [23] experimentally investigated several DE schemes to determine which one is the best to use in DynDE. The schemes investigated were DE/rand/1/bin, DE/rand/2/bin, DE/best/1/bin, DE/best/2/bin, DE/rand-to-best/1/bin, DE/current-to-best/1/bin and DE/current-to-rand/1/bin. It was shown that the most effective scheme to use in conjunction with DynDE is DE/best/2/bin, where each temporary individual is created using

$$\mathbf{v} = \mathbf{x}_{best} + \mathscr{F} \cdot (\mathbf{x}_1 + \mathbf{x}_2 - \mathbf{x}_3 - \mathbf{x}_4) \tag{7.12}$$

with $\mathbf{x}_1 \neq \mathbf{x}_2 \neq \mathbf{x}_3 \neq \mathbf{x}_4$ and $\mathbf{x}_{best}$ being the best individual in the sub-population.

7.5.5 DynDE Discussion

The performance of DynDE was thoroughly investigated by Du Plessis and Engelbrecht [15]. However, to illustrate how effective DynDE is on dynamic environments compared to normal DE, Figure 7.2 depicts the offline error, current error, and diversity of DynDE algorithm on the MPB for 10 changes in the environment. Averages over 30 runs are used.

A comparison between Figures 7.1 and 7.2 shows the considerable improvement of DynDE over DE. Firstly, where the diversity of the DE algorithm sharply declines to a point close to zero, the diversity of DynDE remains high. The average diversity per generation over all repeats was found to be 59.496 for DynDE, compared to the value of 0.422 for DE. The frequent perturbations on the diversity curve in Figure 7.2 show how diversity is perpetually increased by Brownian individuals and sub-populations reinitialized by exclusion.

Secondly, the graphs clearly show considerably lower offline and current errors for DynDE than for DE. In Figure 7.1, the offline error increases after the first few changes in the environment, while DynDE's offline error consistently decreases (see Figure 7.2). The current error of DynDE frequently approaches zero between changes in the environment, while the current error of DE remains high, especially after the first change in the environment.

7.5.6 Competitive Population Evaluation

For most static optimization problems, the effectiveness of an algorithm is measured by the error of the best solution found at the end of the optimization process. In contrast, optimization in dynamic environments implies that a solution is likely to be required at all times (or at least just before changes in the environment occur), not just at the termination of the algorithm. In these situations, it is imperative to find the lowest error value as soon as possible after changes in the environment have occurred. A dynamic optimization algorithm can thus be improved, not only by reducing the error, but also by making the algorithm reach its lowest error value (before a change occurs in the environment) in fewer function evaluations.

The above argument is the motivation for the component of DynPopDE named Competitive Population Evaluation (CPE). The primary goal of CPE is not to decrease the error value found by DynDE, but rather to make the algorithm reach the lowest error value in fewer function evaluations. It is proposed that this can be achieved by initially allocating all function evaluations after a change in the environment to the sub-population that is optimizing the global optimum; thereafter function evaluations are allocated to other sub-populations to locate local optima.

The mechanism used by CPE to allocate function evaluations is based on the performance of sub-populations. The best-performing sub-population is evolved on its own until its performance drops below that of another sub-population. The new best performing sub-population is then evolved on its own until its performance drops below that of another sub-population. This process is repeated until a change occurs in the environment. CPE allows the location of the global optimum to be discovered early, while the sub-optimum peaks are located later. CPE thus differs from DynDE in that peaks are not located in parallel, but sequentially. The CPE process is detailed in Algorithm 7.2.

The performance value, $\mathscr{P}$, of a sub-population depends on two factors: The current fitness of the best individual in the sub-population and the error reduction of the best individual during the last evaluation of the sub-population. Let K be the number of sub-populations, $\mathbf{x}_{best,k}$ the best individual in sub-population k, and $F(\mathbf{x}_{best,k},t)$ the fitness of the best individual in sub-population k during iteration t. The performance $\mathscr{P}(k,t)$ of population k after iteration t is given by:

$$\mathscr{P}(k,t) = (\Delta F(\mathbf{x}_{best,k},t)+1)(R_k(t)+1), \tag{7.13}$$

where

$$\Delta F(\mathbf{x}_{best,k},t) = |F(\mathbf{x}_{best,k},t) - F(\mathbf{x}_{best,k},t-1)| \tag{7.14}$$

For function maximization problems, $R_k(t)$ is calculated as:

$$R_k(t) = |F(\mathbf{x}_{best,k},t) - \min_{q=1,\ldots,K}\{F(\mathbf{x}_{best,q},t)\}| \tag{7.15}$$

Algorithm 7.2. Competitive Population Evaluation

```
while termination criteria not met do
    Allow the standard DynDE algorithm to run for two generations;
    repeat
        for k = 1,...,K;                         // all K sub-populations
        do
            Calculate the performance value, 𝒫(k,t)
        end
        Select a such that 𝒫(a,t) = min_{k=1,...,K} {𝒫(k,t)};
        Evolve only sub-population a using DE;
        t = t + 1;
        Calculate 𝒫(a,t);
        Perform Exclusion;
    until Change occurs in the environment;
end
```

and for function minimization problems,

$$R_k(t) = |F(\mathbf{x}_{best,k},t) - \max_{q=1,\ldots,K}\{F(\mathbf{x}_{best,q},t)\}| \tag{7.16}$$

Therefore, the best performing sub-population is the sub-population with the highest product of fitness and improvement. The motivation for the addition of 1 to the first and second terms in Equation (7.13) is to prevent a population being assigned a performance value of zero. Without the addition in the second term, the least fit population will always be assigned a performance value of zero (since the product of the first and second terms will be zero) and will never be considered for searching for an optimum. Similarly, without the addition in the first term, a good performing population that does not show any improvement during a specific iteration, is assigned a performance value of zero and will never be considered for evaluation again. The addition of 1 to the first and second terms is thus included to ensure that every sub-population could potentially have the highest performance value and subsequently be given function evaluations.

The absolute values of $\Delta F(\mathbf{x}_{best,k},t)$ and $R_k(t)$ are taken to ensure that the performance values are always positive. When a population is reinitialized due to exclusion (see Section 7.5.2), the fitness of the best individual is likely to be lower than before reinitialization (since the sub-population now consists of randomly generated individuals), resulting in a large $\Delta F(\mathbf{x}_{best,k},t)$ value. The population is consequently assigned a relatively large performance value, making it likely that it would be allocated fitness evaluations in the near future.

The CPE approach can be clarified by an example. Figure 7.3 plots the error per function evaluation of the best individual in each of the three sub-populations (labelled Population 1, Population 2 and Population 3) found during an actual run of the DynDE algorithm on the MPB. During the period that is depicted, no changes occurred in the environment. Note that one of the sub-populations (Population 3) converged to the global optimum, as is evidenced by its error value approaching zero, while the others likely converged to local optima.

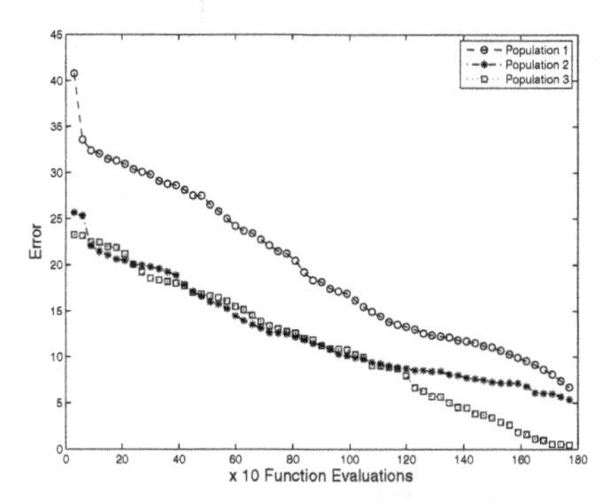

Fig. 7.3 Error profile of three DynDE populations in a static environment

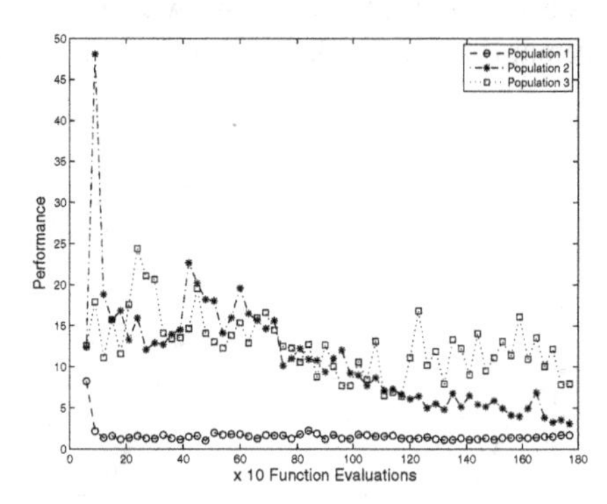

Fig. 7.4 Performance, $\mathscr{P}$, of each population

The performance of each of the populations was calculated using Equation (7.13). A plot of the performance of each of the populations is given in Figure 7.4. For roughly the first half of the period depicted, Population 2 and Population 3 alternated between having the highest performance value. After the first half of the depicted period, Population 3 consistently received the highest performance value, as it converged to the global optimum. Population 3 received, on average, a higher performance value than the other two populations. The population that had the highest overall error received a relatively low average performance value. Note that it was only after about 1 000 function evaluations that Population 2 and Population 3 received a lower performance value than the initial performance value of Population 1.

It is now possible, by using the calculated performance values, to observe the behaviour of each of the populations when the CPE algorithm (refer to Algorithm 7.2) is used to selectively evolve sub-populations on their own based on their respective performance values. The plots of the error of each of the populations per function evaluation when using CPE are given in Figure 7.5. For the first few iterations, Population 2 and Population 3 were alternately evolved, followed by a period where only Population 2 was evolved. After about 250 function evaluations, Population 3 was evolved (except for a few intermittent iterations around 700 function evaluations) until it converged to the global optimum at about 1 000 function evaluations. Population 2 was evolved during the next period which lasted until about 1 200 function evaluations, while Population 1 was evolved during the last period.

Note that the lowest error was reached after 1 000 function evaluations in Figure 7.5, while this point was only reached after 1 800 function evaluations in Figure 7.3. The performance values of the three populations when using CPE are given in Figure 7.6. Observe that for considerable periods the performance values of some of the sub-populations remain constant. This occurs because only one sub-population is evolved at a time, during which the performance value of the other populations remains unchanged.

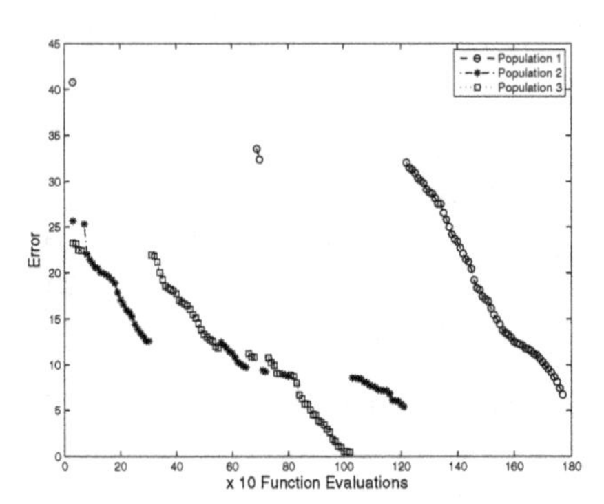

Fig. 7.5 Error profile of three populations when using competitive evaluation

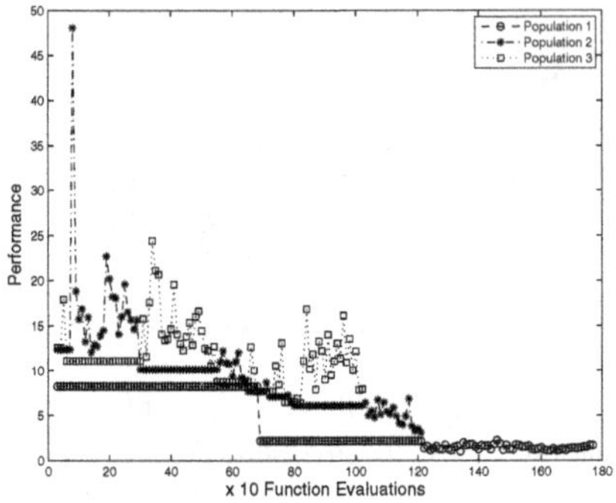

Fig. 7.6 Performance, $\mathscr{P}$, of each population when using competitive evaluation

It is clear how the more successful sub-populations are allocated more function evaluations earlier in the evolution process. Optima are located sequentially by CPE and in parallel by DynDE. Figure 7.7 depicts the effect that using CPE had on the offline error for the three sub-population example. The curve of the offline error when competing sub-populations are used exhibits a steeper downward gradient than the curve of normal DynDE, due to earlier discovery of the global optimum. Overall, the offline error was reduced by more than 30% after 1 800 function evaluations.

By competitively choosing the better performing populations to evolve before other populations, the lowest error value could be reached sooner, thus reducing the average error. This technique has the added advantage that better performing populations will receive more function evaluations that would otherwise have been wasted on finding the maximum of the sub-optimal peaks. The overall error value should consequently also be reduced.

An advantage of CPE is that it only utilizes information that is available in normal DynDE, so that no extra function evaluations are required.

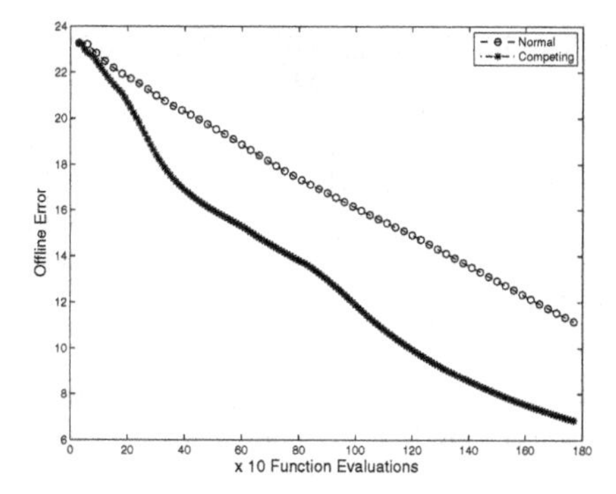

Fig. 7.7 Comparison of offline error for normal and competitive evaluation

7.5.7 *Reinitialization Midpoint Check*

Section 7.5.2 explained how DynDE determines when two sub-populations are located on the same optimum, which results in the weaker sub-population being reinitialized. This approach does not take into account the case when two optima are located extremely close to each other, i.e. within the exclusion threshold from one another. In these situations, one of the sub-populations will be reinitialized, leaving one of the optima unpopulated. It was shown [15] that this problem can be partially remedied by determining whether the midpoint between the best individuals in each sub-population constitutes a higher error value than the best individuals of both sub-populations. If this is the case, it implies that a trough exists between the two sub-populations and that neither should be reinitialized (see Figure 7.8, scenario A). This approach is referred to as the Reinitialization Midpoint Check (RMC) approach. It is apparent that RMC does not work in all cases. Scenarios B and C of Figure 7.8 depict situations where multiple optima within the exclusion threshold are not detected by a midpoint check. Scenario C further constitutes an example where no point between the two optima will give a higher error, thus making it impossible to detect two optima by using any number of intermediate point checks.

This approach is similar to, but simpler, than *hill-valley detection* suggested by [34], since only one point is checked between sub-populations.

The midpoint check approach provides a method of detecting multiple optima within the exclusion threshold without being computationally expensive or using too many function evaluations, since only one point is evaluated.

7.5.8 *Spawning Populations*

It was shown [14] that even if the number of peaks is known, creating an equal number of sub-populations as peaks is not always an effective strategy. When the

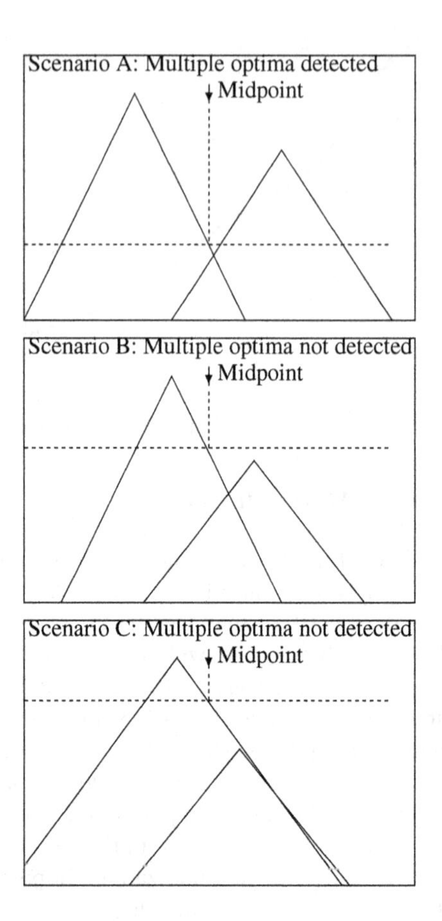

Fig. 7.8 Midpoint checking scenarios

number of peaks is unknown, choosing the number of sub-populations to use would be, at best, an educated guess. It was therefore suggested [14] that the number of sub-populations should not be a parameter of the algorithm, but that sub-populations should be spawned as needed. The question that must be answered is: When should new populations be spawned?

CPE is based on allocating processing time and function evaluations to populations based on a performance value, $\mathscr{P}(k,t)$ (refer to equation (7.13) in Section 7.5.6). Sub-populations are evolved in sequence until all sub-populations converge to their respective optima in the search space. It was proposed that an appropriate time to introduce an additional sub-population is when little further improvement in fitness is found for all the current sub-populations. Introducing new sub-populations earlier would be contrary to the competing population approach of CPE, where inferior sub-populations are deliberately excluded from the evolution process so that optima can be discovered earlier.

A detection scheme is suggested to indicate when evolution has reached a point of little or no improvement in fitness of current sub-populations. This point will be referred to as stagnation. When stagnation is detected, DynPopDE introduces a new population of randomly created individuals so that previously undiscovered optima can be located. CPE calculates the performance value of a sub-population, k, as the product of its current relative fitness, $R_k(t)$, and the improvement that was made in fitness during the previous generation, $\Delta F(\mathbf{x}_{best,k},t)$ (see Equation (7.13)). It is suggested that a meaningful indicator of stagnation is when all of the current sub-populations receive a zero improvement of fitness after their last respective function evaluations. Let $\mathscr{K}$ be the set of current populations. Define a function, $\Upsilon(t)$, that is true if stagnation has occurred, as follows:

$$\Upsilon(t) = \begin{cases} true \text{ if}(\Delta F(\mathbf{x}_{best,k},t) = 0)\ \forall\, k \in \mathscr{K} \\ false \text{ otherwise} \end{cases} \tag{7.17}$$

$\Delta F(\mathbf{x}_{best,k},t)$ is defined in equation (7.14). Note that equation (7.17) does not guarantee that stagnation of all populations has permanently occurred, since the fitness of some of the sub-populations may improve in subsequent generations. However, when $\Upsilon(t)$ is true, it does mean that none of the sub-populations has improved its fitness in the previous generation. Since function evaluations are not effectively used by the current sub-populations, a new sub-population should be created which may lead to locating more optima in the problem space.

After each generation, DynPopDE determines the value of $\Upsilon(t)$. If $\Upsilon(t) = true$ then a new randomly generated sub-population is added to the set of sub-populations. The sub-population spawning approach allows DynPopDE to commence with only a single population and adapt to an appropriate number of populations. The number of sub-populations is thus removed as a parameter from the algorithm.

7.5.9 Removing Populations

The previous section explained how new populations are spawned when necessary. However, it is possible that Equation (7.17) detects stagnation incorrectly since it cannot guarantee that stagnation for all sub-populations has occurred indefinitely. Consequently, more sub-populations than necessary may be created. Furthermore, in problems where the number of optima fluctuates, it would be desirable to remove superfluous sub-populations when the number of optima decreases. It thus becomes necessary to detect and remove redundant sub-populations.

DynDE reinitializes sub-populations through exclusion when the spatial difference between the population and a more fit population drops below the exclusion threshold. It is reasonable to assume that when redundant populations are present (i.e. more sub-populations exist than optima), these sub-populations will perpetually be reinitialized and will not converge to specific optima since exclusion prevents more than one sub-population from converging to the same optima.

Consequently, redundant sub-populations can be detected by finding sub-populations that are successively reinitialized through exclusion without reaching a point of apparent stagnation (i.e. there are no more optima available for the sub-population to occupy).

A sub-population, k, will be discarded when it is flagged for reinitialization due to exclusion and if

$$\Delta F(\mathbf{x}_{best,k},t) \neq 0. \tag{7.18}$$

The RMC exclusion process is thus further adapted to also remove sub-populations (refer to Algorithm 7.3 where $\mathbf{x}_{best,k}$ is the best individual in sub-population $k \in \{1,2,\ldots,K\}$).

Algorithm 7.3. DynPopDE Exclusion

```
for k = 1,...,K do
    for q = 1,...,K do
        if ||x_best,k − x_best,q||_2 < r_excl and k ≠ q then
            Let x_mid = (x_best,k + x_best,q)/2;
            if F(x_mid) < F(x_best,k) and F(x_mid) < F(x_best,q) then
                if F(x_best,k) < F(x_best,q) then
                    if ΔF(x_best,q, t) = 0 then
                        Reinitialize population q
                    else
                        Discard population q
                    end
                else
                    if ΔF(x_best,k, t) = 0 then
                        Reinitialize population k
                    else
                        Discard population k
                    end
                end
            end
        end
    end
end
```

This approach may at times incorrectly classify populations as redundant in that it does not guarantee the removal of populations only when the number of populations outnumbers the number of optima. In some cases a sub-population will be discarded for converging to an optimum, occupied by another sub-population, even when undiscovered optima still exist in the the problem space. However, no optimum is ever left unguarded through the discarding process since a sub-population is discarded only when converging to an optimum already occupied by another sub-population. No information about the search space is thus lost through discarding a sub-population. If all sub-populations have stagnated, a new sub-population will be created through the spawning process. The spawning and the discarding components of DynPopDE thus reach a point of equilibrium where sub-populations are created when function evaluations are not being used effectively by the current sub-populations, and sub-populations are removed when converging to optima that are already guarded by other sub-populations.

7.5.10 Penalty Factor

The detection of stagnation is essential to DynPopDE's population spawning process. The detection strategy described in Section 7.5.8 is based on all sub-populations not improving their respective fitness values in the previous generation. However, in CPE, sub-populations are evolved alone based on performance value. It is possible that a situation could occur where some of the weaker sub-populations are not allocated enough fitness evaluation to reach the stagnation point as detected by Equation (7.17). It was found experimentally that in order to detect stagnation effectively, it is necessary to distribute function evaluations more uniformly among all populations [14]. The final component of DynPopDE is the introduction of a penalty factor into the performance value, $\mathscr{P}(k,t)$ given in Equation (7.13), to penalize populations for successively receiving the highest performance value without showing any improvement in fitness.

Each population, k, maintains a penalty factor $pen_k(t)$ which is calculated as follows:

$$pen_k(t) = \begin{cases} pen_k(t-1)+1 \text{ if}(\Delta F(\mathbf{x}_{best,k},t)=0) \\ 0 \text{ otherwise} \end{cases} . \tag{7.19}$$

The penalty factor is thus reset to zero as soon as an improvement of fitness is found. The performance with penalty $\mathscr{P}_{pen}(k,t)$ is calculated as:

$$\mathscr{P}_{pen}(k,t) = \begin{cases} \frac{\mathscr{P}(k,t)}{pen_k(t)} \text{ if}(pen_k(t)>0) \\ \mathscr{P}(k,t) \text{ otherwise} \end{cases} . \tag{7.20}$$

Populations that stagnate but still receive a large performance value, $\mathscr{P}(k,t)$, will eventually receive a low value for performance with penalty, $\mathscr{P}_{pen}(k,t)$, and will consequently not dominate the evolution process.

The necessity of using a penalty implies that there is an intrinsic cost to detecting stagnation effectively, since it is necessary to waste more function evaluations on weaker sub-populations to ensure a reasonable chance to stagnate.

7.6 Self-Adaptive DynPopDE

The main goal of this study is an investigation into the effect of incorporating self-adaptive control parameters into DynPopDE. This section describes the self-adaptive approach that is incorporated into DynPopDE to form the Self-Adaptive Dynamic Population Differential Evolution (SADynPopDE) algorithm.

7.6.1 The Self-Adaptive DynPopDE Algorithm

In a previous study, it has been shown that CDE can be improved by self-adapting the DE scale and crossover factors [16]. Three different DE-based approaches to self-adaptation were investigated [9, 26, 27], and it was found that the the approach of Brest *et al.* [9] is the most effective of the three when used in conjunction with CDE.

The approach of Brest *et al.* [9] self-adapts the values of both the crossover factor and the scale factor. Each individual, $\mathbf{x}_i$, stores its own value for the crossover factor, Cr_i, and scale factor, $\mathscr{F}_i$. Before Equation (7.1) is used to create a new trial individual, the scale factor and crossover factor of the target individual ($\mathbf{x}_i$ in Equation (7.2)) are used to create new values for the scale factor and crossover factor to be used in equations (7.1) and (7.2). The new scale and crossover factors ($\mathscr{F}_{new}$ and Cr_{new}) are calculated as follows:

$$\mathscr{F}_{new} = \begin{cases} \mathscr{F}_l + U(0,1) \cdot \mathscr{F}_u \text{ if } (U(0,1) < \tau_1) \\ \mathscr{F}_i \text{ otherwise} \end{cases} \tag{7.21}$$

$$Cr_{new} = \begin{cases} U(0,1) \text{ if } (U(0,1) < \tau_2) \\ Cr_i \text{ otherwise} \end{cases}, \tag{7.22}$$

where τ_1 and τ_2 are the probabilities that the factors will be adjusted. Brest *et al.* [9] used 0.1 for both τ_1 and τ_2. Other values used were $\mathscr{F}_l = 0.1$ and $\mathscr{F}_u = 0.9$. $\mathscr{F}_l$ is a constant introduced to avoid premature convergence by ensuring that the scale factor is never too small (see Section 7.4.3), while $\mathscr{F}_u$ determines the range of scale factors that can be explored by the algorithm.

This approach is the basis for the successful *jDE* algorithm [10] which is aimed at dynamic environments. Brest *et al.* [10] used $\mathscr{F}_l = 0.36$ rather than $\mathscr{F}_l = 0.1$ for the dynamic algorithm. Initial values of 0.9 for the crossover factor and 0.5 for the scale factor were used [10]. The *jDE* algorithm makes use of DE/rand/1/bin, but it was found that DE/best/2/bin yielded the best results in the CDE-based version of this algorithm [16].

The self-adaptive approach of Brest *et al.* [9][10] was incorporated into DynPopDE to form SADynPopDE. The following section presents experimental evidence that shows that SADynPopDE is an improvement over DynDE and DynPopDE.

7.6.2 Results and Discussion

This section experimentally compares the performance of SADynPopDE to those of DynDE and DynPopDE, in order to determine whether self-adaptive control parameters yield a more effective algorithm. The modified MPB described in Section 7.2.2 was used to model problems with a fluctuating number of peaks. For these experiments, DynDE was given 10 sub-populations, since this number was shown to yield reasonable performance on this type of problem [14]. The number of individuals in the sub-populations of all algorithms was set to six. For all algorithms, DynDE and DynPopDE were compared to SADynPopDE for different values of maximum number of peaks and percentage change in the number of peaks. Additionally, the effect of varying dimension and change period was investigated. All combinations of settings listed in Table 7.2 were investigated.

Table 7.2 MPB settings for fluctuating number of peaks experiments.

Setting	**Values**
Number of Dimensions	5, 10, 15
Change period	1000, 2000, 3000, 4000, 5000
Maximum Number of Peaks	20, 40, 60, 80, 100, ..., 180, 200
Percentage Change in the Number of Peaks	10, 20, 30, 40, 50

Each experiment was repeated 50 times. Average offline error and 95% confidence interval for DynDE, DynPopDE and SADynPopDE along with the results of Mann-Whitney U tests are summarized in Tables 7.3 and 7.4. For the sake of brevity, some of the results were omitted, but are available from the authors. The tables list the maximum number of peaks (**Max Peaks**), percentage change in number of peaks (**% Change**), change period (**CP**), offline error of DynDE, offline error of DynPopDE, offline error of SADynPopDE, the p-values of Mann-Whitney U tests comparing the differences in offline error between DynDE and DynPopDE (**p-A**), and DynPopDE and SADynPopDE (**p-B**). Values for which the differences were not statistically significant at a 95% confidence level are given in boldface. Cases where the offline error of DynPopDE is lower than that of SADynPopDE are given in italics.

The results show that DynDE generally yields large offline errors when the number of optima fluctuates. Increasing the number of dimensions greatly increases the offline error, because the size of the search space is effectively increased which makes it harder to locate optima. In addition, lowering the number of function evaluations between changes in the environment has a negative effect on the offline error, since fewer generations can be performed between changes in the environment.

Fluctuating the number of optima in the solution space results in severe changes in the environment. When optima vanish from the solution space, entire sub-populations are left with a low fitness value. Similarly, optima are introduced in areas of the search space which is not guarded by individuals. As the maximum number of optima in the environment increases (and subsequently the number of optima that are introduced or removed when a change in the environment occurs), the offline error of DynDE increases.

Increasing the percentage change in the number of peaks negatively affected the performance of DynDE. This trend is expected, since, when peaks are removed, the sub-populations guarding those peaks are reduced to relatively low fitness values.

Table 7.3 Results in 5 dimensions

Max Peaks	% Change	CP	DynDE	DynPopDE	p-A	SADynPopDE	p-B
20	10	1000	15.73 ± 1.22	11.39 ± 0.69	0.00	6.78 ± 0.32	0.00
20	10	3000	3.85 ± 0.33	3.21 ± 0.33	0.00	2.14 ± 0.13	0.00
20	10	5000	2.42 ± 0.19	1.96 ± 0.19	0.00	1.38 ± 0.09	0.00
20	30	1000	28.92 ± 1.46	19.88 ± 0.88	0.00	13.53 ± 0.56	0.00
20	30	3000	7.84 ± 0.64	5.61 ± 0.53	0.00	3.56 ± 0.26	0.00
20	30	5000	3.97 ± 0.45	3.24 ± 0.34	0.01	2.22 ± 0.21	0.00
20	50	1000	37.73 ± 1.15	26.15 ± 0.88	0.00	17.73 ± 0.57	0.00
20	50	3000	11.35 ± 0.86	9.10 ± 0.60	0.00	5.61 ± 0.38	0.00
20	50	5000	6.30 ± 0.59	5.32 ± 0.46	0.04	2.89 ± 0.26	0.00
40	10	1000	16.70 ± 1.44	10.68 ± 0.99	0.00	8.62 ± 0.58	0.01
40	10	3000	5.09 ± 0.48	3.94 ± 0.47	0.00	2.60 ± 0.15	0.00
40	10	5000	3.07 ± 0.20	2.38 ± 0.20	0.00	1.64 ± 0.09	0.00
40	30	1000	28.97 ± 0.99	21.94 ± 1.24	0.00	15.11 ± 0.72	0.00
40	30	3000	10.14 ± 0.92	6.54 ± 0.46	0.00	5.00 ± 0.41	0.00
40	30	5000	6.37 ± 0.71	4.11 ± 0.44	0.00	2.70 ± 0.27	0.00
40	50	1000	38.19 ± 1.22	28.00 ± 0.88	0.00	20.74 ± 0.70	0.00
40	50	3000	15.00 ± 1.18	10.53 ± 0.69	0.00	6.96 ± 0.58	0.00
40	50	5000	8.52 ± 0.89	5.41 ± 0.58	0.00	4.20 ± 0.39	0.00
80	10	1000	16.30 ± 1.70	11.82 ± 1.19	0.00	9.19 ± 0.75	0.00
80	10	3000	6.64 ± 0.80	3.95 ± 0.34	0.00	3.16 ± 0.27	0.00
80	10	5000	3.62 ± 0.22	2.76 ± 0.29	0.00	2.18 ± 0.14	0.00
80	30	1000	29.26 ± 1.35	20.44 ± 0.94	0.00	16.75 ± 0.85	0.00
80	30	3000	13.29 ± 1.13	7.73 ± 0.67	0.00	5.78 ± 0.38	0.00
80	30	5000	6.42 ± 0.58	4.95 ± 0.57	0.00	3.44 ± 0.33	0.00
80	50	1000	37.94 ± 1.17	28.09 ± 0.78	0.00	21.46 ± 0.57	0.00
80	50	3000	17.12 ± 1.12	11.75 ± 0.93	0.00	8.33 ± 0.56	0.00
80	50	5000	9.83 ± 0.87	7.60 ± 0.72	0.00	5.58 ± 0.40	0.00
160	10	1000	16.60 ± 1.55	12.03 ± 1.22	0.00	9.55 ± 0.81	0.00
160	10	3000	7.94 ± 1.29	5.03 ± 0.57	0.00	3.54 ± 0.29	0.00
160	10	5000	4.53 ± 0.46	3.37 ± 0.61	0.00	2.66 ± 0.30	0.00
160	30	1000	28.60 ± 1.31	22.61 ± 1.05	0.00	16.80 ± 0.76	0.00
160	30	3000	14.38 ± 1.26	9.43 ± 0.94	0.00	6.48 ± 0.51	0.00
160	30	5000	8.95 ± 1.04	5.67 ± 0.78	0.00	4.22 ± 0.42	0.01
160	50	1000	36.32 ± 1.18	29.31 ± 0.93	0.00	22.10 ± 0.70	0.00
160	50	3000	18.71 ± 1.20	12.14 ± 0.84	0.00	9.32 ± 0.50	0.00
160	50	5000	12.58 ± 1.10	8.15 ± 0.72	0.00	5.72 ± 0.43	0.00

Similarly, new peaks that are introduced will initially not be located by sub-populations. When the percentage of peaks that are removed or introduced is increased, either, more sub-populations are left not guarding a peak, or more peaks are introduced that are initially unguarded. The detrimental effect of percentage change in number of peaks was more pronounced for larger values of maximum number of peaks, since a greater number of peaks are introduced or removed.

Table 7.4 Results in 15 dimensions

Max Peaks	% Change	CP	DynDE	DynPopDE	p-A	SADynPopDE	p-B
20	10	1000	118.51 ± 5.70	63.59 ± 5.33	0.00	68.97 ± 4.89	***0.14***
20	10	3000	47.52 ± 8.12	19.39 ± 2.74	0.00	16.95 ± 2.36	**0.23**
20	10	5000	25.82 ± 5.15	11.66 ± 2.10	0.00	8.43 ± 1.18	0.02
20	30	1000	146.05 ± 2.30	98.36 ± 4.83	0.00	101.37 ± 3.30	***0.11***
20	30	3000	80.93 ± 6.07	42.26 ± 4.71	0.00	32.34 ± 2.08	0.00
20	30	5000	47.70 ± 5.81	24.73 ± 3.93	0.00	17.64 ± 2.18	0.00
20	50	1000	157.85 ± 2.53	120.04 ± 4.06	0.00	117.58 ± 2.69	**0.72**
20	50	3000	107.02 ± 5.33	51.94 ± 4.49	0.00	41.19 ± 3.12	0.00
20	50	5000	67.92 ± 5.41	33.87 ± 3.07	0.00	26.50 ± 2.26	0.00
40	10	1000	103.93 ± 5.47	64.02 ± 6.34	0.00	58.14 ± 4.89	**0.18**
40	10	3000	37.39 ± 7.52	21.58 ± 3.94	0.00	15.64 ± 2.32	**0.08**
40	10	5000	21.77 ± 4.48	13.66 ± 2.39	0.00	9.62 ± 1.75	0.00
40	30	1000	137.11 ± 3.09	91.94 ± 4.25	0.00	92.80 ± 3.65	***0.60***
40	30	3000	72.50 ± 5.20	40.85 ± 3.86	0.00	32.65 ± 2.94	0.00
40	30	5000	52.98 ± 4.95	24.85 ± 3.93	0.00	20.24 ± 2.45	**0.13**
40	50	1000	146.45 ± 2.19	112.54 ± 3.89	0.00	110.23 ± 3.29	**0.48**
40	50	3000	98.18 ± 4.94	52.51 ± 3.15	0.00	43.75 ± 2.97	0.00
40	50	5000	65.42 ± 5.49	34.49 ± 2.69	0.00	27.34 ± 2.43	0.00
80	10	1000	95.19 ± 6.04	55.47 ± 5.40	0.00	56.74 ± 4.94	***0.49***
80	10	3000	41.76 ± 7.72	27.93 ± 5.55	0.00	16.13 ± 2.43	0.00
80	10	5000	24.61 ± 6.25	12.44 ± 3.29	0.00	7.90 ± 1.06	0.01
80	30	1000	128.25 ± 2.83	89.70 ± 4.70	0.00	84.20 ± 3.49	**0.16**
80	30	3000	78.10 ± 5.63	41.28 ± 4.51	0.00	32.97 ± 3.11	0.01
80	30	5000	43.62 ± 5.92	25.94 ± 3.28	0.00	20.64 ± 2.08	0.03
80	50	1000	138.18 ± 2.17	107.12 ± 3.54	0.00	108.03 ± 2.56	***0.32***
80	50	3000	92.44 ± 4.16	59.88 ± 5.61	0.00	43.56 ± 2.66	0.00
80	50	5000	62.89 ± 4.50	34.59 ± 4.25	0.00	30.44 ± 2.30	**0.29**
160	10	1000	87.46 ± 5.89	56.10 ± 6.78	0.00	48.07 ± 5.54	**0.11**
160	10	3000	33.20 ± 5.60	22.77 ± 3.74	0.00	14.77 ± 2.51	0.00
160	10	5000	25.33 ± 6.24	12.41 ± 3.10	0.00	12.84 ± 2.62	***0.72***
160	30	1000	120.91 ± 2.52	85.79 ± 4.21	0.00	81.53 ± 3.57	**0.22**
160	30	3000	69.23 ± 5.86	38.10 ± 3.93	0.00	34.71 ± 3.42	**0.23**
160	30	5000	47.35 ± 6.68	27.42 ± 3.56	0.00	21.79 ± 3.08	0.02
160	50	1000	132.74 ± 2.26	104.89 ± 3.03	0.00	102.15 ± 2.75	**0.27**
160	50	3000	88.49 ± 4.18	54.33 ± 2.69	0.00	45.54 ± 2.76	0.00
160	50	5000	66.22 ± 6.10	37.94 ± 3.25	0.00	31.54 ± 2.61	0.00

The same general trends that are visible in the DynDE results were also found for both DynPopDE and SADynPopDE, i.e. the offline error deteriorates when the change period is reduced, the percentage change in the number of peaks is increased, or the number of dimensions is increased.

A comparison of the results of DynDE and DynPopDE (refer to Tables 7.3 and 7.4) shows DynPopDE yielded a clear improvement over DynDE. Over all experiments, it was found that 746 of the 750 experiments resulted in a statistically significantly better result for DynPopDE. On average, DynPopDE yielded an improvement over DynDE of 29.22% in 5 dimensions, 37.02% in 10 dimensions, and 40.55% in 15 dimensions. The magnitudes of the improvements are greater when larger values for maximum number of peaks were used, as population spawning ability of DynPopDE is more beneficial when in these regions. Figure 7.9 graphically depicts the surface found when subtracting the average offline error of SA-DynPopDE from the average offline error of DynPopDE for various settings of the MPB when a 50% maximum change in the number of peaks was used. For 5 dimensions almost all the SADynPopDE results were statistically significantly better than those of DynPopDE. For the 10 dimensional experiments the majority of the SADynPopDE results were statistically significantly better than DynPopDE, while in 15 dimensions the majority of the differences were not statistically significant. SADynPopDE did not perform statistically significantly worse than DynPopDE in any of the 5 and 10 dimensional experiments. In 15 dimensions, SADynPopDE was only statistically significantly worse then DynPopDE in 1 of the 250 experiments.

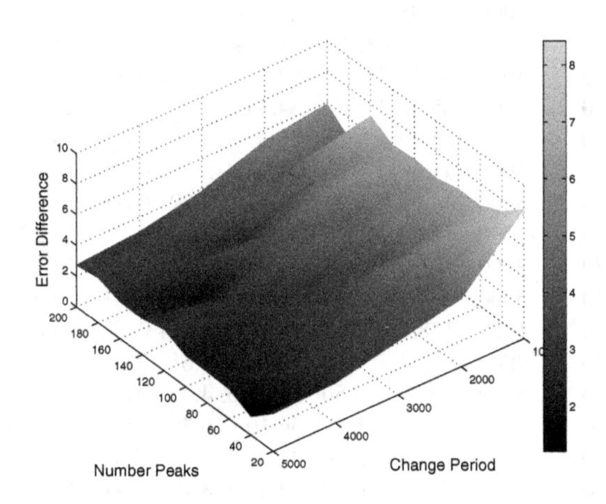

Fig. 7.9 Difference in offline error between DynPopDE and SADynPopDE for various settings of maximum number of peaks and change period in 5 dimensions.

Figure 7.9 shows how the improvement of SADynPopDE over DynPopDE in 5 dimensions becomes more pronounced as the change period becomes smaller. This trend is reversed in 10 and 15 dimensions where the improvement of SADynPopDE over DynPopDE was relatively small for change periods of 1000. The improvements of SADynPopDE over DynPopDE were less frequently significant in the higher dimensions, but differences in offline error are nonetheless often larger than 10. It was found that SADynPopDE was more often better than DynPopDE when the percentage change in the number of peaks is larger then 30%.

From the results it can be concluded that SADynPopDE is an improvement over DynPopDE, especially in low dimensional problems.

7.7 Conclusions

This chapter introduced SADynPopDE, a self-adaptive multi-population DE-based optimization algorithm, aimed at dynamic optimization problems in which the number of optima in the environment fluctuates over time. SADynPopDE was experimentally compared to two algorithms upon which it is based: DynDE and DynPopDE. DynDE extends DE for dynamic environments by utilizing multiple sub-populations which are encouraged to converge to distinct optima by means of exclusion. DynPopDE extends DynDE by: using competitive population evaluation to selectively evolve sub-populations, using a midpoint check during exclusion to determine whether sup-populations are indeed converging to the same optimum, dynamically spawning and removing sub-populations, and using a penalty factor to aid the stagnation detection process. Experimental evidence suggests that DynPopDE is more effective on problems in which the number of optima fluctuates than DynDE. In turn, it was shown that SADynPopDE is more effective than DynPopDE on these types of problems, especially on low dimensional cases.

In conclusion, the incorporation of self-adaptive control parameters into DynPopDE not only yielded a more effective algorithm, but also removed the need to fine-tune the DE crossover and scale factors.

References

[1] A multi-population approach to dynamic optimization problems. In: Adaptive Computing in Design and Manufacturing, pp. 299–308. Springer (2000)

[2] Blackwell, T.: Particle swarm optimization in dynamic environments. In: Evolutionary Computation in Dynamic and Uncertain Environments, pp. 29–49. Springer (2007)

[3] Blackwell, T., Branke, J.: Multiswarm optimization in dynamic environments. Applications of Evolutionary Computing 3005, 489–500 (2004)

[4] Blackwell, T., Branke, J.: Multiswarms, exclusion, and anti-convergence in dynamic environments. IEEE Transactions on Evolutionary Computation 10(4), 459–472 (2006)

[5] Branke, J.: Evolutionary Optimization in Dynamic Environments. Kluwer Academic Publishers, Norwell (2002)

[6] Branke, J.: The moving peaks benchmark (2007), `http://www.aifb.uni-karlsruhe.de/~jbr/MovPeaks/`

[7] Branke, J., Schmeck, H.: Designing evolutionary algorithms for dynamic optimization problems. In: Tsutsui, S., Ghosh, A. (eds.) Theory and Application of Evolutionary Computation: Recent Trends, pp. 239–262. Springer (2002)

[8] Branke, J., Schmeck, H.: Designing evolutionary algorithms for dynamic optimization problems, pp. 239–262 (2003)

[9] Brest, J., Greiner, S., Boskovic, B., Mernik, M., Zumer, V.: Self-adapting control parameters in differential evolution: A comparative study on numerical benchmark problems. IEEE Transactions on Evolutionary Computation 10(6), 646–657 (2006)

[10] Brest, J., Zamuda, A., Boškovic, B., Maučec, M.S., Žumer, V.: Dynamic optimization using self-adaptive differential evolution. In: CEC 2009: Proceedings of the Eleventh Conference on Congress on Evolutionary Computation, Piscataway, NJ, USA, pp. 415–422 (2009)

[11] Carlisle, A., Dozier, G.: Tracking changing extrema with adaptive particle swarm optimizer. In: Proc. World Automation Congress, pp. 265–270 (2002)
[12] Darwin, C.: The origin of species (1859)
[13] du Plessis, M.C., Engelbrecht, A.P.: Improved differential evolution for dynamic optimization problems. In: IEEE Congress on Evolutionary Computation, CEC 2008, pp. 229–234 (June 2008)
[14] du Plessis, M.C., Engelbrecht, A.P.: Differential evolution for dynamic environments with unknown numbers of optima. Submitted to Journal of Global Optimization (2010)
[15] du Plessis, M.C., Engelbrecht, A.P.: Using competitive population evaluation in a differential evolution algorithm for dynamic environments. Submitted to European Journal of Operational Research (2010)
[16] du Plessis, M.C., Engelbrecht, A.P.: Self-adaptive competitive differential evolution for dynamic environments. In: IEEE Symposium Series on Computational Intelligence, SSCI 2011, pp. 1–8 (April 2011)
[17] Engelbrecht, A.P.: Computational Intelligence An Introduction, 2nd edn. John Wiley and Sons (2007)
[18] Hu, X., Eberhart, R.C.: Adaptive particle swarm optimisation: detection and response to dynamic systems. In: Proceedings Congress on Evolutionary Computation, pp. 1666–1670 (2002)
[19] Jin, Y., Branke, J.: Evolutionary optimization in uncertain environments - a survey. IEEE Transactions on Evolutionary Computation 9(3), 303–317 (2005)
[20] Li, C., Yang, S.: A Generalized Approach to Construct Benchmark Problems for Dynamic Optimization. In: Li, X., Kirley, M., Zhang, M., Green, D., Ciesielski, V., Abbass, H.A., Michalewicz, Z., Hendtlass, T., Deb, K., Tan, K.C., Branke, J., Shi, Y. (eds.) SEAL 2008. LNCS, vol. 5361, pp. 391–400. Springer, Heidelberg (2008)
[21] Li, C., Yang, S., Nguyen, T.T., Yu, E.L., Yao, X., Jin, Y., Beyer, H.G., Suganthan, P.N.: University of Leicester, University of Birmingham, Nanyang Technological University, Technical Report (2008)
[22] Li, X., Branke, J., Blackwell, T.: Particle swarm with speciation and adaptation in a dynamic environment. In: GECCO 2006: Proceedings of the 8th Annual Conference on Genetic and Evolutionary Computation, pp. 51–58. ACM, New York (2006)
[23] Mendes, R., Mohais, A.: Dynde: a differential evolution for dynamic optimization problems. In: Congress on Evolutionary Computation, pp. 2808–2815. IEEE (2005)
[24] Mezura-Montes, E., Velázquez-Reyes, J., Coello, C.A.: A comparative study of differential evolution variants for global optimization. In: GECCO 2006: Proceedings of the 8th Annual Conference on Genetic and Evolutionary Computation, pp. 485–492. ACM, New York (2006)
[25] Morrison, R.W.: Designing Evolutionary Algorithms for Dynamic Environments. Springer (2004)
[26] Omran, M.G.H., Salman, A., Engelbrecht, A.P.: Self-adaptive Differential Evolution. In: Hao, Y., Liu, J., Wang, Y.-P., Cheung, Y.-m., Yin, H., Jiao, L., Ma, J., Jiao, Y.-C. (eds.) CIS 2005. LNCS (LNAI), vol. 3801, pp. 192–199. Springer, Heidelberg (2005)
[27] Omran, M.G.H., Engelbrecht, A.P., Salman, A.: Bare bones differential evolution. European Journal of Operational Research 196(1), 128–139 (2009)
[28] Oppacher, F., Wineberg, M.: The shifting balance genetic algorithm: Improving the ga in a dynamic environment. In: Banzhaf, W., et al. (eds.) Genetic and Evolutionary Computation Conference (GECCO), vol. 1, pp. 504–510. Morgan Kaufmann, San Francisco (1999)

[29] Parrott, D., Li, X.: A particle swarm model for tracking multiple peaks in a dynamic environment using speciation. In: Congress on Evolutionary Computation, pp. 98–103. IEEE (2004)
[30] Price, K., Storn, R., Lampinen, J.: Differential evolution - A practical approach to global optimization. Springer (2005)
[31] Storn, R.: On the usage of differential evolution for function optimization. In: Biennial Conference of the North American Fuzzy Information Processing Society, pp. 519–523. IEEE (1996)
[32] Storn, R., Price, K.: Minimizing the real functions of the icec96 contest by differential evolution. In: IEEE Conference on Evolutionary Computation, pp. 842–844. IEEE (1996)
[33] Storn, R., Price, K.: Differential evolution - a simple and efficient heuristic for global optimization over continuous spaces. Journal of Global Optimization 11, 341–359 (1997)
[34] Ursem, R.K.: Multinational GA optimization techniques in dynamic environments. In: Whitley, D., Goldberg, D., Cantu-Paz, E., Spector, L., Parmee, I., Beyer, H.-G. (eds.) Genetic and Evolutionary Computation Conference, pp. 19–26. Morgan Kaufmann (2000)
[35] Zaharie, D., Zamfirache, F.: Diversity enhancing mechanisms for evolutionary optimization in static and dynamic environments. In: 3rd Romanian-Hungarian Joint Symposium on Applied Computational Intelligence, pp. 460–471 (2006)

Chapter 8
Dynamic Multi-Objective Optimization Using PSO

Mardé Helbig and Andries P. Engelbrecht

Abstract. Dynamic multi-objective optimization problems occur in many situations in the real world. These optimization problems do not have a single goal to solve, but many goals that are in conflict with one another - improvement in one goal leads to deterioration of another. Therefore, when solving dynamic multi-objective optimization problem, an algorithm attempts to find the set of optimal solutions, referred to as the Pareto-optimal front. Each dynamic multi-objective optimization problem also has a number of boundary constraints that limits the search space. When the particles of a particle swarm optimization (PSO) algorithm move outside the search space, an approach should be followed to manage violation of the boundary constraints. This chapter investigates the effect of various approaches to manage boundary constraint violations on the performance of the dynamic Vector Evaluated Particle Swarm optimization (DVEPSO) algorithm when solving DMOOP. Furthermore, the performance of DVEPSO is compared against the performance of three other state-of-the-art dynamic multi-objective optimization algorithms.

8.1 Introduction

Many problems in the real-world change over time and require more than one goal to be optimized. However, these goals, or objectives, are normally in conflict with one another, where an improvement in one objective results in deterioration of another objective. Therefore, a single solution does not exist and the goal becomes

Mardé Helbig
CSIR Meraka Institute, Scientia, Meiring Naude Road, 0184,
Brummeria, South Africa
e-mail: mhelbig@csir.co.za

Mardé Helbig · Andries P. Engelbrecht
Department of Computer Science, University of Pretoria, 0002,
Pretoria, South Africa
e-mail: engel@cs.up.ac.za

E. Alba et al. (Eds.): Metaheuristics for Dynamic Optimization, SCI 433, pp. 147–188.
springerlink.com

to find the set of optimal trade-off solutions. These problems are called dynamic multi-objective optimization problem (DMOOP). Each DMOOP has a number of objective functions to optimize and each variable within an objective function has a range of values that are valid, referred to as the search space. These bounds of valid values of the decision variable are called boundary constraints.

A multi-swarm algorithm, called Dynamic Vector Evaluated Particle Swarm optimization (DVEPSO) [10], is presented. The effect that various approaches to manage boundary constraints have on the performance of DVEPSO is investigated. Furthermore, DVEPSO is compared against three other state-of-the-art dynamic multi-objective optimization (DMOO) algorithms.

The rest of the chapter's layout is as follows: Section 8.2 presents theory and background information with regard to particle swarm optimization (PSO) and DMOO. The DVEPSO algorithm is presented in Section 8.3, as well as the approaches that can be used to manage boundary constraints. Section 8.4 provides information about the experiments that were run, including the benchmark functions, performance metrics and statistical analysis that were used to measure the performance of the various algorithms. The results that were obtained from the experiments are discussed in Section 8.5. Conclusions about this research are presented in Section 8.6.

8.2 Background

This section presents background information on PSO, as well as the theory on multi-objective optimization (MOO) and DMOO. Furthermore, some issues when solving DMOOP are presented.

8.2.1 Particle Swarm Optimization

Inspired by the social behaviour of bird flocks, Eberhart and Kennedy introduced PSO [15]. The PSO algorithm maintains a swarm of particles, where each particle represents a solution of the optimisation problem. Each particle moves through the search space and its position is updated based on its own experience (cognitive information), as well as the experience of the its neighbours (social information). The particle's position that produced the best solution so far is referred to as its personal best or *pbest*. The position that leads to the best overall solution by all particles in a pre-defined neighbourhood, is called the neighbourhood best or *nbest*. If the neighbourhood is defined as the whole swarm, the neighbourhood best is referred to as the global best or *gbest*.

In general, the PSO algorithm can be described as indicated in Algorithm 8.1.

Every optimisation problem has boundary constraints and therefore a particle should be prevented from drifting outside the boundary constraints of the problem. In some cases it may be beneficiary to allow a particle to move somewhat outside

Algorithm 8.1. PSO Algorithm

1. Create and initialise a swarm
2. while stopping condition has not been reached
3. for each particle in swarm do
4. set *pbest*
5. set *gbest*
6. for each particle in swarm do
7. calculate new velocity
8. calculate new position

the bounds when the solution is in close proximity of the bounds. However, once a particle has moved outside the bounds, it should not be allowed to roam outside the boundary constraints indefinitely and should be pulled back within the valid bounds of the decision space. Furthermore, if a particle's position is outside the bounds, the position should not be used as the particle's *pbest*.

According to Chu *et al*, there are three basic boundary handling techniques that are widely used, namely [4]:

- Random, where if a particle moves outside the search space, a random value from a uniform distribution between the lower and upper boundaries of the violating dimension is assigned to the violating dimension of the particle's position.
- Absorbing, where if a particle moves outside the search space, the dimension that is violating the bounds are set to the boundary of that dimension, so that it seems as though the particle has been absorbed by the boundary.
- Reflection, where if a particle moves outside the search space, the boundary acts like a mirror that reflects the projection of the particle's displacement.

Recently, studies have been done on the effect of boundary constraint violation approaches on the performance of PSO. Helwig and Wanka investigated four approaches for managing boundary constraints when solving high-dimensional single-objective optimization problem (SOOP) [13]. Chu *et al.* investigated the effect of the three boundary handling techniques mentioned above for high dimensional SOOP and high dimensional composite SOOP. However, in this chapter various boundary handling approaches are investigated to determine their effect on the performance of VEPSO when solving DMOOP.

8.2.2 Multi-Objective Optimization Theory

When dealing with a MOOP, the various objectives are normally in conflict with one another, i.e. improvement in one objective leads to a worse solution for another objective. Therefore, for MOOP, the definition of optimality has to be adjusted from the one that is used for SOOP. When solving a MOOP the goal is to find a set of trade-off solutions where for each of these solutions no objective can be improved without causing a worse solution for at least one of the other objectives. These solutions are referred to as *non-dominated solutions* and the set of such solutions is called the *non-dominated set* or *Pareto-optimal set (POS)*. The corresponding objective vectors in the objective space that lead to the non-dominated solutions are referred to as the *Pareto-optimal front (POF)* or *Pareto-front*.

For MOOP, when one decision vector dominates another, the dominating decision vector is considered as a better decision vector. Therefore, only non-dominated decision vectors are included in the POS. Decision vector domination is defined as follows:

Definition 8.1. Decision Vector Domination: A decision vector $\mathbf{x}_1$ dominates another decision vector $\mathbf{x}_2$, denoted by $\mathbf{x}_1 \prec \mathbf{x}_2$, if and only if

- $\mathbf{x}_1$ is at least as good as $\mathbf{x}_2$ for all the objectives, i.e. $f_m(\mathbf{x_1}) \leq f_m(\mathbf{x_2})$, $\forall m = 1, \ldots, n_m$; and
- $\mathbf{x}_1$ is strictly better than $\mathbf{x}_2$ for at least one objective, i.e. $\exists i = 1, \ldots, n_m \colon f_m(\mathbf{x_1}) < f_m(\mathbf{x_2})$.

The best decision vectors are called Pareto-optimal, defined as follows:

Definition 8.2. Pareto-optimal: A decision vector $\mathbf{x}^*$ is Pareto-optimal if there does not exist a decision vector $\mathbf{x} \neq \mathbf{x}^* \in F$ that dominates it, i.e. $\nexists m \colon f_m(\mathbf{x}) < f_m(\mathbf{x}^*)$. If $\mathbf{x}^*$ is Pareto-optimal, the objective vector, $\mathbf{f}(\mathbf{x}^*)$, is also Pareto-optimal.

Together, all the Pareto-optimal decision vectors form the POS, defined as:

Definition 8.3. Pareto-optimal Set: The POS, P^*, is formed by the set of all Pareto-optimal decision vectors, i.e.

$$P^* = \{\mathbf{x}^* \in F \mid \nexists \mathbf{x} \in F \colon \mathbf{x} \prec \mathbf{x}^*\}, \tag{8.1}$$

The POS contains the best trade-off solutions for the MOOP. The corresponding objective vectors form the POF, which is defined as follows:

Definition 8.4. Pareto-optimal Front: For the objective vector $\mathbf{f}(\mathbf{x})$ and the POS P, the POF, $PF^* \subseteq O$ is defined as

$$PF^* = \{\mathbf{f} = (f_1(\mathbf{x}^*), f_2(\mathbf{x}^*), \ldots, f_{n_m}(\mathbf{x}^*)) \mid \mathbf{x}^* \in P\}, \tag{8.2}$$

Therefore, the POF contains the set of objective vectors that corresponds to the POS, i.e. the set of decision vectors that are non-dominated. The POF can have various shapes, e.g. a convex POF or a concave POF.

8.2.3 Dynamic Multi-Objective Optimisation Theory

Let the n_x-dimensional search space (also referred to as the *decision space*) be represented by $S \subseteq \mathbb{R}^{n_x}$ and the feasible space represented by $F \subseteq S$, where $F = S$ for unconstrained optimisation problems. Let $\mathbf{x} = (x_1, x_2, \ldots, x_{n_x}) \in S$ represent a vector of the decision variables, i.e. the *decision vector*, and let a single objective function be defined as $f_m : \mathbb{R}^{n_x} \to \mathbb{R}$. Then $\mathbf{f}(\mathbf{x}) = (f_1(\mathbf{x}), f_2(\mathbf{x}), \ldots, f_{n_m}(\mathbf{x})) \in O \subseteq \mathbb{R}^{n_m}$ represents an *objective vector* containing n_m objective function evaluations, and O is the *objective space*.

Using the notation above, mathematically, a DMOOP can be defined as:

$$\begin{aligned} &minimise\ \mathbf{f}(\mathbf{x}, \mathbf{W}(t)),\ \mathbf{x} = (x_1, \ldots, x_{n_x}), \mathbf{W}(t) = (\mathbf{w}_1(t), \ldots, \mathbf{w}_{n_m}(t)) \\ &subject\ to\ \ g_i(\mathbf{x}) \leq 0,\ i = 1, \ldots, n_g \\ &\qquad h_j(\mathbf{x}) = 0,\ j = n_g + 1, \ldots, n_h \\ &\qquad \mathbf{x} \in \quad [\mathbf{x}_{min}\, ,\, \mathbf{x}_{max}]^{n_x}, \end{aligned} \tag{8.3}$$

where $\mathbf{W}(t)$ is a vector of time-dependent control parameters of an objective function at time t, n_x is the number of decision variables, $\mathbf{x} \in \mathbb{R}^{n_x}$, n_g is the number of inequality constraints, $\mathbf{g}$, n_h is the number of equality constraints, $\mathbf{h}$, and $\mathbf{x} \in [\mathbf{x}_{min}\, ,\, \mathbf{x}_{max}]^{n_x}$ refers to the boundary constraints.

Unlike DSOOP with only one objective function, DMOOP has many objective functions. Therefore, in order to solve the DMOOP the goal is to track the POF over time, i.e.

$$PF^*(t) = \{\mathbf{f}(t) = (f_1(\mathbf{x}^*, \mathbf{w}_1(t)), f_2(\mathbf{x}^*, \mathbf{w}_2(t)), \ldots, f_{n_m}(\mathbf{x}^*, \mathbf{w}_{n_m}(t))) \mid \mathbf{x}^* \in P\}, \tag{8.4}$$

Farina *et al.* [7] classified dynamic environments for DMOOP into four types, namely:

- **Type I environment** where the POS (optimal set of decision variables) changes, but the POF (corresponding objective function values) remains unchanged.
- **Type II environment** where both the POS and the POF change.
- **Type III environment** where the POS remains unchanged, but the POF changes.
- **Type IV environment** where both the POS and the POF remain unchanged, even though the problem can change.

8.2.4 Dynamic Multi-Objective Optimization Issues

In order to solve a DMOOP, an algorithm has to be able to detect when a change in the environment has occurred and then respond to the change. A change in the environment can be detected through the use of sentry particles [2] where a random number of sentry particles are selected after each iteration. Just before the next iteration is performed, these particles are re-evaluated, and if their current fitness value differs more than a specified value from their fitness value just after the previous iteration, the swarm is alerted that a change has occurred in the environment.

In order to test whether an algorithm can solve DMOOPs, benchmark functions are developed that test an algorithm's ability to manage certain difficulties, such as local POF and a POF that changes shape (such as from convex to concave) over time. Benchmark functions are representative of typical real-world problems. An approach to reformulate a three-objective optimisation test function to define a dynamic two-objective optimisation problem was presented by Jin and Sendhof [14]. Guan *et al.* [11] presented an approach to create DMOOPs by replacing objective functions with new ones at specific times. DMOOPs based on the static MOO two-objective ZDT functions [26] and the scalable DTLZ functions [5] was presented by Farina *et al.* [7]. Some adaptions to these test functions were proposed in [19, 25].

However, when algorithms' performances are compared against each other, performance measures are required [1, 7, 8, 18]. Two main categories of performance metrics for DMOOP exist, namely metrics that require knowledge about the true POF and metrics that do not require any prior knowledge about the DMOOP. Various performance metrics were developed to measure the performance of an algorithm with regard to two main goals when solving a DMOOP, namely finding solutions that are as close as possible to the true POF and finding a diverse set of solutions.

One of the problems when working with DMOOP is that there are no standard benchmark functions or performance metrics that are used when research on an algorithm's performance is presented.

8.3 Dynamic Vector Evaluated Particle Swarm Optimization Approach

This section discusses the Vector Evaluated Particle Swarm Optimization (VEPSO) algorithm and how it has been adapted to solve DMOOPs. One type of constraint that forms part of a DMOOP is the bounds for each decision variable, also referred to as boundary constraints. This section presents approaches that can be used to manage boundary constraint violations when solving DMOOPs.

8.3.1 Vector Evaluated Particle Swarm Optimization

The Vector Evaluated Particle Swarm Optimization (VEPSO) algorithm, inspired by the Vector Evaluated Genetic Algorithm (VEGA) [21], was introduced by Parsopoulos *et al.* [22]. With VEPSO, each swarm solves only one objective function and then shares its knowledge with the other swarms.

$$v_i^j(t+1) = w^j v_i^j(t) + c_1^j r_1 (y_i^j(t) - x_i^j(t)) + c_2^j r_2 (\hat{y}_i^s(t) - x_i^j(t)) \quad (8.5)$$

$$x_i^j(t+1) = x_i^j(t) + v_i^j(t+1), \quad (8.6)$$

where n represents the dimension with $i = 1,\ldots,n$; m represents the number of swarms with $j = 1,\ldots,m$ as the swarm index; $\hat{y}_i^s$ is the global best of the s-th swarm with $s \neq j$; c_1^j and c_2^j are the cognitive and social parameters of the j-th swarm respectively; $r_1, r_2 \in [0,1]$; w^j is the inertia weight of the j-th swarm; and $s \in [1,\ldots,j-1,j+1,\ldots,M]$ represents the index of a respective swarm. The index s can be set up in various ways, affecting the topology of the swarms in VEPSO.

In Equation (8.5) the global best of another swarm (indexed by s) is used to update the velocity of the particles of the j-th swarm. In this way the knowledge of the s-th swarm is shared with the j-th swarm.

8.3.2 Dynamic Vector Evaluated Particle Swarm Optimization

When solving DMOOPs, in order to track the changing POF, an algorithm must be able to detect that a change has occurred in the environment and then respond to the change appropriately. The VEPSO algorithm adapted to solve DMOOPs (DVEPSO) is presented in Algorithm 8.2.

Algorithm 8.2. VEPSO for DMOO problems

```
1.  for number of iterations do
2.      check whether a change has occurred
3.      if change has occurred
4.          respond to change
5.          remove dominated solutions from archive
6.      perform iteration
7.      if new solutions are non-dominated
8.          if space in archive
9.              add new solutions to archive
10.         else
11.             remove solutions from archive
12.             add new solutions to archive
13.     select sentry particles
```

The default configuration of DVEPSO algorithm that is used for this research is as follows:

- Each swarm has 20 particles.
- The non-dominated solutions found so far is stored in an archive and the archive size is set to 100.
- If a particle's new position is non-dominant with regard to its current *pbest*, one of these two positions is randomly selected as the particle's new *pbest*.
- If a particle's new position is non-dominant with regard to the swarm's current *gbest*, one of these two positions is randomly selected as the swarm's new *gbest*.
- Sentry particles are used for change detection (refer to lines 2 and 13 in Algorithm 8.2).
- If a change has been detected, 30% of the particles of the swarm(s) whose objective function changed is re-initialised (refer to line 4 in Algorithm 8.2).

The non-dominated solutions in the archive is re-evaluated and the solutions that have become dominated are removed from the archive (refer to line 5 in Algorithm 8.2).

- If the archive is full, the distance between the solution and the other non-dominated solutions in the archive is calculated, and the one with the lowest average distance is removed. This ensures that a solution from a crowded region in the found POF is removed (refer to line 11 in Algorithm 8.2).
- For knowledge sharing between the various swarms, a ring topology is used. Therefore, s in Equation (8.5) is selected using

$$s = \begin{cases} M & \text{for } j = 1 \\ j-1 & \text{for } j = 2,\ldots,M, \end{cases} \tag{8.7}$$

The next section discusses approaches that can be followed to appropriately respond to a violation of the boundary constraints.

8.3.3 Management of Boundary Constraints

This section presents the various approaches that are used in the experiments to manage boundary constraint violations. Below, $\mathbf{x}_{max}$ and $\mathbf{x}_{min}$ refer to the upper bounds and lower bounds of the decision variables of the DMOOP respectively.

The following approaches to handle boundary constraints are investigated to determine their effect on the performance of DVEPSO when solving DMOOPs:

8.3.3.1 Clamping Approach

With the *clamping* approach, any particle that violates a specific boundary of the search space is placed on or close to the violated boundary of the search space [20]. This approach is used for the default configuration of DVEPSO as discussed in Section 8.3.2. Mathematically, clamping is defined as:

$$\begin{aligned} &\text{if } \mathbf{x}(t+1) > \mathbf{x}_{max} \text{ then } \mathbf{x}(t+1) = \mathbf{x}_{max} - \varepsilon \\ &\text{if } \mathbf{x}(t+1) < \mathbf{x}_{min} \text{ then } \mathbf{x}(t+1) = \mathbf{x}_{min} \end{aligned} \tag{8.8}$$

with ε a very small positive number.

8.3.3.2 Deflection Approach

With the deflection approach, if a particle moves outside the bounds of the search space, the velocity's direction of the violated dimension is inverted, thereby causing a bouncing effect of the bounds. Mathematically, the deflection approach is defined as:

$$
\begin{aligned}
\text{if } x_i(t+1) > x_{max^i} \text{ then } x_i(t+1) &= x^i_{max} - (x_i(t+1) - x^i_{max})\%(x^i_{max} - x^i_{min}) \textit{ and} \\
v_i(t+1) &= -v_i(t) \\
\text{if } x_i(t+1) < x_{min^i} \text{ then } x_i(t+1) &= x^i_{min} + (x^i_{min} - x_i(t+1))\%(x^i_{max} - x^i_{min}) \textit{ and} \\
v_i(t+1) &= -v_i(t), \qquad (8.9)
\end{aligned}
$$

where x_i, x^i_{min} and x^i_{max} are the i-th dimension of $\mathbf{x}$, $\mathbf{x}_{max}$ and $\mathbf{x}_{min}$ respectively.

8.3.3.3 Per Element Re-initialization Approach

With *per element re-initialisation*, if a particle moves outside the search space, each dimension of the particle's position that violates the boundary constraint is re-initialized to a random valid value [20]. Therefore, the dimensions of the position that is valid remain the same. Mathematically, per element re-initialization is defined as:

$$
\begin{aligned}
&\text{if } x_i(t+1) > x^i_{max} \text{ then } x_i(t+1) = rand(x^i_{min}, x^i_{max}) \\
&\text{if } x_i(t+1) < x^i_{min} \text{ then } x_i(t+1) = rand(x^i_{min}, x^i_{max}), \qquad (8.10)
\end{aligned}
$$

8.3.3.4 Periodic Approach

The *periodic approach* is similar to the deflection approach. However, if a particle's position violates the upper boundary for a specific dimension, it is placed near the lower boundary for that dimension and vice versa [24]. Mathematically, the periodic approach is defined as:

$$
\begin{aligned}
&\text{if } x_i(t+1) > x^i_{max} \text{ then } x_i(t+1) = x^i_{min} - (x_i(t+1) - x^i_{max})\%(x^i_{max} - x^i_{min}) \\
&\text{if } x_i(t+1) < x^i_{min} \text{ then } x_i(t+1) = x^i_{max} - (x^i_{min} - x_i(t+1))\%(x^i_{max} - x^i_{min}). \qquad (8.11)
\end{aligned}
$$

8.3.3.5 Random Approach

The *random* approach re-initializes a particle's position to a valid position within the search space if it violates the boundaries of the search space [13, 24]. Therefore, in contrast to the per element re-initialization approach, all dimensions are re-initialized and not only the violating dimensions. Mathematically, it is defined as:

$$
\begin{aligned}
&\text{if } \mathbf{x}(t+1) > \mathbf{x}_{max} \text{ then } \mathbf{x}(t+1) = rand(\mathbf{x}_{min}, \mathbf{x}_{max}) \\
&\text{if } \mathbf{x}(t+1) < \mathbf{x}_{min} \text{ then } \mathbf{x}(t+1) = rand(\mathbf{x}_{min}, \mathbf{x}_{max}). \qquad (8.12)
\end{aligned}
$$

8.3.3.6 Re-initialization Approach

With the *re-initialisation approach*, a particle that violates the bounds of the search space has its position re-initialised to a valid position within the search space, its velocity set to zero and its *pbest* set to the particle's new position [20].

8.3.3.7 Unconstrained Approach

With the *unconstrained* approach, no clamping is performed and particles are free to move outside the search space. However, only valid positions are selected as the *pbest* of a particle.

8.4 Experiments

This section describes experiments that were conducted, using benchmark functions and performance metrics discussed in Sections 8.4.1 and 8.4.2, respectively, to test:

- the effect of various approaches to manage boundary constraints on the performance of DVEPSO (refer to Section 8.3.3); and
- the performance of DVEPSO compared to three other state-of-the-art DMOO algorithms (refer to Section 8.4.3).

All experiments consisted of 30 independent runs and each run consisted of 1,000 iterations. For all benchmark functions the severity of change (n_t) is set to 10 and the frequency of change (τ_t) is set to either 5, 25 or 50. This will cause the DMOOP to change every τ_t iteration with n_t distinct steps in time t.

The PSO parameters were set to values that lead to convergent behaviour [23], namely $w = 0.72$ and $c_1 = c_2 = 1.49$.

All codes are implemented in the Computational Intelligence Library (CIlib) [20]. All simulations were run on the Sun Hybrid System's Nehalem System of the Center for High Performance Computing [3]. The SUN Nehalem system has an Intel Nehalem Processor of 2.93 GHz, 2304 CPU cores, 3465 GB of Memory and produces 24 TFlops at peak performance.

8.4.1 Benchmark Functions

This section presents the benchmark functions that were used to test whether the algorithms can track a POF that changes over time. Three functions presented by Farina *et al.* [7] and three functions of Goh and Tan [8] were used. Additionally, two functions that are based on the ZDT3 function of Deb [26] that were adapted to become DMOOPs were used [12]. Below, τ is the generation counter, τ_t is the number of iterations for which t remains fixed, and n_t is the number of distinct steps in t.

$$\mathsf{FDA1} = \begin{cases} Minimize: f(\mathbf{x},t) = (f_1(\mathsf{x}_\mathsf{I},t), g(\mathbf{x_{II}},t) \cdot h(\mathbf{x_{III}}, f_i(\mathsf{x}_\mathsf{I},t), g(\mathbf{x_{II}},t),t)) \\ f_1(\mathsf{x}_\mathsf{I}) = x_i \\ g(\mathbf{x_{II}}) = 1 + \sum_{x_i \in \mathbf{x_{II}}} (x_i - G(t))^2 \\ h(f_1, g) = 1 - \sqrt{\frac{f_1}{g}} \\ where: \\ G(t) = \sin(0.5\pi t), \ \ t = \frac{1}{n_t} \left\lfloor \frac{\tau}{\tau_t} \right\rfloor \\ \mathsf{x}_\mathsf{I} \in [0,1]; \ \ \mathbf{x_{II}} = (x_2, \ldots, x_n) \in [-1,1] \end{cases}, \tag{8.13}$$

As suggested by [7], the dimension, n, was set to 20. Function FDA1's values in the decision variable space change over time, but its values in the objective space remain the same. Therefore, it is a Type I DMOOP. It has a convex POF with $POF = 1 - \sqrt{f_1}$.

$$\mathsf{FDA2} = \begin{cases} Minimize: f(\mathsf{x},t) = (f_1(\mathbf{x_I},t), g(\mathbf{x_{II}},t) \cdot h(\mathbf{x_{III}}, f_i(\mathbf{x_I},t), g(\mathbf{x_{II}},t),t)) \\ f_1(\mathbf{x_I}) = x_i \\ g(\mathbf{x_{II}}) = 1 + \sum_{x_i \in \mathbf{x_{II}}} x_i^2 \\ h(f_1, g) = 1 - \frac{f_1}{g}^{(H(t) + \sum_{x_i \in \mathbf{x_{III}}} (x_i - H(t))^2)^{-1}} \\ where: \\ H(t) = 0.75 + 0.75 \sin(0.5\pi t), t = \frac{1}{n_t} \left\lfloor \frac{\tau}{\tau_t} \right\rfloor \\ \mathbf{x_I} \in [0,1]; \mathbf{x_{II}}, \mathbf{x_{III}} \in [-1,1] \end{cases}. \tag{8.14}$$

For FDA2 the parameters $|\mathsf{X}_{\mathsf{II}}|$ and $|\mathsf{X}_{\mathsf{III}}|$ were set to: $|\mathsf{X}_{\mathsf{II}}| = |\mathsf{X}_{\mathsf{III}}| = 15$ (as suggested by [7]). Function FDA2 has a POF that changes from a convex to a non-convex shape. It is a Type III DMOOP, since the values in the objective space change while the values in the decision variable space remain the same. For FDA2, $POF = 1 - f_1^{H(t)^{-1}}$.

$$\mathsf{FDA3} = \begin{cases} Minimize: f(\mathsf{x},t) = (f_1(\mathsf{x}_\mathsf{I},t), g(\mathbf{x_{II}},t) \cdot h(\mathbf{x_{III}}, f_i(\mathbf{x_I},t), g(\mathbf{x_{II}},t),t)) \\ f_1(\mathbf{x_I}) = \sum_{x_i \in \mathbf{x_I}} x_i^{F(t)} \\ g(\mathbf{x_{II}}) = 1 + G(t) + \sum_{x_i \in \mathbf{x_{II}}} (x_i - G(t))^2 \\ h(f_1, g) = 1 - \sqrt{\frac{f_1}{g}} \\ G(t) = |sin(0.5\pi t)| \\ F(t) = 10^{2\sin(0.5\pi t)}, t = \frac{1}{n_t} \left\lfloor \frac{\tau}{\tau_t} \right\rfloor \\ \mathbf{x_I} \in [0,1]; \mathbf{x_{II}} \in [-1,1] \end{cases}. \tag{8.15}$$

As suggested by [7], the function parameters $|\mathsf{X}_{\mathsf{II}}|$ and $|\mathsf{X}_{\mathsf{III}}|$ were set to: $|\mathsf{X}_{\mathsf{I}}| = 5$ and $|\mathsf{X}_{\mathsf{II}}| = 25$. Function FDA3 has a convex shaped POF and both the values in the decision variable space, as well as the objective space, change. Therefore, it is called a Type II DMOOP. For FDA3, $POF = (1+G(t))(1-\sqrt{\frac{f_1}{1+G(t)}})$.

$$\mathsf{dMOP1} = \begin{cases} Minimize : f(\mathbf{x},t) = (f_1(\mathsf{x}_{\mathsf{I}},t), g(\mathbf{x_{II}},t)\cdot h(\mathbf{x_{III}}, f_i(\mathsf{x}_{\mathsf{I}},t), g(\mathbf{x_{II}},t),t)) \\ f_1(\mathsf{x}_{\mathsf{I}}) = x_i \\ g(\mathbf{x_{II}}) = 1+9\sum_{x_i\in\mathbf{x_{II}}}(x_i)^2 \\ h(f_1,g) = 1-\frac{f_1}{g}^{H(t)} \\ where : \\ H(t) = 0.75\sin(0.5\pi t)+1.25, \;\; t = \frac{1}{n_t}\left\lfloor\frac{\tau}{\tau_t}\right\rfloor \\ x_i\in[0,1]; \;\; \mathbf{x_I} = (x_1); \;\; \mathbf{x_{II}} = (x_2,\ldots,x_n) \end{cases} \quad (8.16)$$

As suggested by [8], the dimension was set to $n = 10$. Function dMOP1 has a convex POF where the values in the objective space change, but the values in the decision space remain the same. Therefore, it is a Type III problem, with $POF = 1 - f_1^{H(t)}$.

$$\mathsf{dMOP2} = \begin{cases} Minimize : f(\mathbf{x},t) = (f_1(\mathsf{x}_{\mathsf{I}},t), g(\mathbf{x_{II}},t)\cdot h(\mathbf{x_{III}}, f_i(\mathsf{x}_{\mathsf{I}},t), g(\mathbf{x_{II}},t),t)) \\ f_1(\mathsf{x}_{\mathsf{I}}) = x_i \\ g(\mathbf{x_{II}}) = 1+9\sum_{x_i\in\mathbf{x_{II}}}(x_i-G(t))^2 \\ h(f_1,g) = 1-\frac{f_1}{g}^{H(t)} \\ where : \\ H(t) = 0.75\sin(0.5\pi t)+1.25, \\ G(t) = \sin(0.5\pi t) t = \frac{1}{n_t}\left\lfloor\frac{\tau}{\tau_t}\right\rfloor \\ x_i\in[0,1]; \;\; \mathbf{x_I} = (x_1); \;\; \mathbf{x_{II}} = (x_2,\ldots,x_n) \end{cases} \quad (8.17)$$

The dimension, n, was set 10 (as suggested by [8]). Function dMOP2 has a convex POF where the values in both the decision space and objective space change. Therefore, dMOP2 is a Type II problem, with $POF = 1 - f_1^{H(t)}$.

$$\mathsf{dMOP3} = \begin{cases} Minimize : f(\mathbf{x},t) = (f_1(\mathsf{x}_{\mathsf{I}},t), g(\mathbf{x_{II}},t)\cdot h(\mathbf{x_{III}}, f_i(\mathsf{x}_{\mathsf{I}},t), g(\mathbf{x_{II}},t),t)) \\ f_1(\mathsf{x}_{\mathsf{I}}) = x_r \\ g(\mathbf{x_{II}}) = 1+9\sum_{x_i\in\mathbf{x_{II}}\setminus x_r}(x_i-G(t))^2 \\ h(f_1,g) = 1-\sqrt{\frac{f_1}{g}} \\ where : \\ G(t) = \sin(0.5\pi t), \;\; t = \frac{1}{n_t}\left\lfloor\frac{\tau}{\tau_t}\right\rfloor \\ x_i\in[0,1]; r = \mathsf{U}(1,2,\ldots,n) \end{cases} \quad (8.18)$$

As suggested by [8], the dimension, n, was set to 10. Function dMOP3 has a convex POF where the values in the objective space change, but the values in the decision space remain the same, and is therefore a Type I DMOOP, but the spread of the *POF* changes over time. For dMOP3, $POF = 1 - \sqrt{f_1}$.

The following two functions, HE1 and HE2, are based on the function ZDT3 [26], and adapted to be dynamic.

$$\text{HE1} = \begin{cases} Minimize: f(\mathbf{x},t) = (f_1(\mathrm{x_I},t), g(\mathbf{x_{II}},t) \cdot h(\mathbf{x_{III}}, f_i(\mathrm{x_I},t), g(\mathbf{x_{II}},t),t)) \\ f_1(\mathrm{x_I}) = x_i \\ g(\mathbf{x_{II}}) = 1 + \frac{9}{n-1} \sum_{x_i \in \mathbf{x_{II}}} x_i \\ h(f_1, g) = 1 - \sqrt{\frac{f_1}{g}} - \frac{f_1}{g} \sin(10\pi t f_1) \\ where: \\ t = \frac{1}{n_t} \left\lfloor \frac{\tau}{\tau_t} \right\rfloor; \quad \mathrm{x_i} \in [0,1] \\ \mathbf{x_I} = (x_1); \quad \mathbf{x_{II}} = (x_2, \dots, x_n) \end{cases} \tag{8.19}$$

$$\text{HE2} = \begin{cases} Minimize: f(\mathbf{x},t) = (f_1(\mathrm{x_I},t), g(\mathbf{x_{II}},t) \cdot h(\mathbf{x_{III}}, f_i(\mathrm{x_I},t), g(\mathbf{x_{II}},t),t)) \\ f_1(\mathrm{x_I}) = x_i \\ g(\mathbf{x_{II}}) = 1 + \frac{9}{n-1} \sum_{x_i \in \mathbf{x_{II}}} x_i \\ h(f_1, g) = 1 - \sqrt{\frac{f_1}{g}}^{H(t)} - \frac{f_1}{g}^{H(t)} \sin(10\pi f_1) \\ where: \\ H(t) = 0.75 \sin(0.5\pi t) + 1.25; \quad t = \frac{1}{n_t} \left\lfloor \frac{\tau}{\tau_t} \right\rfloor \\ \mathrm{x_i} \in [0,1]; \quad \mathbf{x_I} = (x_1); \quad \mathbf{x_{II}} = (x_2, \dots, x_n) \end{cases} \tag{8.20}$$

The dimension, n, was set to 30 (as suggested by [26]) for both HE1 and HE. Both functions have a discontinuous POF. For HE1, $POF = 1 - \sqrt{f_1} - f_1 \sin(10\pi t f_1)$, and, for HE2, $POF = 1 - \sqrt{f1}^{H(t)} - f_1^{H(t)} \sin(0.5\pi f_1)$.

8.4.2 Performance Metrics

This section discusses the performance metrics that were used to measure the performance of the various algorithms. Each metric is calculated every time just before a change occurs in the environment. The average of all these values is then calculated for each of the runs. However, if it is unknown when a change will occur, the performance metrics can be calculated over all iterations instead of only the iterations just before a change occurs in the environment.

To determine the algorithm with the best performance for a specific function, the algorithm's overall rank is calculated. For each of the performance metrics the algorithm is ranked according to its performance with regards to the specific metric. The algorithm's average rank value is calculated and then the algorithm is ranked accordingly. Two average ranks are calculated, namely: (a) the sum of all ranks divided by the number of performance metrics (indicated in Tables 8.1-8.8 as R_1); and (b) the sum of all ranks (but the ranks of performance metrics that rely on the true POF, namely HV R, VD and MS counted double) divided by the adjusted number of performance metrics (indicated in Tables 8.1-8.8 as R_2).

8.4.2.1 Spacing

Measuring how evenly the non-dominated solutions are distributed along the found POF (POF^*) can be done using the metric of spacing [9], defined as:

$$\overline{S}^i = \frac{1}{n_c}\sum_{j=1}^{n_c} S_j^i, \; S = \frac{1}{n_{PF}}\left[\frac{1}{n_{PF}}\sum_{i=1}^{n_{PF}}(d_i - \overline{d})^2\right]^{\frac{1}{2}}, \; \overline{d} = \frac{1}{n_{PF}}\sum_{i=1}^{n_{PF}} d_i, \tag{8.21}$$

where n_c is the number of changes that occurred in the environment, n_{PF} is the number of non-dominated solutions found at time t and d_i is the Euclidean distance, in the objective space, between non-dominated solution i and its nearest solution in POF^*.

8.4.2.2 Hypervolume Ratio

The hypervolume (HV) or S-metric [26] computes the size of the region that is dominated by a set of non-dominated solutions, based on a reference vector. According to Li *et al.*, comparing the HV averaged over a number of runs may not be as meaningful when dealing with dynamic environments [17]. Therefore, they suggest using the HV ratio (HV R) to overcome this problem, since the HV of the found POF (POF^*) is computed in relation to the HV of the true POF (POF) [17]. Mathematically, HVR is defined as:

$$HVR = \frac{1}{n_c}\sum_{i=1}^{n_c} HVR(t), \; HVR(t) = \frac{HV(POF^*(t))}{HV(POF(t))}. \tag{8.22}$$

Prior knowledge about POF is required to calculate the HVR, POF and the value of the metric will depend on the distribution of sampling points on POF and the selection of the reference vector. For this research the reference vector is selected as the maximum value for each objective.

8.4.2.3 Accuracy

A measure of accuracy that measures the quality of the solutions as a relation between the HV of POF^* and the maximum HV that has been found so far was introduced by Cámara *et al.* [1]. Mathematically, it is defined as:

$$acc = \frac{1}{n_c}\sum_{i=1}^{n_c} acc(t), \quad acc(t) = \frac{HV(POF^*(t))}{HV_{max}(POF^*(t))}, \tag{8.23}$$

8.4.2.4 Stability

The effect of the changes in the environment on the accuracy (*acc* defined above) of the algorithm can be measured by the measure of stability that was introduced by Cámara *et al.* [1]. Mathematically, stability is defined as:

$$stab = \frac{1}{n_c}\sum_{i=1}^{n_c} stab(t), \quad stab(t) = max\{0, acc(t-1) - acc(t)\} \tag{8.24}$$

8.4.2.5 Variable Space Generational Distance

The static generational distance (GD) metric was adapted for dynamic environments by Goh and Tan [8]. It measures the distance between POF^* and POF, i.e. the proximity of POF^* to POF. The variable space GD (VGD) metric calculates the GD just before a change occurs in the environment, and is mathematically expressed as:

$$\begin{aligned} VD(t) &= \frac{1}{\tau}\sum_{t=0}^{\tau} VD(t)I(t) \\ VD(t) &= \frac{1}{n_{POF^*(t)}}\sqrt{n_{POF^*(t)} \sum_{i=1}^{n_{POF^*(t)}} d_i(t)^2} \\ \mathsf{I(t)} &= \begin{cases} 1, & \text{if } t\%\tau_t = 0 \\ 0, & \text{otherwise} \end{cases}, \end{aligned} \tag{8.25}$$

where $n_{POF(t)^*}$ is the number of non-dominated solutions in POF^* at time t and d_i is the Euclidean distance between the i-th solution of POF^* and the nearest solution solution of POF. Goh and Tan calculate d_i in the decision space [8]. However, for this research it is calculated in the objective space.

8.4.2.6 Maximum Spread

Goh and Tan adapted the maximum spread (MS) metric for dynamic environments [8]. MS measures how well POF^* covers the POF, i.e. how well the non-dominated solutions of POF^* are spread along POF. MS for dynamic environments calculates the MS just before a change occurs in the environment, and is defined mathematically as:

$$MS(t) = \frac{1}{\tau}\sum_{t=0}^{\tau} MS(t)I(t)$$

$$MS(t) = \sqrt{\frac{1}{M}\sum_{i=1}^{M}\left[\frac{\min\left[\overline{POF_i^*}(t), \overline{POF_i}(t)\right] - \max\left[\underline{POF_i^*}(t), \underline{POF_i}(t)\right]}{\overline{POF_i}(t) - \underline{POF_i}(t)}\right]}$$

$$\mathsf{I(t)} = \begin{cases} 1, \text{ if } t\%\tau_t = 0 \\ 0, \text{ otherwise} \end{cases} \tag{8.26}$$

where M is the number of objectives, $n_{POF(t)^*}$ is the number of non-dominated solutions in POF^* at time t, $\overline{POF_i^*}$ and $\underline{POF_i^*}$ refer to the maximum and minimum of the i-th objective of non-dominated solutions in POF^* and $\overline{POF_i}$ and $\underline{POF_i}$ refer to the maximum and minimum of the i-th objective of non-dominated solutions in POF respectively.

8.4.3 Comparison

The performance of DVEPSO is compared against three those of the other state-of-the-art DMOO algorithms, namely:

- DNSGA-II-A algorithm, an NSGA-II algorithm adapted for DMOO and proposed by Deb *et al.* [6]. If a change in the environment is detected, a percentage of individuals are randomly selected and replaced with newly created individuals.
- DNSGA-II-B algorithm, an NSGA-II algorithm that selects a percentage of individuals randomly and replaces them with individuals that are mutated from existing individuals when a change is detected. DNSGA-II-B was proposed by Deb *et al.* [6].
- dCOEA algorithm, a dynamic competitive-cooperative coevolutionary algorithm proposed by Goh and Tan [8].

The source code of the dCOEA algorithm was obtained from the first author of [8]. The source code of the static NSGA-II algorithm was obtained from [16] and was adapted for DMOO according to [6].

8.4.4 Statistical Analysis

A Kruskal-Wallis test was performed for each function for each τ_t to determine whether there is a difference in performance with respect to the performance metrics. If this test indicated that there was a difference, pairwise Mann-Whitney U tests were performed.

8.5 Results

This section discusses the results that were obtained from the experiments. The values of the performance metrics that were obtained, are presented in Tables 8.1- 8.8. In all tables, $DVEPSO_c$, $DVEPSO_d$, $DVEPSO_{pe}$, $DVEPSO_p$, $DVEPSO_r$, $DVEPSO_{re}$ and $DVEPSO_u$ refer to the clamping, deflection, per element re-initialisation, periodic, random, re-initialisation and unconstrained approaches respectively (refer to Section 8.3.3 for the definitions of these approaches).

8.5.1 *Managing Boundary Constraints*

This section discusses the results that were obtained by the various boundary constraint management approaches. The values of the performance metrics are presented in Tables 8.1- 8.8.

When comparing the POF that was found by the various approaches to the true POF, the VD and MS metrics provide a good indication of the algorithms' performance. These tables show that for a change frequency of 10, $DVEPSO_c$, $DVEPSO_{pe}$ and $DVEPSO_d$ obtained the best overall VD value for two, one and one function(s) respectively, $DVEPSO_p$, $DVEPSO_u$ and $DVEPSO_{re}$ each obtained the best MS value for one function and $DVEPSO_r$, $DVEPSO_d$ and $DVEPSO_u$ obtained the best rank over all performance measures for one, two and two function(s) respectively.

For a change frequency of 25, $DVEPSO_r$ obtained the best overall VD value for three functions, $DVEPSO_p$, $DVEPSO_d$ and $DVEPSO_{pe}$ each obtained the best MS value for one function, and $DVEPSO_{pe}$, $DVEPSO_c$, $DVEPSO_u$ and $DVEPSO_{re}$ each obtained the best overall rank for one function.

For a change frequency of 50, $DVEPSO_{pe}$, $DVEPSO_u$ and $DVEPSO_r$ obtained the best VD value for two, one and one function(s) respectively, $DVEPSO_u$ and $DVEPSO_r$ obtained the best MS value for one function each and $DVEPSO_u$ obtained the best overall rank for three functions.

Figure 8.1 illustrates the found POF of the various boundary handling approaches for FDA2. Figure 8.1 shows that good results were obtained by $DVEPSO_c$, $DVEPSO_r$, $DVEPSO_u$ and $DVEPSO_{pe}$, but $DVEPSO_d$ and $DVEPSO_p$ struggled to find the POF.

The results obtained by the various boundary handling techniques for dMOP2 can be seen in Figure 8.2. Good results were obtained by all approaches, but the approximated POFs of $DVEPSO_p$ and $DVEPSO_{pe}$ had a worse spread or coverage than the other DVEPSO approaches.

Table 8.9 presents the overall rank that the various algorithms obtained for each performance measure, as well as their overall rank for the various frequencies of change. Table 8.9 shows that with regard to the various boundary constraint management approaches, for a change frequency of 10 the best overall rank for VD was obtained by $DVEPSO_r$ and $DVEPSO_c$, the best MS rank was obtained by $DVEPSO_r$ and the best overall rank for all DVEPSO approaches was obtained by $DVEPSO_c$. For a change frequency of 25 the best overall rank for VD was obtained by $DVEPSO_{cl}$ and $DVEPSO_r$, the best overall rank for MS was obtained by

Table 8.1 Performance Measure Values for FDA1

τ_t	Algorithm	NS	S	HV R	Acc	Stab	VD	MS	R_1	R_2
10	DVEPSO$_c$	99.4	**0.00043**	0.99658	0.9967	0.00154	**0.06593**	0.9761	2	2
10	DVEPSO$_d$	99.4	0.00074	0.99361	0.99373	0.00217	0.12731	0.92471	6	6
10	DVEPSO$_{pe}$	99.2	0.00051	0.99589	0.99601	0.00133	0.08515	0.92806	3.5	3.5
10	DVEPSO$_p$	**99.5**	0.00053	0.99538	0.9955	0.00163	0.08932	0.95041	3.5	3.5
10	**DVEPSO$_r$**	**99.5**	0.00043	**0.99701**	**0.99713**	**0.00116**	0.07035	0.94377	**1**	**1**
10	DVEPSO$_{re}$	99.4	0.00053	0.9953	0.99541	0.00143	0.07855	0.89446	5	5
10	DVEPSO$_u$	99.3	0.00077	0.99391	0.99403	0.00191	0.14115	0.91986	7	7
10	DNSGAII-A	22.8	0.00494	0.97425	0.97436	0.00339	0.83219	0.78693	10	10
10	DNSGAII-B	21.1	0.00612	0.95019	0.9503	0.00543	1.13392	1.19478	9	9
10	dCOEA	33.7	0.00132	0.90528	0.90538	0.01328	1.13184	**2.48561**	8	8
25	DVEPSO$_c$	**99.9**	0.0008	0.99857	**0.99858**	0.00034	0.18913	0.91448	3	4
25	DVEPSO$_d$	**99.9**	**0.00042**	0.98439	0.9763	0.00397	0.12891	0.86929	6	8
25	**DVEPSO$_{pe}$**	**99.9**	0.00046	0.99928	0.99016	0.00032	0.12982	0.90767	**1**	**1**
25	DVEPSO$_p$	**99.9**	0.00045	0.98084	0.97189	0.00485	0.10817	0.89605	9	9
25	DVEPSO$_r$	99.8	0.00047	0.99856	0.98944	0.00049	**0.10446**	0.90257	4.5	3
25	DVEPSO$_{re}$	**99.9**	0.00057	0.99922	0.9901	0.00035	0.13211	0.86428	4.5	5
25	DVEPSO$_u$	98.5	0.00068	**1.00377**	0.99409	0.0013	0.24299	0.88969	7	6
25	DNSGAII-A	37.8	0.00056	0.99903	0.98891	**0.00014**	0.29491	0.9446	8	7
25	DNSGAII-B	38.3	0.00046	0.99913	0.98901	**0.00014**	0.28079	0.94903	2	2
25	dCOEA	39.8	0.00053	0.96001	0.95028	0.00428	1.32408	**2.93453**	10	10
50	DVEPSO$_c$	**100.0**	0.00039	0.99865	**0.99866**	0.00035	0.19331	0.93334	6	7
50	DVEPSO$_d$	99.8	0.00048	0.96771	0.96395	0.00616	0.17621	0.87048	10	10
50	DVEPSO$_{pe}$	99.9	0.00044	0.99915	0.99456	0.0004	**0.09639**	0.86153	7	6
50	DVEPSO$_p$	99.9	0.00037	0.97749	0.97285	0.00541	0.14417	0.83086	9	9
50	DVEPSO$_r$	**100.0**	0.00046	0.99888	0.9941	0.00038	0.24311	0.89013	8	8
50	DVEPSO$_{re}$	**100.0**	0.00033	0.99917	0.99439	0.00041	0.1331	0.87969	4	4
50	**DVEPSO$_u$**	99.9	0.00033	**1.00125**	0.9957	0.00126	0.15148	0.91074	**2**	**1.5**
50	**DNSGAII-A**	40.0	0.00032	0.99985	0.99419	$3.016\text{x}10^{-05}$	0.1716	**0.98858**	**2**	**1.5**
50	**DNSGAII-B**	40.0	0.00033	0.99986	0.9942	**$2.245\text{x}10^{-05}$**	0.17261	0.98778	**2**	3
50	dCOEA	39.9	**0.00026**	0.99965	0.994	0.00017	0.1515	0.95904	5	5

$DVEPSO_r$, and the approach that ranked the best over all performance measures was $DVEPSO_{cl}$. For a change frequency of 50 the best overall rank for VD was obtained by $DVEPSO_{pe}$, the best overall rank for MS was obtained by $DVEPSO_r$ and the approach that ranked the best over all performance measures was $DVEPSO_u$. It is interesting to note that $DVEPSO_r$ consistently provided the best overall MS value. Furthermore, $DVEPSO_c$ and $DEVPSO_r$ obtained the best rank for VD for change frequencies of 10 and 25. Therefore, for the lower change frequencies of 10 and 25, $DVEPSO_c$ and $DVEPSO_r$ outperformed the other approaches and for a change frequency of 50 $DVEPSO_u$ performed the best of the DVEPSO approaches.

Table 8.2 Performance Measure Values for FDA2

τ_t	Algorithm	NS	S	HV R	Acc	Stab	VD	MS	R_1	R_2
10	DVEPSO$_c$	63.3	0.00367	0.99525	**0.99191**	0.00049	0.43937	0.87783	7.5	7
10	DVEPSO$_d$	**73.4**	0.00118	0.99533	0.97848	0.00049	0.45824	0.90878	7.5	8
10	DVEPSO$_{pe}$	63.0	0.00391	0.99905	0.98157	0.00029	**0.43234**	0.88916	5	3
10	DVEPSO$_p$	68.5	0.002	0.99846	0.98098	0.00034	0.45147	**0.91258**	3	2
10	DVEPSO$_r$	68.6	0.00372	0.99634	0.9789	0.00043	0.44453	0.90914	6	5
10	DVEPSO$_{re}$	63.3	0.00297	0.99554	0.97812	0.00037	0.45008	0.87382	9.5	9
10	**DVEPSO$_u$**	71.5	0.00283	1.00171	0.98418	0.00019	0.44998	0.90757	**1**	**1**
10	DNSGAII-A	39.4	0.00044	1.0044	0.98681	$9.565\text{x}10^{-06}$	0.71581	0.77096	4	6
10	DNSGAII-B	39.6	**0.00042**	**1.00441**	0.98683	**$9.206\text{x}10^{-06}$**	0.71681	0.77866	2	4
10	dCOEA	38.4	0.00051	1.00209	0.98454	0.00122	0.70453	0.61923	9.5	10
25	DVEPSO$_c$	78.5	0.0023	0.99644	0.99421	0.00037	0.43181	0.86647	7	7
25	DVEPSO$_d$	77.2	0.00204	0.99354	0.98997	0.00058	0.43196	0.86884	9.5	8.5
25	DVEPSO$_{pe}$	76.7	0.00221	0.99882	**0.99493**	0.00024	0.43695	0.85983	4	4
25	DVEPSO$_p$	**79.3**	0.00166	0.99701	0.9893	0.0004	0.4421	**0.89688**	6	4
25	DVEPSO$_r$	78.0	0.00114	0.9968	0.98855	0.00036	**0.42211**	0.87893	2.5	**1**
25	DVEPSO$_{re}$	78.5	0.00251	0.99684	0.98859	0.00028	0.42642	0.82876	9.5	8.5
25	DVEPSO$_u$	76.0	0.00145	1.00077	0.99249	0.00021	0.43903	0.86418	5.0	4
25	DNSGAII-A	39.7	**0.00043**	**1.00314**	0.99484	$7.579\text{x}10^{-06}$	0.72841	0.78969	2.5	6
25	**DNSGAII-B**	39.7	0.00051	**1.00314**	0.99484	**$6.707\text{x}10^{-06}$**	0.7268	0.83159	**1**	2
25	dCOEA	39.9	0.00099	1.00265	0.99436	0.00017	0.74606	0.78319	8	10
50	DVEPSO$_c$	93.7	0.00031	0.99961	0.9979	0.00017	0.50599	0.95397	3	3
50	DVEPSO$_d$	93.3	**0.00028**	0.99491	0.99166	0.00173	0.49882	0.94	4.5	5
50	DVEPSO$_{pe}$	93.1	0.00031	1.001	0.99732	$7.344\text{x}10^{-05}$	0.4994	0.95325	2	2
50	DVEPSO$_p$	**94.0**	0.00031	0.99524	0.99158	0.00161	0.51161	0.93862	9	9
50	DVEPSO$_r$	93.0	0.00032	1.00035	0.99668	0.00012	0.50096	0.92995	8	7
50	DVEPSO$_{re}$	93.7	0.00036	0.99904	0.99537	0.00012	0.49984	0.95716	6.5	4
50	**DVEPSO$_u$**	91.4	0.00031	1.00155	0.99787	$9.68\text{x}10^{-05}$	**0.49669**	**0.95937**	**1**	**1**
50	DNSGAII-A	40.0	0.0005	1.00287	**0.99918**	$2.804\text{x}10^{-06}$	0.67584	0.75404	4.5	6
50	DNSGAII-B	40.0	0.00039	**1.00287**	**0.99918**	**$2.778\text{x}10^{-06}$**	0.67736	0.74332	6.5	8
50	dCOEA	40.0	0.00207	1.00268	0.999	$4.575\text{x}10^{-05}$	0.69043	0.86612	10	10

8.5.2 Comparison

This section discusses the results that were obtained by the various DMOO algorithms. The results are presented in Tables 8.1- 8.8. These tables show that for a change frequency of 10, *dCOEA* and *DNSGAII-A* each obtained the best overall VD value for 2 functions and with regard to the MS value, *DNSGAII-A* and *dCOEA* obtained the best overall value for two and three functions respectively a change frequency of 25, *dCOEA* obtained the best overall VD value for two functions and *DNSGAII-A* and *DNSGAII-B* each obtained the best overall VD value for one function; *dCOEA* and *DNSGAII-A* obtained the best MS value for two and three

Table 8.3 Performance Measure Values for FDA3

τ_t	Algorithm	NS	S	HV R	Acc	Stab	VD	MS	R$_1$	R$_2$
10	DVEPSO$_c$	**100.0**	0.00109	1.00221	**0.99889**	0.00013	**0.95943**	0.83848	4	**1.5**
10	**DVEPSO$_d$**	**100.0**	0.00084	1.963x10^{+44}	0.00773	0.00334	0.98365	0.82782	**2**	3
10	DVEPSO$_{pe}$	**100.0**	0.00095	1.00169	5.124x10^{-48}	7.62x10^{-52}	0.99045	0.80762	7.5	8
10	**DVEPSO$_p$**	**100.0**	0.00084	5.822x10^{+40}	2.977x10^{-07}	2.96x10^{-07}	1.0308	0.86383	**2**	4
10	DVEPSO$_r$	**100.0**	0.00087	1.00164	5.124x10^{-48}	8.014x10^{-52}	0.99326	0.82882	7.5	7
10	DVEPSO$_{re}$	**100.0**	0.00091	1.0017	5.124x10^{-48}	7.63x10^{-52}	0.99818	0.8472	6	6
10	**DVEPSO$_u$**	**100.0**	0.00081	**4.767x10^{+45}**	0.00068	0.00068	0.97488	0.81679	**2**	**1.5**
10	DNSGAII-A	32.8	0.00318	0.99967	7.171x10^{-50}	2.796x10^{-53}	1.32639	1.09947	9	9.5
10	DNSGAII-B	27.3	0.00498	0.99796	7.158x10^{-50}	5.576x10^{-53}	1.31649	1.18386	10	9.5
10	dCOEA	39.3	**0.00076**	1.00182	7.186x10^{-50}	**1.503x10^{-53}**	1.08503	**1.30535**	5	5
25	**DVEPSO$_c$**	**100.0**	0.00076	1.00045	**0.99981**	2.746x10^{-05}	1.0931	0.95493	**1.5**	2
25	DVEPSO$_d$	**100.0**	0.00087	**7.955x10^{+41}**	0.01251	0.0025	1.14336	1.02693	5	4
25	DVEPSO$_{pe}$	**100.0**	0.00069	1.00037	1.053x10^{-45}	**2.484x10^{-50}**	**1.08436**	0.91634	6.5	6
25	DVEPSO$_p$	**100.0**	0.00071	4.334x10^{+41}	0.00046	0.00045	1.10933	0.96636	4	5
25	DVEPSO$_r$	**100.0**	0.00066	1.00036	1.053x10^{-45}	2.646x10^{-50}	1.11311	0.99296	6.5	7
25	DVEPSO$_{re}$	**100.0**	0.00069	1.00037	1.053x10^{-45}	2.5x10^{-50}	1.10671	0.95784	8	8
25	**DVEPSO$_u$**	**100.0**	0.0008	1.508x10^{+35}	1.588x10^{-10}	1.586x10^{-10}	1.10233	0.97723	**1.5**	**1**
25	DNSGAII-A	38.2	0.00124	1.00039	1.053x10^{-45}	4.373x10^{-50}	1.27408	1.1752	10	9.5
25	DNSGAII-B	39.1	0.0011	1.00041	1.053x10^{-45}	3.612x10^{-50}	1.27814	1.17337	9	9.5
25	dCOEA	39.9	**0.00052**	1.00044	1.053x10^{-45}	3.221x10^{-50}	1.22933	**1.37518**	3	3
50	DVEPSO$_c$	**100.0**	0.00103	1.01768	**0.98517**	0.00231	0.70117	0.98572	6	6
50	DVEPSO$_d$	**100.0**	0.00098	5.573x10^{+41}	0.02758	0.00885	0.68577	0.97358	5	5
50	DVEPSO$_{pe}$	**100.0**	0.00076	1.00645	6.998x10^{-45}	1.587x10^{-47}	0.67082	0.98334	2.5	4
50	DVEPSO$_p$	**100.0**	0.00115	**1.969x10^{+43}**	0.00167	0.00167	0.68958	0.98313	4	2
50	DVEPSO$_r$	**100.0**	0.00077	1.00532	8.548x10^{-47}	2.067x10^{-49}	**0.66911**	0.9844	2.5	3
50	DVEPSO$_{re}$	**100.0**	0.00092	1.00664	8.559x10^{-47}	1.935x10^{-49}	0.70841	0.97215	10	10
50	**DVEPSO$_u$**	**100.0**	0.00088	4.341x10^{+41}	3.674x10^{-05}	3.674x10^{-05}	0.68476	0.98049	**1**	**1**
50	DNSGAII-A	40.0	0.00137	1.02952	8.753x10^{-47}	3.248x10^{-50}	1.15409	**0.99744**	7	7
50	DNSGAII-B	40.0	0.00141	1.02976	8.755x10^{-47}	**2.781x10^{-50}**	1.16742	0.99743	8.5	8
50	dCOEA	40.0	**0.00065**	1.01787	8.654x10^{-47}	2.083x10^{-49}	0.75373	0.9469	8.5	9

functions respectively; and *dCOEA*, *DNSGAII-A* and *DNSGAII-B* obtained the best overall rank for one, two and two functions respectively.

For a change frequency of 50, *dCOEA* and *DNSGAII-B* obtained the best *VD* value for two and three functions respectively; *DNSGAII-A* and *DNSGAII-B* obtained the best *MS* value for four and one function(s) respectively; and *DNSGAII-A* and *DNSGAII-B* obtained the best overall rank for four and three functions respectively.

Figure 8.3 illustrates the found POF of the various DMOO algorithms for FDA2. Figure 8.3 shows that *DVEPSO* was tracking the changing POF well over time, but *DNSGAII-A* and *dCOEA* struggled to track the changing POF once it changed from convex to concave.

Table 8.4 Performance Measure Values for dMOP1

τ_t	Algorithm	NS	S	HV R	Acc	Stab	VD	MS	R_1	R_2
10	DVEPSO$_c$	**99.9**	0.00407	0.99962	**0.99962**	0.00035	0.26344	0.87907	2	3
10	DVEPSO$_d$	**99.9**	0.00452	0.99821	0.99796	7.232x10^{-05}	0.29477	0.89326	8	9.5
10	DVEPSO$_{pe}$	**99.9**	0.00484	0.9991	0.99885	0.00046	0.29445	0.89736	7	6
10	DVEPSO$_p$	**99.9**	0.00405	0.9983	0.99805	6.998x10^{-05}	0.28384	0.89884	5	5
10	DVEPSO$_r$	**99.9**	0.00431	0.99841	0.99816	0.00083	0.29631	0.90964	9	8
10	DVEPSO$_{re}$	**99.9**	0.00365	0.99921	0.99896	0.00041	0.23362	0.88294	4	4
10	DVEPSO$_u$	**99.9**	0.00386	0.99866	0.99817	0.00086	0.23642	0.86045	6	7
10	**DNSGAII-A**	38.8	0.00577	**0.99991**	0.99933	**3.603x10^{-05}**	0.15212	**0.9834**	**1**	**1**
10	DNSGAII-B	38.7	0.00497	**0.99991**	0.99933	5.904x10^{-05}	0.15351	0.93976	3	2
10	dCOEA	39.8	**0.00045**	0.99582	0.99524	0.00253	**0.03892**	0.86235	10	9.5
25	DVEPSO$_c$	**100.0**	0.00361	0.9936	**0.99343**	0.00148	0.68678	0.76746	4	4
25	DVEPSO$_d$	**100.0**	0.00352	0.99097	0.97202	0.00091	0.77566	0.75222	8	8
25	DVEPSO$_{pe}$	**100.0**	0.00395	0.99877	0.96826	0.00055	0.71365	0.73278	6	6.5
25	DVEPSO$_p$	**100.0**	0.00351	0.99056	0.96029	0.00105	0.70929	0.74939	9	9
25	DVEPSO$_r$	**100.0**	0.00358	0.99347	0.96311	0.00177	0.80396	0.76349	10	10
25	DVEPSO$_{re}$	**100.0**	0.00386	0.99892	0.9684	0.00049	0.72382	0.72943	6	6.5
25	DVEPSO$_u$	**100.0**	0.00361	**1.0082**	0.94919	0.00276	0.72882	0.75882	6	5
25	**DNSGAII-A**	39.3	0.0004	0.9998	0.93468	**7.896x10^{-06}**	0.15351	**0.97874**	**1**	**1**
25	DNSGAII-B	39.3	0.0004	0.99976	0.93464	1.998x10^{-05}	0.13231	0.9755	2	2
25	dCOEA	40.0	**0.0003**	0.99887	0.93381	0.00064	**0.0686**	0.95086	3	3
50	DVEPSO$_c$	**100.0**	0.00136	0.97142	**0.9714**	0.00117	1.43242	0.56964	7	9.5
50	DVEPSO$_d$	**100.0**	0.00146	0.97285	0.91286	0.00368	1.48468	0.58482	9	9.5
50	DVEPSO$_{pe}$	**100.0**	0.00164	0.9977	0.90593	0.00074	1.27847	0.5732	6	6
50	DVEPSO$_p$	**100.0**	0.00112	0.97275	0.88327	0.00248	1.40793	0.6043	8	7
50	DVEPSO$_r$	**100.0**	0.00148	0.97523	0.88553	0.00208	1.4258	0.60255	10	8
50	DVEPSO$_{re}$	**100.0**	0.00194	0.99825	0.90643	0.00068	1.26249	0.56443	5	5
50	DVEPSO$_u$	**100.0**	0.00126	**1.01387**	0.91444	0.00885	1.60305	0.66245	4	4
50	DNSGAII-A	40.0	0.00034	0.99967	0.89645	1.349x10^{-05}	0.11787	0.98323	2	2
50	**DNSGAII-B**	40.0	0.00032	0.99967	0.89646	**1.246x10^{-05}**	0.121	**0.98338**	**1**	**1**
50	dCOEA	40.0	**0.00023**	0.99942	0.89624	0.00021	**0.09572**	0.97838	3	3

Although all DMOO algorithms tracked the changing POF of dMOP1 very well over time, Figure 8.4 shows that *DNSGAII-A* and *dCOEA* struggled to track the changing POF of dMOP2 over time. However, *DVEPSO* had no problem tracking the changing POF of dMOP2. The *VD* value that is obtained by the *DVEPSO* approaches for dMOP1 is high compared to the evolutionary algorithms. The *DVEPSO* approaches find much more solutions than the evolutionary algorithms, and most of the these solutions are on or very close to the true POF. However, a few outlier solutions in the archive of the *DVEPSO* approaches lead to the high *VD* values, even though they have tracked the changing POF.

Table 8.9 presents the overall rank that the various algorithms obtained for each performance measure, as well as their overall rank for the various frequencies of

Table 8.5 Performance Measure Values for dMOP2

τ_t	Algorithm	NS	S	HV R	Acc	Stab	VD	MS	R_1	R_2
10	$DVEPSO_c$	**99.9**	0.00073	0.99962	**0.99951**	**0.00027**	0.07904	0.97647	2	2
10	**$DVEPSO_d$**	**99.9**	**0.00062**	1.00667	0.98227	0.00042	**0.07402**	0.97937	**1**	**1**
10	$DVEPSO_{pe}$	**99.9**	0.00083	0.99915	0.9732	0.00039	0.09291	0.97288	5	7
10	$DVEPSO_p$	**99.9**	0.00067	1.00603	0.97887	0.00045	0.07467	0.9744	3	3
10	$DVEPSO_r$	**99.9**	0.00076	0.99891	0.97127	0.00047	0.08269	0.97518	6.5	6
10	$DVEPSO_{re}$	**99.9**	0.00067	0.99903	0.97138	0.00046	0.0855	0.97567	4	4
10	$DVEPSO_u$	**99.9**	0.0008	**1.00709**	0.95911	0.00197	0.08911	0.9689	6.5	5
10	DNSGAII-A	33.5	0.00095	0.99321	0.93715	0.00064	0.90415	**1.45643**	8	8
10	DNSGAII-B	28.7	0.00212	0.99216	0.93616	0.00068	1.03746	1.43973	9.5	9.5
10	dCOEA	33.7	0.00112	0.98988	0.93401	0.00213	0.81297	1.40996	9.5	9.5
25	$DVEPSO_c$	**100.0**	0.00078	0.998	**0.9978**	0.00097	0.17631	0.91634	3	4
25	$DVEPSO_d$	**100.0**	0.00085	0.99396	0.96988	0.00188	0.1772	0.93172	6	5
25	$DVEPSO_{pe}$	**100.0**	0.00076	0.99874	0.96992	**0.00056**	0.18783	0.94799	2	2
25	$DVEPSO_p$	**100.0**	0.00085	0.99719	0.96842	0.00163	0.17535	0.91477	6	6.5
25	$DVEPSO_r$	**100.0**	0.00054	0.99767	0.96888	0.00096	**0.17112**	0.93158	4	3
25	**$DVEPSO_{re}$**	**100.0**	0.00079	0.99867	0.96986	0.00064	0.17207	0.93278	**1**	**1**
25	$DVEPSO_u$	**100.0**	0.00099	**1.0045**	0.96771	0.00241	0.18725	0.91586	9	9
25	DNSGAII-A	39.9	0.00043	0.98884	0.95201	0.00101	0.93768	**1.63537**	8	6.5
25	DNSGAII-B	39.9	0.00041	0.98885	0.95203	0.001	0.94214	1.63414	6	8
25	dCOEA	39.8	**0.0004**	0.98775	0.95096	0.00144	0.93822	1.61199	10	10
50	$DVEPSO_c$	**100.0**	**0.00016**	0.97296	**0.97124**	0.00629	0.19285	0.84654	9	10
50	$DVEPSO_d$	**100.0**	0.00017	1.05717	0.71632	0.01254	0.1688	0.85856	2.5	4
50	$DVEPSO_{pe}$	**100.0**	**0.00016**	0.99637	0.62876	0.00077	0.16929	0.85865	4	6.5
50	$DVEPSO_p$	**100.0**	0.00017	1.13016	0.61318	0.01325	0.20012	0.88857	6	5
50	$DVEPSO_r$	**100.0**	**0.00016**	0.98452	0.49138	0.0024	0.16467	0.87944	6	8
50	$DVEPSO_{re}$	**100.0**	0.00018	0.99619	0.4972	0.00065	0.14661	0.85065	6	6.5
50	$DVEPSO_u$	**100.0**	**0.00016**	**1.23115**	0.61117	0.04407	0.15933	0.87335	2.5	2
50	DNSGAII-A	40.0	0.00032	0.99845	0.45172	0.00014	0.15645	0.9955	8	3
50	**DNSGAII-B**	40.0	0.00032	0.99863	0.45181	**0.00012**	**0.14069**	**0.99639**	**1**	**1**
50	dCOEA	39.8	0.00027	0.98953	0.44769	0.00229	0.15248	0.95434	10	9

change. Table 8.9 shows that for a change frequency of 10 the best overall rank for VD was obtained by $DVEPSO_c$ and $DVEPSO_r$ and the best overall rank for MS was obtained by *DNSGAII-A*. *DNSGAII-B* obtained the best rank over all performance measures and $DVEPSO_c$ obtained the best rank over all performance measures when the measures that use the true POF count more towards the overall rank average.

For a change frequency of 25 the best overall rank for VD was obtained by $DVEPSO_{cl}$ and $DVEPSO_r$, the best overall rank for MS was obtained by *DNSGAII-A* and the approach that ranked the best over all performance measures was *DNSGAII-B*. For a change frequency of 50 the best overall rank for VD was obtained by *DNSGAII-B* and *dCOEA*. The best overall rank for MS was obtained by *DNSGAII-A* and the approach that ranked the best over all performance measures was *DNSGAII-A*.

Table 8.6 Performance Measure Values for dMOP3

τ_t	**Algorithm**	**NS**	**S**	**HV R**	**Acc**	**Stab**	**VD**	**MS**	$\mathbf{R_1}$	$\mathbf{R_2}$
10	DVEPSO$_c$	5.1	0.07368	0.9973	0.99735	**0.00045**	1.35206	1.91012	2.5	2
10	DVEPSO$_d$	5.1	0.0789	0.91942	0.91945	0.01367	1.38248	1.93216	10	10
10	DVEPSO$_{pe}$	5.2	0.07884	0.99545	0.99548	0.00108	1.38861	1.93614	4	4
10	DVEPSO$_p$	5.1	0.07866	0.91629	0.91632	0.01296	1.37957	1.92784	9	9
10	DVEPSO$_r$	5.0	0.07815	0.99515	0.99519	0.00125	1.3581	1.90401	8	7
10	DVEPSO$_{re}$	5.1	0.08445	0.99547	0.9955	0.00107	1.38452	1.93409	5	5.5
10	DVEPSO$_u$	5.2	0.07905	0.99403	0.99406	0.00232	1.38136	**1.94846**	6.5	5.5
10	**DNSGAII-A**	36.7	**0.00075**	**0.99744**	**0.99745**	0.00086	0.84372	1.54286	**1**	**1**
10	DNSGAII-B	28.2	0.00146	0.97038	0.97039	0.00558	0.94898	1.27303	6.5	8
10	dCOEA	**36.9**	0.00078	0.99611	0.99612	0.00159	**0.74777**	1.34982	2.5	3
25	DVEPSO$_c$	5.5	0.07858	0.98458	0.98515	0.00178	1.43742	2.08363	7	5
25	DVEPSO$_d$	5.5	0.07596	0.88596	0.88644	0.01572	1.45526	2.12036	9	10
25	DVEPSO$_{pe}$	5.5	0.08177	0.98197	0.9825	0.002	1.48467	**2.14814**	8	8
25	DVEPSO$_p$	5.5	0.07608	0.89117	0.89164	0.01617	1.45777	2.1446	10	9
25	DVEPSO$_r$	5.8	0.07085	0.98163	0.98215	0.00206	1.42254	2.10335	4	4
25	DVEPSO$_{re}$	5.5	0.07123	0.98211	0.98263	0.00191	1.44345	2.12769	3	3
25	DVEPSO$_u$	5.5	0.07527	0.98564	0.98583	0.00289	1.46414	2.10228	6	6
25	DNSGAII-A	**40.0**	0.00038	0.99017	0.99018	0.0014	0.90791	1.6252	2	2
25	DNSGAII-B	**40.0**	**0.00035**	0.9741	0.97411	0.00812	0.89235	1.49494	5	7
25	**dCOEA**	39.9	0.00042	**0.99104**	**0.99106**	**0.00116**	**0.88038**	1.52265	**1**	**1**
50	DVEPSO$_c$	8.9	0.01957	0.99372	0.99612	0.00105	0.60082	0.83513	5	6
50	DVEPSO$_d$	9.6	0.01643	0.86685	0.86849	0.0281	0.5848	0.86028	6	5
50	DVEPSO$_{pe}$	9.3	0.01741	0.99347	0.99535	0.00123	0.59723	0.84383	4	4
50	DVEPSO$_p$	9.0	0.01997	0.86521	0.86588	0.03018	0.60588	0.83987	9	9
50	DVEPSO$_r$	9.2	0.01932	0.99301	0.99365	0.00154	0.58804	0.83986	7	7
50	DVEPSO$_{re}$	8.9	0.02124	0.99359	0.99423	0.00114	0.61199	0.84182	8	8
50	DVEPSO$_u$	8.7	0.02248	0.98684	0.98649	0.00439	0.61941	0.83323	10	10
50	**DNSGAII-A**	**40.0**	0.00032	**0.99982**	**0.99912**	$3.1\text{x}10^{-05}$	0.11419	**0.99401**	**1**	**1**
50	DNSGAII-B	**40.0**	**0.00029**	0.99561	0.99491	0.00422	**0.09471**	0.97185	3	3
50	dCOEA	39.9	**0.00025**	0.99942	0.99871	0.00028	0.12694	0.968	2	2

With regard to the overall rank presented in Table 8.9 over all frequencies of change, the DVEPSO approaches performed the best with regards to VD and the dynamic NSGA-II approaches performed the best with regards to MS and the overall rank. The DVEPSO approaches obtained the best overall rank for VD on eleven occasions, and the dynamic NSGA-II approaches and $dCOEA$ on six occasions each. With regards to MS, the DVEPSO approaches obtained the highest rank on 8 occasions, the dynamic NSGA-II approaches on eleven occasions and $dCOEA$ on 4 occasions. The dynamic NSGA-II approaches obtained the best overall rank on 15 occasions, the DVEPSO approaches on 12 occasions and $dCOEA$ on no occasion.

Table 8.7 Performance Measure Values for HE1

τ_t	Algorithm	NS	S	HV R	Acc	Stab	VD	MS	R_1	R_2
10	DVEPSO$_c$	13.5	0.01173	0.66388	**0.85439**	0.01399	1.55763	0.78202	7	9
10	DVEPSO$_d$	12.6	0.01359	0.73684	0.67688	0.01324	1.51808	0.77148	8	6
10	DVEPSO$_{pe}$	13.1	0.01518	0.68747	0.61666	0.00955	1.522	0.76341	10.0	10
10	DVEPSO$_p$	13.3	0.01196	0.78513	0.70427	0.01191	1.53101	0.77698	4.5	4.5
10	DVEPSO$_r$	14.4	0.0108	0.67781	0.608	0.00835	1.53891	0.76821	6	7.5
10	DVEPSO$_{re}$	13.5	0.01411	0.70286	0.63047	0.01085	1.54639	0.78341	9	7.5
10	DVEPSO$_u$	13.8	0.01429	0.89687	0.7716	0.01334	1.56948	0.78503	4.5	4.5
10	DNSGAII-A	**40.0**	0.00058	**0.96747**	0.80366	0.00827	**0.13607**	**0.8044**	**1**	**1**
10	DNSGAII-B	**40.0**	**0.00033**	0.90325	0.75031	**0.00317**	0.13917	0.42739	2	2
10	dCOEA	28.6	0.00326	0.92802	0.77089	0.01347	0.18269	0.60639	3	3
25	DVEPSO$_c$	19.9	0.0069	0.66264	**0.88732**	0.01299	1.54781	0.77076	6.5	5.5
25	DVEPSO$_d$	24.2	0.00558	0.8044	0.75121	0.02017	1.55831	0.76052	9.5	10
25	DVEPSO$_{pe}$	18.3	0.00837	0.69068	0.63844	0.00961	1.51748	0.75548	9.5	9
25	DVEPSO$_p$	21.1	0.0075	0.83401	0.74319	0.01696	1.53058	0.76397	8	7
25	DVEPSO$_r$	25.0	0.00543	0.68311	0.60386	0.00915	1.57137	0.76668	4.5	5.5
25	DVEPSO$_{re}$	19.9	0.00543	0.71364	0.63085	0.00988	1.57552	0.76542	6.5	8
25	DVEPSO$_u$	17.3	0.009	0.87798	0.77612	0.01192	1.52789	0.77961	4.5	4
25	**DNSGAII-A**	**40.0**	0.00058	**0.96607**	0.85399	0.01143	**0.15803**	**0.79656**	**1**	**1**
25	DNSGAII-B	**40.0**	**0.00038**	0.90938	0.80388	**0.00648**	0.16232	0.47371	2	2
25	dCOEA	39.7	0.0011	0.94994	0.83974	0.0125	0.18167	0.72581	3	3
50	DVEPSO$_c$	34.2	0.00367	0.71493	**0.91246**	0.01146	1.59845	0.76024	5	8.5
50	DVEPSO$_d$	34.7	0.00401	0.87724	0.79065	0.01461	1.56689	0.7584	9	8.5
50	DVEPSO$_{pe}$	29.0	0.00409	0.72615	0.64561	0.00899	1.56256	0.76148	6.5	5
50	DVEPSO$_p$	29.4	0.00489	0.89416	0.79499	0.01309	1.54536	0.75448	10	6.5
50	DVEPSO$_r$	33.4	0.00357	0.71771	0.63811	0.00863	1.5742	0.75946	6.5	10
50	DVEPSO$_{re}$	34.2	0.00342	0.75603	0.67218	0.00934	1.56573	0.76762	4	4
50	DVEPSO$_u$	32.4	0.00416	0.89113	0.7923	0.01105	1.56451	0.75721	8	6.5
50	**DNSGAII-A**	**40.0**	0.00058	**0.96827**	0.86088	0.01083	0.19626	**0.78734**	**1**	**1**
50	DNSGAII-B	**40.0**	**0.00041**	0.91908	0.81715	**0.00778**	0.19108	0.46374	3	3
50	dCOEA	**40.0**	0.00072	0.96564	0.85854	0.00961	**0.18173**	0.78317	2	2

8.5.3 Statistical Analysis

This section discusses the statistical analysis that was done on the performance metrics values. Kruskal-Wallis tests were performed to determine whether there was a statistically significant difference between the values obtained by the various DMOO algorithms for a performance metric for a specific function at a specific τ_t. The p-values that were obtained from the Kruskal-Wallis tests are presented in Tables 8.10- 8.17. In these tables, p-values that are statistically significant are displayed in bold.

When the p-value of the Kruskal-Wallis test indicated that there was a statistically significant difference, Mann-Whitney U tests were performed to determine

Table 8.8 Performance Measure Values for HE2

τ_t	Algorithm	NS	S	HV R	Acc	Stab	VD	MS	R_1	R_2
10	DVEPSO$_c$	24.4	0.01986	0.48235	0.54748	0.01305	1.52451	0.98783	4.5	3.5
10	DVEPSO$_d$	29.8	0.0115	0.46427	0.51349	0.0134	1.55614	1.02734	6.5	7
10	DVEPSO$_{pe}$	27.8	0.0132	0.47364	0.52271	0.01315	1.52684	1.01328	4.5	5
10	DVEPSO$_p$	28.1	0.01271	0.46458	0.51271	0.01325	1.56867	1.01558	10	10
10	DVEPSO$_r$	23.3	0.02176	0.46844	0.51697	0.01302	1.53622	1.024	8.5	8
10	DVEPSO$_{re}$	24.4	0.01208	0.46198	0.50984	0.01338	1.55797	**1.03365**	8.5	9
10	DVEPSO$_u$	23.9	0.01354	0.47324	0.52226	0.01305	1.5348	1.01992	6.5	6
10	DNSGAII-A	**40.0**	0.00062	0.99071	0.94744	0.00213	0.20337	0.91933	2	2
10	**DNSGAII-B**	**40.0**	**0.00061**	**0.99095**	**0.9474**	**0.00206**	**0.20331**	0.92084	**1**	**1**
10	dCOEA	27.4	0.00452	0.9062	0.89176	0.01591	0.23457	0.6925	3	3.5
25	DVEPSO$_c$	**43.5**	0.00478	0.6976	**0.7411**	0.00911	1.50652	0.89599	5	5
25	DVEPSO$_d$	42.7	0.01078	0.98233	0.45845	0.00316	1.5213	**0.90232**	4	4
25	DVEPSO$_{pe}$	34.1	0.01208	0.71057	0.18521	0.00214	1.48753	0.88314	9	9
25	DVEPSO$_p$	39.5	0.00823	0.98768	0.25744	0.00207	1.49003	0.89328	3	3
25	DVEPSO$_r$	32.0	0.01914	0.69406	0.18091	0.00235	1.50608	0.88741	10	10
25	DVEPSO$_{re}$	**43.5**	0.01553	0.74847	0.19509	0.00181	1.51314	0.89228	8	8
25	DVEPSO$_u$	33.6	0.01342	0.9345	0.24358	0.00187	1.52096	0.89767	7	7
25	DNSGAII-A	**40.0**	0.00065	**0.9896**	0.25794	**0.00022**	0.26994	0.88562	2	2
25	**DNSGAII-B**	**40.0**	**0.00061**	**0.9896**	0.25794	**0.00022**	**0.26682**	0.88501	**1**	**1**
25	dCOEA	39.4	0.00131	0.95101	0.24789	0.00204	0.2783	0.74454	6	6
50	DVEPSO$_c$	33.2	0.0145	0.66815	**0.89051**	0.00519	1.75131	1.16669	9	10
50	DVEPSO$_d$	24.4	0.02218	0.9366	0.72392	0.00342	1.71371	1.14846	7	7
50	DVEPSO$_{pe}$	28.0	0.02009	0.68452	0.48548	0.00278	1.71787	1.18569	8	8
50	DVEPSO$_p$	32.3	0.01708	0.94995	0.67374	0.00244	1.73754	1.1905	4	3.5
50	DVEPSO$_r$	20.3	0.01837	0.67025	0.47536	0.00298	1.74609	**1.1989**	10	9
50	DVEPSO$_{re}$	33.2	0.01405	0.72528	0.51439	0.00193	1.72075	1.1981	5	5
50	DVEPSO$_u$	25.9	0.01376	0.86189	0.61128	0.00301	1.71865	1.17589	6	6
50	**DNSGAII-A**	**40.0**	0.00063	**0.9985**	0.70817	**0.00048**	0.19138	0.91808	**1**	**1**
50	DNSGAII-B	**40.0**	**0.00062**	0.99847	0.70815	**0.00048**	**0.17538**	0.91793	2	2
50	dCOEA	**40.0**	0.00146	0.97275	0.68991	0.00346	0.25389	0.81778	3	3

between which DMOO algorithms' performance metric values there were a statistically significant difference. Both the Kruskal-Wallis tests and the Mann-Whitney U tests were performed using the statistical software package R and testing for a confidence level of 95%. Due to a lack of space all results of the Mann-Whitney U tests are not presented. However, Tables 8.18- 8.25 in the appendix present the results of the Mann-Whitney U tests for the VD performance metric. In all these tables "-" indicates that there was no statistically significant difference and "x" indicates that according to the Mann-Whitney U test, there was a statistically significant difference between the specific performance metric values.

Table 8.10 shows that for FDA1 there is a statistically significant difference between almost all of the algorithms for a change frequency of 10 and for almost half

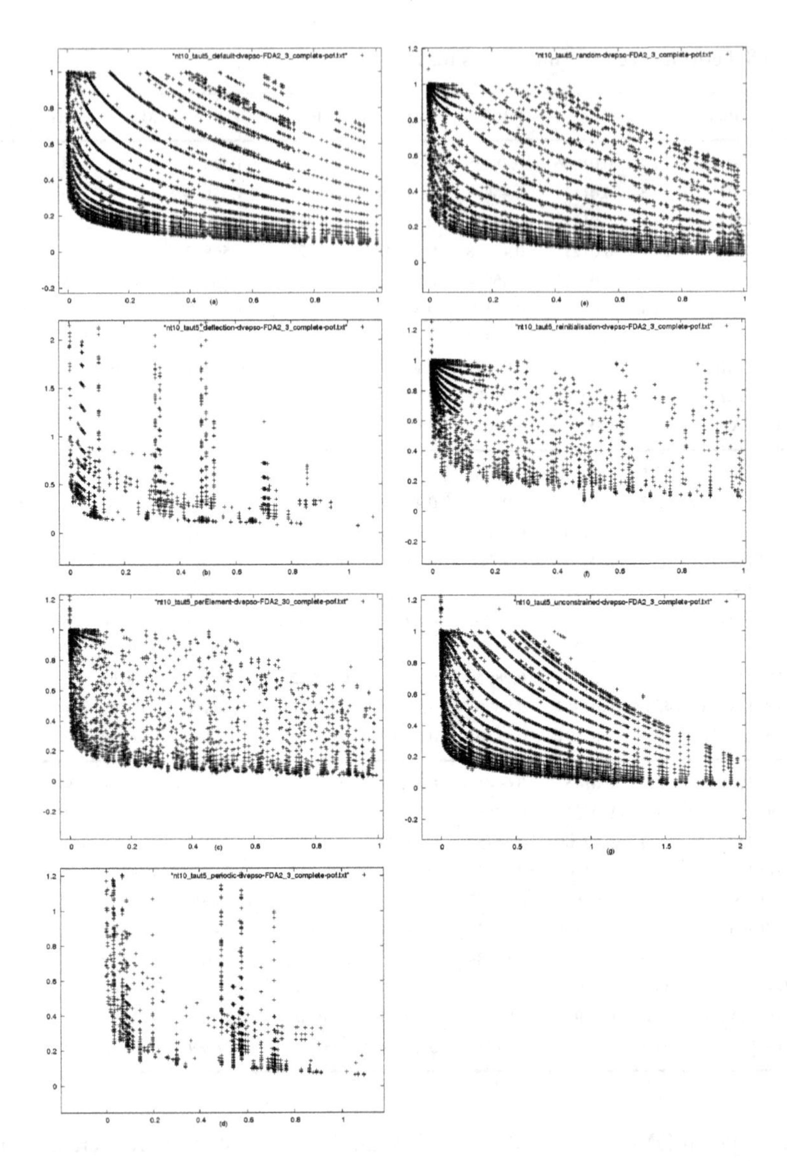

Fig. 8.1 Results of various boundary constraint management approaches solving FDA2, with (a) DVEPSO$_c$, (b) DVEPSO$_d$, (c) DVEPSO$_{pe}$, (d) DVEPSO$_p$, (e) DVEPSO$_r$, (f) DVEPSO$_{re}$ and (g) DVEPSO$_u$. The numbering is from top to bottom on the left, and then from top to bottom on the right.

of the algorithm combinations for a change frequency of 50. However, for a change frequency of 25 there is no statistically significant difference when comparing the evolutionary algorithms against each other, but there is a statistically significant difference for almost all combinations when comparing the evolutionary algorithms against the DVEPSO approaches.

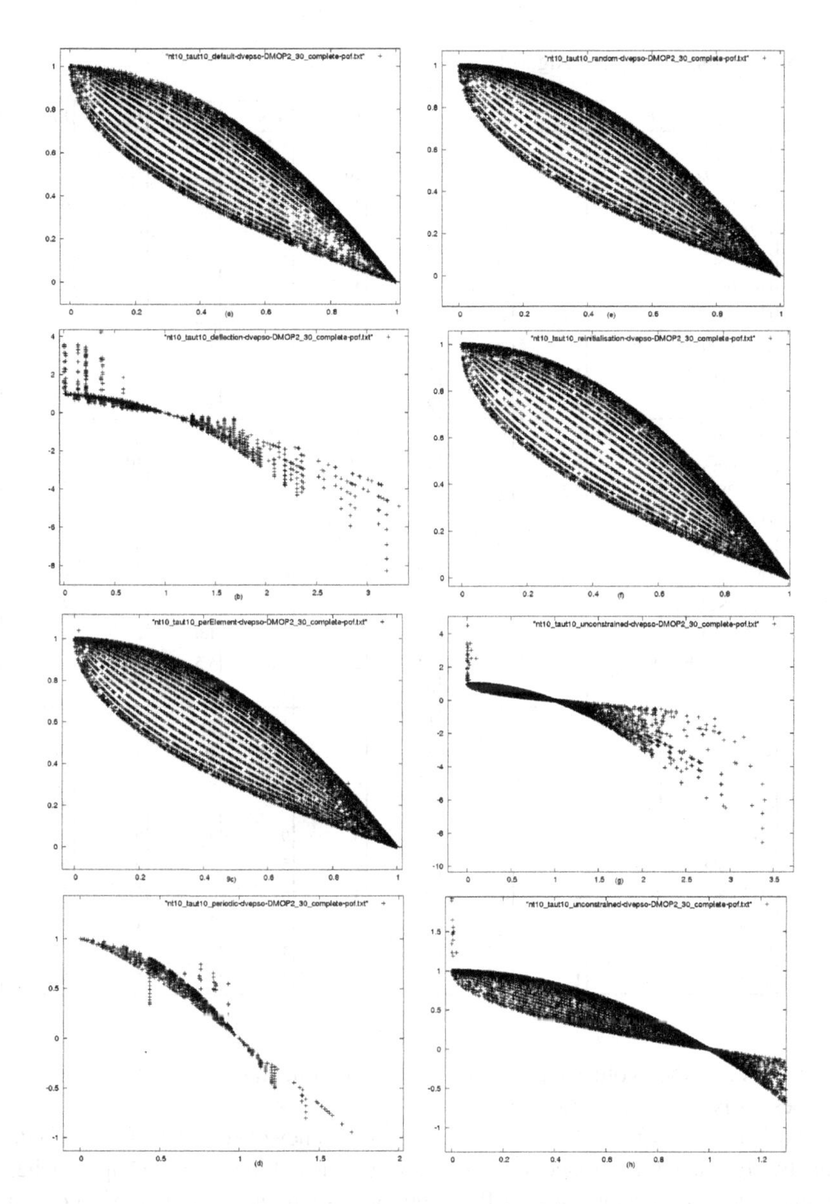

Fig. 8.2 Results of various boundary constraint management approaches solving dMOP2, with (a) DVEPSO$_c$, (b) DVEPSO$_d$, (c) DVEPSO$_{pe}$, (d) DVEPSO$_p$, (e) DVEPSO$_r$, (f) DVEPSO$_{re}$, (g) DVEPSO$_u$. The numbering is from top to bottom on the left, and then from top to bottom on the right.

For FDA2 with a change frequency of 10 and 25, only a few of the DVEPSO approaches have statistically significant differences when compared to other DVEPSO approaches, but almost all combinations of comparisons between DVEPSO

Table 8.9 Overall Ranking of Algorithms

τ_t	**Algorithm**	$\mathbf{R}_{NS}$	$\mathbf{R}_{S}$	$\mathbf{R}_{HVR}$	$\mathbf{R}_{Acc}$	$\mathbf{R}_{Stab}$	$\mathbf{R}_{VD}$	$\mathbf{R}_{MS}$	$\mathbf{R}_{O_1}$	$\mathbf{R}_{O_2}$
10	DVEPSO$_c$	**1**	9	8	**1**	8	**1.5**	4	3	**1**
10	DVEPSO$_d$	3	6	9	2	10	9.5	3	10	9
10	DVEPSO$_{pe}$	4.5	7.5	6	3	4	3.5	9	6	6
10	DVEPSO$_p$	4.5	5	7	8	9	3.5	7	9	4
10	DVEPSO$_r$	6	4	10	10	6	**1.5**	2	6	7
10	DVEPSO$_{re}$	2	7.5	5	6.5	3	6.5	10	8	8
10	DVEPSO$_u$	7	10	**1**	5	7	9.5	5.5	6	2
10	DNSGAII-A	10	3	2	4	**1**	6.5	**1**	2	3
10	**DNSGAII-B**	9	**1.5**	3	2	2	5	5.5	**1**	5
10	dCOEA	8	**1.5**	4	9	5	8	8	4	10
25	DVEPSO$_c$	**1**	9	8	**1**	8	**1.5**	4	3	3
25	DVEPSO$_d$	3	6	9	2	10	9.5	3	10	10
25	DVEPSO$_{pe}$	4.5	7.5	6	3	4	3.5	9	6	6
25	DVEPSO$_p$	4.5	5	7	8	9	3.5	7	9	9
25	DVEPSO$_r$	6	4	10	10	6	**1.5**	2	6	5
25	DVEPSO$_{re}$	2	7.5	5	6.5	3	6.5	10	8	8
25	DVEPSO$_u$	7	10	**1**	5	7	9.5	5.5	6	4
25	DNSGAII-A	10	3	2	4	**1**	6.5	**1.0**	2	2
25	DNSGAII-B	9	**1.5**	3	6.5	2	5.0	5.5	**1.0**	**1**
25	dCOEA	8	**1.5**	4	9	5	8	8	4	7
50	DVEPSO$_c$	**1.5**	5	10	**1**	7	10	8	7	9.5
50	DVEPSO$_d$	**1.5**	9	7	4.5	10	7	10	8	8
50	DVEPSO$_{pe}$	5.5	6	8	6.5	5	3	5.5	4	4
50	DVEPSO$_p$	3.5	9	5	9	9	8	9	10	7
50	DVEPSO$_r$	3.5	7	9	10	6	9	4	9	9.5
50	DVEPSO$_{re}$	5.5	9	6	8	4	5	7	6	6
50	DVEPSO$_u$	9	3	3	2	8	6	5.5	3	3
50	DNSGAII-A	7.5	4	**1**	3	2	4	**1**	**1**	**1**
50	DNSGAII-B	7.5	2	2	4.5	**1**	**1.5**	2	2	2
50	dCOEA	10	**1**	4	6.5	3	**1.5**	3	5	5

approaches and the evolutionary algorithms resulted in statistically significant differences. This is shown in Table 8.11.

Table 8.12 shows that for FDA3 for a change frequency of 10 and 25 there is no statistically significant difference between the VD values of the DVEPSO approaches, but almost all comparisons of DVEPSO approaches with an evolutionary algorithm resulted in a statistically significant difference in VD values. For a change frequency of 50, almost all comparisons resulted in statistically significant differences.

For dMOP1 with a change frequency of 10 most DVEPSO approaches compared against each other resulted in statistically significant differences and for a change frequency of 25 and 50 almost half of the DVEPSO approaches comparisons resulted in statistically significant differences. However, for all three change frequencies all comparisons between the evolutionary algorithms and the DVEPSO approaches resulted in statistically significant differences and the values obtained by

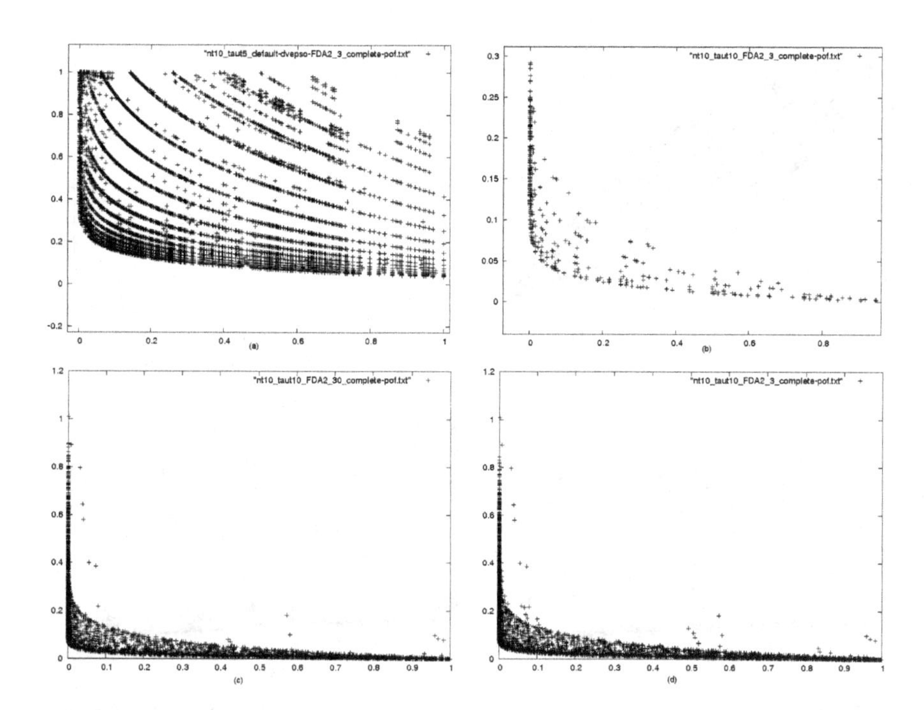

Fig. 8.3 Results of various algorithms solving FDA2, with (a) DVEPSO, (b) dCOEA, (c) DNSGAII-A and (d) DNSGAII-B.

DNSGAII-A and *DNSGAII-B* was statistically significantly different, but the comparison between *DNSGAII-B* and *dCOEA* was not statistically significantly different. This is shown in Table 8.13. Table 8.14 shows that for dMOP2 for a change frequency of 10 all comparisons lead to statistically significant differences and for a change frequency of 50 only the comparison between *DNSGAII-A* and *DNSGAII-B* indicated a statistically significant difference. For a change frequency of 25 all comparisons amongst the evolutionary algorithms, and all comparisons between the evolutionary algorithms and the DVEPSO approaches, resulted in statistically significant differences. However, only a few comparisons amongst the DVEPSO approaches resulted in a statistically significant difference.

For dMOP3 no statistically significant difference was found for any comparisons amongst the DVEPSO approaches for all frequencies of change. For a change frequency of 10, all comparisons amongst the evolutionary algorithms indicated a statistically significant difference, but not for the change frequencies of 25 and 50. All comparisons between the evolutionary algorithms and the DVEPSO approaches indicated a statistically significant difference for all frequencies of change for dMOP3. This is shown in Table 8.15.

Table 8.16 shows that for HE1 for a change frequency of 10, all comparisons lead to a statistically significant difference, except the comparison of *DNSGAII-B*

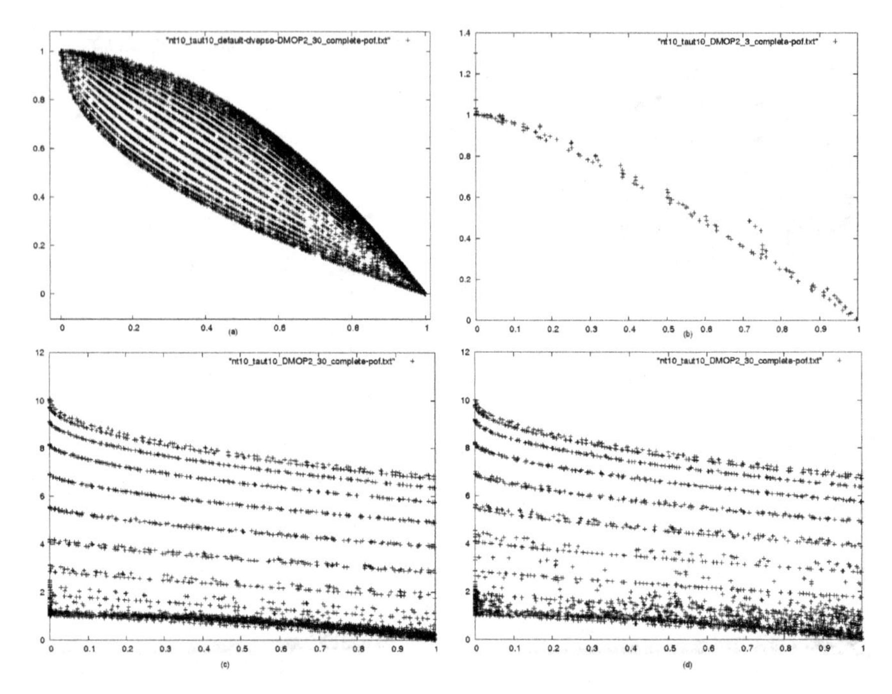

Fig. 8.4 Results of various algorithms solving dMOP2, with (a) DVEPSO, (b) dCOEA, (c) DNSGAII-A and (d) DNSGAII-B.

and *dCOEA*. For a change frequency of 25, almost all DVEPSO comparisons and all comparisons between DVEPSO and evolutionary computation algorithms lead to a statistically significant difference and amongst the evolutionary algorithms only the comparison between *DNSGAII-B* and *dCOEA* indicated *VD* values that were not statistically significantly different. For a change frequency of 50, all comparisons between DVEPSO and the evolutionary algorithms, and a few of the comparisons amongst the DVEPSO approaches, indicated statistically significant different values.

For HE2 for a change frequency of 10 and 50, amongst the evolutionary algorithms only the comparison between *DNSGAII-B* and *dCOEA* indicated *VD* values that were not statistically significantly different and for a change frequency of 25 none of the comparisons amongst the evolutionary algorithms indicated a statistically significant difference. From the comparisons amongst the DVEPSO approaches, approximately half indicated a statistically significant difference for change frequencies of 10 and 50. For a change frequency of 25, none of the comparisons amongst the DVEPSO approaches indicated a statistically significant difference. However, for all frequencies of change, the comparisons between the evolutionary algorithms and the DVEPSO approaches indicated a statistically significant difference.

Table 8.10 p-values of Kruskal-Wallis test for FDA1

τ_t	**S**	**HV R**	**Acc**	**Stab**	**VD**	**MS**
10	**$< 2.2x10^{-16}$**	**$< 2.2x10^{-16}$**	**$< 2.2x10^{-16}$**	**$< 2.2x10^{-16}$**	**$< 2.2x10^{-16}$**	**$< 2.2x10^{-16}$**
25	**0.00045**	**$< 2.2x10^{-16}$**	**$< 2.2x10^{-16}$**	**$< 2.2x10^{-16}$**	**$< 2.2x10^{-16}$**	**$< 2.2x10^{-16}$**
50	**0.01745**	**$< 2.2x10^{-16}$**	**$< 2.2x10^{-16}$**	**$< 2.2x10^{-16}$**	**0.0001837**	**$< 2.2x10^{-16}$**

Table 8.11 p-values of Kruskal-Wallis test for FDA2

τ_t	**S**	**HV R**	**Acc**	**Stab**	**VD**	**MS**
10	**$3.509x10^{-14}$**	**$< 2.2x10^{-16}$**	**$< 2.2x10^{-16}$**	**$< 2.2x10^{-16}$**	**$< 2.2x10^{-16}$**	**$< 2.2x10^{-16}$**
25	**0.003196**	**$< 2.2x10^{-16}$**	**$< 2.2x10^{-16}$**	**$< 2.2x10^{-16}$**	**$< 2.2x10^{-16}$**	**$5.444x10^{-12}$**
50	**$< 2.2x10^{-16}$**	**$< 2.2x10^{-16}$**	**$< 2.2x10^{-16}$**	**$< 2.2x10^{-16}$**	**$< 2.2x10^{-16}$**	**$< 2.2x10^{-16}$**

Table 8.12 p-values of Kruskal-Wallis test for FDA3

τ_t	**S**	**HV R**	**Acc**	**Stab**	**VD**	**MS**
10	**$< 2.2x10^{-16}$**	**$< 2.2x10^{-16}$**	**$< 2.2x10^{-16}$**	**$< 2.2x10^{-16}$**	**$9.382x10^{-14}$**	**0.01549**
25	**$4.898x10^{-08}$**	**$< 2.2x10^{-16}$**	**$< 2.2x10^{-16}$**	**$< 2.2x10^{-16}$**	**0.0182**	0.08228
50	**$1.864x10^{-07}$**	**$< 2.2x10^{-16}$**	**$< 2.2x10^{-16}$**	**$< 2.2x10^{-16}$**	**$< 2.2x10^{-16}$**	**$1.96x10^{-08}$**

Table 8.13 p-values of Kruskal-Wallis test for dMOP1

τ_t	**S**	**HV R**	**Acc**	**Stab**	**VD**	**MS**
10	**$< 2.2x10^{-16}$**	**$< 2.2x10^{-16}$**	**$< 2.2x10^{-16}$**	**$< 2.2x10^{-16}$**	**$< 2.2x10^{-16}$**	**$< 2.2x10^{-16}$**
25	**$< 2.2x10^{-16}$**	**$< 2.2x10^{-16}$**	**$< 2.2x10^{-16}$**	**$< 2.2x10^{-16}$**	**$< 2.2x10^{-16}$**	**$< 2.2x10^{-16}$**
50	**$< 2.2x10^{-16}$**	**$< 2.2x10^{-16}$**	**$< 2.2x10^{-16}$**	**$< 2.2x10^{-16}$**	**$< 2.2x10^{-16}$**	**$< 2.2x10^{-16}$**

Table 8.14 p-values of Kruskal-Wallis test for dMOP2

τ_t	**S**	**HV R**	**Acc**	**Stab**	**VD**	**MS**
10	**$< 2.2x10^{-16}$**	**$< 2.2x10^{-16}$**	**$< 2.2x10^{-16}$**	**$< 2.2x10^{-16}$**	**$< 2.2x10^{-16}$**	**$7.723x10^{-16}$**
25	0.9811	**$< 2.2x10^{-16}$**	**$< 2.2x10^{-16}$**	**$3.127x10^{-16}$**	**$2.888x10^{-08}$**	**$7.932x10^{-08}$**
50	**$2.564x10^{-15}$**	**$< 2.2x10^{-16}$**	**$< 2.2x10^{-16}$**	**$< 2.2x10^{-16}$**	0.9032	**$< 2.2x10^{-16}$**

Table 8.15 p-values of Kruskal-Wallis test for dMOP3

τ_t	S	HV R	Acc	Stab	VD	MS
10	$< 2.2\text{x}10^{-16}$	$< 2.2\text{x}10^{-16}$	$< 2.2\text{x}10^{-16}$	$< 2.2\text{x}10^{-16}$	$< 2.2\text{x}10^{-16}$	$1.07\text{x}10^{-12}$
25	$< 2.2\text{x}10^{-16}$	$< 2.2\text{x}10^{-16}$	$< 2.2\text{x}10^{-16}$	$< 2.2\text{x}10^{-16}$	$1.573\text{x}10^{-07}$	0.1925
50	$< 2.2\text{x}10^{-16}$	$< 2.2\text{x}10^{-16}$	$< 2.2\text{x}10^{-16}$	$< 2.2\text{x}10^{-16}$	$< 2.2\text{x}10^{-16}$	$< 2.2\text{x}10^{-16}$

Table 8.16 p-values of Kruskal-Wallis test for HE1

τ_t	S	HV R	Acc	Stab	VD	MS
10	$< 2.2\text{x}10^{-16}$	$< 2.2\text{x}10^{-16}$	$< 2.2\text{x}10^{-16}$	$< 2.2\text{x}10^{-16}$	$< 2.2\text{x}10^{-16}$	$< 2.2\text{x}10^{-16}$
25	$< 2.2\text{x}10^{-16}$	$< 2.2\text{x}10^{-16}$	$< 2.2\text{x}10^{-16}$	$1.231\text{x}10^{-15}$	$< 2.2\text{x}10^{-16}$	$< 2.2\text{x}10^{-16}$
50	$< 2.2\text{x}10^{-16}$	$< 2.2\text{x}10^{-16}$	$< 2.2\text{x}10^{-16}$	0.009434	$< 2.2\text{x}10^{-16}$	$1.158\text{x}10^{-06}$

Table 8.17 p-values of Kruskal-Wallis test for HE2

τ_t	S	HV R	Acc	Stab	VD	MS
10	$< 2.2\text{x}10^{-16}$	$< 2.2\text{x}10^{-16}$	$< 2.2\text{x}10^{-16}$	$< 2.2\text{x}10^{-16}$	$< 2.2\text{x}10^{-16}$	$2.682\text{x}10^{-10}$
25	$< 2.2\text{x}10^{-16}$	$< 2.2\text{x}10^{-16}$	$< 2.2\text{x}10^{-16}$	$3.902\text{x}10^{-15}$	$< 2.2\text{x}10^{-16}$	0.003448
50	$< 2.2\text{x}10^{-16}$	$< 2.2\text{x}10^{-16}$	$< 2.2\text{x}10^{-16}$	$2.815\text{x}10^{-05}$	$< 2.2\text{x}10^{-16}$	$< 2.2\text{x}10^{-16}$

8.6 Conclusions

This chapter discussed DMOO and issues that should be addressed when solving DMOOP. The DVEPSO algorithm was presented and the effect that various boundary handling approaches have on the performance of DVEPSO was investigated. It could clearly be seen that the deflection and periodic boundary handling approaches lead to bad performance with especially the FDA2 problem.

The performance of DVEPSO were compared against those of three other state-of-the-art DMOO algorithms. DVEPSO performed quite well with regards to the VD metric that measures the closeness of the approximated POF to the true POF and the MS metric that measures the spread of the found non-dominated solutions. The DNSGAII approaches and dCOEA struggled to track the changing POF of the FDA2 and dMOP2 problems, but DVEPSO had no problem to track the changing POF for these problems. However, the DNSGAII approaches and dCOEA outperformed DVEPSO with the problems that have a discontiuous POF.

Acknowledgement. The authors would like to thank the Centre for High Performance Computing (CHPC) for the use of their infrastructure for this research. Furthermore, they would like to thank C.-K. Goh for sharing his source code of the dCOEA algorithm, and Kalyanmoy Deb for making the code of the static NSGA-II available on his website.

Appendix

Tables 8.18- 8.25 present the results that were obtained with the Mann-Whitney U tests that were performed on the performance metric values. In all tables below, D_d, D_{pe}, D_p, D_r, D_{re}, D_u, *N-A*, *N-B* and *C* refers to $DVEPSO_d$, $DVEPSO_{pe}$, $DVEPSO_p$, $DVEPSO_r$, $DVEPSO_{re}$, $DVEPSO_u$, *DNSGAII-A* and *DNSGAII-B* respectively. In all tables "-" indicates that there was no statistically significant difference and "x" indicates that according to the Mann-Whitney U test, there was a statistically significant difference between the specific performance metric values.

Table 8.18 Results of Mann-Whitney U test for VD metric for FDA1

τ_t	Algorithm	Algorithm									
		D_c	D_d	D_{pe}	D_p	D_r	D_{re}	D_u	N-A	N-B	C
10	D_c	n/a									
10	D_d	x	n/a								
10	D_{pe}	x	x	n/a							
10	D_p	x	x	–	n/a						
10	D_r	x	x	–	x	n/a					
10	D_{re}	x	x	–	x	–	n/a				
10	D_u	x	–	x	x	x	x	n/a			
10	N-A	x	x	x	x	x	x	x	n/a		
10	N-B	x	x	x	x	x	x	x	x	n/a	
10	C	x	x	x	x	x	x	x	x	x	n/a
25	D_c	n/a									
25	D_d	x	n/a								
25	D_{pe}	x	–	n/a							
25	D_p	x	–	–	n/a						
25	D_r	x	x	–	–	n/a					
25	D_{re}	x	–	–	–	x	n/a				
25	D_u	–	x	x	x	x	x	n/a			
25	N-A	x	x	x	x	x	x	x	n/a		
25	N-B	x	x	x	x	x	x	x	–	n/a	
25	C	x	x	x	x	x	x	–	–	–	n/a
50	D_c	n/a									
50	D_d	–	n/a								
50	D_{pe}	x	x	xn/a							
50	D_p	–	–	x	n/a						
50	D_r	–	x	–	x	n/a					
50	D_{re}	–	–	–	–	x	n/a				
50	D_u	–	–	x	–	x	–	n/a			
50	N-A	–	–	x	–	x	–	–	n/a		
50	N-B	–	–	x	x	x	x	–	x	n/a	
50	C	–	–	x	x	–	x	–	x	–	n/a

Table 8.19 Results of Mann-Whitney U test for VD metric for FDA2

τ_t	Algorithm	D_c	D_d	D_{pe}	D_p	D_r	D_{re}	D_u	N-A	N-B	C
10	D_c	n/a									
10	D_d	x	n/a								
10	D_{pe}	–	x	n/a							
10	D_p	–	–	x	n/a						
10	D_r	–	–	–	–	n/a					
10	D_{re}	–	–	x	–	–	n/a				
10	D_u	–	–	–	–	–	–	n/a			
10	N-A	x	x	x	x	x	x	x	n/a		
10	N-B	x	x	x	x	x	x	x	–	n/a	
10	C	x	x	x	x	x	x	x	–	–	n/a
25	D_c	n/a									
25	D_d	–	n/a								
25	D_{pe}	–	–	n/a							
25	D_p	–	–	–	n/a						
25	D_r	–	–	–	x	n/a					
25	D_{re}	–	–	–	x	–	n/a				
25	D_u	–	–	–	–	–	–	n/a			
25	N-A	x	x	x	x	x	x	x	n/a		
25	N-B	x	x	x	x	x	x	x	–	n/a	
25	C	x	x	x	x	x	x	x	–	–	n/a
50	D_c	n/a									
50	D_d	x	n/a								
50	D_{pe}	x	–	n/a							
50	D_p	x	x	x	n/a						
50	D_r	x	–	–	x	n/a					
50	D_{re}	–	–	–	x	–	n/a				
50	D_u	x	–	–	x	–	–	n/a			
50	N-A	x	x	x	x	x	x	x	n/a		
50	N-B	x	x	x	x	x	x	x	x	n/a	
50	C	x	x	x	x	x	x	x	–	–	n/a

Table 8.20 Results of Mann-Whitney U test for VD metric for FDA3

τ_t	Algorithm	Algorithm									
		D_c	**D_d**	**D_{pe}**	**D_p**	**D_r**	**D_{re}**	**D_u**	**N-A**	**N-B**	**C**
10	D_c	n/a									
10	D_d	–	n/a								
10	D_{pe}	–	–	n/a							
10	D_p	–	–	–	n/a						
10	D_r	–	–	–	–	n/a					
10	D_{re}	–	–	–	–	–	n/a				
10	D_u	–	–	–	–	–	–	n/a			
10	N-A	–	–	–	–	–	–	–	n/a		
10	N-B	x	x	x	x	x	x	x	x	n/a	
10	C	x	x	x	x	x	x	x	x	–	n/a
25	D_c	n/a									
25	D_d	–	n/a								
25	D_{pe}	–	–	n/a							
25	D_p	–	–	–	n/a						
25	D_r	–	–	–	–	n/a					
25	D_{re}	–	–	–	–	–	n/a				
25	D_u	–	–	–	–	–	–	n/a			
25	N-A	–	–	–	–	–	–	–	n/a		
25	N-B	x	x	x	x	x	x	x	–	n/a	
25	C	x	x	x	x	x	x	x	–	–	n/a
50	D_c	n/a									
50	D_d	x	n/a								
50	D_{pe}	x	–	n/a							
50	D_p	x	–	–	n/a						
50	D_r	x	–	x	–	n/a					
50	D_{re}	x	x	x	x	x	n/a				
50	D_u	x	x	x	x	x	–	n/a			
50	N-A	x	x	x	x	x	x	x	n/a		
50	N-B	x	x	x	x	x	x	x	–	n/a	
50	C	x	x	x	x	x	x	x	x	–	n/a

Table 8.21 Results of Mann-Whitney U test for VD metric for dMOP1

τ_t	Algorithm	Algorithm									
		D_c	D_d	D_{pe}	D_p	D_r	D_{re}	D_u	N-A	N-B	C
10	D_c	n/a									
10	D_d	x	n/a								
10	D_{pe}	x	–	n/a							
10	D_p	x	–	–	n/a						
10	D_r	x	–	–	x	n/a					
10	D_{re}	x	x	x	x	x	n/a				
10	D_u	x	x	x	x	x	–	n/a			
10	N-A	x	x	x	x	x	x	x	n/a		
10	N-B	x	x	x	x	x	x	x	x	n/a	
10	C	x	x	x	x	x	x	x	x	–	n/a
25	D_c	n/a									
25	D_d	x	n/a								
25	D_{pe}	–	–	n/a							
25	D_p	–	x	–	n/a						
25	D_r	x	–	x	x	n/a					
25	D_{re}	–	x	–	–	x	n/a				
25	D_u	–	–	–	–	x	–	n/a			
25	N-A	x	x	x	x	x	x	x	n/a		
25	N-B	x	x	x	x	x	x	x	x	n/a	
25	C	x	x	x	x	x	x	x	x	–	n/a
50	D_c	n/a									
50	D_d	–	n/a								
50	D_{pe}	x	x	n/a							
50	D_p	–	x	x	n/a						
50	D_r	–	–	x	–	n/a					
50	D_{re}	x	x	–	x	x	n/a				
50	D_u	x	x	x	x	x	x	n/a			
50	N-A	x	x	x	x	x	x	x	n/a		
50	N-B	x	x	x	x	x	x	x	x	n/a	
50	C	x	x	x	x	x	x	x	x	–	n/a

Table 8.22 Results of Mann-Whitney U test for VD metric for dMOP2

τ_t	Algorithm	Algorithm									
		D_c	D_d	D_{pe}	D_p	D_r	D_{re}	D_u	N-A	N-B	C
10	D_c	n/a									
10	D_d	x	n/a								
10	D_{pe}	x	x	n/a							
10	D_p	x	x	x	n/a						
10	D_r	x	x	x	x	n/a					
10	D_{re}	x	x	x	x	x	n/a				
10	D_u	x	x	x	x	x	x	n/a			
10	N-A	x	x	x	x	x	x	x	n/a		
10	N-B	x	x	x	x	x	x	x	x	n/a	
10	C	x	x	x	x	x	x	x	x	x	n/a
25	D_c	n/a									
25	D_d	–	n/a								
25	D_{pe}	–	x	n/a							
25	D_p	–	–	x	n/a						
25	D_r	–	–	x	–	n/a					
25	D_{re}	–	–	–	–	–	n/a				
25	D_u	x	–	–	–	–	–	n/a			
25	N-A	x	x	x	x	x	x	x	n/a		
25	N-B	x	x	x	x	x	x	x	x	n/a	
25	C	x	x	x	x	x	x	x	x	x	n/a
50	D_c	n/a									
50	D_d	–	n/a								
50	D_{pe}	–	–	n/a							
50	D_p	–	–	–	n/a						
50	D_r	–	–	–	–	n/a					
50	D_{re}	–	–	–	–		n/a				
50	D_u	–	–	–	–	–	–	n/a			
50	N-A	–	–	–	–	–	–	–	n/a		
50	N-B	–	–	–	–	–	–	–	x	n/a	
50	C	–	–	–	–	–	–	–	–	–	n/a

Table 8.23 Results of Mann-Whitney U test for VD metric for dMOP3

τ_t	Algorithm	Algorithm									
		D_c	D_d	D_{pe}	D_p	D_r	D_{re}	D_u	N-A	N-B	C
10	D_c	n/a									
10	D_d	–	n/a								
10	D_{pe}	–	–	n/a							
10	D_p	–	–	–	n/a						
10	D_r	–	–	–	–	n/a					
10	D_{re}	–	–	–	–	–	n/a				
10	D_u	–	–	–	–	–	–	n/a			
10	N-A	x	x	x	x	x	x	x	n/a		
10	N-B	x	x	x	x	x	x	x	x	n/a	
10	C	x	x	x	x	x	x	x	x	x	n/a
25	D_c	n/a									
25	D_d	–	n/a								
25	D_{pe}	–	–	n/a							
25	D_p	–	–	–	n/a						
25	D_r	–	–	–	–	n/a					
25	D_{re}	–	–	–	–	–	n/a				
25	D_u	–	–	–	–	–	–	n/a			
25	N-A	x	x	x	x	x	x	x	n/a		
25	N-B	x	x	x	x	x	x	x	–	n/a	
25	C	x	x	x	x	x	x	x	–	–	n/a
50	D_c	n/a									
50	D_d	–	n/a								
50	D_{pe}	–	–	n/a							
50	D_p	–	–	–	n/a						
50	D_r	–	–	–	–	n/a					
50	D_{re}	–	–	–	–	–	n/a				
50	D_u	–	–	–	–	–	–	n/a			
50	N-A	x	x	x	x	x	x	x	n/a		
50	N-B	x	x	x	x	x	x	x	–	n/a	
50	C	x	x	x	x	x	x	x	–	–	n/a

Table 8.24 Results of Mann-Whitney U test for VD metric for HE1

τ_t	Algorithm	Algorithm									
		D_c	D_d	D_{pe}	D_p	D_r	D_{re}	D_u	N-A	N-B	C
10	D_c	n/a									
10	D_d	x	n/a								
10	D_{pe}	x	x	n/a							
10	D_p	x	x	x	n/a						
10	D_r	x	x	x	x	n/a					
10	D_{re}	x	x	x	x	x	n/a				
10	D_u	x	x	x	x	x	x	n/a			
10	N-A	x	x	x	x	x	x	x	n/a		
10	N-B	x	x	x	x	x	x	x	x	n/a	
10	C	x	x	x	x	x	x	x	x	–	n/a
25	D_c	n/a									
25	D_d	x	n/a								
25	D_{pe}	x	x	n/a							
25	D_p	x	x	x	n/a						
25	D_r	x	x	x	x	n/a					
25	D_{re}	x	x	x	x	–	n/a				
25	D_u	x	x	–	–	x	x	n/a			
25	N-A	x	x	x	x	x	x	x	n/a		
25	N-B	x	x	x	x	x	x	x	x	n/a	
25	C	x	x	x	x	x	x	x	x	–	n/a
50	D_c	n/a									
50	D_d	x	n/a								
50	D_{pe}	x	–	n/a							
50	D_p	x	x	x	n/a						
50	D_r	x	–	–	x	n/a					
50	D_{re}	x	–	–	x	–	n/a				
50	D_u	x	–	–	x	–	–	n/a			
50	N-A	x	x	x	x	x	x	x	n/a		
50	N-B	x	x	x	x	x	x	x	x	n/a	
50	C	x	x	x	x	x	x	x	–	–	n/a

Table 8.25 Results of Mann-Whitney U test for VD metric for HE2

τ_t	Algorithm	D_c	D_d	D_{pe}	D_p	D_r	D_{re}	D_u	N-A	N-B	C
10	D_c	n/a									
10	D_d	x	n/a								
10	D_{pe}	–	x	n/a							
10	D_p	x	–	x	n/a						
10	D_r	–	x	–	x	n/a					
10	D_{re}	x	–	x	–	x	n/a				
10	D_u	–	x	–	x	–	x	n/a			
10	N-A	x	x	x	x	x	x	x	n/a		
10	N-B	x	x	x	x	x	x	x	x	n/a	
10	C	x	x	x	x	x	x	x	x	–	n/a
25	D_c	n/a									
25	D_d	–	n/a								
25	D_{pe}	–	–	n/a							
25	D_p	–	–	–	n/a						
25	D_r	–	–	–	–	n/a					
25	D_{re}	–	–	–	–	–	n/a				
25	D_u	–	–	–	–	–	–	n/a			
25	N-A	x	x	x	x	x	x	x	n/a		
25	N-B	x	x	x	x	x	x	x	–	n/a	
25	C	x	x	x	x	x	x	x	–	–	n/a
50	D_c	n/a									
50	D_d	x	n/a								
50	D_{pe}	x	–	n/a							
50	D_p	–	x	–	n/a						
50	D_r	–	x	x	–	n/a					
50	D_{re}	x	–	–	–	–	n/a				
50	D_u	x	–	–	–	x	–	n/a			
50	N-A	x	x	x	x	x	x	x	n/a		
50	N-B	x	x	x	x	x	x	x	x	n/a	
50	C	x	x	x	x	x	x	x	x	–	n/a

References

[1] Cámara, M., Ortega, J., Toro, J.: Parallel Processing for Multi-objective Optimization in Dynamic Environments. In: Proc. of IEEE International Parallel and Distributed Processing Symposium, p. 243 (2007)

[2] Carlisle, A., Dozier, G.: Adapting Particle Swarm Optimization to Dynamic Environments. In: Proc. of International Conference on Artificial Intelligence (ICAI 2000), pp. 429–434 (2000)

[3] CHPC. Sun hybrid system, `http://www.chpc.ac.za/sun` (last accessed online on March 15, 2011)

[4] Chu, W., Gao, X., Sorooshian, S.: Handling boundary constraints for particle swarm optimization in high-dimensional search space. Information Sciences (2010) (in press)

[5] Deb, K., Thiele, L., Laumanns, M., Zitzler, E.: Scalable multi-objective optimization test problems. In: Proc. of Congress on Evolutionary Computation (CEC 2002), vol. 1, pp. 825–830 (2002)

[6] Deb, K., Udaya Bhaskara Rao, N., Karthik, S.: Dynamic Multi-objective Optimization and Decision-Making Using Modified NSGA-II: A Case Study on Hydro-thermal Power Scheduling. In: Obayashi, S., Deb, K., Poloni, C., Hiroyasu, T., Murata, T. (eds.) EMO 2007. LNCS, vol. 4403, pp. 803–817. Springer, Heidelberg (2007)

[7] Farina, M., Deb, K., Amato, P.: Dynamic multiobjective optimization problems: test cases, approximations, and applications. IEEE Transactions on Evolutionary Computation 8(5), 425–442 (2004)

[8] Goh, C.-K., Tan, K.C.: A competitive-cooperative coevolutionary paradigm for dynamic multiobjective optimization. IEEE Transactions on Evolutionary Computation 13(1), 103–127 (2009)

[9] Goh, C.K., Tan, K.C.: An Investigation on Noisy Environments in Evolutionary Multiobjective Optimization. IEEE Transactions on Evolutionary Computation 11(3), 354–381 (2007)

[10] Greeff, M., Engelbrecht, A.P.: Solving dynamic multi-objective problems with vector evaluated particle swarm optimisation. In: Proc. of IEEE World Congress on Evolutionary Computation: IEEE Congress on Evolutionary Computation, Hong Kong, pp. 2917–2924 (June 2008)

[11] Guan, S.-U., Chen, Q., Mo, W.: Evolving Dynamic Multi-Objective Optimization Problems with Objective Replacement. Artificial Intelligence Review 23(3), 267–293 (2005)

[12] Helbig, M., Engelbrecht, A.P.: Archive management for dynamic multi-objective optimisation problems using vector evaluated particle swarm optimisation. Submitted for Review

[13] Helwig, S., Wanka, R.: Particle swarm optimization in high-dimensional bounded search spaces. In: Proc. of IEEE Swarm Intelligence Symposium, Honululu (HI), pp. 198–205 (2007)

[14] Jin, Y., Sendhoff, B.: Constructing Dynamic Optimization Test Problems Using the Multi-objective Optimization Concept. In: Raidl, G.R., Cagnoni, S., Branke, J., Corne, D.W., Drechsler, R., Jin, Y., Johnson, C.G., Machado, P., Marchiori, E., Rothlauf, F., Smith, G.D., Squillero, G. (eds.) EvoWorkshops 2004. LNCS, vol. 3005, pp. 525–536. Springer, Heidelberg (2004)

[15] Kennedy, J., Eberhart, R.C.: Particle Swarm Optimization. In: Proc. of IEEE International Conference on Neural Networks, vol. IV, pp. 1942–1948 (1995)

[16] Deb, K.: Kanpur Genetic Algorithms Laboratory (2011), `http://www.iitk.ac.in/kangal/codes.shtml` (last accessed online on March 6, 2011)
[17] Li, X., Branke, J., Blackwell, T.: Particle Swarm with Speciation and Adaptation in a Dynamic Environment. In: Proc. of 8th Conference on Genetic and Evolutionary Computation (GECCO 2006), pp. 51–58 (2006)
[18] Li, X., Branke, J., Kirley, M.: On Performance Metrics and Particle Swarm Methods for Dynamic Multiobjective Optimization Problems. In: Proc. of Congress of Evolutionary Computation (CEC 2007), pp. 1635–1643 (2007)
[19] Mehnen, J., Wagner, T., Rudolph, G.: Evolutionary Optimization of Dynamic Muli-Objective Test Functions. In: Proc. of 2nd Italian Workshop on Evolutionary Computation and 3rd Italian Workshop on Artificial Life (2006)
[20] Pampara, G., Engelbrecht, A.P., Cloete, T.: Cilib: A collaborative framework for computational intelligence algorithms - part i. In: Proc. of IEEE World Congress on Computational Intelligence (WCCI), Hong Kong, June 1-8, pp. 1750–1757 (2011), Source code available at, `http://www.cilib.net` (last accessed on March 6, 2011)
[21] Parsopoulos, K.E., Tasoulis, D.K., Vrahatis, M.N.: Multiobjective Optimization using Parallel Vector Evaluated Particle Swarm Optimization. In: Proc. of IASTED International Conference on Artificial Intelligence and Applications, Innsbruck Austria (2004)
[22] Parsopoulos, K.E., Vrahatis, M.N.: Recent Approaches to Global Optimization Problems through Particle Swarm Optimization. Natural Computing 1(2-3), 235–306 (2002)
[23] Bergh, F.V.D.: An analysis of particle swarm optimizers. PhD thesis, Department of Computer Science, University of Pretoria (2002)
[24] Zhang, W.-J., Xie, X.-F., Bi, D.-C.: Handling boundary constraints for numerical optimization by particle swarm flying in periodic search space. In: IEEE Congress on Evolutionary Computation, vol. 2, pp. 2307–2311 (June 2004)
[25] Zheng, B.: A New Dynamic Multi-Objective Optimization Evolutionary Algorithm. In: Proc. of third International Conference on Natural Computation (ICNC 2007), vol. V, pp. 565–570 (2007)
[26] Zitzler, E., Deb, K., Thiele, L.: Comparison of Multiobjective Evolutionary Algorithms: Emperical Results. Evolutionary Computation 8(2), 173–195 (2000)

Chapter 9
Ant Colony Based Algorithms for Dynamic Optimization Problems

Guillermo Leguizamón and Enrique Alba

Abstract. The use of metaheuristic approaches to deal with dynamic optimization problems has been largely studied, being evolutionary techniques the more widely used and assessed techniques. Nevertheless, successful applications coming from other nature-inspired metaheuristics, e.g., ant algorithms, have also shown their applicability in dynamic optimization problems, but received a limited attention until now. Different from perturbative techniques, ant algorithms use a set of agents which evolve in an environment to construct one solution. They cooperate by means of asynchronous communications based on numerical information laid on an environment. This environment is often modeled by a graph which constitutes a formalism with a great expressiveness, specially well-suited for dynamic optimization problems. A solution could be a structure like a subgraph, a route, a short path, a spanning tree, or even a partition of vertices. In this chapter we present a general overview of the more relevant works regarding the application of ant colony based algorithms for dynamic optimization problems. We will also highlight the mechanisms used in different implementations found in the literature, and thus show the potential of this kind of algorithms for research in this area.

9.1 Introduction

Metaheuristic techniques have widely proved to be suitable approaches for dynamic environments. In this regard, it should be noticed that Evolutionary Algorithms (AEs) are without any doubt the pioneer and more widely used metaheuristic [30].

Guillermo Leguizamón
Universidad Nacional de San Luis,
Av. Ejército de Los Andes 950 (5700), San Luis, Argentina
e-mail: legui@unsl.edu.ar

Enrique Alba
Universidad de Málaga, Campus de Teatinos (3.2.12)
Málaga - 29071, Spain
e-mail: eat@lcc.uma.es

E. Alba et al. (Eds.): Metaheuristics for Dynamic Optimization, SCI 433, pp. 189–210.
springerlink.com

However, Ant Colony Optimization (ACO) metaheuristic has also shown to be an efficient candidate to deal with this kind of problem [35]. In a broad sense, ACO [17] refers to a class of algorithms whose design is mainly based on the foraging behavior of real ants, being the more representative ACO algorithms those designed for solving a certain type of combinatorial optimization problem: those problems for which a solution is obtained by simulating a walk through a construction graph. One of the more studied problems under the above (original) vision of ACO algorithms is the Traveling Salesperson Problem (TSP), a well-known and classical NP-complete problem that owns important features that can be easily exploited to show the applicability of this metaheuristic. Several ACO algorithms were designed since the publication of the first ACO, the so-called Ant System (AS) by Dorigo et al. [16]. These algorithms include (see Figure 9.1 where each algorithm name includes the main keyword that respectively defines it) [17]: elitist-AS (an AS with an elitist strategy for updating the pheromone trail levels), AS rank (a rank-based version of Ant System), *MAX-MIN* Ant System (an AS that incorporates a mechanism to control the pheromone levels), the Ant Colony System (ACS) (a more advanced ACO algorithm with a modified transition rule, with local and global pheromone update), the Best-Worst Ant System (an extended AS characterized by integrating some components taken from evolutionary computation), ANTS (an AS that uses lower bounds on the completion of a partial solution to derive the heuristic desirability), and ANT-Q (a version of an AS that combines concepts of the reinforcement learning theory). Many of them were initially applied to TSP and also to the Quadratic Assignment Problem (QAP) [34]. After that, many other variants of these algorithms have been proposed in the literature. However, not all of them strictly follow the standard design principles given for the ACO metaheuristic in [17]. Instead, these algorithms follow in many different ways the metaphor of the ants behavior (foraging or some other) as a general framework which let the researcher broaden the field of application of this nature-inspired approach. In a more general sense, algorithms that follow in some way the above mentioned metaphor could be called Ant Colony Based algorithms (ACB).

Ant colony based algorithms have proven to be successfully applied to many different real world and academic problems that include combinatorial optimization problems, continuous domains, and also dynamic optimization problems (DOPs), the kind of problems and applications considered in the current chapter.

The rest of the chapter is organized as follows. The next section describes the ACO metaheuristic as the more representative ant colony based algorithm. Section 9.3 gives a general description of the type of dynamic problems usually found in the literature, as well as in real-world applications. Section 9.4 presents the mechanisms used in ACB algorithms to deal with dynamic problems, and a survey of the works in the literature. The last section contains some conclusions and outlines some future challenges.

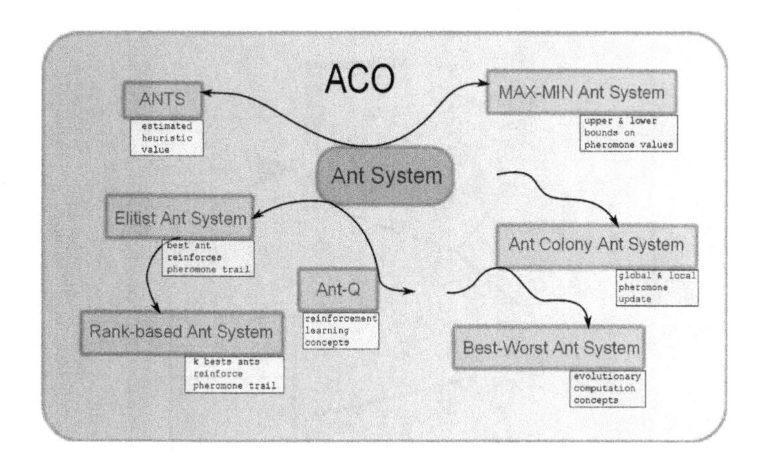

Fig. 9.1 The most representative and widely used algorithms based on the ACO metaheuristic.

9.2 Ant Colony Optimization

As mentioned earlier in this chapter, ant colony based algorithms involve a broad class of optimization algorithms designed under the metaphor of real ants behavior. More precisely, ant colonies are social insect societies that can be considered distributed systems composed of simple interacting individuals. From the interaction between those individuals, a complex and highly structured social organization arises. Different types of ant colonies self-organize according to a particular behavior, e.g., foraging, division of labor, brood sorting, and cooperative transport. In the case of foraging ants, as well for the remaining behaviors, the activities are coordinated through indirect communication known as stigmergy [23]. A foraging ant deposits a chemical substance (pheromone trail) on the ground to communicate to other ants the desirability of following a particular path. At the same time, as more intense is the pheromone trail sensed on the ground, more amount of the chemical substance is deposited by a particular ant. From this autocatalytic or positive feedback process emerges a self-organized system in which the shortest paths connecting the nest and the food source remain candidates to follow by the whole colony. In addition to the above, it is worth noticing that an evaporation process occurs in the environment which helps the colony to keep the exploration capabilities during the search of alternative paths to the food source.

The foraging ants behavior just described as well as other types of behavior are taken as useful metaphors to design optimization and search algorithms under different names. In this work we adopted the definition of the Ant Colony Optimization (ACO) metaheuristic, as given by Dorigo and Stützle [17], which is the classical and more widely used formulation for this type of algorithms. Nevertheless, other ant colony based algorithms that not strictly follow the definition of the ACO

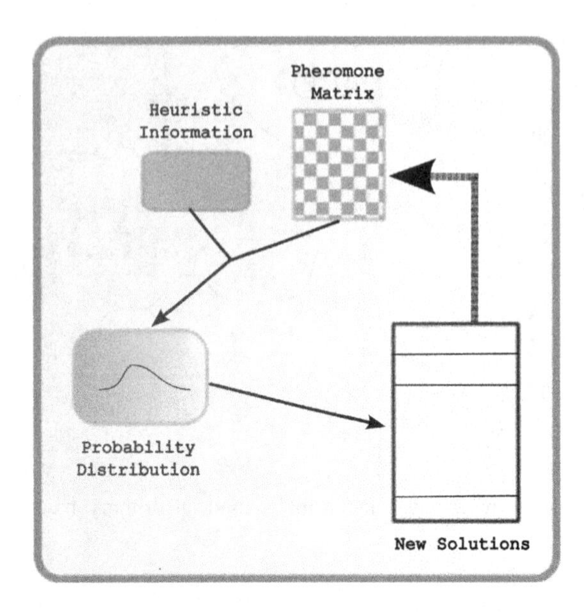

Fig. 9.2 A general overview of the behavior of an ACO algorithm: pheromone trail plus heuristic information are used to find the probabilistic distribution to generate new solutions. These new solutions are then used to update pheromone trails to bias the search for the next iterations.

metaheuristic will be also described in further sections. It is well-known that ACO is one of the most representative metaheuristics derived from the broad concept known as swarm intelligence, where the behavior of social insects is the main source of inspiration. As a typical swarm intelligence approach, the ACO metaheuristic is mainly characterized by its distributiveness, flexibility, capacity of interaction among simple agents, and its robustness.

ACO algorithms (in the standard version) generate solutions for an optimization problem by a construction mechanism where the selection of the solution component to be added at each step is probabilistically influenced by pheromone trails and (in most of the cases) heuristic information. Thus, the solution construction process is mainly influenced by pheromone trails and heuristic information[1] from which a probabilistic model is evolved to guide to exploration of the search space.

This means that the construction process probabilistically builds step by step the problem solutions and the probabilistic model has a feedback for its modification based on the solutions found. Figure 9.2 displays a general overview of an ACO algorithm. Heuristic information plus pheromone values are used to find a probability distribution over the search space. Initial pheromone values are in general set to a constant value, thus, at the first iterations the algorithm is highly explorative.

[1] The use of heuristic information is not mandatory, if used, this information is also taken into account to build a probabilistic model.

The probability distribution is then used to sample new solutions to the problem at hand. According to some criterion, all or a subset of these new solutions are involved in the pheromone update process. As the cycle displayed in Figure 9.2 is repeated, the pheromone values will bias the search to specific regions of the search space as it directly influences the probability distribution.

Although the application domains of ACO algorithms are certainly diverse, the more and well-known field is that related to combinatorial optimization problems like TSP [15, 16], QAP [21, 44], and Vehicle Routing Problem (VRP) [12, 41]. Thus, we present in the following some important considerations when applying this kind of algorithm to discrete problems as the mentioned above.

First of all, it is necessary to define an appropriate problem representation. In the jargon of ACO metaheuristic, this means to properly define:

(i) the construction graph and the way this represents the different problem components and connections among them,
(ii) the definition (if any) of the problem information to be exploited,
(iii) the behavior of the artificial ants, in the sense of how each ant will walk through the construction graph to build the corresponding solutions.

Algorithm 9.1. ACO algorithm

```
1: Init();
2: while not (termination-condition) do
3:     Build-Sols-Step-by-Step();
4:     Pheromone-Update();
5:     Daemon-Actions(); // Optional step
6: end while
```

A general design of an ACO algorithm (as showed in Algorithm 9.1) includes a set of four main activities (or steps) that define this iterative search technique. It must be noticed that variations of the way these activities are implemented define the kind of ACO Algorithm. For example, a variation on the activity Pheromone-Update() will mainly define Algorithm 9.1 as an Ant System (AS), elitist-AS, AS-rank, *MAX-MIN* Ant System, or an ACS (in this last case, Build-Sols-Step-by-Step() activity also involves a local pheromone update step). Also, any other ant colony based algorithm not necessarily fitting exactly in the canonical definition of those algorithms could be included in the family of ACO algorithms. Nevertheless, the activities in Algorithm 9.1 can be described in a general way in the following. To do that we assume an Ant System applied to TSP, thus the problem components are cities and connections (routes) between them. The connections have an associated value, e.g., distance between cities or cost to travel from one city to another:

- **Init():** As in any typical population-based algorithm, some basic tasks need to be done before starting the exploration of the search space. In this case, the initialization of pheromone trail matrix which at time 0 is:

$$\tau_{ij}(0) = \tau_0, \text{ for } i,j \in \{1,\ldots,n\}, \tag{9.1}$$

where n represents the problem size and τ_0 is a small constant value (e.g., the following value is suggested for TSP in [17]: $\tau_0 = m/C^{nn}$, where m is the number of ants, and C^{nn} is the length of a tour generated by the nearest-neighbor heuristic). Also the heuristic values (represented by η symbol) are calculated (when available and used) and any other structure, necessary to complete the problem representation, is initialized.

- **Build-Sols-Step-by-Step():** This activity involves the release of an independent colony of artificial ants in charge of incrementally building a solution to the problem. Each ant, at each step of the construction process, makes a local stochastic decision about the next component to be included in the solution under construction. For example, for an Ant System applied to TSP, the decision of adding city j (problem component) to the solution under construction when city i was the last visited is given by:

$$p_{ij}^k = \begin{cases} \frac{\tau_{ij}(t)^\alpha \eta_{ij}^\beta}{\Sigma_{h \in \mathcal{N}^k(i)} \tau_{ih}(t)^\alpha \eta_{ih}^\beta} & \text{if } j \in \mathcal{N}^k(i) \\ 0 & \text{otherwise,} \end{cases} \tag{9.2}$$

where α and β are the parameters that, respectively represent the importance of the pheromone trail ($\tau_{ij}(t)$) and the heuristic information (η_{ij}), and $\mathcal{N}^k(i)$ represents the set of cities that can be visited by ant k, i.e., the feasible or unvisited cities.

- **Pheromone-Update():** The acquired experience achieved at each iteration by the colony is considered in this activity. High quality solutions will positively affect the amount of pheromone trail, i.e., those edges that are part of solutions found will receive an increased amount of pheromone trail according to the goodness of these solutions. This is known as the global pheromone update.

  As in Nature [22], a process of pheromone evaporation takes place (usual implementations of this metaheuristic decrease the amount of pheromone trail for all edges in the construction graph) [9]. Thus, the amount of pheromone corresponding to those edges that are not part of any solution at the current iteration will show a gradually diminishing pheromone intensity. It should be noticed that some ACO algorithms, such as ACS [15], apply a local pheromone update rule which does not depend on the solution quality. Instead, a fixed amount is deposited as soon as an edge in the construction graph is selected to make the move (the next component added to the solution under construction). The following equation is a possible way of (global) updating pheromone values:

$$\tau_{ij}(t+1) = (1-\rho) \cdot \tau_{ij}(t) + \Delta\tau_{ij}, \tag{9.3}$$

  where ρ is the evaporation rate and $\Delta\tau_{ij}$ represents the amount of pheromone trail deposited in edge (i,j). That amount is calculated according to the quality of the solutions, found by the whole colony, that include edge (i,j).

- **Daemon-Actions():** As single ants cannot carry out centralized actions, many ACO algorithms include some specific activities called *daemon actions*. Examples of these activities are: activation of a local search procedure or a collection of global information (e.g., use of a set of the best ranked solutions) that could be employed to modify some entries in the pheromone trail matrix.

Based on the above description, several types of ACO algorithms can be obtained [17]. In addition, many other algorithms that not necessarily follow the standard design can be considered either for discrete problems [9], continuous domains [43], and dynamic optimization problems [28] as we will show in the next sections.

9.3 Dynamic Optimization Problems (DOPs)

Dynamic optimization problems (DOPs) involve any problem definition for which at least one of its components varies with time. Thus, we can find problems where the objective function changes over time or some problem constraints depend on environmental conditions. These situations include many real-world tasks for which changes in the environment affect the applied optimization process, as this has to react to the new environmental conditions.

In this section we present the more relevant concepts involved in DOPs that are usually considered when applying metaheuristic techniques[2] for solving them. A classical reference to the use of Evolutionary Algorithms (EAs) for dynamic problems is given in Branke [6] in which a widely referenced classification of dynamic problems is presented, and the fundamentals of possible mechanisms to deal with are also analyzed.

In order to obtain a fairly self-contained chapter we succinctly describe some important concepts regarding the mechanism to deal with certain dynamic optimization problems. The interested reader can find a good source of information in this regard in Branke [6, 30] and Morrison [37]. In addition, a complementary source of information can be found in Leguizamón et al. [33] where alternative metaheuristic techniques to deal with dynamic optimization problems under a unified perspective are described.

Any change in a dynamic problem can be seen as the activation of a new optimization problem (replacing the previous one) for which a new solution must be provided as the quality of the current solution could be no longer acceptable for the new environment. Therefore, the adaptation of the current solution to the new problem will be the main objective when a change occurs. The most primitive mechanism to deal with dynamic problems is restarting from scratch, i.e., the optimization algorithm does not consider any previous experience or information that could be helpful under the new conditions. However, this approach is impractical for many reasons and alternative mechanisms should be taken into account that consider in different ways the past experience on the search. This is particularly desirable when

[2] We are assuming through this chapter population-based metaheuristic techniques.

the solution for the new problem should be not too different from the solution previously found for the (probably related) old problem. Also, the past experience is a valuable element, as many dynamic problems are defined as a sequence of static instances of a basic problem with slight variations from one instance to another, resulting in a sequence that defines the complete dynamic problem. In that regard, it is worth noticing that the quantity and quality of the past experience considered for an optimization algorithm will determine the capacity of such algorithm to adapt to a possibly continuously changing environment. In addition, the control of the population diversity is a key factor, as the adaptation to the changes could be harder if the algorithm rapidly losses diversity.

The use of explicit or implicit memory to remember past experience is the typical approach implemented in metaheuristic techniques to guide the exploration of the search space. However, some metaheuristics implement by definition some sort of memory of the past experience as a mechanism to bias the search during the incoming iterations; also this "natural" memory can be used to deal with dynamic problems. This is the case of ant colony based algorithms, as will be described in the next section.

Many types of dynamic features can be found in real-world problems. From earlier application of evolutionary algorithms to dynamic problems (see for example, Abdunnaser [1], Bianchi [7], Branke [6], and Psaraftis [31]) the following elements and features are commonly considered when dealing with DOPs:

- the problem can change with time in such a way that future scenarios are not completely known, yet the problem is completely known up to the current moment;
- a solution that is optimal or near optimal at a certain time may reduce its quality in the future, or may even become infeasible;
- the goal of the optimization algorithm is to track the shifting optima through time as closely as possible;
- solutions cannot be determined ahead of time but should be found in response to the incoming information; and
- solving the problem entails setting up a strategy that specifies how the algorithm should react to environmental changes, e.g., to solve the problem from scratch or adapt some parameters of the algorithm at every change.

Besides the described features, DOPs can be classified in different ways depending on the sources of dynamism and its effects on the objective function. Simple questions will help us to determine the nature of the change: i) *what?* (i.e., aspects of change), ii) *when?* (i.e., frequency of change), and iii) *how?* (i.e., severity of the change, effect of the algorithm over the scenario, and presence of patterns).

Different descriptions for dynamic problems are given in the literature [6, 37] and some of them were proposed having in mind certain type of metaheuristic algorithm that could be used for solving them [8, 11, 28]. Nevertheless, before applying any algorithm to solve a particular dynamic problem, the dynamic nature of the problem must be taken into account as well as the capability of the chosen algorithm to react to the changes. Next section discusses different ant colony based algorithms and the way they were applied to dynamic problems.

9.4 Solving DOPs with ACB algorithms

Different metaheuristic approaches have been applied to dynamic optimization problems, being EAs the more widely used search technique to deal with this kind of problems [6, 8, 28]. However, other well-known metaheuristics have also been increasingly and successfully applied to DOPs. Some of them include Nature inspired metaheuristics like the ACO and Particle Swarm Optimization (PSO). For example, a recent short survey by Hendtlass et al. [28] examines some representative works and methodologies to deal with DOPs by using Ant Colony Optimization, Particle Swarm Optimization, and Extremal Optimization. Besides describing the applications based in these Nature inspired metaheuristics, the authors also analyzed some limitations of the presented algorithms. Other short review in this regard (see the technical report from Angus [2]) focuses on the ACO metaheuristic and describes some relevant and related works. Besides the previous surveys on this topic, we give here a unified and broad perspective of the different ant colony based algorithms to deal with DOPs.

There are two concepts closely related in any algorithm dealing with DOPs: 1) the mechanisms implemented to avoid stagnation and hence 2) the capacity of the algorithm *to react to the changes*. Particularly in the ant colony based algorithms the pheromone structure τ presented in Section 9.2 is the key algorithm component that should mainly be taken into account, as this represents, on one hand, the memory of the algorithm. On the other hand, the pheromone structure is built in the solution search space by a graph and that dynamics modify this graph valuated by variable amounts of pheromone. Thus, in any ant based algorithm an appropriate strategy must be defined to let the algorithm adapt to the changes by modifying the pheromone values, e.g., by resetting all or part of the pheromone values to reduce, in some extent, the learnt experience. This is a classical strategy for increasing an explorative behavior as the new scenario has been detected.

The seminal works of the ACO metaheuristic to deal with dynamic optimization problems (see [24, 26, 27, 36]) were mostly devoted to the TSP, QAP, and VRP. All the dynamic versions of these problems have the following characteritic: the changes are produced by adding/eliminating problem components, i.e., cities in TSP, locations in QAP, and orders in VRP. However, other alternatives are also possible like changing the problem data, e.g., distance between cities in TSP, cost of assignment in QAP, or cost of delivery between the depot and customers in VRP.

In the following we present in three sections a short review of literature and a global description of the respective mechanism used to deal with certain types of DOPs in the past. The sections are divided according to the following criteria: Section 9.4.1 presents the so-called standard versions of ACO algorithms, Section 9.4.2

describes the class of population-based ACO (P-ACO) algorithms, and finally in Section 9.4.3 some general algorithms based in the metaphor of ants behavior are presented.

9.4.1 *Standard ACO Algorithms*

One of the seminal works regarding this type of algorithm is the proposal by Guntsch et al. [27]. Authors there investigate several strategies for pheromone updating in reaction to changes on the problem instance. In this case, the problem instance is changed by adding/eliminating cities.

Regarding dynamic TSP, Angus and Hendtlass [3, 4] study another strategy to modify the pheromone matrix when a change occurs. In this case, the strategy consists in normalizing the pheromone values in a way that the past memory is maintained but avoiding extreme values.

Another proposal to solve dynamic TSP is presented by Eyckelhof and Snoek [18]. They study two different ways of modifying pheromone trails: local and global. In the dynamic TSP studied, the distance between cities is seen as the time to travel from one city to another one. Thus, the changes are obtained by modifying the traveling times and hence, traffic jams could be produced in the paths as the ants are walking while solving the problem. The strategy proposed logarithmically smooths the pheromone values (called *shaking* process) maintaining the relative ordering among them before and after the modification. The global shaking produces a modification on all over the edges while the local one only changes the pheromone values around a certain distance where the traffic jam was produced. The strategy also avoids to assign pheromone values below a certain lower bound.

Montemanni et al. [36] investigate the Dynamic Vehicle Routing Problem (DVRP) through an Ant Colony System based algorithm. In the studied version of the DVRP new orders can arrive at any time and they have to be dynamically incorporated in the constantly evolving schedule. In the particular case of [36] new orders can be assigned after the vehicle left the depot. The mechanism for pheromone updating is inspired by the work by Guntsch and Middendorf [24, 25]. The strategy adopted evaporates the old pheromone values and at the same time increases the amount of pheromone values by a constant amount.

The Binary Ant Algorithm (BAA) proposed by Fernandes et al. [20] is designed by using a particular construction graph to work on binary dynamic environments. The main characteristic of BAA is that it stresses the role of the negative feedback (i.e., give more relevance to the evaporation processes). BAA was tested on two dynamic continuous functions: Oscillatory Royal Road and Dynamic Schaffer.

In summary, several strategies have been proposed in the literature to deal with DOPs by applying the classical ACO algorithms (i.e., those which more closely follow the principle of using a construction graph as earlier defined in [17]):

- Global pheromone modification:
  - Increase the values proportionally to their difference to the maximum pheromone value.
  - The new pheromone values are a combination of the old values and an increment of a constant small pheromone value. Those values are regulated by a parameter $0 \leq \gamma_r \leq 1$ that controls the relative importance of both values:

$$\tau_{i,j}^{new} = (1 - \gamma_r) \cdot \tau_{i,j}^{old} + \gamma_r \cdot \tau_0, \tag{9.4}$$

  where τ_0 is a small constant value (usually used in the pheromone matrix initialization process); and $\tau_{i,j}^{new}$ and $\tau_{i,j}^{old}$ represent, respectively the pheromone value on edge (i, j) before and after the change in the environment.

- Local pheromone modification:
  - η-strategy (based on the heuristic information) and τ-strategy (based on pheromone information). They both refer to connection problems, thus, problem components can be inserted/deleted. Consequently, only the edges connecting the problem components must be added/eliminated from the construction graph will influence (locally) the pheromone values.
  - Combined strategies based on η-strategy and τ-strategy.
  - Normalize the pheromone values regarding the maximum pheromone value all over the current edges, i.e., $\tau_{ij}(t+1) = \tau_{ij}(t)/\tau_{i,max}(t)$. The term "current edges" is used here as the edges directly connecting component i with some other component and $\tau_{i,max}(t)$ indicates the maximum pheromone value in the neighborhood of component i. Thus, the normalization is local with respect to each component i of the problem instance.

- Local and global pheromone modification:
  - Using a mechanism to limit the lower pheromone values (similar to the *MAX-MIN* Ant System) and smoothing the pheromone values keeping the relative order before and after the change (this promote exploration without losing information of the past experience). The modification can be applied locally or globally (depending on an *ad-hoc* parameter).

- Other:
  - Keeping elitist ant: in this case the best-so-far ant is modified after a change has occurred (e.g., adding/eliminating solution components). Thus, the eliminated components are deleted from the elitist solution, whereas the new added components (if any) are located in the places left in the solution. The new components could be added, for example, by using some heuristic procedure. This is a possible way of remembering the past experience that will influence the pheromone values when the usual step of pheromone updating takes place.
  - Choosing only the ants with a quality value equal to or below[3] the population's average fitness to generate new solutions. This strategy is combined with a quick evaporation of pheromone trails.

[3] For a maximization problem the value is above or equal to the average.

Please notice that the above strategies could be adapted or enhanced in many ways to deal with unseen dynamic problems. In this regards, the approach to use and adapt the pheromone matrix is the more important issue in order to achieve an efficient mechanism to react to the changes.

9.4.2 Population Based ACO Algorithms

Population based ACO algorithms (P-ACO) were first proposed by Guntsch and Middendorf [26] as an alternative ACO in which a set of solutions is transferred between the current iteration and the next one instead of the pheromone trail values (or pheromone matrix). Thus the set of transferred solutions is used to calculate the new pheromone values used later for building the new set of solutions. Although not applied to dynamic problems, FIFO-Queue ACO ([26]) could be considered as a preliminary work of P-ACO in this direction as the authors claimed about its applicability to create new metaheuristic algorithms, as well as to solve dynamic problems, since P-ACO could rapidly adapt the pheromone values as the environment changes in comparison with standard ACO algorithms. In Guntsch and Middendorf [25] FIFO-Queue was studied on dynamic versions of QAP and TSP, where different strategies for updating the population were investigated. More recently, Ho and Ewe [29] proposed a P-ACO algorithm where three different mechanisms to adapt the pheromone values are investigated to solve the dynamic load-balanced clustering problem in ad hoc networks. Although the proposed mechanisms are novel, they only represent slight variations of the original P-ACO [26]. For that reason, we describe first the main features of P-ACO, as proposed by Guntsch and Middendorf [26] and then we give some highlights of the variations of P-ACO according to Ho and Ewe [29].

It is important to note that P-ACO algorithms also adopt the principle of the construction graph as in the standard ACO described in the previous section. However, we have decided to describe them in a section apart. The reason for that is because they differ from the standard ACO algorithms. On one side, ACO algorithms include problem-based strategies in the pheromone updating process. On the other side, P-ACO algorithms use the solutions themselves to define the strategies to calculate the new pheromone values. Another interesting feature of P-ACO is that this algorithm does not use any problem information to handle the changes. This makes it a convenient approach for different types of dynamic as the basic information needed is that provided by the fitness function.

The following strategies are the alternatives implemented in [26] to manipulate the set of solutions maintained (i.e., the population) in order to influence in different ways the pheromone values:

1. *Age strategy:* Add the best solution found in the current iteration and remove the oldest one.
2. *Quality strategy:* Add the best solution found in the current iteration and remove the worst one.

3. *Prob strategy:* Add the best solution found in the current iteration and remove a solution randomly chosen. To do that a distribution probability is created in a way that bad solutions have more chances to be removed; however, all solutions are candidate to be eliminated.

Combinations of two of the above three strategies also exist, as used and tested in [26]. It must be noticed that the above strategies were designed for controlled changes where the number of cities (problem components for the general case) that are added is the same as the number of deleted ones, i.e., the problem size is kept constant.

Under the population-based ACO the *KeeepElite* strategy, first defined in [27], can also be applied to repair the solutions in the population regarding the changes that took place in the environments, i.e., deleted problem components are eliminated from the elite solution whereas the new inserted ones are added based on some criterion.

Instances considered in [26] include two additional parameters regarding the dynamic features:

- c, which indicates the severity of the change, i.e., the number of components deleted (respectively inserted) from (to) the problem instance.
- t, the time window that controls the occurrence of the change.

No overall definitive conclusion about the use of the different strategies in P-ACO algorithms can be achieved according to the results presented in [26]. However, strategy *Prob* was the worst performer for all the scenarios considered in the experimental study. Also, a small number of solutions in the population (size of the population $k = 3$) was enough to perform fairly well on all the problems (TSP and QAP) and considered instances.

Three variations of P-ACO (see Ho and Ewe [29]) were studied when solving the dynamic load-balanced clustering problem in ad hoc networks. The changes here operate on the problem structure. The variations presented are intended to adapt the pheromone values as closely as possible to reflect the new problem structure after a change has occurred by:

i) Applying a repairing process (based on the new problem structure) to the solutions in the population.
ii) Adapting the parameters that control the importance of the pheromone and heuristic values: parameters α and β as used in standard ACO algorithms (see Equation 9.2). Thus, more importance is given to the heuristic values (they are recalculated when the problem changes) and less importance to the pheromone values (forgetting past experience). The adapted parameters are used by a percentage of ants of the whole colony.
iii) Combining two approaches to build new solutions: greedy and pheromone based. Thus, a percentage of ants have a greedy behavior since they disregard the accumulated pheromone values and only consider the new problem structure. The remaining ants follow the usual steps to find a solution based on the accumulated pheromone values.

The above P-ACO variations were called, respectively P-ACO, PAdapt, and GreedyAnts. These algorithms were tested on 12 instances representing 4 ad hoc networks. The best performer algorithms were P-ACO and GreedyAnts. The authors highlight that all the algorithms experience some difficulties to react to the first change, however, they improved the respective capacity of reaction in subsequent changes of the problems structure.

To finish this section we would like to remark that only a few applications of P-ACO for dynamic problems were found in the literature. Nevertheless, it is interesting to note, as claimed by Guntsch and Middendorf [26], that P-ACO has a potential to solve other dynamic problems.

9.4.3 Other Ants Based Algorithms

In addition to the standard ACO algorithms (Section 9.4.1) and Population-based ACO algorithms (Section 9.4.2), there exist other alternatives to implement algorithms based on the metaphor of ants behavior which do not completely fit in the mentioned two classes. In this section we describe some representative works in this regard in order to show the reader the potential of following the ant behavior metaphor to solve unseen dynamic problems. An earlier ant-based method to deal with dynamic problems is the proposal by Schoonderwoerd et al. [42] where an ant-based load balancing approach is applied to telecommunication networks. This methodology considers the ants as mobile agents that can travel though the network with similar abilities as the real ants, as they can deposit certain amount of pheromone according to the distance between two pair of nodes and the congestion found during the journey.

The implemented mechanism to route the call is fairly simple as the pheromone value deposited on the route connecting two nodes is used to calculate a probability distribution that will influence the decision maker to route a call. Based on the experimental study accomplished, the authors claim that good load balance can be reached due to the emergent organization of the proposed ant-based methodology (thus, other related problems could also be solved).

Another methodology where the ants are seen as mobile agents is the AntNet [14], maybe one of the more widely known ant based algorithms different from standard ACO algorithms. This algorithm was designed as an alternative approach to the adaptive learning tables in communication networks, an intrinsically dynamic problem. AntNet follows the core ideas of the ACO metaheuristic: a set of independent ants (agents) try to find an optimal or near-optimal path by indirect communications.

Interestingly, there are two types of ants (*forward* and *backward*) which are distributed on the network (represented by small packages) to find paths from a source node to a destination node (the task of forward ants) and propagate the collected information on the routing tables (the task of backwards ants). In short, during their travel ants collect information about the network traffic which is later used to adapt certain values (resembling pheromone values) used by the decision maker at the time of routing the actual data packages. AntNet was thoroughly tested on real and artificial IP datagram networks and achieved superior performance with respect to the other algorithms tested for comparison. In the same mobile agent approaches as AntNet and Schoonderwoerd's ABC, Pigné and Guinand [39] consider the problem of mobile ad hoc networks where, prior to the throughput and the quality of service, the problem of energy consuming is the biggest issue in these wireless networks, composed of small handheld devices. The authors propose a bi-objective model where the length of communication paths is minimized and the selection of robust (unlikely to fail) links is maximized. The approach is decentralized and only relies on local rules for heuristics and pheromone updates.

In Cicirello and Smith [13] the so-called Ant Colony Control (AC^2) is proposed to an adaptive and dynamic shop floor scheduling problem. AC^2 is a decentralized algorithm conformed by a set of artificial ants which use indirect communication to make all shop routing decisions. This is done by altering and reacting to a dynamically changing common environment through the use of simulated pheromone trails. Thus, the amount of pheromones will be used to control the reaction of the algorithm to the changes, i.e., the decision about to which processing shop a job should be assigned. Another interesting feature of AC^2 is that, as the shop can process job of different types, an ant associated to that job will have different pheromone type from an ant associated to a different type of job. According to this, AC^2 manages many pheromone matrices such as the number of different types of jobs the shop floor can process. From the experimental study (applied to different shop floors) the authors claim that, for complex problems, AC^2 evolves local decision making policies that lead to near-optimal solutions with respect to its global performance.

A recent work by Fernandes et al. [19] proposed an extension of the Univariate Marginal Distribution Algorithm (UMDA) (see Mühlenbein and Paas [38]) called Reinforcement-Evaporation UMDA (RE UMDA). This new algorithm includes a different update strategy for the probability model based on the equation of the transition probability equations found in ACO algorithms. To do that, RE UMDA uses two real vectors τ^0 and τ^1 that, respectively, represent the pheromone values associated with the desirability of having a 0 or 1 at a particular position in the solution (a binary search space it is assumed). These two vectors are then used to calculate the probability of assigning 1 (respectively 0) to a particular solution component, i.e., to generate new solutions. As mentioned before, the two pheromone trail vectors are updated (reinforcement stage) according to the solutions found in the current iteration before obtaining the probability to generate new solutions. The evaporation stage, which uses an evaporation parameter as in the traditional ACO algorithms, takes place when the new population has been completely generated. This new approach delays (or avoids) the complete convergence of the population which increases the chances to adaptation to a new environment when a change occurs. RE UMDA was tested on dynamic versions of Onemax and Royal Road

(R1) functions. In addition, several variations of specific parameters of UMDA were studied under the dynamic functions considered.

Xi et al. [45] propose an Ant Colony System (ACS) based methodology to deal with the curing of polymeric coating process, a complex and dynamic optimization problem of great interest in the automotive industry. This is a large-scale multi-stage dynamic optimization problem that involves a time variant objective function (energy consumption for the coating process) and also time dependent linear and nonlinear constraints. More precisely, the ACS-based methodology is called *a dynamic model-embedded ACS-based optimization methodology.* The solution search space is represented by N trees (N is the problem dimension) where each of them is traversed by a set of M ants in charge of cooperatively finding a complete solution by building a tour on each of the N trees, i.e., when ant j traverses tree i, a value for dimension i will be found by ant j. Pheromone and heuristic values are associated to each branch tree to bias the probability values that will guide the ants when traveling the respective trees from the root to the leaves. Each tree node is assumed to have L possible branches (i.e., L possible values for each problem dimension). As in a standard ACS, global and local pheromone updates are applied, as well as similar transition rule as the originally defined in ACS. The authors claim that by building the solutions in this way the algorithm can easily adapt the solutions to the current state of the problem environment to reach the minimum energy consumption. The proposed ACS-based methodology was successfully compared with a genetic algorithm (GA) as the first one was capable of decreasing in about 9% the energy consumption with respect a Genetic Algorithm. Although the reported results are encouraging, it is still necessary to study in more detail the changes produced in the values associated to the edges in the trees when a change occurs in the environment.

In a recent work, an ant-stigmergy based algorithm to solve dynamic optimization problems (DASA) was proposed by Korošec and Šilc [31]. Interestingly, DASA was first proposed to solve (static) problems in continuous domains (see [32]). More precisely, it was applied on a benchmark suite from the Special Session on Real Parameter Optimization of the International Congress on Evolutionary Computation (CEC) 2005. Interestingly, the same algorithm DASA was applied without any modification on the set of benchmark problems provided for CEC' 2009 Special Session on Evolutionary Computation in Dynamic and Uncertain Environments with encouraging results. The main idea behind the DASA design is the construction of a special graph called *differential graph* used by ants to build a solution (vector of real numbers) starting from a given solution called temporary best solution which is initially chosen at random. Each edge in the graph represents either an increment or decrement (Δ value) that have to be applied to a particular dimension as the ant walks through the graph. The decision to choose the next vertex to visit is based on the pheromone values which are initialized according to the Cauchy distribution. Although no exhaustive discussion is provided in this work concerning the features of DASA, the results showed a natural capacity of the algorithm when dealing with the kind of tested dynamic problems. In [5, 6] the authors propose to distribute agents or entities-based applications on a grid or network of computers using several distinct colonies of ants. Each colony represents a distinct computing resource identified by a color. The colored ants compete to detect and colonize evolving communities, or organizations in a dynamic graph representing the set of entities (nodes) and their interactions one with another (edges). The importance of the interaction can be used to weight the edges. Communities are commonly defined as areas of a graph where nodes are more connected one with another than with the other parts of the graph. Organizations are evolving communities, as the underlying graph changes during time. Indeed, often, although individual nodes and edges of a community appear, change, or disappear, the community remains stable for a longer period of time. The proposed algorithm (called AntCO^2) shows two important differences with standard ACB: it does not use an explicit objective function, and it uses several colonies of ants in competition one with another. The ants detect organizations of the evolving graph by laying down "colored" pheromone corresponding to their colony. Pheromone of the same color as an ant probabilistically attracts it, whereas other colors repulse it. Furthermore, the larger the weight on an edge, the more this edge attracts ants, and strongly connected areas capture ants. These mechanisms act as positive feedback to create "colonized" areas on organizations.

Both the evolution of the network (disappearance of edges and node) and the evaporation of pheromone act as negative feedback to remove old solutions (old colonized areas) that are no more valid when the environment changes, therefore providing adaption to dynamics of the graph. Indeed, there is no need to evaluate any objective function to use this algorithm, which allows it to be easily distributed, since it uses only local information. Colonies can be of distinct sizes (number of ants) to accommodate the difference in power of the corresponding computing resources, and therefore colonize larger areas to distribute more entities or agents on more powerful computers. Furthermore, colonies can be added or removed as computing resources appear or disappear, therefore providing another level of dynamism.

9.4.4 Summary of ACB Applications on DOPs

To finish the main section on ACB for DOPs, we show in Table 9.1 a summary of the main applications commented in this chapter. Table 9.1 displays in the first column the type of application considered (Application), the name and reference of the ant colony based algorithm used (ACB), and finally some remarks about the algorithm are given in the third column (Remarks). In column ACB the proposal is called as: Standard ACO (Section 9.4.1), P-ACO (Section 9.4.2), and the respective names found in the literature for other ant colony based algorithms (Section 9.4.3).

Table 9.1 Summary of the reviewed literature indicating the application, type of ACB or algorithms' names, and some additional remarks about the respective proposal.

Application	ACB	Remarks
Dynamic TSP	Standard ACO [3, 4, 18, 24] P-ACO [25, 26]	Mostly aimed to investigate different strategies for pheromone updating on dynamic problems where components are added/eliminated at certain times.
Dynamic QAP	P-ACO [25, 26]	Like the previous row.
Dynamic VRP	Standard ACO [36]	Idem above with new orders arriving when vehicles have already started their tours.
Load balancing in telecommunication newtorks	Ant-Based Control [42]	Use of simple mobile agents (ants) with abilities to laid pheromone trails. Pheromone tables are used to balance the load generated by calls between nodes.
Ad hoc networks	P-ACO, PAdapt, and GreedyAnts [29]	Variations of P-ACO are studied to manage in different ways the modification of the pheromone matrix. The algorithms use knowledge of the problem structure (dynamic component) to carry on the pheromone updating process.
Dynamic load balancing in individual-based simulations	AntCO^2 [5, 6]	Use of several ant colonies in competition to colonize communities in an evolving network of interacting entities.
Continuous functions	DASA [31]	An ant-stigmergy based algorithm originally designed for static continuous functions is successfully applied on a benchmark of dynamic continuous functions.
Shop floor scheduling problem	AC^2 [13]	Ants use only the stigmergy principle to make all shop routing decisions by altering and reacting to their dynamically changing common environment through the use of simulated pheromone trails.
Oscillatory Royal Road & dynamic Schaffer's function	Standard ACO [20]	Optimization of dynamic binary landscapes by stressing the role of negative feedback when modifying pheromone values.
Routing tables in communication networks	AntNet [14]	Tiny packages (ants) are used to collect and distribute information from the network to modify the routing tables.
Communications paths in wireless mobile ad hoc networks	Ant-based algorithm [39]	An ant colony constructs and maintains communication paths trying to minimize both, the length of the constructed paths and the number of link reconnections.
Curing of Polymeric Coating	ACS-based algorithm [45]	Utilization of tree structures to find quality values for each problem dimension. The traversing from the root to the leaves in the respective trees is governed by the deposited pheromone values. The pheromone updating is made by rules resembling those used in Ant Colony Systems.
OneMax & Royal Road	RE UMDA [19]	Use of principles of pheromone trail to keep diversity.

9.5 Conclusions

Many real-world problems are dynamic by definition and the use of metaheuristic techniques to solve them seems to be a good alternative, as those kind of algorithms are robust and flexible. Ant colony based algorithms share these particular features as they can be easily adapted to deal with dynamic problems.

In this chapter we presented a general perspective of the more relevant works regarding the application of ant colony based algorithms for dynamic optimization problems. The main mechanisms used in different implementations found in the literature were described. Interestingly, the metaphor of ant colony behavior could potentially be used in many different ways, which make ACB algorithms good candidates to solve known and unseen DOPs.

Promising research areas seem to be related with applications in which the metaphor of ants behavior (basically stigmergy by pheromone trail) can be used as a source of information to rapidly react to the changes. In that regard, it could be interesting to define and thoroughly study general strategies to adapt pheromone values on classes of dynamics problems as well as comparisons with other, more studied and applied, metaheuristics for solving DOPs, e.g., evolutionary algorithms.

Acknowledgments. The first author acknowledges funding from Universidad Nacional de San Luis (UNSL), Argentina and the National Agency for Promotion of Science and Technology, Argentina (ANPCYT). The second author acknowledges funds from the Junta de Andalucía (CICE), under contract P07-TIC-03044 (DIRICOM project) and Spanish Ministry of Sciences and Innovation (MICINN) and FEDER under contracts TIN2008-06491-C04-01 (M* project) and TIN2011-28194 (roadME project).

References

[1] Abdunnaser, Y.: Adapting Evolutionary Approaches for Optimization in Dynamic Environments. PhD thesis, University of Waterloo, Waterloo, Ontario, Canada (2006)

[2] Angus, D.: The current state of ant colony optimisation applied to dynamic problems. Technical Report TR009, Centre for Intelligent Systems & Complex Processes, Faculty of Information & Communication Technologies Swinburne University of Technology, Melbourne, Australia (2006)

[3] Angus, D., Hendtlass, T.: Ant Colony Optimisation Applied to a Dynamically Changing Problem. In: Hendtlass, T., Ali, M. (eds.) IEA/AIE 2002. LNCS (LNAI), vol. 2358, pp. 618–627. Springer, Heidelberg (2002)

[4] Angus, D., Hendtlass, T.: Dynamic ant colony optimisation. Applied Intelligence 23(1), 33–38 (2005)

[5] Bertelle, C., Dutot, A., Guinand, F., Olivier, D.: Organization Detection Using Emergent Computing. International Transactions on Systems Science and Applications 2(1), 61–70 (2006)

[6] Bertelle, C., Dutot, A., Guinand, F., Olivier, D.: Organization Detection for Dynamic Load Balancing in Individual-Based Simulations. Multi-Agent and Grid Systems 3(1), 42 (2007)

[7] Bianchi, L.: Notes on dynamic vehicle routing - the state of the art. Technical Report IDSIA 05-01, Istituto Dalle Molle di Studi sull' Intelligenza Artificiale (IDSIA), Manno-Lugano, Switzerland (2000)
[8] Blackwell, T.: Particle Swarm Optimization in Dynamic Environments. In: Yang, S., Ong, Y., Jin, Y. (eds.) Evolutionary Computation in Dynamic and Uncertain Environments. SCI, vol. 51, pp. 29–49. Springer, Heidelberg (2007)
[9] Blum, C.: Ant colony optimization: Introduction and recent trends. Physics of Life Reviews 2(4), 353–373 (2005)
[10] Branke, J.: Evolutionary Optimization in Dynamic Environments. Kluwer Academic Publishers, Norwell (2002)
[11] Branke, J., Schmeck, H.: Designing evolutionary algorithms for dynamic optimization problems. In: Advances in Evolutionary Computing, pp. 239–262. Springer-Verlag New York, Inc. (2003)
[12] Bullnheimer, B., Hartl, R.F., Strauss, C.: An improved ant system algorithm for the vehicle routing problem. Annals of Operations Research 89, 319–328 (1999)
[13] Cicirello, V.A., Smith, S.F.: Ant Colony Control for Autonomous Decentralized Shop Floor Routing. In: International Symposium on Autonomous Decentralized Systems, pp. 383–390. IEEE Computer Society, Dallas (2001)
[14] Di Caro, G., Dorigo, M.: AntNet: distributed stigmergetic control for communications networks. J. Artif. Int. Res. 9, 317–365 (1998)
[15] Dorigo, M., Gambardella, L.M.: Ant Colony System: a Cooperative Learning Approach to the Traveling Salesman Problem. IEEE Transactions on Evolutionary Computation 1(1), 53–66 (1997)
[16] Dorigo, M., Maniezzo, V., Colorni, A.: Ant system: Optimization by a colony of cooperating agents. IEEE Trans. on Systems, Man, and Cybernetics–Part B 26(1), 29–41 (1996)
[17] Dorigo, M., Stützle, T.: Ant Colony Optimization. MIT Press (2004)
[18] Eyckelhof, C.J., Snoek, M.: Ant Systems for a Dynamic TSP. In: Proceedings of the Third International Workshop on Ant Algorithms, ANTS 2002, pp. 88–99. Springer, London (2002)
[19] Fernandes, C.M., Lima, C., Rosa, A.C.: UMDAs for dynamic optimization problems. In: Proceedings of the 10th Annual Conference on Genetic and Evolutionary Computation, GECCO 2008, pp. 399–406. ACM, New York (2008)
[20] Fernandes, C.M., Rosa, A.C., Ramos, V.: Binary ant algorithm. In: Proceedings of the 9th Annual Conference on Genetic and Evolutionary Computation, GECCO 2007, pp. 41–48. ACM, New York (2007)
[21] Gambardella, L.M., Taillard, E.D., Dorigo, M.: Ant colonies for the quadratic assignment problem. Journal of the Operational Research Society 50(2), 167–176 (1999)
[22] Goss, S., Aron, S., Deneubourg, J.L., Pasteels, J.M.: Self-organized shortcuts in the argentine ant. Naturwissenschaften 76, 579–581 (1989)
[23] Grassé, P.P.: La reconstruction du nid et les coordinations interindividuelles chez bellicositermes natalensis et cubitermes sp. la theéorie de la stigmergie: Essai dfinterprétation du comportement des termites constructeurs. Insectes Sociaux 6(1), 41–48 (1959)
[24] Guntsch, M., Middendorf, M.: Pheromone Modification Strategies for Ant Algorithms Applied to Dynamic TSP. In: Boers, E.J.W., Gottlieb, J., Lanzi, P.L., Smith, R.E., Cagnoni, S., Hart, E., Raidl, G.R., Tijink, H. (eds.) EvoIASP 2001, EvoWorkshops 2001, EvoFlight 2001, EvoSTIM 2001, EvoCOP 2001, and EvoLearn 2001. LNCS, vol. 2037, pp. 213–222. Springer, Heidelberg (2001)

[25] Guntsch, M., Middendorf, M.: Applying population based aco to dynamic optimization problems. In: Proceedings of the Third International Workshop on Ant Algorithms, ANTS 2002, pp. 111–122. Springer (2002)
[26] Guntsch, M., Middendorf, M.: A Population Based Approach for ACO. In: Cagnoni, S., Gottlieb, J., Hart, E., Middendorf, M., Raidl, G.R. (eds.) EvoIASP 2002, EvoWorkshops 2002, EvoSTIM 2002, EvoCOP 2002, and EvoPlan 2002. LNCS, vol. 2279, pp. 72–81. Springer, Heidelberg (2002)
[27] Guntsch, M., Middendorf, M., Schmeck, H.: An Ant Colony Optimization Approach to Dynamic TSP. In: Spector, L., Goodman, E.D., Wu, A., Langdon, W.B., Voigt, H.-M., Gen, M., Sen, S., Dorigo, M., Pezeshk, S., Garzon, M.H., Burke, E. (eds.) Proceedings of the Genetic and Evolutionary Computation Conference, GECCO 2001, pp. 860–867. Morgan Kaufmann, San Francisco (2001)
[28] Hendtlass, T., Moser, I., Randall, M.: Dynamic Problems and Nature Inspired Meta-Heuristics. In: Proceedings of the Second IEEE International Conference on e-Science and Grid Computing, E-SCIENCE 2006, pp. 111–116. IEEE Computer Society, Washington, DC (2006)
[29] Ho, C.K., Ewe, H.T.: Ant Colony Optimization Approaches for the Dynamic Load-Balanced Clustering Problem in Ad Hoc Networks. In: Swarm Intelligence Symposium, SIS 2007, pp. 76–83. IEEE (April 2007)
[30] Jin, Y., Branke, J.: Evolutionary optimization in uncertain environments - A survey. IEEE Transactions on Evolutionary Computation 9(3), 303–318 (2005)
[31] Korošec, P., Šilc, J.: The differential ant-stigmergy algorithm applied to dynamic optimization problems. In: Proceedings of the Eleventh Conference on Congress on Evolutionary Computation, CEC 2009, pp. 407–414. IEEE Press, Piscataway (2009)
[32] Korošec, P., Šilc, J., Oblak, K., Kosel, F.: The differential ant-stigmergy algorithm: an experimental evaluation and a real-world application. In: IEEE Congress on Evolutionary Computation, CEC 2007, pp. 157–164 (September 2007)
[33] Leguizamón, G., Ordóñez, G., Molina, S., Alba, E.: Canonical Metaheuristics for Dynamic Optimization Problems. In: Alba, E., Blum, C., Isasi, P., León, C., Gómez, J.A. (eds.) Optimization Techniques for Solving Complex Problems, pp. 83–100. John Wiley & Sons, Inc. (2008)
[34] Maniezzo, V.: Exact and Approximate Nondeterministic Tree-Search Procedures for the Quadratic Assignment Problem. Informs Journal on Computing 11(4), 358–369 (1999)
[35] Monmarché, N., Guinand, F., Siarry, P.: Artificial Ants. Wiley-ISTE (2010)
[36] Montemanni, R., Gambardella, L.M., Rizzoli, A.E., Donati, A.V.: A new algorithm for a Dynamic Vehicle Routing Problem based on Ant Colony System. In: Second International Workshop on Freight Transportation and Logistics, pp. 27–30 (2003)
[37] Morrison, R.W.: Designing Evolutionary Algorithms for Dynamic Environments. Natural Computing Series. Springer (2004)
[38] Mühlenbein, H., Paass, G.: From Recombination of Genes to the Estimation of Distributions I. Binary Parameters. In: Ebeling, W., Rechenberg, I., Voigt, H.-M., Schwefel, H.-P. (eds.) PPSN 1996. LNCS, vol. 1141, pp. 178–187. Springer, Heidelberg (1996)
[39] Pigné, Y., Guinand, F.: Short and Robust Communication Paths in Dynamic Wireless Networks. In: Dorigo, M., Birattari, M., Di Caro, G.A., Doursat, R., Engelbrecht, A.P., Floreano, D., Gambardella, L.M., Groß, R., Şahin, E., Sayama, H., Stützle, T. (eds.) ANTS 2010. LNCS, vol. 6234, pp. 520–527. Springer, Heidelberg (2010)

[40] Psaraftis, H.N.: Dynamic vehicle routing: Status and prospect. Annals Operations Research 61, 143–164 (1995)
[41] Reimann, M., Doerner, K., Hartl, R.F.: D-ants: Savings based ants divide and conquer the vehicle routing problem. Computers & Operations Research 31(4), 563–591 (2004)
[42] Schoonderwoerd, R., Holland, O.E., Bruten, J.L., Rothkrantz, L.J.M.: Ant-based load balancing in telecommunications networks. Adaptive Behavior 2, 169–207 (1996)
[43] Socha, K., Dorigo, M.: Ant colony optimization for continuous domains. European Journal of Operational Research 185(3), 1155–1173 (2008)
[44] Stützle, T., Dorigo, M.: ACO algorithms for the quadratic assignment problem. In: Corne, D., Dorigo, M., Glover, F. (eds.) New Ideas in Optimization, pp. 33–50. McGraw Hill, London (1999)
[45] Xiao, J., Li, J., Xu, Q., Huang, W., Lou, H.: ACS-based Dynamic Optimization for Curing of Polymeric Coating. The American Institute of Chemical Engineers (AIChE) Journal 52(2), 1410–1422 (2005)

Chapter 10
Elastic Registration of Brain Cine-MRI Sequences Using MLSDO Dynamic Optimization Algorithm

Julien Lepagnot, Amir Nakib, Hamouche Oulhadj, and Patrick Siarry

Abstract. In this chapter, we propose to use a dynamic optimization algorithm to assess the deformations of the wall of the third cerebral ventricle in the case of a brain cine-MR imaging. In this method, an elastic registration process is applied to a 2D+t cine-MRI sequence of a region of interest (*i.e.* lamina terminalis). This registration process consists in optimizing an objective function that can be considered as dynamic. Thus, a dynamic optimization algorithm based on multiple local searches, called MLSDO, is used to accomplish this task. The obtained results are compared to those of several well-known static optimization algorithms. This comparison shows the efficiency of MLSDO, and the relevance of using a dynamic optimization algorithm to solve this kind of problems.

10.1 Introduction

Hydrocephalus is a medical condition in which there is an abnormal accumulation of cerebrospinal fluid in the ventricles, or cavities, of the brain. This may cause increased intracranial pressure inside the skull and progressive enlargement of the head, convulsion, tunnel vision, and mental disability. Hydrocephalus can also cause death. Hydrocephalus may be suggested by symptoms; however, imaging studies of the brain are the mainstay of diagnosis. In this paper, we focus on a method based on cine-MRI sequences to facilitate this diagnosis, and to assist neurosurgeons in the treatment of hydrocephalus. This method makes use of the dynamic optimization paradigm.

Julien Lepagnot · Amir Nakib · Hamouche Oulhadj · Patrick Siarry
Université Paris-Est Créteil (UPEC)
LISSI, E.A. 3956
61 avenue du Général de Gaulle,
94010 Créteil, France
e-mail: siarry@u-pec.fr

E. Alba et al. (Eds.): Metaheuristics for Dynamic Optimization, SCI 433, pp. 211–224.
springerlink.com

Recently, optimization in dynamic environments has attracted a growing interest, due to its practical relevance. Almost all real-world problems are time dependent or dynamic, *i.e.* their objective function changes over the time. For dynamic environments, the goal is not only to locate the global optimum, but also to track it as closely as possible over the time. Then, a dynamic optimization problem can be expressed as in (10.1), where $f(\mathbf{x},t)$ is the objective function of a minimization problem, $h_j(\mathbf{x},t)$ denotes the jth equality constraint and $g_k(\mathbf{x},t)$ denotes the kth inequality constraint. All of these functions may change over time (iterations), as indicated by the dependence on the time variable t.

$$\begin{aligned} &\min\ f(\mathbf{x},t) \\ &\text{s.t.}\ \ h_j(\mathbf{x},t) = 0 \text{ for } j = 1,2,...,u\,, \\ &\qquad g_k(\mathbf{x},t) \leq 0 \text{ for } k = 1,2,...,v \end{aligned} \tag{10.1}$$

In this chapter, we focus on a dynamic optimization problem with time constant constraints. We propose to apply the *Multiple Local Search algorithm for Dynamic Optimization* (MLSDO) [5] to the registration of sequences of images. Image registration is the process of overlaying two or more images of the same scene taken at different times, from different viewpoints, and/or by different sensors. It is a critical step in all image analysis tasks in which the final information is gained from the combination of various data sources like in image fusion or change detection.

It geometrically aligns two images: the source and the target images. It is done by determining a transformation that maps the target image to the source one. Thus, registering a sequence of images consists in determining, for each couple of successive images, the transformation that makes the first image of the couple match the following image.

Comprehensive surveys of the registration approaches are available in the literature, we can cite [6, 11]. Registration approaches can be roughly based on:

- geometric image features (geometric registration), such as points, edges and surfaces;
- measures computed from the image gray values (intensity based registration), such as mutual information.

In many cases, a satisfactory solution can be found by using a rigid or an affine transformation (deformation model applied to the target image), *i.e.* the target image is only translated, rotated, and scaled to match the source image [7]. Elastic registration is required to register inter-patient images or regions containing non-rigid objects. The goal is to remove structural variation between the two images to be registered. As stated in [6], most applications represent elastic transformations in terms of a local vector displacement (disparity) field or as polynomial transformations in terms of the old coordinates. In the problem at hand, each image of the region of interest (*i.e.* lamina terminalis) is extracted from a brain cine-MRI

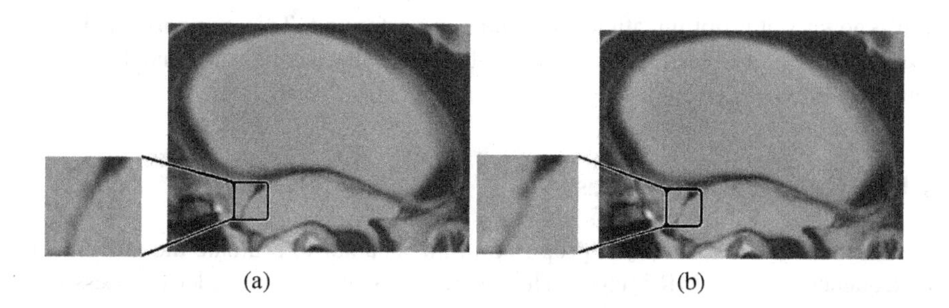

Fig. 10.1 Two images from a brain cine-MRI sequence: (a) first image of the sequence, (b) sixth image of the sequence.

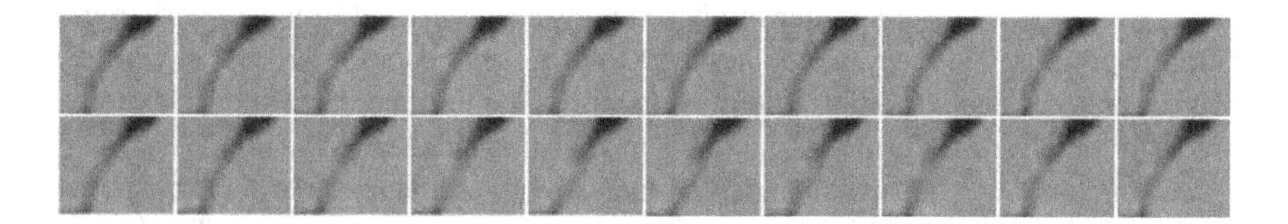

Fig. 10.2 A sequence of cine-MR images of the region of interest.

sequence of 20 images. This sequence corresponds to 80% of a R-R cardiac cycle, more details about the acquisition procedure are given in [7]. An example of two images extracted from a brain cine-MRI sequence is presented in Figure 10.1. Hence, each sequence is composed of 20 MR images. An example of sequence is illustrated in Figure 10.2. The goal is to register each couple of successive images of the sequence. Hence, for a sequence of 20 images, 19 couples of successive images have to be registered. Then, the transformations that result from this matching operation can be used to assess the deformation movements of the third cerebral ventricle.

Several papers are proposed in the literature about the analysis and quantification of cardiac movements, we can cite those recently published [1, 2, 10]. In our case, the single approach that deals with the problem at hand is [7], because of the recent appearance of the acquisition method of the images. The main difference between the problem at hand and the cardiac problem lies in the amplitude of the movements of the ventricles. Indeed, the amplitude of the cardiac ventricle movements is higher than the amplitude of the cerebral ventricle movements. In this chapter, we propose a method inspired from [7] to assess the movements of a region of interest (ROI). Besides, another contribution of the present work is to show the importance of the use of dynamic optimization algorithms for brain cine-MRI registration.

The rest of this chapter is organized as follows. In section 10.2, the method proposed to register sequences of images is described. In section 10.3, the MLSDO algorithm and its use for the problem at hand are presented. In section 10.4, a comparison of the results obtained by MLSDO on this problem to the ones of several well-known static optimization algorithms is performed. This comparison shows the relevance of using MLSDO on this problem. Finally, a conclusion and the works under progress are given in section 10.5.

10.2 Registration Process

A method inspired from [7] is proposed in this chapter to evaluate the movement in sequences of cine-MR images. This operation is required in order to assess the movements in the ROI over time. In [7], a segmentation process is performed on each image of the sequence, to determine the contours (as a set of points) of the walls of the third cerebral ventricle. Then, a geometric registration of each successive contours is performed, based on an affine deformation model. In the present work, we propose to use an intensity based registration instead of a geometric registration process. This way, we do not have to use a segmentation process anymore. Moreover, to evaluate the pulsatile movements of the third cerebral ventricle more precisely, an elastic deformation model is used in this chapter.

Let Im_1 and Im_2 be two successive images of the sequence. Let the transpose of a matrix A be denoted by A^{T}. Then, we assume that a transformation T_{Φ} allows to match Im_1 with $Im'_1 = T_{\Phi}(Im_2)$ and, for every pixel $o_2 = (x_2\ y_2)^{\mathrm{T}}$ of Im_2, it is defined by:

$$\begin{array}{c} x'_1 = c_1\, x_2^2 + c_2\, y_2^2 + c_3\, x_2\, y_2 + (c_4\, |c_4| + 1)\, x_2 + c_5\, |c_5|\, y_2 + (c_6)^3 \\ y'_1 = c_7\, x_2^2 + c_8\, y_2^2 + c_9\, x_2\, y_2 + c_{10}\, |c_{10}|\, x_2 + (c_{11}\, |c_{11}| + 1)\, y_2 + (c_{12})^3 \end{array}, \tag{10.2}$$

where $o'_1 = (x'_1\ y'_1)^{\mathrm{T}} = T_{\Phi}(o_2)$. The set of parameters $\Phi = \{c_1, c_2, ..., c_{12}\}$ is estimated through the maximization of the following criterion:

$$C(\Phi) = \frac{NMI(\Phi)}{P(\Phi) + 1}, \tag{10.3}$$

where $NMI(\Phi)$ computes the normalized mutual information [9] of Im_1 and Im'_1 ; $P(\Phi)$ is part of a regularization term that penalizes large deformations of Im_2, as we are dealing with slight movements in the ROI. Besides, as the size of the ROI is not constant, we have to normalize the coordinates of the pixels. Then, we make the pixels in the ROI range in the interval $[-0.5, 0.5]$. The use of this interval transforms discrete coordinates of the pixels into continuous ones. This interval was determined empirically, and it is well fitted to the regularization term, and to the transformation model used. $NMI(\Phi)$ and $P(\Phi)$ are defined as follows:

$$NMI(\Phi) = \frac{H(Im_1) + H(Im'_1)}{H(Im_1,\, Im'_1)}, \tag{10.4}$$

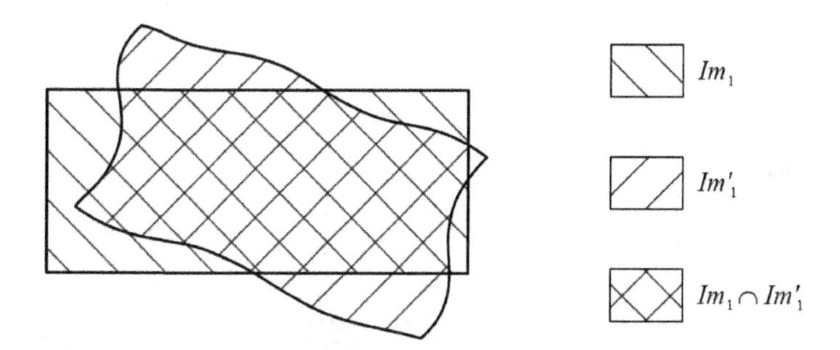

Fig. 10.3 Overlapping area ($Im_1 \cap Im'_1$) of the source image (Im_1) and the transformed target image (Im'_1) in the registration of a couple of successive images of a sequence.

$$P(\Phi) = \max_{o_2 \in Im_1 \cap Im'_1} \left(o_2 - o'_1\right)^{\mathrm{T}} \left(o_2 - o'_1\right), \tag{10.5}$$

where $Im_1 \cap Im'_1$ is the overlapping area of Im_1 and Im'_1 (see Figure 10.3); $H(Im_1)$ and $H(Im'_1)$ compute the Shannon entropy of Im_1 and Im'_1, respectively, in their overlapping area; $H(Im_1, Im'_1)$ computes the joint Shannon entropy of Im_1 and Im'_1, in their overlapping area. They are defined as follows:

$$H(Im_1) = -\sum_{i=0}^{L-1} p(i) \log_2 (p(i)), \tag{10.6}$$

$$H(Im'_1) = -\sum_{j=0}^{L-1} p'(j) \log_2 (p'(j)), \tag{10.7}$$

$$H(Im_1, Im'_1) = -\sum_{i=0}^{L-1} \sum_{j=0}^{L-1} p(i,j) \log_2 (p(i,j)), \tag{10.8}$$

where L is the number of possible gray values that a pixel can take; $p(i)$, $p'(j)$ and $p(i,j)$ are the probability of the pixel intensity i in Im_1, the probability of the pixel intensity j in Im'_1, and the joint probability of having a pixel intensity i in Im_1 and j in Im'_1, respectively. They are defined as follows:

$$p(i) = \frac{g(i)}{\sum_{k=0}^{L-1} g(k)}, \tag{10.9}$$

$$p'(j) = \frac{g'(j)}{\sum_{l=0}^{L-1} g'(l)}, \tag{10.10}$$

$$p(i,j) = \frac{g(i,j)}{\sum_{k=0}^{L-1} \sum_{l=0}^{L-1} g(k,l)}, \tag{10.11}$$

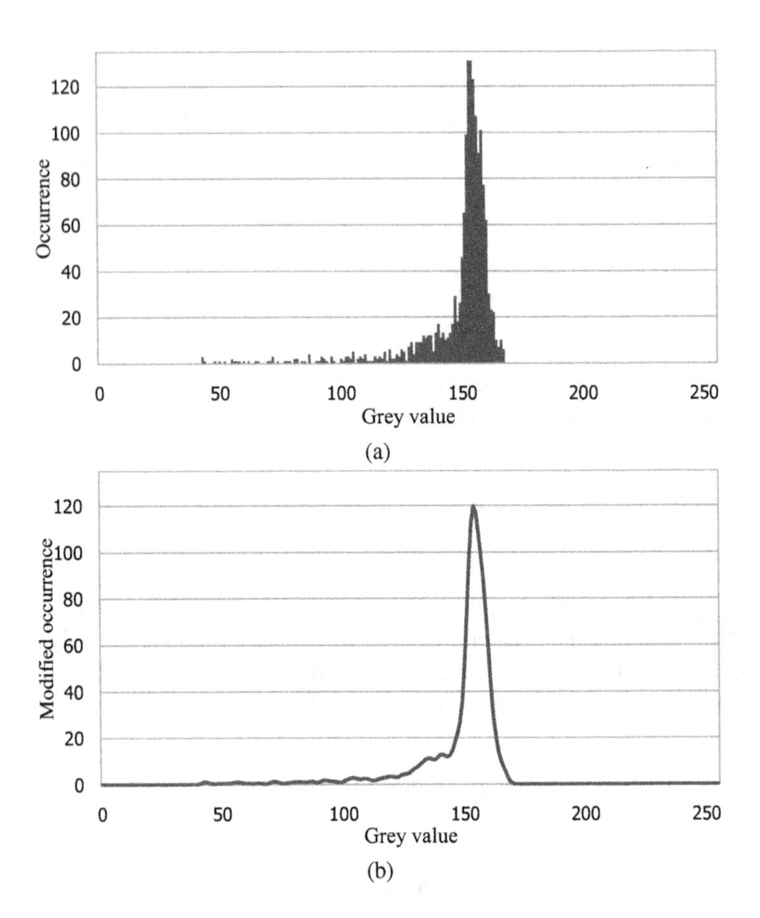

Fig. 10.4 Illustration of the histogram of an MR image: (a) original histogram, (b) smoothed histogram used to accelerate the optimization process.

where $g(i)$ is the histogram of the overlapping area of Im_1 (occurrence of gray level i in Im_1); $g'(j)$ is the histogram of the overlapping area of Im'_1 (occurrence of gray level j in Im'_1); $g(i,j)$ is the joint histogram of the overlapping area of Im_1 and Im'_1 (occurrence of having a grey value equal to i in Im_1 and to j in Im'_1, see Equation (10.12)). However, in this work, we apply a low-pass filter to these histograms, using a convolution with a Gaussian function, in order to accelerate the convergence of the optimization process. Applying this filter reduces indeed the number of local optima in the objective function, by smoothing it. An illustration of the histogram of an MR image from a sequence, and of its corresponding smoothed histogram, is shown in Figure 10.4.

In (10.12), the cardinal function is denoted by *card*, and the functions $Im_1(o)$ and $Im'_1(o)$ return the gray values of a given pixel o in Im_1 and Im'_1, respectively.

$$g(i,j) = \text{card}\left\{o \in Im_1 \cap Im'_1,\ Im_1(o) = i \wedge Im'_1(o) = j\right\}. \tag{10.12}$$

The registration problem can be formulated as an optimization problem defined by:

$$\max\ C(\Phi). \tag{10.13}$$

10.3 The MLSDO Algorithm

In this section, MLSDO and its use on the problem at hand are described. At first, the algorithm is presented. Then, the dynamic objective function proposed for the problem at hand is described. Afterwards, the parameter fitting of MLSDO is given to solve this problem.

10.3.1 Description of the Algorithm

MLSDO uses several local searches, each one performed in parallel with the others, to explore the search space and to track the found optima over the changes in the objective function. These local searches consist in moving step-by-step in the search space, from a current solution to its best neighbor one, until a stopping criterion is satisfied, reaching thus a local optimum. Each local search is performed by an agent, and all the agents are coordinated by a dedicated module (the coordinator). Two types of agents exist in MLSDO: the exploring agents (to explore the search space in order to discover the local optima) and the tracking agents (to track the found local optima over the changes in the objective function). The strategies used to coordinate these local search agents enable the fast convergence to well diversified optima, in order to quickly react to a change and find the global optimum. Furthermore, the local optima found during the optimization process are archived, to accelerate the detection of the global optimum after a change in the objective function. The overall scheme of MLSDO is illustrated in Figure 10.5, where the local search agents are depicted by the numbered black-filled circles in the search space S, and the neighborhood of the ith agent is denoted by N_i. More details about this algorithm are given in [5].

10.3.2 Cine-MRI Registration as a Dynamic Optimization Problem

The registration of a cine-MRI sequence can be seen as a dynamic optimization problem. Then, the dynamic objective function optimized by MLSDO changes according to the following rules:

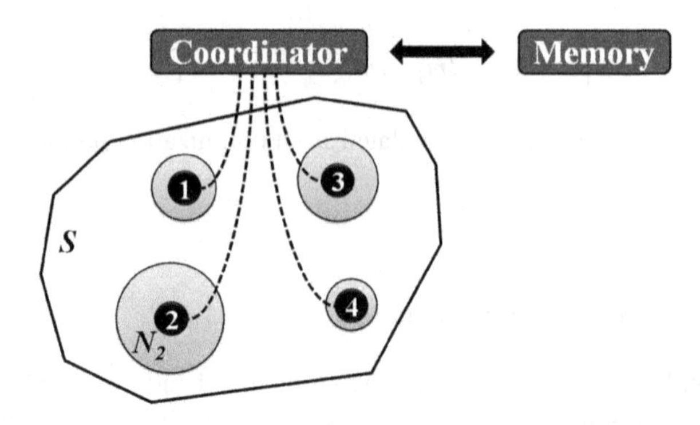

Fig. 10.5 Overall scheme of MLSDO.

- The criterion in (10.3) has to be minimized for each couple of successive images, as we are in the case of a sequence, then the optimization criterion becomes:

$$C(\Phi(t)) = \frac{NMI(\Phi(t))}{P(\Phi(t))+1}, \tag{10.14}$$

  where t is the index of the current couple of images in the sequence. $\Phi(t)$, $NMI(\Phi(t))$, and $P(\Phi(t))$ are the same as Φ, $NMI(\Phi)$, and $P(\Phi)$ defined before, respectively, but here are dependent on the couple of images.
- Then, the dynamic optimization problem is defined by:

$$\max\ C(\Phi(t)), \tag{10.15}$$

- If the current best solution (transformation) found for the couple t cannot be improved anymore (according to a stagnation criterion), the next couple $(t+1)$ is treated.
- The stagnation criterion of the registration of a couple of successive images is satisfied if no significant improvement (higher than 1E-5) in the current best solution is observed during 5000 successive evaluations of the objective function.
- Thus, the end of the registration of a couple of images and the beginning of the registration of the next one constitute a change in the objective function.

10.3.3 Parameter Fitting of MLSDO

Table 10.1 summarizes the six parameters of MLSDO that the user has to define. In this table, the values given are suitable for the problem at hand, and they were fixed experimentally. These values will be used to perform the experiments reported in the following section.

Table 10.1 MLSDO parameter setting for the problem at hand.

Name	Type	Interval	Value	Short description
r_l	real	$(0, r_e)$	0.005	initial step size of tracking agents
r_e	real	$(0, 1]$	0.1	exclusion radius of the agents, and initial step size of exploring agents
δ_{ph}	real	$[0, \delta_{pl}]$	1E-5	highest precision parameter of the stopping criterion of the agents local searches
δ_{pl}	real	$[\delta_{ph}, +\infty]$	1E-4	lowest precision parameter of the stopping criterion of the agents local searches
n_a	integer	$[1, 10]$	1	maximum number of exploring agents
n_c	integer	$[0, 20]$	2	maximum number of tracking agents created after the detection of a change

10.4 Experimental Results and Discussion

The registrations of two couples of slightly different images are illustrated in Figures 10.6 and 10.7, and the registrations of two couples of significantly different images are illustrated in Figures 10.8 and 10.9. As we can see in Figures 10.6(e) and 10.6(f), as well as in Figures 10.7(e) and 10.7(f), if the movements in the ROI are not significant, then only noise appears in the difference images. Hence, the transformation used to register the couple of images (Figures 10.6(d) and 10.7(d)) does not deform the second image of the couple significantly. On the other hand, significant movements in the ROI leave an important white trail in the difference images, as illustrated in Figures 10.8(e) and 10.9(e). Then, a significant transformation (Figure 10.8(d) and 10.9(d)) has to be applied in order to eliminate the white trail (see Figure 10.8(f) and 10.9(f)).

A comparison between the results obtained by MLSDO and those obtained by several well-known static optimization algorithms is presented in this section. These algorithms, and their parameter setting, empirically fitted to the problem at hand, are defined below (see references for more details on these algorithms and their parameter fitting):

- CMA-ES (*Covariance Matrix Adaptation Evolution Strategy*) [4] using the recommended parameter setting, except for the initial step size σ, set to $\sigma = 0.5$. The population size λ of children and the number of selected individuals μ are set to $\lambda = 11$ and $\mu = 5$;
- SPSO-07 (*Standard Particle Swarm Optimization* in its 2007 version) [3] using the recommended parameter setting, except for the number S of particles ($S = 12$) and for the parameter K used to generate the particles, neighborhood ($K = 8$) ;

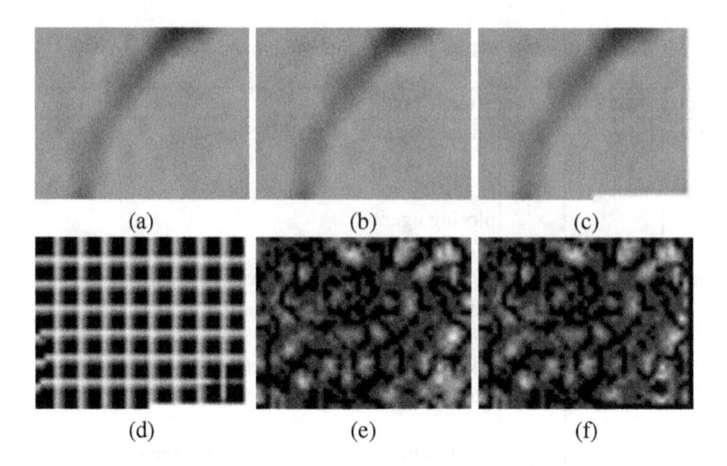

Fig. 10.6 Illustration of the registration of a couple of slightly different images of a sequence: (a) the first image of the couple, (b) the second image of the couple, (c) the second image after applying the found transformation to it, (d) illustration showing the transformation applied on the second image of the couple to register it, (e) illustration showing the difference, in the intensity of the pixels, between the two images of the couple: a black pixel indicates that the intensities of the corresponding pixels in the images are the same, and a white pixel indicates the highest difference between the images, (f) illustration showing the difference, in the intensity of the pixels, between the first image and the transformed second image.

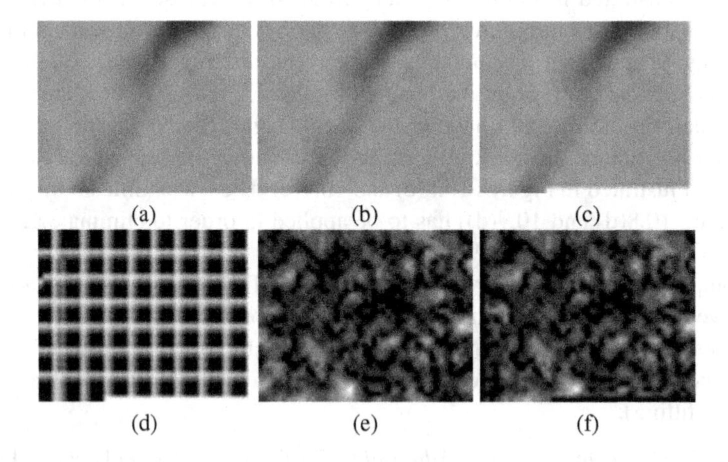

Fig. 10.7 Illustration of the registration of another couple of slightly different images of a sequence, in the same way as in Figure 10.6.

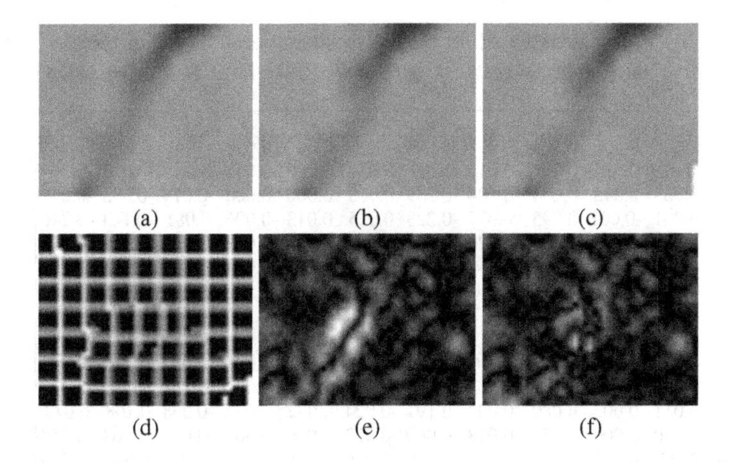

Fig. 10.8 Illustration of the registration of a couple of significantly different images of a sequence: (a) the first image of the couple, (b) the second image of the couple, (c) the second image after applying the found transformation to it, (d) illustration showing the transformation applied on the second image of the couple to register it, (e) illustration showing the difference, in the intensity of the pixels, between the two images of the couple: a black pixel indicates that the intensities of the corresponding pixels in the images are the same, and a white pixel indicates the highest difference between the images, (f) illustration showing the difference, in the intensity of the pixels, between the first image and the transformed second image.

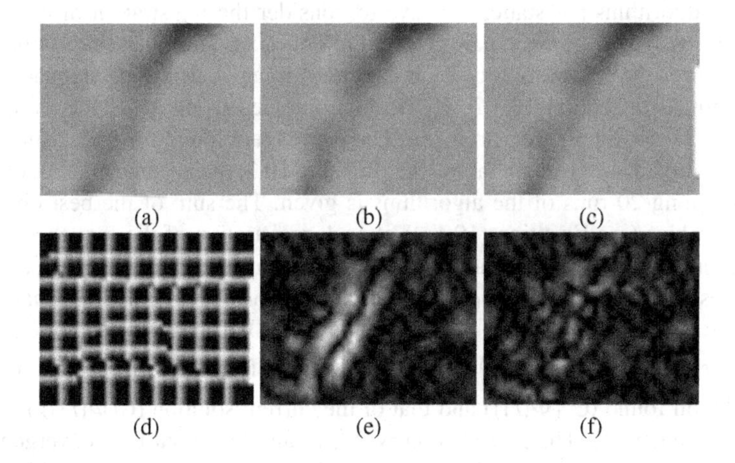

Fig. 10.9 Illustration of the registration of another couple of significantly different images of a sequence, in the same way as in Figure 10.8.

Table 10.2 Transformations found for the registration of each couple of images. The value of the objective function of the best solution found, denoted by $C^*(\Phi(t))$, is also given.

t	c_1	c_2	c_3	c_4	c_5	c_6	c_7	c_8	c_9	c_{10}	c_{11}	c_{12}	$C^*(\Phi(t))$
1	0.039	-0.022	0.005	0.105	-0.034	0.139	-0.039	0.017	0.025	0.091	0.132	0.090	1.199
2	-0.005	-0.029	0.025	0.085	-0.014	0.203	0.077	0.055	0.051	0.068	-0.077	-0.264	1.201
3	0.055	0.063	0.048	0.094	-0.104	-0.239	0.068	0.000	0.000	-0.074	-0.083	-0.256	1.192
4	0.021	0.031	-0.001	0.095	-0.077	-0.223	0.025	0.013	0.006	0.081	-0.144	-0.246	1.195
5	0.063	0.000	0.003	-0.074	-0.026	-0.089	-0.026	0.041	0.011	0.100	0.145	-0.128	1.218
6	0.002	-0.063	-0.033	-0.115	0.034	0.224	-0.019	-0.027	0.024	0.015	0.087	0.258	1.209
7	0.013	-0.092	0.016	0.036	0.080	0.253	-0.060	-0.045	-0.033	-0.077	0.131	0.247	1.208
8	0.003	-0.068	-0.004	-0.023	0.117	0.238	-0.069	-0.047	-0.032	-0.078	0.131	0.247	1.195
9	0.065	-0.020	-0.007	0.044	0.061	-0.046	-0.064	-0.047	-0.023	-0.081	0.131	0.251	1.201
10	0.050	-0.004	-0.017	0.072	0.056	-0.061	0.051	0.005	0.011	-0.052	0.135	-0.043	1.216
11	0.050	0.000	-0.012	-0.004	0.073	-0.053	-0.059	0.047	0.002	0.099	0.164	-0.178	1.216
12	0.060	0.011	0.003	0.080	-0.033	-0.191	-0.024	0.032	0.036	-0.068	0.108	0.048	1.225
13	0.042	0.000	0.000	0.050	-0.018	-0.060	-0.023	0.016	0.002	-0.085	-0.064	-0.218	1.232
14	0.064	-0.005	0.000	0.094	-0.021	-0.199	-0.016	0.075	0.065	-0.039	0.065	-0.210	1.232
15	0.025	-0.008	0.042	0.049	-0.072	0.172	0.037	0.029	0.000	0.104	0.107	-0.037	1.235
16	0.060	0.007	0.003	0.082	-0.026	-0.191	-0.024	0.032	0.034	-0.063	0.111	-0.049	1.216
17	0.050	-0.005	0.000	0.021	0.010	-0.071	-0.025	0.047	0.052	0.018	0.080	-0.170	1.226
18	0.052	-0.005	-0.017	0.083	0.108	-0.121	-0.018	0.042	-0.001	0.071	0.075	0.149	1.225
19	-0.006	0.056	-0.011	-0.080	0.072	-0.210	-0.025	0.076	0.033	-0.057	0.084	-0.158	1.214

- DE (*Differential Evolution*) [8] using the "DE/target-to-best/1/bin" strategy, a number of parents equal to $NP = 30$, a weighting factor $F = 0.8$, and a crossover constant $CR = 0.9$.

As these algorithms are static, we have to consider the registration of each couple of successive images as a new problem to optimize. Thus, these algorithms are restarted after the registration of each couple of images, using the stagnation criterion defined in section 10.3.2. The results obtained using MLSDO, as a static optimization algorithm, are also given. The parameters found for the elastic transformation model are given in Table 10.2. In Table 10.3, the average number of evaluations among 20 runs of the algorithms is given. The sum of the best objective function values (see Equation (10.14)) of each registration of the sequence is also given, averaged on 20 runs of the algorithms. The convergence of MLSDO, and that of the best performing static optimization algorithm on the problem at hand, *i.e.* CMA-ES, are illustrated by the curves in Figure 10.10. It shows the evolution of the relative error $\left(\frac{C^*(\Phi(t))-C(\Phi(t))}{C^*(\Phi(t))}\right)$ between the value of the objective function of the best solution found $(C^*(\Phi(t)))$ and that of the current solution $(C(\Phi(t)))$ for each couple of images (t). The presented curves give an idea about the convergence of the algorithms to an optimal value. It can also be seen as a stagnation metric of the algorithms. In this figure, the number of evaluations per registration of a couple of images is fixed to 5000, in order to enable the comparison of the convergence of the algorithms. For readability, a logarithmic scale is used on the ordinates.

As we can see, the average sum of objective function values given in Table 10.3 shows that the algorithms have a similar average precision. However, we can see in Table 10.3 that the number of evaluations of the objective function performed by

Table 10.3 Average number of evaluations to register all couples of images, and average sum of $C^*(\Phi(t))$, obtained by each algorithm.

	Algorithm	Evaluations	$\sum_{t=1}^{19} C^*(\Phi(t))$
Dynamic optimization	MLSDO	6880.68 ± 585.92	1.21 ± 7.0E-4
Static optimization	CMA-ES	7709.14 ± 467.75	1.21 ± 9.1E-4
	SPSO-07	8007.21 ± 364.24	1.21 ± 8.8E-4
	DE	9131.25 ± 279.20	1.21 ± 9.3E-4
	MLSDO	9522.76 ± 648.87	1.21 ± 1.7E-3

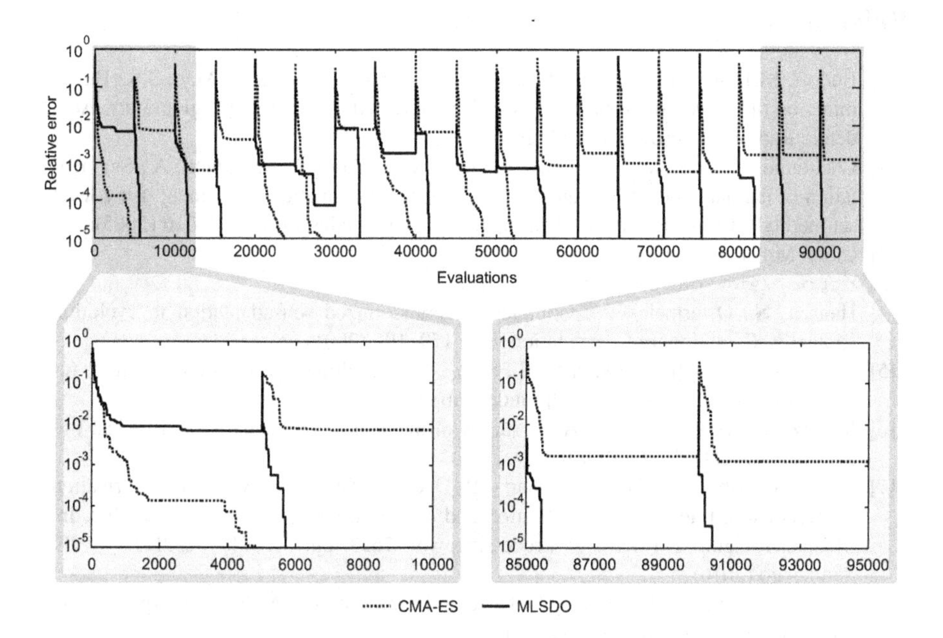

Fig. 10.10 Convergence graph of MLSDO and CMA-ES on the problem at hand.

MLSDO, used as a dynamic optimization algorithm, is significantly lower than the ones of the static optimization algorithms. A Wilcoxon-Mann-Whitney statistical test has been applied on the numbers of evaluations performed by MLSDO and CMA-ES, the best ranked algorithms in terms of the number of evaluations. This test confirms at a 99% confidence level that there is a significant difference between their performances. It can be seen also in Figure 10.10 that the convergence of MLSDO to an acceptable solution is faster than CMA-ES for the registration of most of the couples of contours, especially for the last ones. MLSDO needs indeed to learn from the first registrations in order to accelerate its convergence on the next ones. Thus, this comparison shows the efficiency of MLSDO and the significance of using a dynamic optimization algorithm on the problem at hand.

10.5 Conclusions

In this chapter, a registration process based on a dynamic optimization algorithm is proposed to register quickly all the images of a cine-MRI sequence. It takes profit from the effectiveness of the dynamic optimization paradigm. The process is sequentially applied on all the 2D images. The entire procedure is fully automated and provides an accurate assessment of the ROI deformation throughout the entire cardiac cycle. Our work under progress consists in the parallelization of the MLSDO algorithm using Graphics Processing Units.

References

[1] Budoff, M.J., Ahmadi, N., Sarraf, G., Gao, Y., Chow, D., Flores, F., Mao, S.S.: Determination of left ventricular mass on cardiac computed tomographic angiography. Academic Radiology 16(6), 726–732 (2009)

[2] Chenoune, Y., Deléchelle, E., Petit, E., Goissen, T., Garot, J., Rahmouni, A.: Segmentation of cardiac cine-MR images and myocardial deformation assessment using level set methods. Computerized Medical Imaging and Graphics 29(8), 607–616 (2005)

[3] Clerc, M., et al.: The Particle Swarm Central, `http://www.particleswarm.info`

[4] Hansen, N., Ostermeier, A.: Completely derandomized self-adaptation in evolution strategies. Evolutionary Computation 9(2), 159–195 (2001)

[5] Lepagnot, J., Nakib, A., Oulhadj, H., Siarry, P.: A multiple local search algorithm for continuous dynamic optimization (under submission)

[6] Maintz, J.B.A., Viergever, M.A.: A survey of medical image registration. Medical Image Analysis 2(1), 1–36 (1998)

[7] Nakib, A., Aiboud, F., Hodel, J., Siarry, P., Decq, P.: Third brain ventricle deformation analysis using fractional differentiation and evolution strategy in brain cine-MRI. In: Medical Imaging 2010: Image Processing, vol. 7623, pp. 76232I–76232I–10. SPIE, San Diego (2010)

[8] Price, K., Storn, R., Lampinen, J.: Differential Evolution - A Practical Approach to Global Optimization. Springer (2005)

[9] Studholme, C., Hill, D.L.G., Hawkes, D.J.: An overlap invariant entropy measure of 3D medical image alignment. Pattern Recognition 32(1), 71–86 (1999)

[10] Sundar, H., Litt, H., Shen, D.: Estimating myocardial motion by 4D image warping. Pattern Recognition 42(11), 2514–2526 (2009)

[11] Zitová, B., Flusser, J.: Image registration methods: a survey. Image and Vision Computing 21(11), 977–1000 (2003)

Chapter 11
Artificial Immune System for Solving Dynamic Constrained Optimization Problems

Victoria S. Aragón, Susana C. Esquivel, and Carlos A. Coello

Abstract. In this chapter, we analyze the behavior of an adaptive immune system when solving dynamic constrained optimization problems (DCOPs). Our proposed approach is called Dynamic Constrained T-Cell (DCTC) and it is an adaptation of an existing algorithm, which was originally designed to solve static constrained problems. Here, this approach is extended to deal with problems which change over time and whose solutions are subject to constraints. Our proposed DCTC is validated with eleven dynamic constrained problems which involve the following scenarios: dynamic objective function with static constraints, static objective function with dynamic constraints, and dynamic objective function with dynamic constraints. The performance of the proposed approach is compared with respect to that of another algorithm that was originally designed to solve static constrained problems (*SMES*) and which is adapted here to solve DCOPs. Besides, the performance of our proposed DCTC is compared with respect to those of two approaches which have been used to solve dynamic constrained optimization problems (RIGA and dRepairRIGA). Some statistical analysis is performed in order to get some insights into the effect that the dynamic features of the problems have on the behavior of the proposed algorithm.

Victoria S. Aragón · Susana C. Esquivel
Laboratorio de Investigación y Desarrollo en Inteligencia Computacional (LIDIC),
Universidad Nacional de San Luis - Ejército de los Andes 950 (5700)
San Luis, Argentina
e-mail: vsaragon@unsl.edu.ar

Carlos A. Coello
CINVESTAV-IPN (Evolutionary Computation Group) - Computer Science Department,
Av. IPN No. 2508, Col. San Pedro Zacatenco,
México D.F. 07300, México
e-mail: ccoello@cs.cinvestav.mx

E. Alba et al. (Eds.): Metaheuristics for Dynamic Optimization, SCI 433, pp. 225–263.
springerlink.com

11.1 Introduction

Three are the main scenarios that we could have in dynamic constrained optimization problems (DCOPs): 1) we could have a dynamic objective function and static constraints (if the objective function changes over time, this could affect the location of the optimum, which could move, for example, from a disconnected feasible region to another one), 2) we could have a static objective function and dynamic constraints (in this case, new local optima (or even a new global optimum) could appear as the infeasible region changes), and 3) we could have a dynamic objective function and dynamic constraints (this is perhaps the most difficult case, since there are changes in both the location of the optimum and the infeasible region).

Regardless of the scenario that we consider, DCOPs are, clearly, very difficult problems [17]. When the objective function is moving, it is necessary to have good mechanisms to track it. When the dynamic components of the problem are given by the constraints, then the changes could vary the shape, ratio (with respect to the entire search space), and/or structure of the feasible region [17].

DCOPs are not simply an academic challenge, since there are several real-world problems with such features. For example: the cargo movement problem in metropolitan areas adjacent to marine ports. In particular, truck scheduling and route planning, where ISO (International Standards Organization) containers need to be transferred between marine terminals, intermodal facilities, and end customers. In such an application, the objective is to reduce empty miles, and to improve customer service. A dynamic component is given in this case for incorporating the information of new customers after the set of routes has been determined [10]. Another example is the assembly of a schedule for transport ships, where the ships transport liquified natural gas from different ports around the world to one destination port. In this case, after finding a valid schedule, a recalculation may be needed because some ships got delayed (for example, due to a storm or some mechanical damage) [14]. Another example are the hydro-thermal power generation systems, in which both the hydroelectric and the thermal generating units are utilized to meet the total power demand. The optimum power scheduling problem involves the allocation of power to all concerned units, but the total fuel cost of thermal generation and emission properties has to be minimized, while satisfying all the constraints imposed by the hydraulic and power system networks. The problem is dynamic due to the changing nature of power demand over time. Thus, ideally, the optimal power scheduling problem should be considered as an online dynamic optimization problem in which solutions must be found when there is a change in the power demand [9]. All of these problems have a great industrial impact, and their efficient solution can, therefore, be very profitable. Surprisingly enough, however, the literature on the solution of DCOPs is relatively scarce. In this chapter, we propose the use of an algorithm based on an adaptive immune system model for solving DCOPs. The proposed approach has been inspired by the immune responses mediated by the T cells, and constitutes an extension of an algorithm (developed by the authors of this chapter) that was originally designed for solving static constrained optimization problems.

The remainder of this chapter is organized as follows. Section 11.2 formally defines the problem of our interest. In Section 11.3, we provide a review of the previous related work. In Section 11.4, we describe our proposed approach. Section 11.5 presents the test problems and performance measures adopted to evaluate our proposed approach. This section also describes the algorithm that we used to compare our results and the experimental design that we adopted. Finally, Sections 11.6 and 11.7 present the results that we obtained from the experiments performed and our conclusions, respectively.

11.2 Problem Statement

We are interested in solving problems of the form[1]:
minimize

$$f(X;t) \tag{11.1}$$

subject to:

$$g_j(X;t) \leq 0 \qquad j = 1,\ldots,m \tag{11.2}$$

$$h_k(X;t) = 0 \qquad k = 1,\ldots,p \tag{11.3}$$

$$x_i^l \leq x_i \leq x_i^u \qquad i = 1,\ldots,n. \tag{11.4}$$

In Equation (11.1), f designates the objective function, $X = (x_1, x_2, \ldots, x_n)^T$ is a vector containing the design variables, and t is a positive integer which denotes time. The remaining functions correspond to inequality constraints g (Equation (11.2)), equality constraints h (Equation (11.3)), and side constraints with lower and upper limits indicated by the superscripts l and u (Equation (11.4)), respectively. Both the objective function and the constraints could be linear or nonlinear.

When an inequality constraint takes a zero value at the optimum, we say that it is **active**. By definition, all equality constraints are active at all points of the search space.

11.3 Previous Related Work

The literature on the solution of DCOPs using artificial immune systems is very scarce and is briefly reviewed next. Schiex et al. [29] defined the maintenance solution problem in dynamic constraint satisfaction problems (CSPs) and underlined that the iterative application of commonly used constraint satisfaction algorithms would result in redundant search and inefficiency. They also indicated that a complete description of the space explored justified in terms of the set of constraints may

[1] Without loss of generality, we will assume only minimization problems.

grow exponentially in space. Thus, they proposed a class of nogood recording algorithms to solve the satisfaction problem and simultaneously offered a polynomially bounded space compromise between both of these approaches. They outperformed all algorithms considered for the solution maintenance problem in dynamic CSPs, and also provided very good results for static CSPs.

Modi et al. [16] proposed a formalization of distributed resource allocation that, its authors claim to be expressive enough to represent both dynamic and distributed aspects of the problem. They defined different categories of difficulties of the problem and presented complexity results for them. Also, they defined the notion of Dynamic Distributed Constraint Satisfaction Problem (DyDCSP) and presented a generalized mapping from distributed resource allocation to DyDCSP. Through both theoretical analysis and experimental verifications, they showed that this approach to dynamic and distributed resource allocation is powerful and can be applied to real-world problems such as the Distributed Sensor Network Problem.

Mailler [11] presented two protocols for solving dynamic distributed constraint satisfaction problems which are based on the classical Distributed Breakout Algorithm (DBA) and the Asynchronous Partial Overlay (APO) algorithm. These two new algorithms are compared on a broad class of problems varying their overall difficulty as well as the rate at which they change over time. The results obtained by the author indicate that neither of the algorithms completely dominates the other on all problem classes, but that depending on environmental conditions and the needs of the user, one method may be preferable over the other.

Richter and his collaborators have investigated the use of evolutionary algorithms in dynamic environments in a number of papers [19–28]. Some of these works will be briefly reviewed next.

In [19], Richter studied the behavior of an evolutionary algorithm in dynamic environments that change chaotically. Additionally, he analyzed the concept of dynamic severity when applied to chaotic changes, as well as the relationship between severity, change frequency, and predictability of the changes. Richter et al. [20] proposed a memory scheme based on abstraction for evolutionary algorithms with the aim of solving dynamic optimization problems. In this scheme, the memory does not store good solutions as themselves but as their abstraction, i.e., their approximate location in the search space. Thus, when the environment changes, the stored abstraction information is extracted to generate new individuals into the population. The authors argued that their results show the efficiency of their proposed approach. In a further paper, Richter [25] proposed a memory design for solving constrained dynamic optimization problems using an evolutionary algorithm. Based on ideas from abstract memory, Richter introduced and tested two schemes: blending and censoring. Through some experiments he showed that such a memory can be used to solve certain types of constrained dynamic problems. Richter's work also involves a study of the automatic detection of changes through population-based and sensor-based schemes [22]. In another paper, Richter [24] employed a negative selection algorithm to detect changes in order to solve dynamic optimization problems.

His numerical experiments showed that the use of an immunological approach can be successfully used to solve the change detection problem for dynamic fitness landscapes. In Richter et al. [26], the authors proposed a method for generating variable-sized detectors in the framework of negative selection based artificial immune systems. The method is inspired by the idea of interpreting the feature space as a potential field. The authors used a divide-and-conquer algorithm in order to accelerate the generation of detectors. They also generalized the idea of overlapping detectors by introducing multiple detector layers compromising different geometric structures for the used detectors. In [27], Richter et al. proposed a method for solving the change detection problem for constrained dynamic optimization using evolutionary algorithms. The basis for this detection is the use of both fitness values and corrected fitness values of the individuals, without requiring any additional data from the fitness landscape. The fitness distribution of the individuals of one generation is analyzed using simple statistical measures and by hypothesis test based classifiers. The authors argued that good results could be found for the constraint change and the landscape change detection, for classifiers based on the Kolmogorov-Smirnov test whereas simple statistical methods failed to detect changes with high robustness. Richter et al. [28] considered optimization problems with a dynamic fitness landscape and dynamic constraints that may change in an independent manner. The authors argued that this situation can lead to asynchronous change patterns with the possibility of occasional synchronization points. So, they presented a framework for describing such a dynamical setting and for performing numerical experiments on the algorithm's behavior. The DynCOAA algorithm was proposed by Mertens et al. [14]. It is based on the Ant Colony Optimization (ACO) metaheuristic and was designed for solving DCOPs. Its authors compared this approach with respect to two other algorithms in two different types of problems: artificial graph coloring problems and real-world ship scheduling problems. The main conclusions from the authors were the following: DynCOAA is well suited for solving DCOPs, as it beats DynAWC [16, 29] in the two types of problems that they studied and DynDBA [11] in one of them. Their second conclusion was that choosing the right algorithm for a specific problem is very important because the performance of the algorithms greatly depends on the type of problem being solved. The main difference between DynCOAA and the approach proposed by us in this paper resides on the biological metaphors they are based on: DynCOAA is based on the behavior of ants when foraging for food and ours is based on the behavior of the immune system when receiving an external attack. There is also another difference worth emphasizing. DynCOAA builds a solution and then it takes the best solution as a guideline when a change occurs. In contrast, our approach does not follow the direction of the best solution found so far but, instead, the activities of each cell are influenced by another random cell from the same population to which the cell belongs.

Deb et al. [9] modified the NSGA-II so that it could track down a new Pareto optimal front, as soon as there was a change in the problem. The authors investigated the introduction of a few randomly generated solutions or a few mutated solutions. The proposed approaches were tested and compared on a benchmark problem and on the real-world optimization of a hydro-thermal power scheduling problem. This systematic study was able to find a minimum frequency of change allowed in the problem for two dynamic EMO procedures to adequately track down the Pareto optimal frontiers on-line. Based on their results, the authors suggested an automatic decision-making procedure for arriving at a single optimal solution on-line.

Nguyen et al. [17] studied the characteristics that might make dynamic constrained problems difficult to solve by some of the existing dynamic optimization and constraint-handling algorithms. They also introduced a set of (numerical) dynamic benchmark problems with the features analyzed in the paper. They tested several dynamic and constrained optimization strategies on the proposed benchmark problems, including the use of two canonical algorithms, a triggered hyper-mutation Genetic Algorithm (GA), a random-immigrant GA named RIGA (*RIGA-elit* is an algorithm derivated from the GA (Genetic Algorithm), but after applying the mutation operator on each generation, a fraction of the population is replaced by randomly generated individuals, in order to maintain diversity.), and a GA plus repair, named *dRepairGA*. The authors stated that their results confirm that dynamic constrained problems have special characteristics that might significatively affect the performance of the algorithms traditionally used for static problems. At the end of the paper, they proposed a list of possible requirements that an algorithm should meet to solve dynamic constrained problems effectively. Our algorithm differs from the ones used by Nguyen et al. [17] in the following aspects: in our proposed DCTC, the mutation probability is not increased when a change occurs, but instead, it remains fixed during all the search process. Additionally, any of the randomly generated cells is inserted into any population, except when virgin cells are initializated. Finally, our proposed DCTC works over feasible as well as over infeasible solutions and, therefore, it does not require any repair algorithm. Nguyen et al. [19] proposed a new approach to solving DCOPs, combining existing dynamic optimization techniques with constraint-handling techniques in order to handle objective-function changes and constraint-function changes separately and differently. They modified an existing repair method to track down the moving constraints and combined it with existing random-immigrant and hyper-mutation operators, in order to handle objective-function changes. They also used different techniques to detect objective function changes and constraint-function changes separately and differently. This proposed approach was used to derive two new algorithms. The first of them was called *dRepairGA* and is based on *dRepairRIGA*, an algorithm which integrates the characteristics of a GA, a repair method (to transform infeasible solutions into feasible ones, if possible) and the dynamic optimization strategy *RIGA-elit*). The second approach was called *dGenocop* and is based on Genocop III. They also proposed variants of these two algorithms. The authors validated their algorithms using 18 test problems and argued that their proposed algorithms were able to significantly outperform GA/RIGA/HyperM and GA+Repair/Genocop III in solving DCOPs while still maintaining equal or better overall performance in solving other groups of problems except for the static cases.

11.4 Our Proposed Approach

Our proposed approach consists of an adaptive immune system model based on the immune responses mediated by the T cells. Originally, this approach was used to solve static optimization problems (see [3]). Then, it was extended to solve dynamic (unconstrained) problems and, later on, to solve (static) constrained problems (see [1, 2, 4, 5]).

The model that we developed is called TCELL, and it considers many of the processes that T cells suffer from their origin in the hematopoietic stem cells in the bone marrow until they become memory cells. T cells belong to a group of white blood cells known as lymphocytes. They play a central role in cell-mediated immunity. They present special receptors on their cell surface called T cell receptors (TCR[2]). All T cells originate from hematopoietic stem cells in the bone marrow. Hematopoietic progenitor derived from hematopoietic stem cells populate the thymus and expand by cell division to generate a large population of immature thymocytes [30].

Several subsets of the T cells have been discovered, each with a distinct function. Thus, they can be classified in different populations according to the antigen receptor they express. These antigens receptors could be TCR-1 or TCR-2. Additionally, TCR-2 cells express CD4 or CD8.[3]

Also, T cells can be divided into three groups according to their maturation or development level (phylogenies of the T cells [8]): virgin, effector, and memory cells. Virgin cells are those which have never been activated (i.e., they have not suffered proliferation or differentiation). At the beginning, these cells do not express CD4, nor CD8. However, later on, they develop and express both marks, CD4 and CD8. Finally, virgin cells mature and express only one mark, either CD4 or CD8. Before these cells release the thymus, they are subject to both positive selection [12] and negative selection [12]. Positive selection guarantees that the only survivors are the cells with TCRs that present a moderate affinity with respect to the self MHC. Negative selection eliminates the cells with TCRs that recognize self components unrelated to the MHC.

Effector cells are a type of cells that express only one mark, CD4 or CD8. They can be activated by co-stimulating signals plus their ability to recognize an antigen [7, 13]. The immune cells interact through the secretion of cytokines.[4] Cytokines allow cellular communication. Thus, an immune cell c_i influences the activities (proliferation and differentiation) of another cell c_j through the secretion of cytokines, modulating the production and secretion of cytokines by c_j [8]. In order to activate an effector cell, a co-stimulated signal is necessary. Such signal corresponds to the cytokines secreted from another effector cell. The activation of an effector cell im-

[2] TCRs are responsible for recognizing antigens bound to major histocompatibility complex (MHC) molecules.

[3] Lymphocytes express a large number of surface molecules that can be used to mark different cellular populations. CD means *Cluster Denomination* and indicates the group to which lymphocytes belong.

[4] Proteins act as signal transmitters between cells, and also induce growth, differentiation, activation, etc.

plies that it will be replicated and differentiated. Thus, the proliferation process has as its goal to replicate the cells and the differentiation process changes the clones so that they acquire specialized functional properties.

Finally, the memory cells are those that remain in the host even when the infection or danger has been overtaken, so that in the future, they are able to get stimulated by the same or by a similar antigen. Usually, they respond (through proliferation and differentiation) faster with a low dosage of antigens than the B memory cells. It is worth noting that, although the effector and memory cells are replicated, they are not subject to somatic hypermutation. For the effector cells, the differentiation process is subject to the cytokines released by another effector cell. In our model, the differentiation process of the memory cells relies on their own cytokines. The immune response consists of two phases: the first (called *recognizing phase*) involves the processes that suffer only the virgin cells and the second (called *effector phase*) is related to the processes that suffer the effector and memory cells. The *recognizing phase* has to provide some diversity so that the next phase can produce a cell to eliminate the antigen. Meanwhile, the *effector phase* is in charge of doing this job.

Summarizing the features of the natural immune system that inspired our model, we can highlight that TCELL considers that T cells react when the system is invaded by an external pathogen as well as the presence of co-stimulating signals, sent by the own T cells, according to the Danger Theory. Additionally, TCELL uses the Self-Non-self concept (in the *recognizing phase*) in order to remove undesirable cells which can be considered dangerous to the host. Finally, TCELL also considers the interaction among the T cells through the secretion of cytokines, as a communication mechanism.

11.4.1 Proposed Algorithm Based on T-CELL

DCTC (Dynamic Constrained T-Cell) is an algorithm inspired on the TCELL model [4], which we propose here to solve dynamic constrained optimization problems. DCTC operates on four populations, corresponding to the groups in which the T-cells are divided: (1) Virgin Cells (VC), (2) Effector Cells with cluster denomination CD4 (CD4), (3) Effector Cells with cluster denomination CD8 (CD8), and (4) Memory Cells (MC). Each population is composed by a set of T cells whose characteristics are subject to the population to which they belong.

Virgin Cells (VC) do not suffer the activation process. They have to provide diversity. This is reached through the random acquisition of TCR receptors. Virgin cells are represented by: 1) a *TCR* represented by a bitstring using Gray coding (called TCR_b) and 2) a *TCR* represented by a vector of real numbers (called TCR_r).

Into the natural immune system, the positive and negative selections have to remove the potentially harmful cells. Thus, in our proposed algorithm, positive selection is in charge of eliminating the cells that recognize the antigen with a low matching. On the other hand, negative selection has to eliminate the cells that have a similar TCR, according to a Hamming or a Euclidean distance, depending on whether the TCR is represented by a TCR_b or by a TCR_r.

Effector Cells are composed by: 1) a TCR_b or TCR_r, if they belong to CD4 or CD8, respectively, 2) a proliferation level, and 3) a differentiation level. The goal of this type of cell is to explore in a global way the search space. Thus, CD4 explores the search space, taking advantage of the Gray coding properties (there is only one bit of difference between two consecutive numbers), while CD8 uses real numbers representation (big or small jumps).

The goal of the memory cells is to explore the neighborhood of the best found solutions. These cells are represented by the same components as CD8.

In our proposal, the TCR identifies the decision variables of the problem, independently of the TCR representation. The proliferation level indicates the number of clones that will be assigned to a cell and the differentiation level indicates the number of bits or decision variables (depending on the TCR representation adopted) that will be changed, when the differentiation process is applied.

The activation of an effector cell, called ce_i, implies the random selection of a set of potential activator (or stimulating) cells. The closest cell to ce_i (using Hamming or Euclidean distance), according to the TCR in the set, is chosen to become the stimulating cell, say ce_j. Then, ce_i proliferates and differentiates.

At the beginning, the proliferation level of each stimulated cell, ce_i, is given by a random value within [1, 3],[5] but then, it is determined taking into account the proliferation level of its stimulating cell (ce_j). If the ce_i is better than ce_j, then ce_i keeps its own proliferation level; otherwise, ce_i receives a level which is 10% lower than the level of ce_j.

Memory cells proliferate and differentiate according to their proliferation level (randomly between 1 and the size of MC[6]) and differentiation level (number of decision variables,[7]) respectively. Both levels are independent from the other memory cells.

In our proposed DCTC algorithm, the constraint-handling method needs to calculate, for each cell (solution), regardless of the population to which it belongs, the following: 1) the sum of constraint violations (sum_res)[8] and 2) the value of the objective function (only if the cell is feasible).

We consider that a ce_i cell is better than a ce_j cell if: 1) TCR's ce_i is feasible and TCR's ce_j is infeasible, 2) both cells have feasible TCRs but the objective function value of ce_i is lower than the objective function value of ce_j, and 3) both cells have infeasible TCRs but sum_res' of ce_i is lower than sum_res' of ce_j. This criterion is used to sort the population. Each type of cell has its own differentiation process, which is blind to their representation and population:

Differentiation for CD4: the differentiation level of ce_i is determined by the Hamming distance between the stimulated (ce_i) and stimulating (ce_j) cells. It indicates the number of bits to be changed. Each decision variable and the bit to

---

[5] This value was derived after numerous experiments.

[6] This is an arbitrary value in order to avoid overloading the number of required parameters.

[7] This value was set thinking on performing an intensive local search.

[8] This is a positive value determined by $g_i(x)^+$ for $i = 1, \ldots, m$ and $|h_k(x)|$ for $k = 1, \ldots, p$.

be inverted are chosen in a random way. The bits change according to a probability $prob_{diff-CD4}$. The pseudo-code for the proliferation and differentiation of cell ce_i is shown in Algorithm 11.1.

Algorithm 11.1. Differentiation CD4

```
for np = 1 to Proliferation Level of ce_i do
  | clone^np ← ce_i ;
end
for nd = 1 to Differentiation Level of ce_i do
  | if prob_diff-CD4 then
  |   | k ← U(1, | vd |);
  |   | l ← U(1, | bits_k |);
  |   | Invert the l^th-bit of vd_k of the clone^np;
  | end
end
```

where $U(1,w)$ refers to a random number with a uniform distribution in the range (1,w), $\mid vd \mid$ is the number of decision variables of the problem, $\mid bits_k \mid$ is the number of bits to represent the kth decision variable, and vd_k indicates the kth decision variable.

Differentiation for CD8: the differentiation level for cell ce_i is related to its stimulating cell (ce_j). If the TCR_r of the ce_j is better than the TCR_r of the stimulated cell ce_i (according to the objective function value), then the level (for ce_i) is a random number within [1, $\mid dv \mid$[9]]; otherwise, it is a random value within [1, $\mid dv \mid /2$], where $\mid dv \mid$ is the number of decision variables of the problem. Each variable to be changed is chosen in a random way and it is modified according to Equation (11.5):

$$x' = x \pm \frac{U(0, lu - ll)^{U(0,1)}}{10^7 iter}, \tag{11.5}$$

where x and x' are the original and the mutated decision variables, respectively. lu and ll are the upper and lower bounds of x, respectively. $iter$ indicates the number of iterations until reaching the maximum number of evaluations for a change. At the moment of the differentiation of a cell (ce_i), the value of the objective function of its stimulating cell (ce_j) is taken into account. In order to determine if $r = \frac{U(0,lu-ll)^{U(0,1)}}{10^7 iter}$, will be added or subtracted to x, the following criteria are considered: if ce_j is better than ce_i and the decision variable value of ce_j is less than the value of ce_i, or if ce_i is better than ce_j and the decision variable value of ce_i is less than the value of ce_j, then r is subtracted from x; otherwise, r is added to x. Both criteria aim to guide the search towards the best solutions found so far. The pseudo-code for the proliferation and differentiation of the cell ce_i with the stimulating cell ce_j is shown in Algorithm refCD8:

[9] If the stimulating cell is better, then ce_i should change more decision variables

Algorithm 11.2. Differentiation CD8

for $np = 1$ *to Proliferation Level of* ce_i **do**
  Clone$^{np} \leftarrow ce_i$;
  for $nd = 1$ *to Differentiation Level of* ce_i **do**
    $k \leftarrow U(1, \mid vd \mid)$;
    $r \leftarrow \frac{U(0,lu-ll)}{10^7 iter}^{U(0,1)}$;
    if $f(ce_{jTCR_r})$ *is better than* $f(ce_{iTCR_r})$ *and* $ce_{jTCR_{r_k}} < ce_{iTCR_{r_k}}$ *o* $f(ce_{iTCR_r})$ *is better than* $f(ce_{jTCR_r})$ *and* $ce_{jTCR_{r_k}} > ce_{iTCR_{r_k}}$ **then**
      Clone$^{np}TCR_{r_k} \leftarrow ce_{iTCR_{r_k}} - r$
    else
      else if $f(ce_{jTCR_r})$ *is better than* $f(ce_{iTCR_r})$ *and* $ce_{jTCR_{r_k}} > ce_{iTCR_{r_k}}$ *o* $f(ce_{iTCR_r})$ *is better than* $f(ce_{jTCR_r})$ *and* $ce_{jTCR_{r_k}} < ce_{iTCR_{r_k}}$ **then**
        Clone$^{np}TCR_{r_k} \leftarrow ce_{iTCR_{r_k}} + r$;
      end
      Add or subtract r with probability 50%;
    end
  end
end

where $U(w_1, w_2)$ refers to a random number with a uniform distribution in the range (w_1,w_2), $\mid vd \mid$ is the number of decision variables of the problem, lu_x and ll_x are the upper and lower bounds of x, respectively. $iter$ indicates the number of iterations until reaching the maximum number of evaluations for a change. $f(ce_{hTCR_r})$ is the objective function value for the TCR$_r$ of the cell ce_h, and $ce_{hTCR_{r_k}}$ indicates the kth decision variable of the cell h. If after ten trials, the procedure cannot find an x' in the allowable range, a random number with a uniform distribution is assigned to it.

Differentiation for MC: the number of decision variables to be changed is determined by the differentiation level of the cell to be differentiated. Each variable to be changed is chosen in a random way and it is modified according to Equation (11.6):

$$x' = x \pm \left(\frac{U(0, lu_x - ll_x)}{10^7 iter}\right)^{U(0,1)}, \tag{11.6}$$

where x and x' are the original and the mutated decision variables, respectively. $U(0,w)$ refers to a random number with a uniform distribution in the range (0,w). lu_x and ll_x are the upper and lower bounds of x, respectively. $iter$ indicates the number of iterations until reaching the maximum number of evaluations for a change. In a random way, we decide if $r = \left(\frac{U(0,lu_x-ll_x)}{10^7 iter}\right)^{U(0,1)}$ will be added or subtracted to x. If after ten trials, the procedure cannot find an x' in the allowable range, then a random number with a uniform distribution is assigned to it.

The general structure of our proposed algorithm for dynamic constrained optimization problems is given in Algorithm 11.3.

Algorithm 11.3. DCTC Algorithm

```
 1: Initialize_VC();
 2: Evaluate_VC();
 3: Assign_Proliferation();
 4: Divide_CDs();// take into account feasibility of the solutions
 5: Positive_Selection_CD4();// eliminate the worst cells in CD4
 6: Positive_Selection_CD8();// eliminate the worst cells in CD8
 7: Negative_Selection_CD4();// eliminate the most similar cells in CD4
 8: Negative_Selection_CD8();// eliminate the most similar cells in CD8
 9: while A predetermined number of changes has not been reached do
10:    while A predetermined number of evaluations has not been performed do
11:       Proliferate_CD4();
12:       Differentiate_CD4();
13:       Sort_CD4();
14:       Proliferate_CD8();
15:       Differentiate_CD8();
16:       Sort_CD8();
17:       Insert_CDs_en_MC();
18:       for i = 1 to rep_MC do
19:          Proliferate_MC();
20:          Differentiate_MC();
21:       end for
22:       Sort_CM();
23:    end while
24:    Statistics();
25:    Change_Function();
26:    Re-evaluate_Populations();
27: end while
```

The algorithm works in the following way. At the beginning, the TCR_b and TCR_r from the virgin cells are initialized in a random way, according to the TCR's encoding (step 1). Then, each TCR of a virgin cell is evaluated (step 2). In step 3, the proliferation levels are assigned. Then, in step 4, the virgin cells are divided taking into account their feasibility. Next, the feasible cells, TCR_b and TCR_r, from VC are selected to form CD4 and CD8, respectively. If it is not possible to complete the required number of cells, then the infeasible TCRs needed to reach such a value are selected. Each effector cell will inherit the proliferation level of the virgin cell which received the TCR.

The negative and positive selections are applied to each effector population (CD4 and CD8). The first selection eliminates 10% of the worst cells and the second selection eliminates cells that are similar among them (keeping the best from them). This mechanism works in the following way: for each effector cell, we search inside its population the closest cell (using Hamming or Euclidean distance according to the TCR's cell) and the worst from them is eliminated. This process reduces the effector's population sizes.

The first iteration (line 9) is controlled by the number of changes of the objective function.

Furthermore, for each change, a maximum number of objective function evaluations is allowed (line 10)[10]. The steps inside the last iteration are: first, to activate the CD4 population; in other words, to perform proliferation and differentiation of all the cells from CD4 (lines 11 and 12). Then, these cells are sorted (line 13). Next, the CD8 population is activated. This means that we perform proliferation and differentiation of all the cells from CD8 (lines 14 and 15), which are sorted in (line 16).

The best solutions from CD4 and CD8 are inserted or are used to replace the 50% of the worst solutions in MC (depending on whether or not, MC is empty) (lines 17). Since the representation schemes of the TCR, for CD4 and MC, are different, before the insertion of the best cell from CD4 (with TCR_b) into MC, the receptor has to be converted into a real-values vector (TCR_r). For this process, we use Equation (11.7), which takes as input a bitstring generated with Gray coding and returns a real number (this process is applied as many times as decision variables has the problem):

$$dv_j = l_{lj} + \frac{\sum_{i=0}^{L_j} 2^{L_j - i} dv'_{ij}(lu_j - ll_j)}{2^L_j - 1}, \tag{11.7}$$

where dv_j is the jth decision variable with $j = 1, \ldots,$ number of decision variables of the problem, L_j is the number of bits for the jth decision variable, lu_j and ll_j are the upper and lower limits for the decision variable dv_j, respectively. And dv'_{ij} is the ith bit of the bitstring that represents dv_j. Also, Equation (11.7) is used when a cell from CD4 has to be decoded in order to be evaluated. Next, the cells from MC are activated a certain (predefined) number of times, rep_{MC} (lines 19 and 20).

The algorithm is notified about the existence of a change in the environment (line 25), since that information is required in order to re-evaluate the populations (step 26). Even when some literature about change detection exists (see for example, [22, 24, 26, 27]), dealing with this (rather difficult) topic is beyond the scope of this chapter. Here, we only focus on the mechanisms to react to any incoming changes. In fact, when using metaheuristics, it is normally assumed that the search engine will be informed whenever a change has occurred.

[10] Since it is not known *a priori* how many clones will be assigned to each cell, it is possible to exceed the maximum number of evaluations per change in $3 \mid feasible(CD4) \mid + \mid feasible(CD8) \mid + rep_{MC} \mid feasible(MC) \mid^2$, where $feasible(x)$ indicates the number of feasible solutions in population x.

11.5 Experiments

In this section, we describe our experimental setup. This includes a description of the set of test problems used to validate our proposed DCTC, the performance measures used to evaluate it, the corresponding parameters settings, and the description of the algorithm chosen to compare our results. Some statistical analysis is performed in order to determine the effect of the dynamic features of the problems on the behavior of the proposed algorithm.

11.5.1 Dynamic Constrained Benchmark

In order to validate our proposed approach, we adopted eleven dynamic constrained optimization problems from a set that was originally proposed by Nguyen et al. [17]. The subset of problems chosen presents some kind of dynamism, either in the objective function, in the constraint or both. Table 11.1 summarizes the main features of the test problems adopted.

11.5.2 Performance Measures

Here, we describe the performance measures adopted for our experimental study. One of them is relatively popular in the literature [6]: the **offline error** (oe), which represents the average of the best error at each iteration. This measure is calculated here, only for feasible solutions. If an infeasible solution is found then nothing is added, as defined by Equation (11.8).

$$oe = \frac{1}{N_c} \sum_{j=1}^{N_c} \sum_{i=1}^{iter} (f_j^* - f_{ji}^*)), \tag{11.8}$$

where N_c is the total number of changes within an experiment, $iter$ is the current iteration number, f_j^* is the value of the optimum solution for the jth state[11] and f_{ji}^* is the current best fitness value found for the jth state.

The ideal value for oe is zero, which would mean that the optimum was found at the very beginning of each state.

As oe is calculated only for feasible solutions, it is possible that this value becomes zero or a value close to it, but without finding any feasible solution. For this reason, we use this measure along with the measure rf, which calculates the percentage of runs in which at least one feasible solution was found for all the changes that took place. Thus, the ideal value of rf is 100%, which would mean that, in all runs, feasible solutions were found, for all changes.

[11] We call *state* to the time period where objective function and constraints remain fixed.

Table 11.1 Main features of the test problems adopted

Problem	ObjFunc	Constr.	DFR	Parameters Setting
G24_1	f^1 Dynamic	g^1 g^2 Fixed	2	$p_2(t) = r_i(t) = 1; q_i(t) = s_i(t) = 0; p_1(t) = sin(k\pi t + \frac{\pi}{2})$
G24_2	f^1 Dynamic	g^1 g^2 Fixed	2	$if(t\,mod\,2 = 0) = \begin{cases} p_1(t) = sin(\frac{k\pi t}{2} + \frac{\pi}{2}) \\ p_2(t) = \begin{cases} p_2(t-1) & if\,t>0 \\ p_2(0) = 0 & if\,t=0 \end{cases} \end{cases}$ $if(t\,mod\,2 \neq 0) = \begin{cases} p_1(t) = sin(\frac{k\pi t}{2} + \frac{\pi}{2}) \\ p_2(t) = sin(\frac{k\pi(t-1)}{2} + \frac{\pi}{2}) \end{cases}$ $r_i(t) = 1; q_i(t) = s_i(t) = 0;$
G24_3	f^1 Fixed	g^1 g^2 Dynamic	2-3	$p_i(t) = r_i(t) = 1; q_i(t) = s_1(t) = 0; s_2(t) = 2 + t\frac{x_2 \max - x_2 \min}{S}$
G24_3b	f^1 Dynamic	g^1 g^2 Dynamic	2-3	$p_1(t) = sin(k\pi t + \frac{\pi}{2}); p_2(t) = r_i(t) = 1; q_i(t) = s_1(t) = 0; s_2(t) = 2 + t\frac{x_2 \max - x_2 \min}{S}$
G24_4	f^1 Dynamic	g^1 g^2 Dynamic	2-3	$p_1(t) = sin(k\pi t + \frac{\pi}{2}); p_2(t) = r_i(t) = 1; q_i(t) = s_1(t) = 0; s_2(t) = t\frac{x_2 \max - x_2 \min}{S}$
G24_5	f^1 Dynamic	g^1 g^2 Dynamic	2-3	$if(t\,mod\,2 = 0) = \begin{cases} p_1(t) = sin(\frac{k\pi t}{2} + \frac{\pi}{2}) \\ p_2(t) = \begin{cases} p_2(t-1) & if\,t>0 \\ p_2(0) = 0 & if\,t=0 \end{cases} \end{cases}$ $if(t\,mod\,2 \neq 0) = \begin{cases} p_1(t) = sin(\frac{k\pi t}{2} + \frac{\pi}{2}) \\ p_2(t) = sin(\frac{k\pi(t-1)}{2} + \frac{\pi}{2}) \end{cases}$ $r_i(t) = 1; q_i(t) = s_1(t) = 0; s_2(t) = t\frac{x_2 \max - x_2 \min}{S}$
G24_6a	f^1 Dynamic	g^3 g^6 Fixed	2	$p_1(t) = sin(\pi t + \frac{\pi}{2}); p_2(t) = r_i(t) = 1; q_i(t) = s_i(t) = 0;$
G24_6c	f^1 Dynamic	g^3 g^4 Fixed	2	$p_1(t) = sin(\pi t + \frac{\pi}{2}); p_2(t) = r_i(t) = 1; q_i(t) = s_i(t) = 0;$
G24_6d	f^1 Dynamic	g^5 g^6 Fixed	2	$p_1(t) = sin(\pi t + \frac{\pi}{2}); p_2(t) = r_i(t) = 1; q_i(t) = s_i(t) = 0;$
G24_7	f^1 Fixed	g^1 g^2 Dynamic	2	$p_i(t) = r_i(t) = 1; q_i(t) = s_1(t) = 0; s_2(t) = t\frac{x_2 \max - x_2 \min}{S}$
G24_8b	f^2 Dynamic	g^1 g^2 Fixed	2	$p_i(t) = -1; q_i(t) = -(1.4706 + 0.859cos(k\pi t)); q_2(t) = -(3.442 + 0.859sin(k\pi t)); r_i(t) = 1; s_i(t) = 0$

Fixed - There is no change
Dynamic - The function is dynamic

$f^1 = -(X_1(x_1;t) + X_2(x_2;t))$

$f^2 = -3\exp\left(-\sqrt{\sqrt{(X_1(x_1;t))^2 + (X_2(x_2;t))^2}}\right)$

$g^1 = -2Y_1(x_1;t)^4 + 8Y_1(x_1;t)^3 - 8Y_1(x_1;t)^2 + Y_2(x_2;t) - 2$

$g^2 = -4Y_1(x_1;t)^4 + 32Y_1(x_1;t)^3 - 88Y_1(x_1;t)^2 + 96Y_1(x_1;t) + Y_2(x_2;t) - 36$

$g^3 = 2Y_1(x_1;t) + 3Y_2(x_2;t) - 9$

$g^4 = \begin{cases} -1 \; if\,(0 \leq Y_1(x_1;t) \leq 1)\,or\,(2 \leq Y_1(x_1;t) \leq 3) \\ 1 \quad otherwise \end{cases}$

$g^5 = \begin{cases} -1 \; if\,(0 \leq Y_1(x_1;t) \leq 0.5)\,or\,(2 \leq Y_1(x_1;t) \leq 2.5) \\ 1 \quad otherwise \end{cases}$

$g^6 = \begin{cases} -1 \; if\,[(0 \leq Y_1(x_1;t) \leq 1)\,and\,(2 \leq Y_2(x_2;t) \leq 3)]\,or\,(2 \leq Y_1(x_1;t) \leq 3) \\ 1 \quad otherwise \end{cases}$

where $X_i(x;t) = p_i(t)(x + q_i(t))$, $Y_i(x;t) = r_i(t)(x + s_i(t))$, $0 \leq x_1 \leq 3$, $0 \leq x_2 \leq 4$, $p_i(t)$, $q_i(t)$, $r_i(t)$ and $s_i(t)$ are the dynamic parameters. The first two of them determine how the objective function changes over time and the rest determine how the constraint functions change
DFR - Number of Disconnected Feasible Regions
In all problems, except for G24_3 and G24_7, the global optimum switches between disconnected regions
Only in problem G24_3 a new optimum appears without changing the existing one.

From [17], we took the measures *ARR* and *RR*, which indicate how quickly does the algorithm converge to the global optimum before the next change occurs, and how quickly does the algorithm recover from an environmental change and starts converging to a new solution before a change occurs, respectively. *ARR* and *RR* are defined by equations (11.9) and (11.10):

$$ARR = \frac{1}{N_c} \sum_{i=1}^{N_c} \frac{\sum_{j=1}^{p(i)} [f_{ij}^* - f_{i1}^*]}{p(i)[f_i^* - f_{i1}^*]} \tag{11.9}$$

$$RR = \frac{1}{N_c} \sum_{i=1}^{N_c} \frac{\sum_{j=1}^{p(i)} [f_{ij}^* - f_{i1}^*)]}{p(i)[f_{ip(i)}^* - f_{i1}^*]} \tag{11.10}$$

where f_{ij}^* is the objective function value of the best feasible solution found since the last change until the jth iteration of the algorithm of the state i, N_c is the number of changes, $p(i)$ is the maximum number of iterations performed by the algorithm for the state i, and f_i^* is the optimum value for the state i. Both, *ARR* and *RR* have their ideal values in 1. Both *ARR* and *RR* would be 1 when the algorithm is able to recover and converge to a solution (the optimal solution for *ARR*) immediately after a change, and would be equal to zero in case the algorithm is completely unable to recover from the change.

Nguyen et al. [17] proposed how to analyze the convergence behavior/recovery speed of an algorithm through a plot of the *RR*/*ARR* scores. If a point is:

1. on the thick diagonal line, the algorithm has recovered and has converged to the optimum;
2. at the top right corner, the algorithm has recovered quickly and is having a good performance;
3. at the bottom right corner, it is likely that the algorithm has converged to a local optimum;
4. at the bottom left corner, the algorithm has recovered slowly and has not converged yet.

11.5.3 Parameters Settings

Since the literature on dynamic constrained optimization using artificial immune systems is scarce, in order to validate our proposed approach, we decided to adapt an algorithm that was originally proposed to solve (static) constrained optimization problems. This approach was proposed in [15], and it consists of a simple multi-membered evolution strategy, called *SMES*. This approach does not require the use of penalty factors (or a penalty function at all). Instead, it uses a diversity mechanism based on allowing infeasible solutions to remain in the population. It also uses a comparison mechanism based on feasibility to guide the process towards the feasible region of the search space. Also, the initial step size of the evolution strategy is reduced in order to perform a finer search and a combined (discrete/intermediate) panmictic recombination technique improves its exploitation capabilities. The approach was tested with a well-known benchmark, obtaining very competitive results. Its source code was taken from `http://www.cs.cinvestav.mx/~EVOCINV/SES/principal.html`. We modified this code by adding a mechanism for re-evaluating populations after a change occurs. We called this new version *SMESD*. Table 11.2 highlights the main differences between DCTC and *SMESD*. Additionally, our results are indirectly compared to two approaches which are known to perform well in dynamic optimization, namely *RIGA-elit* and *dRepairRIGA* [19].

Table 11.2 Differences and similarities between DCTC and *SMESD*

	DCTC	*SMESD*
Search engine	Artificial Immune System based on T cells behavior	Multimembered Evolution Strategy
Population size	4	1
Number of mutation operators	3	1
Mutation rate	Fixed	It is decreased during the search process
Recombination operator	No	Yes
Constraint-handling mechanism	Discrimination between feasible and infeasible solutions. It uses the sum of constraint violations. No penalty function is required.	Discrimination between feasible and infeasible solutions. It uses the sum of constraint violations. No penalty function is required.
Extra diversity mechanism	No	It allows that the best infeasible solution which is closest to the boundary with the feasible region remaining into the population with some probability (given by the user).

The following experiments were performed for our proposed DCTC and for *SMESD* for validation purposes and in order to compare the performance of these two approaches. Both algorithms were implemented in C and the experiments were performed on a PC having an Intel Pentium P6000 processor, running at 1.87 GHz, and with 3 GB in RAM.

11.5.3.1 Benchmark Problems Setting

Table 11.3 indicates the parameters settings adopted for our proposed DCTC, for *SMESD*, for *RIGA-elit*, and for dRepairRIGA. The dynamic parameters were set as follows:

- Number of runs: 50
- Number of changes: 5/*k*

- Change frequency: 250, 500, and 1000 objective function evaluations per change
- Objective function severity of the changes (k): 0.25 (small), 0.5 (medium) and 1.0 (large). For G24_6a, G24_6c, and G24_6d only, k=1.0
- Constraints severity of the changes (S): 10 (small), 20 (medium) and 50 (large)

It is worth noting that optimal values (necessary to calculate the offline errors), for each period through functions, were not provided in [17]. Thus, we obtained them by executing DCTC and *SMESD* with a budget of 350000 objective function evaluations pro period (for each dynamic parameters setting). Then, we took the best solution for each period (choosing from the union of the solutions obtained by both DCTC and *SMESD*). Therefore, the offline errors for DCTC and *SMESD* were calculated using these optimal values. Since the comparison of the results of DCTC with respect to those of *RIGA-elit* and *dRepairRIGA* is indirect and, considering that in [19], the authors do not describe how the optimal values were found, we cannot guarantee that we used the same values adopted by them. Additionally, in [19], only the results for medium severity and when using 1000 objective function evaluations pro period are reported.

Table 11.3 Parameters settings for DCTC, *SMESD*, *RIGA-elit*, and *dRepairRIGA*

Parameter	DCTC	Parameter	*SMESD*	Parameter	*RIGA-elit*	Parameter	dRepairRIGA
VC	20	parents	10	popsize	25	popsize	20
CD4/ CD8	10	children	20	elitism	Yes	elitism	Yes
CM	5	Apply diversity mecha- nism	Yes	selection method	non- linear ranking	selection method	non-linear ranking
rep_{MC}	2	Selection ratio	0.97	mutation operator	uniform (P=0.15)	mutation operator	uniform (P=0.15)
$prob_{mut}$	0.9			crossover operator	arithmetic (P=0.1)	crossover operator	arithmetic (P=0.1)
clones	3			rand- inming	rate (P=0.3)	rand- inming	rate (P=0.3)
						reference popsize	5
						replace rate	0

In order to statistically determine if when we increase the change frequency, the objective function severity, the constraints severity or if when vary the dynamic features of the problems, our proposed DCTC produces results with significant differences, we performed an analysis of variance (ANOVA) taking into account the offline errors attained by our proposed DCTC from each run of all the experiments performed. Thus, the hypotheses considered were the following:

Null Hypothesis: there is no significant difference among the averages of the offline errors (oe). If there are differences, they are due to random effects.

Alternative Hypothesis: there is a combination of factor values for which the averages of the offline errors (oe) are significatively different and these differences are not due to random effects.

As the results (offline errors) do not follow a normal distribution, we applied the Kruskal-Wallis test, to perform the ANOVA and then the Turkey method in order to determine the experimental conditions for which significant differences exist. The results obtained by the ANOVA proved the Null Hypothesis for several combinations of parameters. However, the Alternative Hypothesis was proved, too. Tables 11.9 to 11.11 summarize the values of severity for which significant different results were detected.

11.6 Discussion of Results

The running time depends on the number of objective function evaluations and the test function itself. For instance, for G24_1, the running time taken by one run ranges from 9 to 200 milliseconds when 250 and 1000 objective function evaluations pro period are performed, respectively. For G24_3, the execution time taken by one run ranges from 12 to 243 milliseconds when 250 and 1000 objective function evaluations pro period are performed, respectively. Finally, for G24_3b, the running time taken by one run ranges from 12 to 239 milliseconds when 250 and 1000 objective function evaluations are performed, respectively.

Table 11.4 shows the results obtained for problems with both a dynamic objective function and dynamic constraints. If we fix the number of objective function evaluations pro period as well as the constraint severity values and increase the objective function severity values, we can see how for G24_3b, in general, the offline errors deteriorate. But there are significant differences only between the results produced when adopting low and medium values of k with respect to those obtained when k is large. Furthermore, when the constraint severity is large, the results which show significant differences are those produced with low values of k with respect to those obtained with medium and large values of k.

For G24_4, an increase in the objective function severity value gives rise to worse offline errors with our proposed DCTC. In this case, we obtain results with significant differences when the number of objective function evaluations pro period is equal to 250 and the constraint severity value is low. For 500 and 1000 evaluations, the results that show significant differences are those produced with low and medium values of k with respect to those obtained with large values of k.

For G24_5, an increase in the objective function severity value also deteriorates the offline errors produced by our proposed DCTC. In this case, we obtain significant differences when, in general, the constraint severity values are low and medium and k grows from low to medium. In general, when the constraint severity value is

Table 11.4 Offline errors (the standard deviation is shown in parentheses) for problems with dynamic objective function and dynamic constraints

			Dynamic Parameters								
			k=0.25			k=0.5			k=1.0		
Ev.	Probl.	Alg.	S=10	S=20	S=50	S=10	S=20	S=50	S=10	S=20	S=50
250	G24_3b	DCTC	**0.59**	**0.54**	**0.56**	**0.55**	**0.57**	**1.15**	1.67	**1.09**	**1.68**
			(0.15)	(0.15)	(0.14)	(0.13)	(0.13)	(0.13)	(0.26)	(0.29)	(0.20)
		SMESD	0.85	0.85[1]	0.87	0.78	0.83	1.44	**1.49**	1.31	2.40
			(0.00)	(0.00)	(0.00)	(0.00)	(0.04)	(0.00)	(0.25)	(0.14)	(0.00)
250	G24_4	DCTC	**0.43**	**0.41**	**0.28**	**0.71**	**0.62**	**0.28**	**1.53**	**1.33**	**1.33**
			(0.07)	(0.05)	(0.04)	(0.08)	(0.05)	(0.06)	(0.13)	(0.09)	(0.11)
		SMESD	0.82	0.80	0.69	1.05	0.99	0.68	1.91	2.17	2.14
			(0.00)	(0.00)	(0.00)	(0.00)	(0.01)	(0.00)	(0.15)	(0.00)	(0.00)
250	G24_5	DCTC	**0.21**	**0.18**	**0.11**	**0.34**	**0.31**	**0.12**	0.46	0.28	**0.31**
			(0.02)	(0.02)	(0.02)	(0.04)	(0.04)	(0.04)	(0.08)	(0.11)	(0.09)
		SMESD	0.38	0.44	0.45	0.68	0.53	0.22	**0.43**	**0.24**	0.35
			(0.01)	(0.11)	(0.03)	(0.10)	(0.03)	(0.00)	(0.00)	(0.01)	(0.00)
500	G24_3b	DCTC	**0.54**	**0.49**	**0.49**	**0.49**	**0.51**	**1.03**	**1.59**	**0.93**	**1.54**
			(0.16)	(0.13)	(0.10)	(0.14)	(0.11)	(0.10)	(0.23)	(0.18)	(0.16)
		SMESD	0.84	0.81	0.84[2]	0.74	0.81	1.40	1.72	1.79	2.36
			(0.00)	(0.00)	(0.01)	(0.00)	(0.00)	(0.01)	(0.00)	(0.06)	(0.00)
500	G24_4	DCTC	**0.36**	**0.35**	**0.23**	**0.63**	**0.55**	**0.20**	**1.41**	**1.26**	**1.20**
			(0.05)	(0.02)	(0.03)	(0.05)	(0.04)	(0.04)	(0.06)	(0.08)	(0.07)
		SMESD	0.81	0.81[3]	0.67	0.83[4]	0.94	0.61	2.23	2.05	2.03
			(0.00)	(0.02)	(0.00)	(0.04)	(0.00)	(0.00)	(0.00)	(0.00)	(0.01)
500	G24_5	DCTC	**0.18**	**0.15**	**0.07**	**0.28**	**0.26**	**0.07**	0.38	**0.20**	**0.25**
			(0.01)	(0.01)	(0.01)	(0.02)	(0.03)	(0.03)	(0.04)	(0.05)	(0.07)
		SMESD	0.47	0.58	0.52	0.47	0.45	0.17	**0.32**	0.37	0.73
			(0.08)	(0.19)	(0.16)	(0.02)	(0.03)	(0.00)	(0.00)	(0.28)	(0.00)
1000	G24_3b	DCTC	**0.47**	**0.43**	**0.39**	**0.41**	**0.45**	**0.98**	**1.44**	**0.83**	**1.45**
			(0.12)	(0.08)	(0.06)	(0.12)	(0.09)	(0.09)	(0.18)	(0.14)	(0.08)
		SMESD	0.87[5]	0.82[6]	0.82	0.73	0.81	1.49[7]	2.23	1.65	2.00
			(0.03)	(0.02)	(0.03)	(0.00)	(0.00)	(0.09)	(0.00)	(0.01)	(0.29)
1000	G24_4	DCTC	**0.32**	**0.32**	**0.19**	**0.57**	**0.50**	**0.15**	**1.36**	**1.17**	**1.13**
			(0.02)	(0.02)	(0.01)	(0.02)	(0.02)	(0.02)	(0.05)	(0.05)	(0.05)
		SMESD	0.81	0.80[8]	0.68[9]	1.06[10]	0.87[11]	0.61	2.07	1.57	2.01
			(0.00)	(0.02)	(0.02)	(0.06)	(0.09)	(0.00)	(0.22)	(0.22)	(0.00)
1000	G24_5	DCTC	**0.17**	**0.12**	**0.06**	**0.25**	**0.23**	**0.03**	**0.34**	**0.15**	**0.20**
			(0.02)	(0.02)	(0.01)	(0.01)	(0.01)	(0.02)	(0.03)	(0.03)	(0.03)
		SMESD	0.56	0.73	1.00	0.57	0.72	0.83	0.78	0.64	0.72
			(0.15)	(0.07)	(0.17)	(0.30)	(0.24)	(0.05)	(0.12)	(0.00)	(0.00)

[1] rf=32.0 - [2] rf=68.0 - [3] rf=44.0 - [4] rf=6.0 - [5] rf=30.0 - [6] rf=70.0 - [7] rf=46.0 - [8] rf=80.0 - [9] rf=60.0- [10] rf=54.0 - [11] rf=68.0

large, the results show significant differences only between those obtained with low and medium values k and those corresponding to large values of k.

On the other hand, if we fix the number of objective function evaluations pro period as well as the objective function severity values and we increase the constraint severity value, we can see how G24_3b offline errors with low k improve but without significant differences. If we use medium values of k, the results get worse but show significant differences only between the results produced with low and medium values of S with respect to those obtained when S is large. In general, with large values of k, the best results are obtained with medium values of S showing significant differences with respect to the results obtained with low and medium values of S.

For G24_4, if we adopt either a low or a large value of k, an increase of S improves the results but not with significant differences. With a medium value of k, the results also improve, showing significant differences only between the results produced with low and medium values of S with respect to those obtained when S is large.

For G24_5, an increase in S improves the results produced by our proposed DCTC. For low values of k, the results show significant differences only between the results produced with either a low or a large S. For medium values of k, the results show significant differences between the results produced with low and medium values of S with respect to those obtained with large values of S. For large values of k, the best results are obtained with medium values of S, showing significant differences only with respect to those obtained with low values of S, but not with respect to those produced with large values of S.

An increase in the number of objective function evaluations per change does not produce results with significant differences for G24_3b and G24_4. For G24_5, the results obtained for 250 evaluations present significant differences with respect to those found for 1000 evaluations with a large value of k and either a low or a medium S.

Our proposed DCTC always outperforms *SMESD*, when compared on problems with dynamic objective function and dynamic constraints, except for four cases as shown in Table 11.4. Furthermore, in one case, (the eleventh experiment), *SMESD* fails to find feasible solutions in all changes for all runs, while our proposed DCTC had success in the same task.

Tables 11.5 and 11.6 show the results obtained for problems with a dynamic objective function and fixed constraints. If we fix the number of objective function evaluations pro period and we increase the objective function severity value, we can see how for G24_1 and G24_2 the offline errors get worse. But the results show significant differences only between the results produced when using low and large values of k, for 250 and 500 evaluations per change.

For G24_6a, G24_6c, and G24_8b, in general, offline errors get worse when k grows. But the results show significant differences only when they are produced with low values of k with respect to those produced with medium and large values of k.

For G24_6d, an increase in the objective function severity value deteriorates the offline errors but not with significant differences.

An increase in the number of objective function evaluations pro period when k is fixed produces better results with significant differences.

Table 11.5 Offline errors (the standard deviation is shown in parentheses) for problems with dynamic objective function and fixed constraints

Ev.	Probl.	Alg.	Dynamic Parameters *k*=0.25	*k*=0.5	*k*=1.0
250	G24_1	DCTC	**0.03** (0.01)	**0.05** (0.03)	**0.12** (0.04)
		SMESD	1.66 (0.00)	1.58 (0.00)	2.39 (0.03)
250	G24_2	DCTC	**0.08** (0.03)	**0.12** (0.03)	0.18 (0.10)
		SMESD	0.77 (0.01)	0.38 (0.00)	**0.16** (0.00)
250	G24_8b	DCTC	**0.11** (0.04)	**0.25** (0.08)	**0.58** (0.19)
		SMESD	0.55 (0.01)	0.74 (0.00)	0.76 (0.00)
500	G24_1	DCTC	**0.00** (0.00)	**0.01** (0.01)	**0.03** (0.02)
		SMESD	1.65 (0.00)	1.58 (0.01)	1.74 (0.00)
500	G24_2	DCTC	**0.05** (0.02)	**0.06** (0.03)	**0.12** (0.07)
		SMESD	0.84 (0.08)	0.26 (0.07)	0.57 (0.00)
500	G24_8b	DCTC	**0.04** (0.02)	**0.12** (0.06)	**0.29** (0.13)
		SMESD	0.51 (0.00)	0.69 (0.09)	1.07 (0.00)
1000	G24_1	DCTC	**0.00** (0.00)	**0.00** (0.00)	**0.00** (0.00)
		SMESD	1.65 (0.00)	1.57 (0.00)	2.32 (0.00)
1000	G24_2	DCTC	**0.03** (0.01)	**0.03** (0.02)	**0.04** (0.04)
		SMESD	1.31 (0.12)	0.79 (0.12)	0.49 (0.20)
1000	G24_8b	DCTC	**0.01** (0.01)	**0.03** (0.02)	**0.07** (0.07)
		SMESD	0.51 (0.00)	0.72 (0.01)	1.03 (0.00)

Table 11.6 Offline errors (the standard deviation is shown in parentheses) for problems with dynamic objective function and fixed constraints

Ev.	Probl.	Algorithms DCTC	*SMESD*
250	G24_6a	**0.26** (0.38)	1.76 (0.00)
250	G24_6c	0.12 (0.05)	**0.11** (0.00)
250	G24_6d	**0.14** (0.18)	0.55 (0.00)
500	G24_6a	**0.06** (0.12)	1.75 (0.00)
500	G24_6c	**0.06** (0.03)	0.10 (0.04)
500	G24_6d	**0.04** (0.14)	0.50 (0.00)
1000	G24_6a	**0.02** (0.02)	1.75 (0.00)
1000	G24_6c	**0.04** (0.03)	0.13 (0.00)
1000	G24_6d	**0.00** (0.00)	0.13 (0.00)

Our proposed DCTC always outperforms *SMESD*, when compared on problems with dynamic objective function and static constraints, except for one case, as shown in Table 11.5. In these problems both approaches found, for all runs, feasible solutions, for all changes.

Table 11.7 shows the results obtained for problems with a static objective function and dynamic constraints. If we fix the number of objective function evaluations pro period and increase the constraint severity value we can see how, for G24_3, the offline errors improve. In general, the results show significant differences only between the results produced with low values of S with respect to those obtained with medium and large values of S.

Table 11.7 Offline errors (the standard deviation is shown in parentheses) for problems with static objective function and dynamic constraints

			Dynamic Parameters		
Ev.	Probl.	Alg.	S=10	S=20	S=50
250	G24_3	DCTC	0.16 (0.15)	0.15 (0.21)	0.12 (0.07)
		SMESD	**0.12** (0.00)	**0.04** (0.01)	**0.01** (0.00)
250	G24_7	DCTC	0.15 (0.02)	0.11 (0.03)	0.10 (0.03)
		SMESD	**0.14** (0.00)	**0.03** (0.01)	**0.08** (0.05)
500	G24_3	DCTC	0.13 (0.14)	0.10 (0.13)	0.10 (0.11)
		SMESD	**0.10**[1] (0.00)	**0.02** (0.00)	**0.00** (0.00)
500	G24_7	DCTC	**0.12** (0.02)	0.07 (0.02)	0.06 (0.02)
		SMESD	**0.12** (0.01)	**0.04** (0.03)	**0.00** (0.00)
1000	G24_3	DCTC	0.11 (0.03)	0.05 (0.03)	0.05 (0.04)
		SMESD	**0.09** (0.01)	**0.02** (0.00)	**0.00** (0.00)
1000	G24_7	DCTC	**0.10** (0.02)	0.05 (0.01)	0.04 (0.01)
		SMESD	0.11 (0.01)	**0.02** (0.00)	**0.00** (0.00)

[1] rf=44.0

Finally, an increase in the number of objective function evaluations per change produces results with significant differences for G24_3 and medium values of S as well as for a number of evaluations of 250 and 1000. For G24_7, the results that present significant differences are those found for 250 and 1000 evaluations with low values of S, as well as the results produced with 250 evaluations with respect to those obtained with 500 and 1000 evaluations, using a medium value of S. For those two problems, the results show significant differences when S is large.

SMESD outperforms our proposed DCTC in all problems with static objective function and dynamic constraints, except for one case and, in another case (see Table 11.7), *SMESD* fails to find feasible solutions in all changes for some runs, whereas our proposed DCTC found feasible solutions for all changes in all the runs performed.

Table 11.8 shows the results for DCTC vs *RIGA-elit* and dRepairRIGA. Here we can note that *RIGA-elit* outperforms DCTC only in one test case while DCTC is superior to dRepairRIGA in seven of the eleven test cases adopted.

Table 11.8 Offline errors (the standard deviation is shown in parentheses) for DCTC vs *RIGA-elit* and dRepairRIGA

	Algorithms		
Probl.	DCTC	*RIGA-elit*	*dRepairRIGA*
G24_1	**0.00** (0.00)	0.40 (0.04)	0.08 (0.01)
G24_2	**0.03** (0.02)	0.28 (0.02)	0.16 (0.02)
G24_3	0.05 (0.03)	0.34 (0.04)	**0.02** (0.00)
G24_3b	0.45 (0.09)	0.47 (0.05)	**0.05** (0.00)
G24_4	0.50 (0.02)	0.49 (0.07)	**0.14** (0.02)
G24_5	0.23 (0.01)	0.25 (0.03)	**0.15** (0.01)
G24_6a	**0.02** (0.02)	0.45 (0.05)	0.36 (0.03)
G24_6c	**0.04** (0.03)	0.41 (0.04)	0.32 (0.03)
G24_6d	**0.00** (0.00)	0.42 (0.02)	0.31 (0.02)
G24_7	**0.05** (0.01)	0.45 (0.05)	0.15 (0.03)
G24_8b	**0.03** (0.02)	1.08 (0.11)	0.34 (0.05)

In order to determine if DCTC is able to recover and converge to a solution immediately after a change, we analyze the plot of *RR*/*ARR* scores displayed in Figures 11.1, 11.2, 11.3, 11.4, and 11.5.

For G24_1 and G24_2 (see Figures 11.1 (a) and (b)), our proposed DCTC found, on the median run, solutions close to the optimum. As the number of objective function evaluations grows, the algorithm recovers faster and gets closer to the new optimum. Also, objective function severity has a negative impact on convergence when it is increased.

For those problems in which only the constraints change (see Figures 11.1 (c) and 11.3), the algorithm found solutions close to the optimum when the constraint severity was larger. When constraint severity was low, the algorithm normally converged to local optima.

For G24_6a, G24_6c, and G24_6d with 500 and 1000 evaluations per change (see Figure 11.2 (a)) the algorithm had a perfect and an almost perfect performance regarding convergence behavior and recovery speed. But, with 250 evaluations per change it presents moderate convergence behavior and recovery speed.

For G24_3b and G24_4, with 250 evaluations per change (see Figures 11.3 (a) and 11.4 (a)), our proposed DCTC presented relatively moderate convergence behavior and recovery speed.

For G24_3b with 500 and 1000 evaluations per change and G24_5, with 250 and 500 evaluations per change (see Figures 11.3 (b) and (c) and Figures 11.5 (b) and (c)), our proposed DCTC presented good convergence behavior and recovery speed.

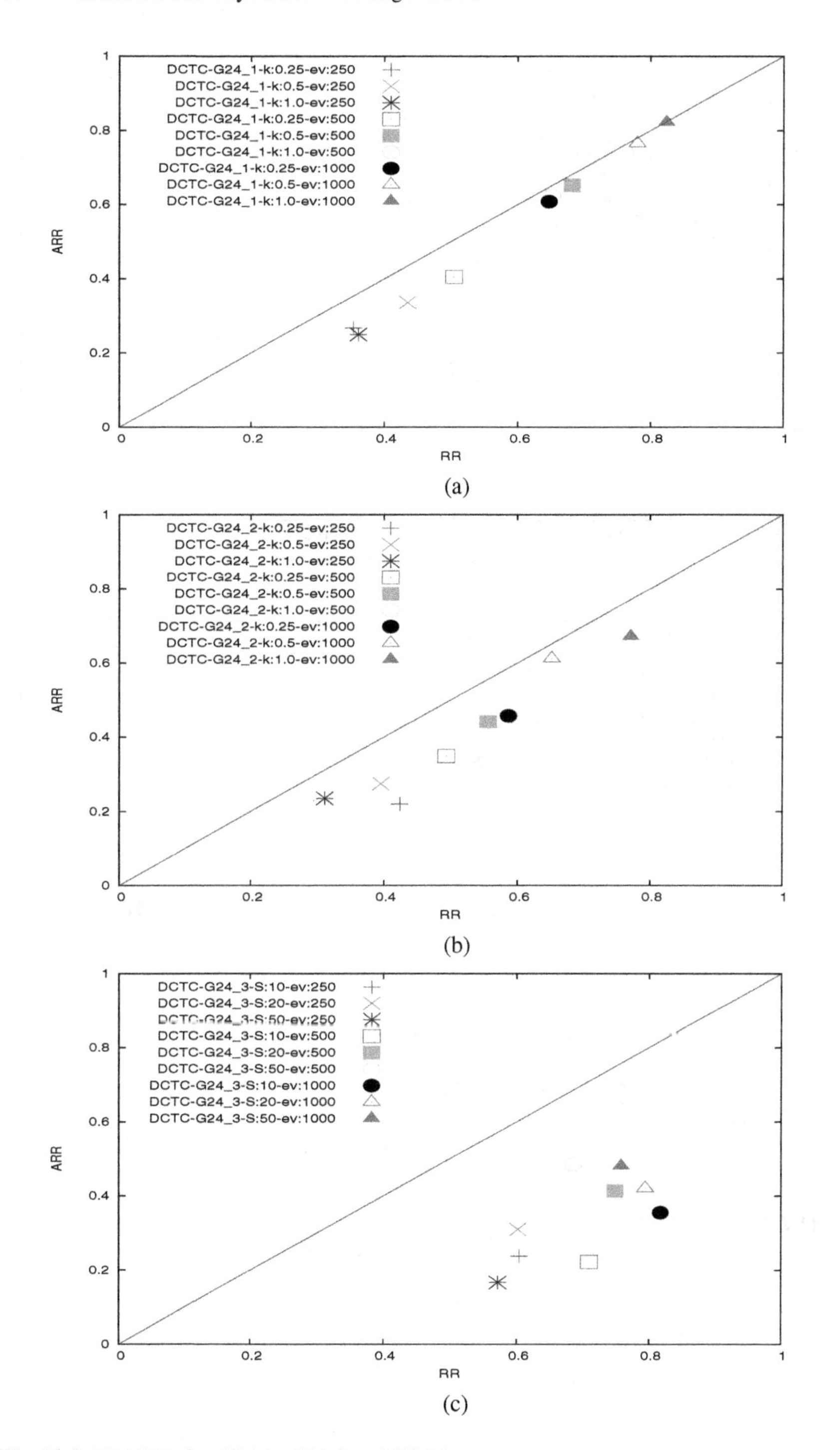

Fig. 11.1 RR/ARR for G24_1, G24_2 and G24_3

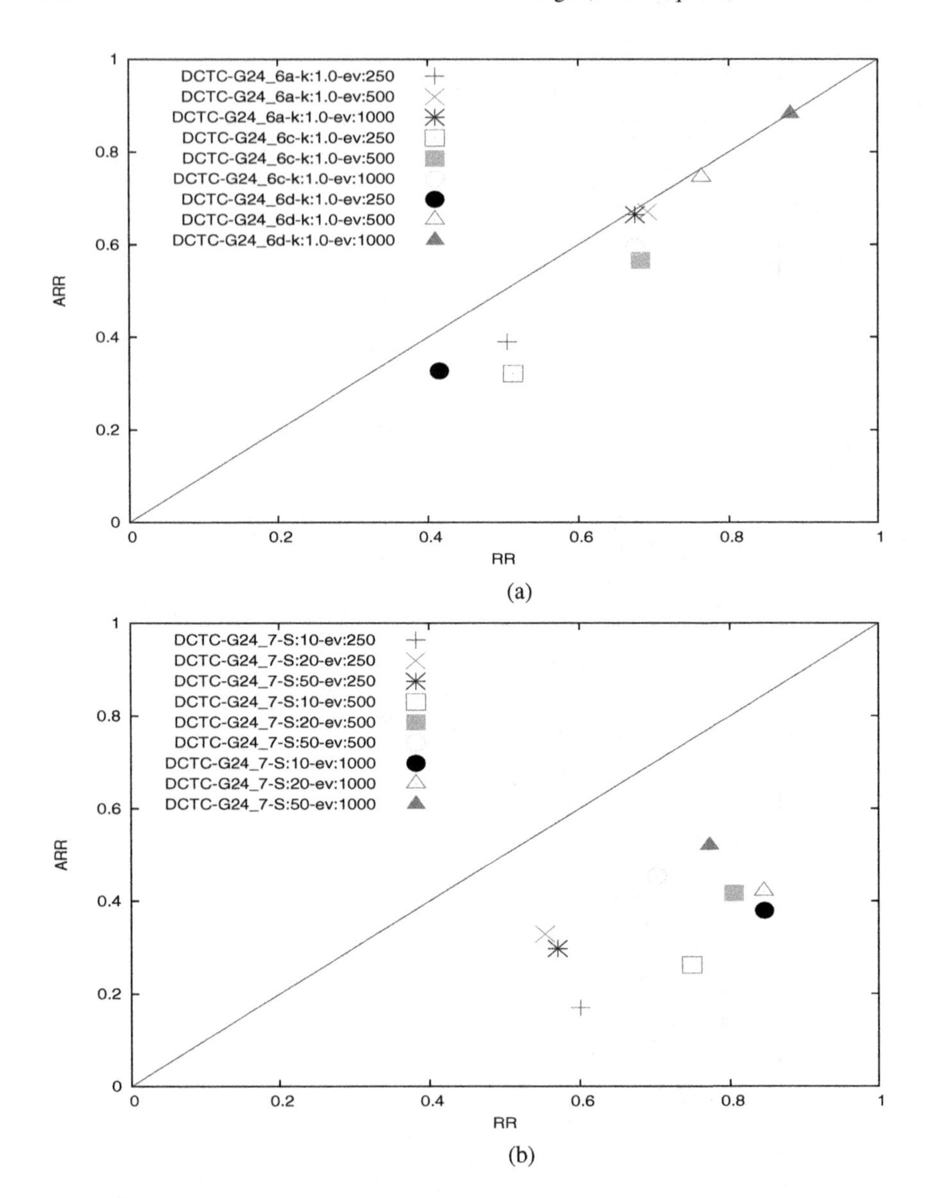

Fig. 11.2 RR/ARR for G24_6a and G24_7

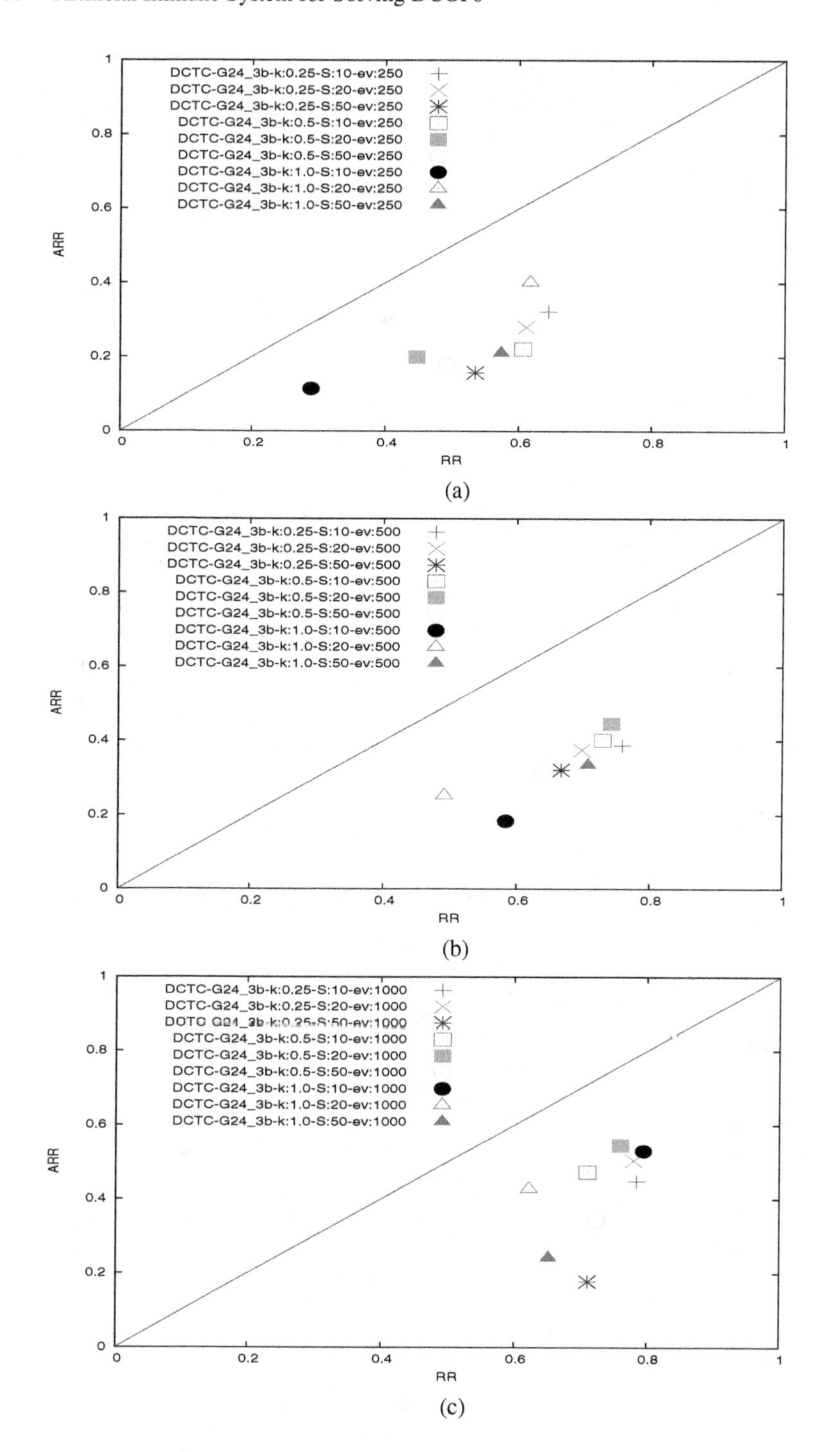

Fig. 11.3 RR/ARR for G24_3b

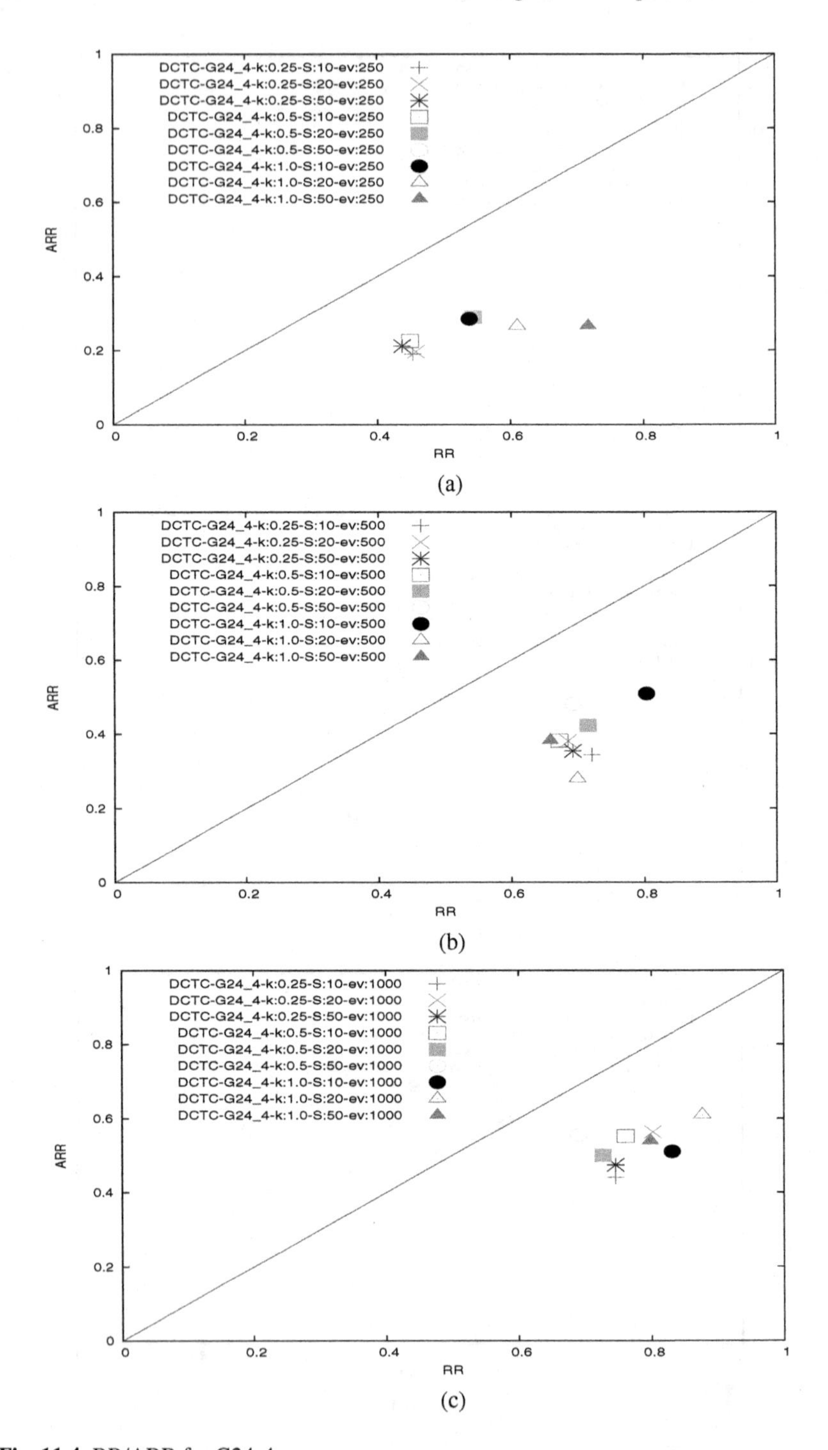

Fig. 11.4 RR/ARR for G24_4

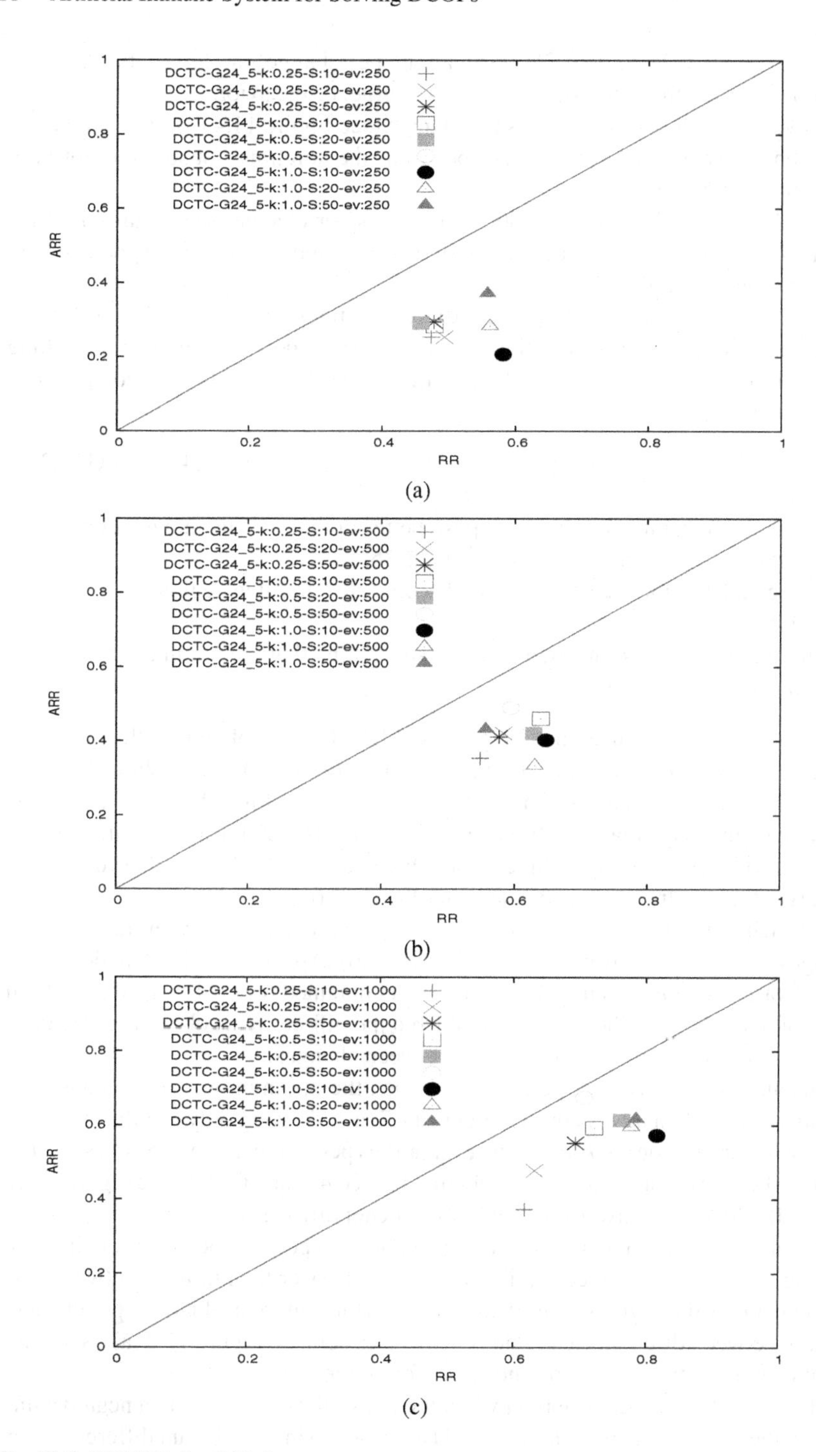

Fig. 11.5 RR/ARR for G24_5

Particularly, for G24_5, with 250 evaluations per change, the larger the objective function severity, the better becomes the convergence behavior.

For G24_4 with 500 evaluations per change (see Figure 11.4 (b)), in general, the algorithm presented a fast recovery speed but the solutions found were not very close to the optimum.

For G24_4 and G24_5, with 1000 evaluations per change (see Figures 11.4 (c) and 11.4 (c)), the convergence behavior of our proposed DCTC is very good and the recovery speed is high.

In order to compare the effect that different features of the dynamic constrained problems had on performance, Nguyen et al. [17] proposed to contrast the offline errors produced over pairs of problems. In this work, the following comparisons were made:

- Static constraints versus dynamic constraints - G24_1 vs G24_4 and G24_2 vs G24_5.
- Moving constraints that do not expose better optima versus moving constraints that expose better optima - G24_3 vs G24_3b.
- Connected feasible regions versus disconnected feasible regions - G24_6c vs G24_6d.
- Optima in the constraints boundary versus optima that are not in the constraints boundary - G24_4 vs G24_5.

The offline errors produced by our proposed DCTC for problems with static constraints and dynamic constraints (see Figures 11.6 (a) to (c)) clearly show the negative impact on performance when constraints change over time. A statistical analysis of variance indicates that the offline errors obtained for G24_1 have significant differences with respect to the offline errors obtained for G24_4. Note that for G24_4, the larger the constraint severity, the better the performance.

For G24_2 versus G24_5, the offline errors also deteriorate when the problem changes its constraints over time (see Figures 11.6 (d) to (f)), but the statistical analysis of variance indicates that the offline errors obtained for G24_5, with a medium value of S, are not significantly different from the offline errors obtained for G24_2, regardless of the number of evaluations between changes.

The algorithm had better performance when the optimum was in the constraint boundary than when it was not (see Figure 11.7 (a)), showing only significant differences with large values of k and a few evaluations per change. The opposite situation occurs when we compare the results obtained for G24_4 and G24_5 (see Figures 11.7 (b) to (d)). Also, they always presented significant differences.

When the algorithm had only a few evaluations to perform per state (or change), moving inside the connected feasible regions was easier than moving between disconnected feasible regions. But, if we could perform more evaluations per change, moving between disconnected regions became easier (see Figure 11.8 (a)), showing significant differences in the offline errors for all the tested cases.

The exposition of better optima when the constraints change, had a negative impact on the performance of our proposed DCTC, showing significant differences for all the tested cases (see Figures 11.8 (b) to (d)).

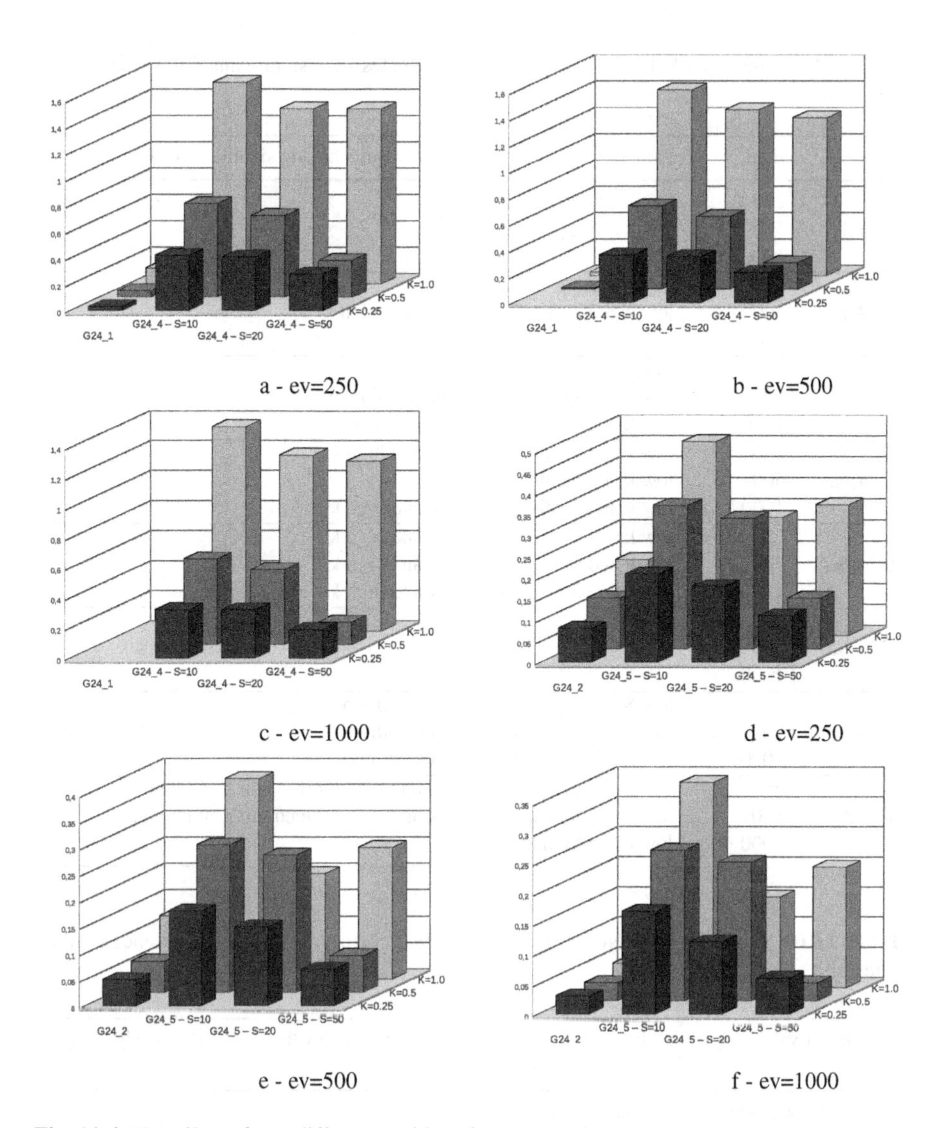

Fig. 11.6 The effect of two different problem features on the performance of DCTC. G24_1 versus G24_4 and G24_2 versus G24_5= Static constraints versus dynamic constraints. Performance is evaluated based on the offline error

Table 11.9 Summary of the ANOVA results. - indicates that significant differences were detected

Probl.	Eval.	Pairs of severity values for the results having significant differences
G24_1	250	low k and high k
	500	low k and high k
	1000	-
G24_2	250	low k and high k
	500	low k and high k / medium k and high k
	1000	-
G24_3	250	low S and medium S
	500	low S and high S / low S and medium S
	1000	low S and high S / low S and medium S
G24_6a	250	low k and medium k / low k and high k
	500	low k and medium k
	1000	low k and medium k / low k and high k
G24_6c	250	low k and medium k / low k and high k
	500	low k and medium k / low k and high k
	1000	low k and medium k / low k and high k
G24_6d	250	-
	500	-
	1000	-
G24_7	250	low S and high S / low S and medium S
	500	low S and high S / low S and medium S
	1000	low S and high S / low S and medium S
G24_8b	250	low k and medium k / low k and high k / medium k and high k
	500	low k and medium k / low k and high k / medium k and high k
	1000	low k and high k

Table 11.10 Summary of ANOVA results. - indicates that significant differences were detected

Probl.	Eval.	k	Pairs of severity values for the results having significant differences
G24_3b	250 / 500 / 1000	small	-
	250/ 500 / 1000	medium	low S and high S / medium S and high S
	250 / 500 / 1000	large	low S and medium S / medium S and high S
G24_4	250 / 500 / 1000	small	-
	250 / 500 / 1000	medium	low S and high S / medium S and high S
	250 / 500 / 1000	large	-
G24_5	250 / 500 / 1000	small	low S and high S
	250 / 500 / 1000	medium	low S and high S / medium S and high S
	250 / 500 / 1000	large	low S and medium S / low S and high S

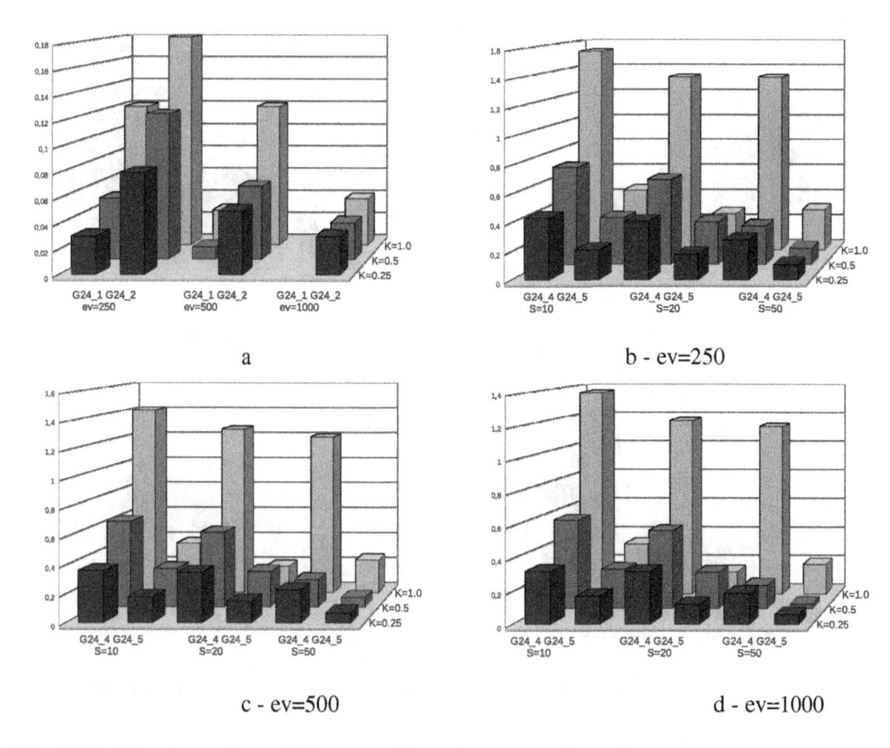

Fig. 11.7 The effect of two different problem features on the performance of DCTC. G24_1 versus G24_2 and G24_4 versus G24_5= Optimum in the constraint boundary versus Optimum not in the constraint boundary. Performance was evaluated based on the offline error

11.6.1 Increasing the Number of Changes per Run

In order to determine if the performance of DCTC and *SMESD* gets affected when more than five changes occur, in the case in which severity is high (k= 0.01 and S= 50) and the minimum time pro period is granted (only 250 objective function evaluations), we ran these two algorithms allowing fifty changes (in 50 independent runs) for each test problem. The results of these experiments are shown in Table 11.12.

First, it can be seen that DCTC outperforms *SMESD* in all the test cases, except for G24_7 but here, *SMESD* could not find feasible solutions for every period (rf = 62.0). It is worth noting that, for G24_3 and G24_3b, even when the offline errors of *SMESD* are zero, their rf values are zero, as well. This means that *SMESD* could not find, in any run, a feasible solution for each period while DCTC could do it.

On the other hand, when we consider the results obtained by DCTC for 5 against 50 changes (with the highest severity), in general (8 from 11 cases) the offline errors improved when more changes were allowed, while, regarding *SMESD*, on 6 of the

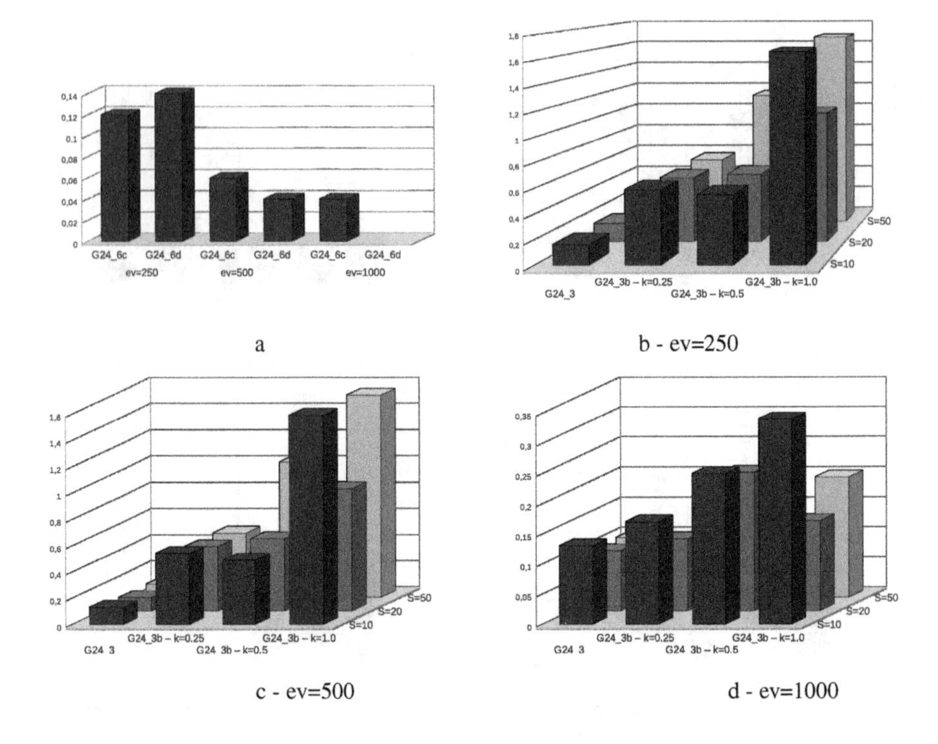

Fig. 11.8 The effect of four different problem features on the performance of DCTC. G24_6c versus G24_6d = Connected feasible regions versus disconnected feasible regions and G24_3 versus G24_3b= Moving constraints do not expose a better optimum versus moving constraints expose a better optimum. Performance was evaluated based on the offline error

11 test cases the results were worst. Thus, we can think that *SMESD* loses its ability to react to changes when these are increased, while DCTC properly maintains diversity during the search process.

11.7 Conclusions

In this chapter, we have analyzed the behavior of an adaptive immune system called Dynamic Constrained T-Cell (DCTC) for solving dynamic constrained optimization problems. One of the strengths we can highlight about DCTC is the few number of parameters that it requires. Furthermore, and analogously to other techniques that do not rely on a penalty function to handle constraints, DCTC does not need to define a penalty factor, which normally has to take a specific value for each problem at hand. An adaptation of an existing algorithm, which was originally used to solve static constrained optimization problems, was used to compare the results obtained by our proposed DCTC on eleven constrained optimization problems which present

Table 11.11 Summary of ANOVA results for G24_1 vs G24_4, G24_2 vs G24_5, G24_4 vs G24_5, G24_1 vs G24_2 and G24_6c vs G24_6d. - indicates that significant differences were detected

First function vs Second Function	Eval.	Did the results obtained for the first function have significant differences with respect to the results obtained for the second function with any dynamic parameter?
G24_1 vs G24_4	250	Always
	500	Always
	1000	Always
G24_2 vs G24_5	250	Yes, when S is small and when S is medium
	500	Yes, when S is small and when S is medium
	1000	Always
G24_4 vs G24_5	250	Always
	500	Always
	1000	Always
G24_3 vs G24_3b	250	Always
	500	Always
	1000	Always
G24_1 vs G24_2	250	Yes, when k is small and when k is medium
	500	Always
	1000	Always
G24_6c vs G24_6d	250	Always
	500	Always
	1000	Always

Table 11.12 Offline errors (the standard deviation is shown in parentheses) for dynamic constrained problems performing 50 changes

	Algorithms	
Probl.	DCTC	*SMESD*
G24_1	**0.06** (0.02)	2.32 (0.00)
G24_2	**0.20** (0.04)	0.55 (0.04)
G24_3	**0.19** (0.05)	0.00 (0.00)[1]
G24_3b	**0.49** (0.17)	0.86 (0.01)[2]
G24_4	**0.16** (0.04)	1.03 (0.07)[3]
G24_5	**0.11** (0.02)	0.32 (0.03)
G24_6a	**0.08** (0.04)	1.80 (0.02)
G24_6c	**0.09** (0.02)	0.23 (0.00)
G24_6d	**0.08** (0.06)	0.97 (0.00)
G24_7	0.05 (0.01)	**0.00** (0.00)[4]
G24_8b	**0.20** (0.07)	0.64 (0.13)

[1] *rf=0.0* - [2] *rf=0.0* - [3] *rf=54.0* - [4] *rf=62.0*

several forms of dynamism (both in the objective function and in the constraints). Additionally, DCTC was also indirectly compared to two approaches used to solve dynamic constrained optimization problems: *RIGA-elit* and dRepairRIGA.

For problems with a dynamic objective function and dynamic constraints, an increase in the objective function severity produces a poorer performance of our proposed DCTC. However, in general, the results that show significant differences are those found with low severity with respect to those found with large severity. An increase in the constraints severity improves the offline errors, showing, in some cases, significant differences, generally between the results found with low severity with respect to those found with a large severity. In general, an increase in the number of evaluations per change improves the offline errors but not with significant differences.

For those problems in which the objective function is dynamic and the constraints are static, in general, an increase in the severity has a negative impact on the behavior of our proposed DCTC. In this case, there are significant differences between the results obtained with low severity with respect to those obtained with a large severity. In this type of problems, an increase in the number of evaluations per change improves the offline errors with significant differences.

In problems in which the objective function is static and the constraints change over time, an increase in the severity improves the results. In this case, there are significant differences when using a low severity with respect to the use of a medium and a large severity. An increase in the number of evaluations per change improves the offline errors but significant differences are detected on results found with low severity, with respect to those obtained with a large severity. Regarding the poor behavior of DCTC in problems that present dynamic constraints, our hypothesis is the following. When the constraint severity is low, it is likely that many of the feasible solutions found so far keep their feasibility. However, in this case, the algorithm has converged to a local optimum and it remains trapped there. On the other hand, when the constraint severity is large, feasible solutions will become infeasible and viceversa. This causes the search to be redirected to the new feasible regions. This situation can be observed in both DCTC and *SMESD*.

When a global optimum switches between disconnected feasible regions and the constraints change, for our proposed DCTC it is more difficult to solve the problem than when the constraints are static.

For all the test problems adopted, we could see, in the median run, that a larger number of objective function evaluations allowed us to keep improving the solutions in that period.

When the problem presented a dynamic objective function and dynamic constraints and the number of objective function evaluations per change was low, the results obtained were not very good. But, we could see that if we increased this number, the results improved, showing significant differences in some cases. This leads us to believe that our proposed approach is able to adapt well to dynamic environments but requires a minimum number of evaluations in order to reach some stability.

Our proposed DCTC was found to be superior to *SMESD* in problems with a dynamic objective function and dynamic constraints and in problems with a dynamic objective function and static constraints. There were only five cases in which *SMESD* outperformed our proposed DCTC when using such types of problems. On the other hand, *SMESD* showed a better behavior than our proposed DCTC in problems having a static objective function and dynamic constraints. When we compared our results against those of *RIGA-elit*, we could note the superior performance of DCTC in all but one test case. On the other hand, DCTC showed to be competitive with respect to *dRepairRIGA*, overcoming it in seven of the eleven test cases adopted. As part of our future work, we aim to improve the mechanisms to maintain diversity of our approach, mainly when dealing with problems in which a change in the constraints gives rise to a new optimum. It is also desirable to improve the exploratory capabilities of our proposed algorithm so that it can be more effective in the test problems in which it was outperformed by *SMESD*. Thus, taking into account the performed experiments and the results obtained from them, we can suggest that DCTC should be suitable for solving problems having a dynamic objective function and either static or dynamic constraints. However, our aim is to improve the behavior of our proposed DCTC in problems having a static objective function and dynamic constraints. Finally, we would also like to extend our approach for solving multi-objective dynamic constrained optimization problems.

Acknowledgments. The first two authors acknowledge support from the Universidad Nacional de San Luis, the Agencia Nacional para promover la Ciencia y Tecnología (ANPCYT) and project no. PROICO 317902. The third author acknowledges support from CONACyT project no. 103570 and from the UMI LAFMIA 3175 CNRS at CINVESTAV-IPN.

References

[1] Aragón, V., Esquivel, S., Coello Coello, C.: Optimizing Constrained Problems through a T-Cell Artificial Immune System. Journal of Computer Science & Technology 8(3), 158–165 (2008)

[2] Aragón, V., Esquivel, S., Coello Coello, C.: Solving constrained optimization using a t-cell artificial immune system. Revista Iberoamericana de Inteligencia Artificial 12(40), 7–22 (2008)

[3] Aragón, V., Esquivel, S., Coello Coello, C.: Artificial Immune System for Solving Global Optimization Problems. Revista Iberoamericana de Inteligencia Artificial (AEPIA) 14(46), 3–16 (2010) ISSN: 1137-3601

[4] Aragón, V., Esquivel, S., Coello Coello, C.: A Modified Version of a T-Cell Algorithm for Constrained Optimization Problems. International Journal for Numerical Methods in Engineering 84(3), 351–378 (2010)

[5] Aragón, V.: Optimización de Problemas con Restricciones a través de Heurísticas BioInspiradas. PhD Tesis

[6] Branke, J.: Evolutionary Optimization in Dynamic Environments. Kluwer Academic Publishers (2002)

[7] Bretscher, P., Cohn, M.: A theory of self-nonself discrimination. Science 169, 1042–1049 (1970)

[8] Dasgupta, D., Nino, F.: Immunological Computation: Theory and Applications. Auerbach Publications, Boston (2008)
[9] Deb, K., Udaya Bhaskara Rao, N., Karthik, S.: Dynamic Multi-objective Optimization and Decision-Making Using Modified NSGA-II: A Case Study on Hydro-thermal Power Scheduling. In: Obayashi, S., Deb, K., Poloni, C., Hiroyasu, T., Murata, T. (eds.) EMO 2007. LNCS, vol. 4403, pp. 803–817. Springer, Heidelberg (2007)
[10] Jula, H., Dessouky, M., Ioannou, P., Chassiakos, A.: Container movement by trucks in metropolitan networks: modeling and optimization. Transportation Research Part E 41, 235–259 (2005)
[11] Mailler, R.: Comparing two approaches to dynamic, distributed constraint satisfaction. In: Proceedings of the 4th International Joint Conference on Autonomous Agents and Multiagent Systems, pp. 1049–1056. ACM, New York (2005), doi:10.1145/1082473.1082632
[12] Male, D., Brostoff, J., Roth, D., Roitt, I.: Inmunology. Mosby, 7th edn. (2006)
[13] Matzinger, P.: Tolerance, danger and the extend family. Annual Review of Immunology 12, 991–1045 (1994)
[14] Mertens, K., Holvoet, T., Berbers, Y.: The DynCOAA algorithm for dynamic constraint optimization problems. In: Proceedings of the 5th International Joint Conference on Autonomous Agents and Multiagent Systems (AAMAS 2006), pp. 1421–1423. ACM, New York (2006), doi:10.1145/1160633.1160898
[15] Mezura Montes, E., Coello Coello, C.: A Simple Multi-Membered Evolution Strategy to Solve Constrained Optimization Problems. IEEE Transactions on Evolutionary Computation 9(1), 1–17 (2005)
[16] Modi, P.J., Jung, H., Tambe, M., Shen, W.-m., Kulkarni, S.: A Dynamic Distributed Constraint Satisfaction Approach to Resource Allocation. In: Walsh, T. (ed.) CP 2001. LNCS, vol. 2239, pp. 685–700. Springer, Heidelberg (2001)
[17] Nguyen, T., Yao, X.: Continuous Dynamic Constrained Optimisation - The Challenges. IEEE Transactions on Evolutionary Computation, 321–354 (2010)
[18] Nguyen, T., Yao, X.: Solving dynamic constrained optimisation problems using repair methods (2011)
[19] Richter, H.: A study of dynamic severity in chaotic fitness landscapes. The 2005 IEEE Congress on Evolutionary Computation 3, 2824–2831 (2005)
[20] Richter, H., Yang, S.: Memory Based on Abstraction for Dynamic Fitness Functions. In: Giacobini, M., Brabazon, A., Cagnoni, S., Di Caro, G.A., Drechsler, R., Ekárt, A., Esparcia-Alcázar, A.I., Farooq, M., Fink, A., McCormack, J., O'Neill, M., Romero, J., Rothlauf, F., Squillero, G., Uyar, A.Ş., Yang, S. (eds.) EvoWorkshops 2008. LNCS, vol. 4974, pp. 596–605. Springer, Heidelberg (2008)
[21] Richter, H., Yang, S.: Learning in Abstract Memory Schemes for Dynamic Optimization. In: Proceedings of the 2008 Fourth International Conference on Natural Computation, vol. 1, pp. 86–91. IEEE Computer Society, Washington, DC (2008)
[22] Richter, H.: Detecting change in dynamic fitness landscapes. In: Proceedings of the Eleventh Conference on Congress on Evolutionary Computation (CEC 2009), pp. 1613–1620. IEEE Press, Piscataway (2009)
[23] Richter, H., Yang, S.: Learning behavior in abstract memory schemes for dynamic optimization problems. Soft Comput. 13(12), 1163–1173 (2009)
[24] Richter, H.: Change detection in dynamic fitness landscapes: An immunological approach. In: World Congress on Nature Biologically Inspired Computing, pp. 719–724 (2009)

[25] Richter, H.: Memory Design for Constrained Dynamic Optimization Problems. In: Di Chio, C., Cagnoni, S., Cotta, C., Ebner, M., Ekárt, A., Esparcia-Alcazar, A.I., Goh, C.-K., Merelo, J.J., Neri, F., Preuß, M., Togelius, J., Yannakakis, G.N. (eds.) EvoApplicatons 2010. LNCS, vol. 6024, pp. 552–561. Springer, Heidelberg (2010)
[26] Schulze, R., Dietel, F., Jandkel, J., Richter, H.: Using an artificial immune system for classifying aerodynamic instabilities of centrifugal compressors. In: World Congress on Nature Biologically Inspired Computing, pp. 31–36 (2010)
[27] Richter, H., Dietel, F.: Change detection in dynamic fitness landscapes with time-dependent constraints. In: Second World Congress on Nature Biologically Inspired Computing, pp. 580–585 (2010)
[28] Richter, H., Dietel, F.: Solving Dynamic Constrained Optimization Problems with Asynchronous Change Pattern. In: Di Chio, C., Cagnoni, S., Cotta, C., Ebner, M., Ekárt, A., Esparcia-Alcázar, A.I., Merelo, J.J., Neri, F., Preuss, M., Richter, H., Togelius, J., Yannakakis, G.N. (eds.) EvoApplications 2011, Part I. LNCS, vol. 6624, pp. 334–343. Springer, Heidelberg (2011)
[29] Schiex, T., Verfaillie, G.: Nogood Recording for Static and Dynamic Constraint Satisfaction Problems. International Journal of Artificial Intelligence Tools 3, 48–55 (1993)
[30] Schwarz, B., Bhandoola, A.: Trafficking from the bone marrow to the thymus: a prerequisite for thymopoiesis. N. Immunol. Rev., 209–247 (2006)
[31] Yang, S., Richter, H.: Hyper-learning for population-based incremental learning in dynamic environments. In: Proceedings of the Eleventh Conference on Congress on Evolutionary Computation (CEC 2009), pp. 682–689. IEEE Press, Piscataway (2009)

[25] Roberts, H., [illegible]: Designing [illegible] Dynamic Optimization Problems. In: Di Chio, C., [illegible], Esparcia-Alcázar, A.I., [illegible], K., Merelo, J.J., [illegible], F., Preuss, M., [illegible], T., Yannakakis, G.N. (eds.) EvoApplications 2010, LNCS, vol. [illegible], pp. [illegible]. Springer, Heidelberg (2010)
[26] Schulte, R., [illegible], Bradfield, J., [illegible]: [illegible] system for [illegible] computation. In: World Congress on Nature [illegible] Computing, pp. [illegible] (2009)
[27] [illegible]
[28] [illegible]
[29] [illegible]
[30] [illegible]

Chapter 12
Metaheuristics for Dynamic Vehicle Routing

Mostepha R. Khouadjia, Briseida Sarasola, Enrique Alba,
El-Ghazali Talbi, and Laetitia Jourdan

Abstract. Combinatorial optimization problems are usually modeled in a static fashion. In this kind of problems, all data are known in advance, i.e. before the optimization process has started. However, in practice, many problems are dynamic, and change while the optimization is in progress. For example, in the Dynamic Vehicle Routing Problem (DVRP), which is one of the most challenging combinatorial optimization tasks, the aim consists in designing the optimal set of routes for a fleet of vehicles in order to serve a given set of customers. However, new customer orders arrive while the working day plan is in progress. In this case, routes must be reconfigured dynamically while executing the current simulation. The DVRP is an extension of the conventional routing problem, its main interest being the connection to many real-word applications (repair services, courier mail services, dial-a-ride services, etc.). In this chapter, the DVRP is examined, and a survey on solving methods such as population-based metaheuristics and trajectory-based metaheuristics is exposed. Dynamic performances measures of different metaheuristics are assessed using dedicated indicators for the dynamic environment.

Enrique Alba · Briseida Sarasola
Departamento de Lenguajes y Ciencias de la Computación,
Universidad de Málaga,
E.T.S.I. Informática, Campus de Teatinos,
29071 Málaga, Spain
e-mail: eat@lcc.uma.es, briseida@lcc.uma.es

Mostepha R. Khouadjia · El-Ghazali Talbi · Laetitia Jourdan
INRIA Lille Nord-Europe, Parc Scientifique de la Haute-Borne,
Bâtiment A, 40 Avenue Halley,
Park Plaza, 59650 Villeneuve d'Ascq Cedex, France
e-mail: mostepha-redouane.khouadjia@inria.fr,
El-talbi@lifl.fr,
laetitia.jourdan@inria.fr

E. Alba et al. (Eds.): Metaheuristics for Dynamic Optimization, SCI 433, pp. 265–289.
springerlink.com

12.1 Introduction

Thanks to recent advances in information and communication technologies, vehicle fleets can now be managed in real-time. When jointly used, techniques like geographic information systems (GIS), global positioning systems (GPS), traffic flow sensors, and cellular telephones are able to provide real-time data, such as current vehicle locations, new customer requests, and periodic estimates of road travel times. If suitably processed, this large amount of data can be used to reduce the cost and improve the service level of a modern company. To do that, revised routes have to be timely generated as soon as new events occur [28].

In this context, Dynamic Vehicle Routing Problems (DVRPs) are getting increasingly important [31, 32, 49, 55]. The VRP [17] is a well-known combinatorial problem which consists in designing routes for a fleet of capacitated vehicles to service a set of geographically dispersed points (customers, stores, schools, cities, warehouses, etc.) at the least cost (distance, time, or any other desired measure). It is possible to define several dynamic features which introduce dynamism in the classical VRP: roads between two customers could be blocked off, customers could modify their orders, the travel time for some routes could be increased due to bad weather conditions, etc. This implies that Dynamic VRPs constitute in fact a set of different problems, which are of crucial importance in today's industry, accounting for a significant portion of many distribution and transportation systems.

In this chapter, we first present an overview of different metaheuristics (from trajectory to population-based algorithms) for solving the DVRP. Second, we evaluate these algorithms according to dynamic performance measures.

The rest of this chapter is organized as follows. Section 12.2 describes the dynamic VRP, its interests in practical applications, and its specific characteristics. An overview on the problem representation as well as solving trajectory/population based metaheuristics is given in Section 12.3. In order to measure the dynamic performances of the metaheuristics. Section 12.4 presents certain measures that can be used to evaluate the performance. The performance evaluation of different metaheuristics: Genetic Algorithm (GA), Ant Colony System (AS), Multi-Particle Swarm Optimization (MAPSO), and Tabu Search (TS) is analyzed in Section 12.5. Finally Section 12.6 presents conclusions and opens some lines for further research.

12.2 Dynamic Vehicle Routing Problem

In this section, we present a formal description of the problem (Section 12.2.1) and a brief state of the art on common interests of the problem in the literature and its variants (Sections 12.2.3 and 12.2.4).

12.2.1 Formal Description

The conventional VRP can be mathematically modeled by using an undirected graph $G = (V, E)$, where V is a vertex set, and E is an edge set. They are expressed as $V = \{v_0, v_1, ..., v_n\}$, and $E = \{(v_i, v_j) | v_i, v_j \in V, i < j\}$. D is a matrix of non-negative distances $d_{i,j}$ between customers v_i and v_j. Furthermore, a set of l homogeneous vehicles each with capacity Q originate from a single depot, represented by the vertex v_0, and must service all the customers, represented by the set $V' = V \setminus \{v_0\}$. The quantity of goods q_i requested by each customer i $(i > 0)$ is associated with the corresponding vertex. The goal is to find a feasible set of tours with the minimum total traveled distance. The VRP thus consists in determining a set of m vehicle routes of minimal total cost, $m \leq l$, starting and ending at a depot, such that every vertex in V' is visited exactly once by one vehicle. The total demand of all customers supplied by each vehicle cannot exceed the vehicle capacity Q. The capacity means the quantity of items (goods) that the vehicle can carry during its travel. Let be $S = \{R_1, \ldots, R_m\}$ a partition of V representing the routes of the vehicles to service all the customers. The cost of a given route $R_j = (r_0, r_1, \ldots, r_{k+1})$, where $r_i \in V$ and $r_0 = r_{k+1}$ (denoting the depot), is given by:

$$Cost(R_j) = \sum_{i=0}^{k} d_{r_i, r_{i+1}} \tag{12.1}$$

and the cost of the problem solution S is:

$$F_{VRP}(S) = \sum_{j=1}^{m} Cost(R_j) \tag{12.2}$$

with a constraint on the vehicle capacity:

$$\sum_{i=1}^{k} q_{r_i} \leq Q, \tag{12.3}$$

where q_{r_i} is the associated quantity of the customer at r_i (items to be delivered/picked up).

We will consider a service time δ_i (time needed to unload/load all goods), required by a vehicle to load the quantity q_i at v_i. It is required that the total duration of any vehicle route (travel plus service times) may not surpass a given bound T, so, a route $R_j = (r_0, r_1, \ldots, r_{k+1})$ is feasible if the vehicle stops exactly once in each customer and the travel time of the route does not exceed a prespecified bound T corresponding to the end of the working day.

$$\sum_{i=0}^{k} d_{r_i, r_{i+1}} + \sum_{i=1}^{k} \delta_{r_i} \leq T \tag{12.4}$$

There may exist some restrictions such as the total traveling distance allowed for each vehicle, time windows to visit the specific customers, and so forth. The basic VRP deals with customers which are known in advance; all other information such as the driving time between the customers and the service times at the customers are also usually known prior to the planning.

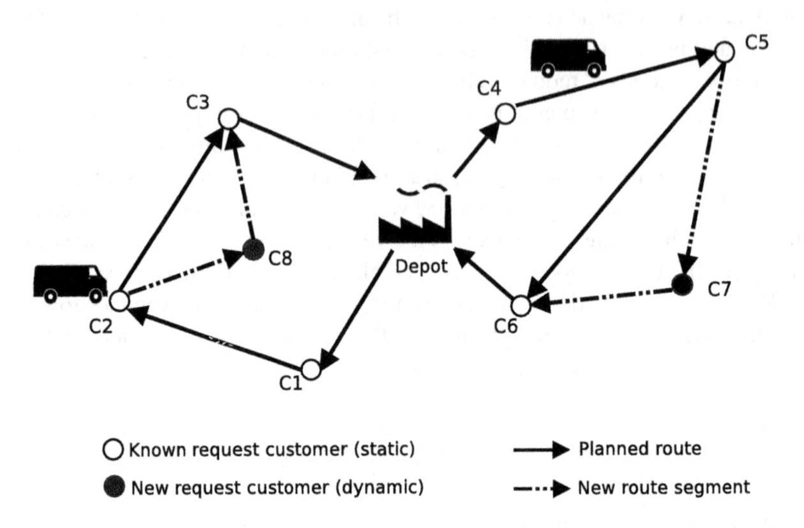

Fig. 12.1 A dynamic vehicle routing case.

The Dynamic Vehicle Routing Problem (DVRP) [55] is strongly related to the static VRP, as it can be described as a routing problem in which information about the problem can change during the optimization process. As conventional static VRPs are NP-hard, DVRP also belongs to the class of NP-hard problems. It is a discrete-time dynamic problem, and can be viewed as a sequence of P instances; each instance is a static problem, which starts at time t and must be solved within a specific time interval Δ_t. We summarize that as follows:

$$P = \{(P_i, t_i, \Delta_t) \ / \ i = 0, 1, \ldots, i_{max}\}. \tag{12.5}$$

With this information the duration of the instance i is $t_{i+1} - t_i$. The maximum number of instances i_{max} can be infinite if the problem is open-ended. A new instance P_{i+1} is generated by the action of the environment change ρ_i on the instance i. This is expressed by $P_{i+1} = \rho_i \oplus P_i$. This change in the environment can be due to several factors; for example, travel times can be time [29] or traffic [66] dependent, orders may be withdrawn or changed [62], some clients may be unknown when the execution begins [47], etc.

One standard approach to deal with this change is to consider the entire problem as a sequence of instances related to the events that happen in the environment. Each change corresponds to the arrival of new optimization problem that has to be solved. The time devoted to solve each instance depends on the frequency of changes [12]. The aim of the present study is to design an optimization algorithm that is able of continuously adapting the solution to a changing environment. This technique is now commonly followed by the community that works on the DVRP domain [31, 39, 49]. Therefore, a partial static VRP has to be solved each time a new request is received. A simple example of a dynamic vehicle routing situation is shown in Figure 12.1. In the example, two uncapacitated vehicles must service both known and new customer requests. Designing a real-time routing algorithm depends to a large extent on how much the problem is dynamic. To quantify this concept, [46] and [45] have defined the degree of dynamism of a problem (*dod*). Without loss of generality, we assume that the planning horizon is a given interval $[0,T]$, possibly divided into a finite number of smaller intervals. Let n_s and n_d be the number of static and dynamic requests, respectively. Moreover, let $t_i \in [0,T]$ be the occurrence time of request i. Static requests are such that $t_i = 0$ while dynamic ones have $t_i \in]0,T]$. The degree of dynamism is defined as:

$$dod = \frac{n_d}{n_s + n_d} \tag{12.6}$$

which may vary between 0 and 1. If it is equal to 0, all requests are known in advance (static problem), while if it is equal to 1, all requests are dynamic (completely dynamic problem). Larsen [42] generalizes the definition proposed by Lund *et al.* [46] in order to take into account both dynamic request occurrence times and possible time windows. He observed that a system in which dynamic requests are received late over the planning horizon $[0,T]$ is more dynamic than another one in which the requests occur at the beginning of the working day. Thus, he introduces a new measure of dynamism:

$$edod = \frac{\sum_{i=1}^{n_s+n_d} (t_i/T)}{n_s + n_d}$$

The effective degree of dynamism then represents an average of how late the requests are received compared to the latest possible time the requests could be received. It is possible to easily see *edod* ranges between 0 and 1. It is equal to 0 if all user requests are known in advance while it is equal to 1 if all user requests occur at time T. Finally, Larsen extends the definition of *edod* to take into account possible time windows on user service time. Let $[a_i, b_i]$ be the interval time of the client i referred to as time windows, with a_i and b_i corresponding to the earliest and the latest possible times when the service should begin, respectively.

$$edod_{tw} = \frac{\sum_{i=1}^{n_s+n_d} [T - (b_i - t_i)]/T}{n_s + n_d}$$

.

It is also obvious that $edod_{tw}$ varies between 0 and 1. Moreover, if no time windows are imposed (i.e. $a_i = t_i$ and $b_i = T$), then $edod_{tw} = edod$.

Larsen *et al.* [43] describe and test several dynamic policies to minimize routing costs for the Partially Dynamic Traveling Repairman Problem (PDTRP) with various degrees of dynamism.

12.2.2 DVRP Interests

There are several important problems that must be solved in real-time. In [27, 28, 42], the authors list a number of real-life applications that motivate the research in the domain of dynamic vehicle routing problems.

- *Supply and distribution companies*: In seller-managed systems, distribution companies estimate customer inventory level in such a way as to replenish them before stock depletion. Hence, demands are known beforehand in principle and all customers are static. However, since the actual demand quantity is uncertain, some customers might run out of their stock and have to be serviced urgently.
- *Courier Services*: It refers to the international express mail services that must respond to customer requests in real-time. The load is collected at different customer locations and has to be delivered at another location. The package to be delivered is brought back to a remote terminal for further processing and shipping. The deliveries form a static routing problem since recipients are known by the driver. However, most pickup requests are dynamic because neither the driver nor the planner knows where the pickups are going to take place.
- *Rescue and repair service companies*: Repair services usually involve a utility firm (broken car rescue, electricity, gas, water and sewer, etc) that responds to customer requests for maintenance or repair of its facilities.
- *Dial-a-ride systems*: Dial-a-ride systems are mostly found in demand-responsive transportation systems aimed at servicing small communities or passengers with specific requirements (elderly, disabled). These problems are of the many-to-many when any node can serve as a source or destination for any commodity or service. Customers can book a trip one day in advance (static customers) or make a request at short notice (dynamic customers) [3, 21, 60].
- *Emergency services*: They cover the police, firefighting, and ambulance services [30, 61]. By definition, the problem is pure dynamic since all customers are unknown beforehand and arrive in real-time. In most situations, routes are not formed because the requests are usually served before a new request appears. The problem then is to assign the best vehicle (for instance, the nearest) to the new request. Solving methods are based on location analysis for deciding where to dispatch the emergency vehicles or to escape the downtown traffic jam.
- *Taxi cab services*: Managing taxi cabs is still another example of a real-life dynamic routing problem. In most taxi cab systems the percentage of dynamic customers is very high, i.e. only very few customers are known by the planner before the taxi cab leaves the central at the beginning of its working day [20].

12.2.3 *Objectives*

Depending on the nature of the system, the objective to be optimized is often a combination of different measures. DVRP inherits the classical objectives defined in the conventional VRP. Moreover, the dynamic nature of the problem leads to the emergence of new objectives. For instance, in weakly dynamic systems the focus is on minimizing routing cost [31, 49]. However, in strongly dynamic systems such as emergency services, the interest is to minimize the expected response time (i.e. the expected time lag between the moment when the user request occurs and its service time) [25, 43, 48]. Furthermore, there are other objectives such as maximizing the expected number of requests serviced during a given period of time [4, 5].

12.2.4 *Related Works*

In this section, we present a classification and an overview on the state of the art of dynamic vehicle routing problems. Different surveys have been proposed in scientific articles on DVRPs [8, 11, 28]. Psaraftis [54] defines a VRP to be dynamic when some input to the problem is revealed during the execution of the algorithm. Solutions to the problem should change as new information is revealed to the algorithm and to the decision maker. Possible information attributes may include the evolution of information (static / dynamic), the quality of information (known-deterministic / forecast / probabilistic / unknown), the availability of information (local / global), and the processing of information (centralized / decentralized).

We propose to classify the DVRPs according to the degree of knowledge that we have on the input data of the problem. A dynamic problem can be either deterministic or stochastic (see Figure 12.2). DVRP is deterministic if all data related to the customers are known when the customer demands arrive, otherwise it is stochastic. Both of these classes can be subject to different factors such as service time window, traffic jam, road maintenance, weather changes, breakdown of vehicles and so on. These factors often change the speed of vehicles, and the travel time of arriving at the depot. Consequently, they lead to other sub-variants of the problem (see Table 12.1):

1. **Deterministic:** In the deterministic case, all the data related to the inputs are known. For instance, when a new customer demand appears, the customer location and the quantity of its demand are known. Different types of deterministic DVRP can be found in the literature as:

    a. **Dynamic Capacitated Vehicle Routing Problem with Dynamic Requests (DCVRP):** An important number of works exist on this variant [26, 40, 49] which represents the conventional definition of the problem, and where the existence of all customers and their localizations are deterministic, but their order can arrive at any time. The objective is to find a set of routes with the lowest traveled distance, observing the vehicle capacity limit.

b. **Dynamic Vehicle Routing Problem with Time Window (DVRPTW):** It is one of the most well-studied variants of the DVRP [1, 18, 19, 32, 44, 48, 66]. Besides the possibility of requiring services in real time, the time window associated to each customer must be respected. The DVRPTW is closely related to the Dynamic Traveling Repairman Problem (DTRP) [6, 7], in which m identical vehicles must service the upcoming demands. At each location, the vehicle serving the demand must spend some amount of time in on-site service. This service time is a random variable that is realized only when the service is completed. The objective is to find service policies that minimize the expected waiting time of the demands. Larsen *et al.* [43] proposed on-line policies for the Partially Dynamic Traveling Salesman Problem with Time Windows (PDTSPTW) that could be considered as an instance of DVRPTW with a single vehicle. The objective is to minimize the total or maximum lateness over the set of customers. A simple policy consists in requiring the vehicle to wait at the current customer location until it can service another customer without being early. Other policies may suggest repositioning the vehicle at a location different from that of the current customer, based on prior information on future requests.
c. **Dynamic Vehicle Routing Problem with Time-Dependent Travel Times (DVRPTT):** Described in [29], it assumes that the travel times from the customer i to the customer j are variable over time. This variation could occur due to the type of the road, weather, and traffic conditions that may strongly influence the speed of vehicles and hence travel times.
d. **Dynamic Pickup and Delivery Vehicle Routing Problem (DPDVRP):** It is based on the conventional Pickup and Delivery Vehicle Routing Problem (PDVRP) [59]. The problem consists of determining a set of optimal routes for a fleet of vehicles in order to serve customer requests. The objective is to minimize the total route length, i.e. the sum of the distances traveled by all the vehicles, under the following constraints: all requests must be served, each request must be served entirely by one vehicle (pairing constraint), and each pickup location has to be served before its corresponding delivery location (precedence constraint). The dynamic version arises when not all requests are known in advance [48].

   Attanasio *et al.* present in [2] parallel implementations of a tabu search method developed previously by Cordeau and Laporte for the Dial-a-Ride Problem (DARP) [16]. Gendreau *et al.* [25] developed a tabu search heuristic where the neighborhood structure is based on ejection chains heuristic. Yang *et al.* [69] introduce a real-time multi-vehicle truckload pickup and delivery problem. They propose a mixed-integer programming formulation for the off-line version of the problem and propose a new rolling horizon re-optimization strategy for a dynamic version.

2. **Stochastic:** In stochastic dynamic problems (also known as probabilistic dynamic problems) uncertain data are related to customer demands and are represented by random variables.

   a. **Dynamic and Stochastic Capacitated Vehicle Routing Problem (DSCVRP):** It considers customer requests are unknown and revealed over time. In addition, customer locations and service times are random variables and are realized dynamically during the plan execution. Bent and Van Hentenryck [4, 5] considered dynamic DVRPTW with stochastic customers. They proposed a multiple scenario approach that continuously generates routing plans for scenarios including known and immediate requests to maximize the number of serviced customers. The approach was adapted from Solomon benchmarks with varying degree of dynamism. Hvattum *et al.* [33] addressed this variant of the problem. The authors consider that both customer locations and demands may be unknown in advance. They formulate the problem as a multi-stage stochastic programming problem, and a heuristic method was developed to generate routes by exploiting the information gathered on future customer demand.
   b. **Dynamic and Stochastic Vehicle Routing Problem with Time Window (DSVRPTW):** It has been introduced in [51]. In this problem, each service request is generated according to a stochastic process; once a service request appears, it remains active for a certain deterministic amount of time, and then expires. The objective is to minimize the number of possible vehicles and ensure that each demand is visited before its expiration.
   c. **Dynamic Vehicle Routing Problem With Stochastic Travel Times (DVRPSTT):** It assumes that the problem is subject to stochastic travel times. The travel times may change from one period to the next one. Some works present this version of the problem as in [52], where the travel time to the next destination is perturbed by adding a value generated with a normal probability law. This perturbation represents any unforeseen events that may occur along the current travel journey. It is known to the dispatching system only when the vehicle arrives at its planned destination.
   d. **Dynamic and Stochastic Pickup and Delivery Vehicle Routing Problem (DSPDVRP):** In this version of the problem, the stochastic process concerns the demand quantity that the vehicle must pick or delivery to each customer. Thus, we have uncertain quantities to pick up or deliver at the customers' location [68]. This distribution can be modelled by using a probabilistic law, such as a normal law for example, or by using fuzzy logic.

12.3 Solving Methods

In this part we present a common solution representation of the problem in the Section 12.3.1, and the major classes of metaheuristics proposed to solve this problem in Section 12.3.2 and Section 12.3.3.

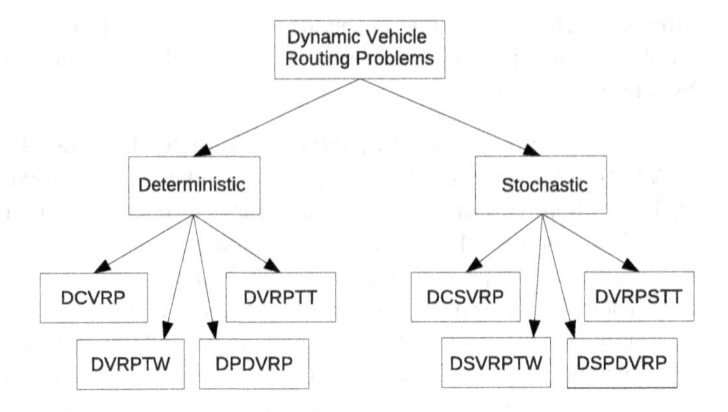

Fig. 12.2 Classification of DVRPs according to deterministic and stochastic information related to customer requests.

Table 12.1 Major publications on different variants of Dynamic Vehicle Routing problems.

DVRPs	Authors	Class	Characteristics	Objectives
Deterministic	Psaraftis *et al.* [54] Kilby *et al.* [40] Montemanni *et al.* [49] Hanshar *et al.* [31] Branching *et al.* [11] Khouadjia *et al.* [38] Sarasola *et al.* [58]	DCVRP	Dynamic requests Capacitated vehicles	Minimize the total traveled distance
	Oliveira *et al.* [18] Gendreau *et al.* [26] Mitrović-Minić *et al.* [48] Larsen *et al.* [44] Housroum *et al.* [32] Alvarez *et al.* [1]	DVRPTW	Dynamic requests Time windows	Minimize the total traveled distance and minimize the total lateness at the customer
	Haghani *et al.* [29] Kritzinger *et al.* [41]	DVRPTT	Variable travel times	Minimize the total traveled distance Minimize the total lateness/tardiness at the customer
	Gendreau *et al.* [25] Mitrović-Minić *et al.* [48]	DPDVRP	Dynamic requests Pickup and Delivery	Minimize the total travel time, tardiness over all pick-up and delivery locations, and sum of overtime over all vehicles
Stochastic	Bent *et al.* [4, 5] Hvattum *et al.* [33]	DSCVRP	Random customer locations Random service times	Maximize the number of serviced customers Minimize the total traveled distance
	Pavone *et al.* [51] Bertsimas *et al.* [6, 7]	DSVRPTW	Random customer locations Random service times Time window	Minimize the number of vehicles and the traveled distance Minimize the wait for completion of service

12.3.1 *Solution Representation*

The solution representation in dynamic vehicle routing problems takes its source from representations that have already been proposed in the literature for the conventional static version of the problem [53, 56, 63]. Different representations for the DVRP solutions have been proposed in the literature either for trajectory-based metaheuristics or population-based metaheuristics [26, 31, 38, 49].

The difference between a dynamic representation and a static representation of the vehicle routing problem consists in the fact that given the dynamic nature of the problem, a representation with a variable length is required. It is justified by the fact that demands arrive over time and have to be inserted in the existing routes or by creating new ones. This representation is encoded as a set of vehicle routes. Each route includes some information on committed customers that have been visited by a given vehicle as well as those that are waiting for completion of service, or new customers that have been added to the day's service, but not yet assigned to any vehicle. Another important point is the information related to vehicles. The current vehicle position in the service area must be known by the dispatcher at any moment of the day's service. This allows to redirect the vehicle when new requests arrive into the system.

In [49], Montemanni *et al.* propose a representation for their Ant Colony System (ACS) algorithm. The authors consider v dummy depots (one for each vehicle of the fleet) and they refer to them as $d_1, \ldots, d_v$. Solutions retrieved by ants will be represented as long, single tours. In this context, nodes contained within two consecutive dummy depots d_a and d_b (with $d_a, d_b \in \{1,...,v\}$) form the (partial) tour associated with vehicle a. The partial tour associated with vehicle b will start from the dummy depot d_b, which corresponds to the location of the last customer committed to vehicle b. The starting time from d_b corresponds to the end of the serving time for the last customer committed to vehicle b, while the capacity of b will be equal to the residual capacity of b, i.e. Q_b minus the quantity ordered by customers already committed to vehicle b.

Another representation is proposed by Hanshar *et al.* [31] for a Genetic Algorithm. Their chromosomal representation consists of two types of nodes: a node with a positive integer number representing a single customer (which has not yet been assigned to a vehicle) and a node depicted with a negative integer number representing a group of clustered customers that have already been committed to a given vehicle. Thus, the chromosome consists of integers, where new customers are directly represented on a chromosome with their corresponding positive index number and each committed customer is indirectly represented within one of the groups representing a given deployed vehicle. When the chromosome is decoded, new customers could be added to these pre-existing vehicles (i.e. groups) if they still have the capacity to accommodate new customer orders.

Garrido *et al.* [24] have tackled the DVRP using Evolutionary Hyper-Heuristics (EH-DVRP). The authors propose a chromosome representation for the low-level heuristics composed by two main data structures: a list of new unassigned customers represented by their identifier, and a set of routes which represents a set of partial solutions or states of the problem, formed by committed and uncommitted requests.

In [38], Khouadjia *et al.* propose a representation for a Particle Swarm Optimization (PSO) Algorithm. It consists in a discrete representation which expresses the route of m vehicles over the n customers to serve. The encoding allows the insertion of dynamic customers in the already planned routes. The representation of each route R_k is a permutation of n customers $R_k : (v_0, v_1, v_2, ..., v_i, ..., v_n, v_{n+1})$. This representation handles the static and dynamic aspects of the problem. On the one side, it allows the insertion of dynamic customers in the already planned routes. On the other side, if the customer is served, it cannot be shifted from its current route to another one. For the simulation process, the authors keep some information, such as the state of each customer (served / not served) and its time of service, the state of vehicles as their current position in the service region, their remaining capacity, the traveled distance, and their condition (committed / not committed).

12.3.2 Trajectory-Based Metaheuristics

Many works are related to trajectory-based metaheuristics for solving the DVRP (see Table 12.2). Gendreau *et al.* in [26] propose a parallel tabu search heuristic with an adaptive memory. The adaptive memory stores previously found elite solutions and uses them to generate new starting points for the tabu search. This is achieved by combining routes taken from different solutions in the memory. Any new solution produced by the tabu search is included in the memory if it is not filled yet. Otherwise, the new solution replaces the worst solution in memory, if it is better. The parallelization of the procedure was achieved at two different levels: (1) different tabu search threads run in parallel, each of them starting from a different initial solution; and (2) within each search thread, many tabu searches run independently of subproblems obtained through a decomposition procedure of the whole problem. For the parallel implementation a masterâslave scheme was chosen to implement the procedure. The master process manages the adaptive memory and creates initial solutions for the slave processes that run the tabu search. Ichoua *et al.* in [34] reused the same algorithm with some enhancement related to the strategy for assigning customer requests to vehicles.

Mitrović -Minić *et al.* [48] dealt with the Dynamic Pickup and Delivery Problem with Time Windows (DPDVRPTW) and applied the cheapest insertion procedure in order to determine the overall best insertions for the locations of a request before its insertion. The improvement procedure is based on Tabu Search (TS). It is applied after the reinsertion procedure and it runs while new requests are being received.

Hanshar *et al.* have implemented a basic tabu search in [31]. Two operators are employed as neighborhood structure procedures: an inversion operator and a λ-exchange operator [50], each one applied according to some probability. Furthermore, Montemanni *et al.* [49] have implemented a GRASP (Greedy Randomized Adaptive Search Procedure) for dealing with the DVRP. Basically, initial tours are generated by iteratively selecting the next customers to visit. The procedure is repeated until a complete solution is built. Sarasola *et al.* [58] designed a flexible VNS for the VRP with dynamic requests. The flexibility strategy is based on the relaxation of the maximum tour length constraint.

Attanasio *et al.* present in [2] parallel implementations of a tabu search method developed previously by Cordeau and Laporte for the static Dial-a-Ride Problem (DARP) [16]. In this latter the requests are received throughout the day and the primary objective is to accept as many requests as possible as possible with the available fleet of vehicles. Furthermore, the routes are constructed under the constraint that users specify pick-up and drop-off requests between origins and destinations.

12.3.3 Population-Based Metaheuristics

Several population-based metaheuristics have been proposed in the literature (see Table 12.2). Next, we outline the major works that follow this research line.

12.3.3.1 Ant Colony Optimization

Ant System (AS) has been applied to tackle a large variety of Dynamic Vehicle Routing Problems [23, 37, 49, 57, 64].

Tian *et al.* [64] present a hybrid Ant System to handle the dynamism by means of modifying the pheromone matrix in order to take advantage of the old information gathered during the previous search. They propose a new pheromone initialization for new demands, which works better than a re-start optimization. Furthermore, they use a simple strategy that consists in grouping new requests at every fixed interval-time before their introduction into the system. In addition, they make further improvements on vehicle routes with the local search 2-Opt heuristic.

Jun *et al.* [37] addressed a hybrid multi-objective ant colony algorithm for solving DVRPTW. They consider two sub-objectives such as the vehicle number and the time cost. In their Ant Colony Algorithm, an Evolutionary Algorithm (EA) is embedded to increase the pheromone update. They explain that EA participates to speed up the convergence of their algorithm.

Montemanni *et al.* [49] exploit some features of the Ant Colony System optimization paradigm to smoothly save information about promising solutions when the optimization problem evolves because of the arrival of new orders. One of these characteristics is the pheromone conservation procedure, which contains information about good solutions features. In particular, pairs of customers that have been visited in sequence in good solutions will have high values in the corresponding entries of the pheromone matrix. In the dynamic context, it is used to pass on information about the properties of good solutions from previously obtained results in the new changing environment, since the two problems are potentially very similar. This operation avoids restarting the optimization from scratch. Based on the Montemanni's algorithm, Rizzoli *et al.* [57] discuss the applications of ACO on a number of real-world problems. They propose some results obtained by their algorithm on an on-line VRP for fuel distribution in the city of Lugano (Switzerland).

Oliveira et al. [18] propose an Ant Colony Algorithm for the DVRPTW with two different forms of attractiveness (time windows and distance) for building the vehicles routes. According to their experiments, the higher the degree of dynamism, the fewer customers will be served.

Chitty *et al.* [14] introduce a hybrid dynamic programming-ant colony optimization approach to solve bi-criterion Vehicle Routing Problems. The aim is to find routes that have both shortest overall travel time and smallest variance in travel time. The hybrid approach uses the principles of dynamic programming to first solve simple problems using ACO (routing from each adjacent node to the end node), and then builds on this preliminary solution to eventually provide solutions (i.e. Pareto fronts) for routing between each node in the network and the destination node. However, the hybrid technique updates the pheromone concentrations only along the first edge visited by each ant. This technique is shown to provide the overall solution faster than an established bi-criterion ACO technique that is concerned only with routing between the start and destination nodes, allowing re-routing vehicle to dynamic changes within the road network.

12.3.3.2 Evolutionary Algorithms

Hanshar [31] proposes a Genetic Algorithm (GA) that handles the optimization of the static VRP-like instances that correspond to the whole dynamic optimization problem. The GA is launched at each fixed duration and must run within an efficient amount of time. The fitness evaluation involves the vehicle routes obtained after the translation of the chromosome representation. It returns the total travel distance / cost of the routes. The Best-Cost Route Crossover (BCRC) is used as the crossover operator and the inversion operator is used as the mutation operator.

Housroum *et al.* [32] deal with the Dynamic Vehicle Routing Problem with Time Windows (DVRPTW). The authors propose an approach based on genetic algorithms. For their algorithm, they use the PMX crossover and different mutation operators such as Or-Opt, 1-Opt, or swap. They validate their approach on modified Solomon's benchmarks which have been proposed by Gendreau *et al.* [26]. Zhao *et al.* in [70] use a GA similar to Housroum's algorithm [32] for solving the Dynamic Vehicle Routing Problem with time-dependent Travel Times (DVRPTT).

Alvarenga *et al.* propose in [1] a hybrid GA with Column Generation Heuristic for the DVRPTW. The authors propose a specific crossover that, at the first step, makes a random choice of routes from each parent involved. After all feasible routes have been inserted in the offspring, remaining customers are inserted into existing routes, if possible (second step). New routes are created if some customers remain after this insertion step. Eight different operators are used as mutation operators.

Branke *et al.* [13] propose a GA with different waiting strategies of vehicles for the DCVRP. A two-point crossover is chosen and the mutation is done by adding to each value a normally distributed random value.

For their Evolutionary Hyper-Heuristics (EH-DVRP) [24], Garrido *et al.* propose a high-level algorithm which evolves and combines different types of low-level heuristics (constructive, perturbation, and noise heuristics) to solve the problem. Each individual of the population refers to a sequence of genes that corresponds to a constructive and improvement heuristics which gradually inserts customers and repairs the set of routes created so far. These dedicated heuristics are applied to construct and improve partial states of the problem. The hyper-heuristic uses four operators to find new individuals: one recombination and three mutation-like operators. The recombination operator performs a one point crossover to generate two new offsprings. For the mutation operators, the first one randomly selects and copies one of the heuristics to another position in the chromosome which allows us to include new heuristics in different steps of the algorithm. The second operator selects and replaces a gene by one single heuristic. The authors' idea is to give an alternative heuristic which may perform better in cooperation with existing ones. The last operator deletes a gene from the chromosome and discards some heuristics which cannot be useful to improve candidate solutions.

Wang *et al.* [66] have proposed an EA for solving the DVRPTW. For the algorihtm's reproduction phase, the authors used a two-point crossover operator and a mutation operator that consists in changing the assignment of unserved customers to another vehicle. In order to enhance their algorithm, the authors propose to hybridize their algorithm with a modified Dijkstra's algorithm for finding real-time shortest paths.

Jih *et al.* [36] address a hybrid genetic algorithm for solving single-vehicle pickup and delivery problem with time windows and capacity constraints (DPDVRPTW). Their approach enables dynamic programming to achieve real-time performance and genetic algorithms to approximate optimal solutions. The initial population is created by the dynamic programming instead of generating it randomly. The dynamic programming passes the unfinished routes to the genetic algorithm in order to produce final solutions. The authors compare the performance of four crossover operators. These operators are order crossover (OX), uniform order-based crossover (UOX), merge cross #1 (MX1), and merge cross #2 (MX2) [9]. In addition, they consider three mutation operators:(i) two genes are selected randomly, and their positions are interchanged (swap operator); (ii) two break points are selected randomly and the order of the sub-route specified by the genes is inverted (inverse operator); (iii) if the vehicle arrives at the i_{th} stop and violates the constraints, the order of the genes within the first i_{th} sub-route is disturbed (rearrangement operator).

Haghani and Jung [29], deal with the pick-up and delivery vehicle routing problem with soft time windows, where they consider multiple vehicles with different capacities, real-time service requests, and real-time variations in travel times between demand nodes. This algorithm includes a vehicle merging operator in addition to the generic genetic operators, namely the crossover and the mutation operators.

Bosman *et al.* [10] introduce a probabilistic model to describe the behavior of the load announcements. This allows the routing to be informed about customer positions where loads are expected to arrive shortly. This approach outperforms the EA that only considers currently available loads. Only mutation is considered. In the mutation of an individual, two vehicles are chosen randomly (could be the same), and two customers from their respective routes are chosen randomly, and are swapped. This operator allows visits to customers to be exchanged between vehicles or to be re-ordered in the route of a single vehicle directly.

Van Hemertand and La Poutré [65] present an evolutionary algorithm that is able to provide solutions in real-time for the DVRP. The authors analyze the benefit of anticipatory vehicle moves within regions that have a high potential of generating loads (fruitful regions). Only mutation is considered. Two vehicles, possibly the same one, are chosen uniform randomly. In both vehicles two nodes are selected uniform randomly. If only one vehicle is chosen, these nodes are chosen to be distinct. Then, the nodes are swapped.

12.3.3.3 Particle Swarm Optimization

Khouadjia *et al.* [39] have proposed a Particle Swarm Optimization (PSO) for solving DCVRP. The authors suggest a discrete optimization of the problem and some adaptive mechanisms. Since PSO is intrinsically a memory-based approach, due to the memorization by each particle of its current and best position in the search space, they propose to reuse the best positions gathered in the past to face the changing environment. At each new sub-problem, the algorithm selects the positions with the best solution cost in the new search landscape. From these positions, the particles are re-positioned (re-initialized) for the new optimization process. The velocity vector of each particle corresponds to the likely routes in which a customer could belong. The updating of the position vector is the application of the velocity vector. It is summarized in shifting customers from their respective route to another one according to the velocity vector and with the cheapest strategy insertion (i.e. by minimizing the cost of the insertion). The updating process is very similar to the *ejection chain* method that has been applied successfully to vehicle routing [56].

Khouadjia *et al.* [38] have enhanced their algorithm, particularly against the early well-known convergence of PSO algorithm. They propose in [38], a multi-swarm approach called MAPSO (Multi-Adaptive Particle Swarm Optimization) to investigate whether a multi-population metaheuristic might be beneficial in dynamic vehicle routing environments. The aim is to place different swarms on the search space to counterbalance the loss of diversity population and to provide better reactivity to the arrival of new customers.

12.4 Dynamic Performance Measures

The goal of optimization in dynamic environments is not only to find an optimum within a given number of generations, but rather a perpetual adjustment to changing environmental conditions. Besides the accuracy of an approximation at time t, the stability of the algorithm is also of interest as well as the recovery time to reach

Table 12.2 State of the art metaheuristics for DVRP and its variants.

Metaheuristics		Authors	Problem	Operators or Neighborhood
Trajectory-Based	Tabu Search	Hanshar *et al.* [31]	DCVRP	λ-interchange and inversion
		Gendreaul *et al.* [26] Ichoua *et al.* [34, 35]	DVRPTW	CROSS exchange
		Mitrović-Minić *et al.* [48] Attanasio *et al.* [2]	DPDVRP	
	GRASP	Montemanni *et al.* [49]	DCVRP	Greedy insertion
	VNS	Sarasola *et al.* [58]	DCVRP	Swap, insertion, 2-Opt, 2-Opt*
Population-Based	Ant Colony Optimization	Montemanni *et al.* [49]	DCVRP	Greedy heuristic
		Rizzoli *et al.* [57]		–
		Tian *et al.* [64]		2-opt heuristic
		Chitty [14]		Dynamic programming
		Jun *et al.* [37]	DVRPTW	Cooperation with EA
		Oliveira *et al.* [18]		Greedy heuristic
	Evolutionary Algorithms	Hanshar *et al.* [31]	DCVRP	BCRC crossover Mutation (inversion)
		Branke *et al.* [13]		Two-point crossover Mutation (replacement)
		Garrido2010 *et al.* [24]		One-point crossover 3 mutation operators (replacement,insertion,deletion)
		Van Hemertand and La Poutré [65]		mutation (CROSS exchange)
		Housroum *et al.* [32]	DVRPTW	PMX crossover 3 mutations (Or-Opt, 1-Opt, swap)
		Alvarenga *et al.* [1]		Specific crossover 8 mutations (insertion, exchange, ...)
		Wang *et al.* [66]		Two-point crossover Mutation (Insertion)
		Jih *et al.* [36]	DPDVRPTW	3 crossovers (OX, UOX MX1, MX2) 3 mutations (rearrangement, swap, 2-Opt)
		Bosman *et al.* [10]		CROSS exchange mutation
		Zhao *et al.* [70]	DVRPTT	PMX crossover 3 mutations (Or-Opt, 1-Opt, swap)
	Particle Swarm Optimization	Khouadjia *et al.* [38, 39]	DCVRP	2-Opt heuristic Cheapest insertion heuristic

again a certain approximation quality. We report here some measures that could be used for evaluating the performance of an algorithm designed for the DVRP.

Weicker [67] proposes three features for describing the goodness of a dynamic adaptation process: accuracy, stability, and ε-reactivity.

The optimization *accuracy* at time t for a fitness function F and optimization algorithm A is defined as

$$accuracy^t_{F,A} = \frac{Min^t_F}{F(best^t_A)}, \tag{12.7}$$

where $best^t_A$ is the best candidate solution in the population at time t and Min^t_F is the best fitness value in the search space (best known solution). The optimization accuracy ranges between 0 and 1, where accuracy 1 is the best possible value.

As a second goal, stability is an important issue in optimization. In the context of dynamic optimization, an adaptive algorithm is called stable if changes in the environment do not affect the optimization accuracy severely. Even in the case of drastic changes an algorithm should be able to limit the respective fitness drop. The stability at time, t is defined as

$$stability^{t}_{F,A} = max\{0, accuracy(t) - accuracy(t-1)\} \tag{12.8}$$

and ranges between 0 and 1. A value close to 0 implies a high stability.

Finally, another aspect to be considered is the ability of the algorithm to react quickly to changes. This is measured by the ε-reactivity, which ranges in $[1, maxgen]$ (a smaller value implies a higher reactivity):

$$\varepsilon - reactivity_i = \min\{i' - i | i < i' \leq maxgen, i \in \mathbb{N}, \frac{accuracy_{i'}}{accuracy_i} \geq (1-\varepsilon)\}$$

12.5 Performance Assessment

This section is devoted to the performance evaluation of different recent metaheuristics proposed in the literature [31, 38, 39, 49]. We justify this choice by the fact that these approaches follow the same experimental protocol, from the simulation framework to the set of benchmarks. Thus, it is easy to have an idea about the performances of these algorithms. Besides, all the classes of population-based metaheuristics described in Section 12.3 are represented.

Several benchmarks have been used. The most used ones are those of Kilby [40]. They were derived from publicly available VRP benchmark data from three separate VRP sources, namely Taillard [63] (13 instances), Christofides and Beasley [15] (7 instances), and Fisher *et al.* [22] (2 instances). These instances were organized and extended by Kilby *et al.* [40]. Kilby *et al.* organized the instances into two groups, pickup and delivery, and gave each request an available time which signifies when the order was placed in the system and a duration, which represents the minimum amount of time a vehicle waits at a customer. In [31, 39, 49] authors use the *dod* described in the Section 12.2 in order to determine the percentage of dynamic requests over the entire working day. The degree of dynamism was fixed at 0.5; this means that a half of the customers is considered as static, while the other half is dynamic. The optimization begins to plan routes with the known static customers at the beginning of the working day.

We report in Table 12.3 the best found solutions from the literature on metaheuristics; Adaptive Particle Swarm Optimization (APSO) [39], Multi-Adaptive Particle Swarm Optimization (MASPO) [38], Genetic Algorithm(GA) [31], Tabu Search (TS) [31], and Ant System(AS) [49] on Kilby's instances. These metaheuristics deal with pickup instances. In this case, the driver of the vehicle is not concerned with what is being transported, but only the quantity that has to be picked from the customer.

We highlight the best found solutions in dark shaded cells and the average results in light shaded cells. For each instance, 30 runs of the algorithms have been considered. We can see that the multi-swarm *MAPSO* is able to provide higher

Table 12.3 Numerical results obtained by the state-of-the-art metaheuristics on Kilby's instances

Instances	Metaheuristics									
	APSO [39]		MAPSO [38]		AS [49]		GA [31]		TS [31]	
	Best	Average	Best	Average	Best	Average	Best	Average	Best	Average
c50	575.89	647.75	571.34	610.67	631.30	681.86	570.89	593.42	603.57	627.90
c75	970.45	1046.25	931.59	965.53	1009.36	1042.39	981.57	1013.45	981.51	1013.82
c100	988.27	1087.96	953.79	973.01	973.26	1066.16	961.10	987.59	997.15	1047.60
c100b	924.32	970.66	866.42	882.39	944.23	1023.60	881.92	900.94	891.42	932.14
c120	1276.88	1450.82	1223.49	1295.79	1416.45	1525.15	1303.59	1390.58	1331.22	1468.12
c150	1371.08	1499.54	1300.43	1357.71	1345.73	1455.50	1348.88	1386.93	1318.22	1401.06
c199	1640.40	1751.63	1595.97	1646.37	1771.04	1844.82	1654.51	1758.51	1750.09	1783.43
f71	279.52	339.08	287.51	296.76	311.18	358.69	301.79	309.94	280.23	306.33
f134	15875	16477.4	15150.5	16193	15135.51	16083.56	15528.81	15986.84	15717.90	16582.04
tai75a	1816.07	1978.51	1794.38	1849.37	1843.08	1945.20	1782.91	1856.66	1778.52	1883.47
tai75b	1447.39	1489.24	1396.42	1426.67	1535.43	1704.06	1464.56	1527.77	1461.37	1587.72
tai75c	1481.35	1555.36	1483.1	1518.65	1574.98	1653.58	1440.54	1501.91	1406.27	1527.72
tai75d	1414.28	1481.05	1391.99	1413.83	1472.35	1529.00	1399.83	1422.27	1430.83	1453.56
tai100a	2249.84	2378.26	2178.86	2214.61	2375.92	2428.38	2232.71	2295.61	2208.85	2310.37
tai100b	2238.42	2426.58	2140.57	2218.58	2283.97	2347.90	2147.70	2215.93	2219.28	2330.52
tai100c	1532.56	1612.1	1490.40	1550.63	1562.30	1655.91	1541.28	1622.66	1515.10	1604.18
tai100d	1955.06	2092.31	1838.75	1928.69	2008.13	2060.72	1834.60	1912.43	1881.91	2026.76
tai150a	3400.33	3581.66	3273.24	3389.97	3644.78	3840.18	3328.85	3501.83	3488.02	3598.69
tai150b	3013.99	3391.08	2861.91	2956.84	3166.88	3327.47	2933.40	3115.39	3109.23	3215.32
tai150c	2714.34	2859.97	2512.01	2671.35	2811.48	3016.14	2612.68	2743.55	2666.28	2913.67
tai150d	3025.43	3143.16	2861.46	2989.24	3058.87	3203.75	2950.61	3045.16	2950.83	3111.43
Total	50190.87	53260.37	48104.13	50349.66	50876.23	53794.02	49202.73	51089.37	49987.8	52725.85

quality solutions than the other algorithms. It outperforms the other metaheuristics, and gives 18 new best solutions out of the 21 Kilby's instances. *MAPSO* algorithm also provides the shortest total traveled distance over all instances. The improvement brought about by *MAPSO* ranges between $[2.23 - 5.76]$ compared to the other metaheuristics on the total traveled distance. As to the dynamic performance measures, we have computed the accuracy at the end of the working day. Table 12.4 shows the accuracy of the previous algorithms. It reports the best obtained distances and the bounds Min_F^T (best known solutions) found by an (ideal) off-line algorithm which had access to the entire instance, including dynamic requests, beforehand. These solutions can be found in the literature[1] over the 21 Kilby's instances.

These best known solutions consider all customers to be static, and then are not feasible solutions for the DVRP. They work as a bound for the algorithms. From Table 12.4, we see that MAPSO has the best accuracy average at the end of the simulation. This accuracy is equal to 0.89 (being 1.0 the best value) which denotes that the algorithm is able to produce good solutions on the conventional dynamic

[1] `http://neo.lcc.uma.es/radi-aeb/WebVRP/`

benchmarks. We do not report the reactivity and stability because we would need all the minimum traveled distances (exact optimal cost) at each arrival of a new customer.

Table 12.4 Accuracy of different metaheuristics on the Kilby's instances

Instance	Min_P^T	Metaheuristics									
		APSO [39]		MAPSO [38]		AS [49]		GA [31]		TS [31]	
		Best	Accu.	Best	Accu.	Best	Accu.	Best	Accu.	Best	Accu.
c50	521	575.89	0.90	571.34	0.91	631.3	0.83	570.89	0.91	603.57	0.86
c75	832	970.45	0.86	931.59	0.89	1009.36	0.82	981.57	0.85	981.51	0.85
c100	817	988.27	0.83	953.79	0.86	973.26	0.84	961.1	0.85	997.15	0.82
c100b	820	924.32	0.89	866.42	0.95	944.23	0.87	881.92	0.93	891.42	0.92
c120	1042.11	1276.88	0.82	1223.49	0.85	1416.45	0.74	1303.59	0.8	1331.22	0.78
c150	1028.42	1371.08	0.75	1300.43	0.79	1345.73	0.76	1348.88	0.76	1318.22	0.78
c199	1291.45	1640.4	0.79	1595.97	0.81	1771.04	0.73	1654.51	0.78	1750.09	0.74
f71	237	279.52	0.85	287.51	0.82	311.18	0.76	301.79	0.79	280.23	0.85
f134	11620	15875	0.73	15150.5	0.77	15135.51	0.77	15528.81	0.75	15717.9	0.74
tai75a	1618.36	1816.07	0.89	1794.38	0.90	1843.08	0.88	1782.91	0.91	1778.52	0.91
tai75b	1344.64	1447.39	0.93	1396.42	0.96	1535.43	0.88	1464.56	0.92	1461.37	0.92
tai75c	1291.01	1481.35	0.87	1483.1	0.87	1574.98	0.82	1440.54	0.90	1406.27	0.92
tai75d	1365.42	1414.28	0.97	1391.99	0.98	1472.35	0.93	1399.83	0.98	1430.83	0.95
tai100a	2041.33	2249.84	0.91	2178.86	0.94	2375.92	0.86	2232.71	0.91	2208.85	0.92
tai100b	1940.61	2238.42	0.87	2140.57	0.91	2283.97	0.85	2147.7	0.90	2219.28	0.87
tai100c	1406.2	1532.56	0.92	1490.4	0.94	1562.3	0.9	1541.28	0.91	1515.1	0.93
tai100d	1581.25	1955.06	0.81	1838.75	0.86	2008.13	0.79	1834.6	0.86	1881.91	0.84
tai150a	3055.23	3400.33	0.90	3273.24	0.93	3644.78	0.84	3328.85	0.92	3488.02	0.88
tai150b	2656.47	3013.99	0.88	2861.91	0.93	3166.88	0.84	2933.4	0.91	3109.23	0.85
tai150c	2341.84	2714.34	0.86	2512.01	0.93	2811.48	0.83	2612.68	0.90	2666.28	0.88
tai150d	2645.39	3025.43	0.87	2861.46	0.92	3058.87	0.86	2950.61	0.90	2950.83	0.90
Average	1976.03	2390.04	0.86	2290.67	0.89	2422.68	0.83	2342.99	0.87	2380.37	0.86

12.6 Conclusions and Future Work

The Dynamic Vehicle Routing Problem (DVRP) has been surveyed in this chapter. This problem is important both in research and industrial domains due to its many real-world applications. The state of the art presented in this chapter covers the problem representation as well as the existing solving metaheuristics. In addition, a practical study of several metaheuristics in terms of the solution quality is reported. Besides, according to dynamic performance measures, the accuracy of different algorithms is calculated.

This study shows that the multi-population-based metaheuristics are able to find high quality solutions comparatively to the rest of metaheuristics. This is easy to understand since they offer a rich diversity in the exploration, which allows the algorithm to easily track the moving optima throughout the search space.

Different issues remain open. One of them is the landscape analysis of the DVRP and the severity of the changes that can occur. Landscape study techniques will allow the algorithm to locate better the current optimum and anticipate its movements, given that, unless the change in the problem is strong, the new problem can be similar to the old one. Concerning the severity, if the change is strong and frequent, trajectory-based metaheuristics usually fail to react and to track the optima. Enhancing the diversity within this class of metaheuristics is an inescapable issue.

Another prospect is the flexibility and robustness of the solutions. The best solution in terms of fitness quality may not be the most flexible or robust one when it comes to updating it when the problem changes. The underlying idea is searching for robust solutions or the manner to obtain them. Robust solutions are those that promise high quality even if the environment changes. One way to ensure robustness is to introduce flexibility in these solutions. Through anticipating the changes in the environment, we will be able to provide solutions that not only have a high quality, but that allow the adaptation to high quality solutions after the environment has changed. To preserve flexibility, we could construct initial solutions being aware about the potential arrival of new orders; in order to do so, we can imagine to adjust dynamically the length of the working day, making it smaller at the beginning of the optimization process and letting it increase until it reaches the value defined by the problem instance. In this way, we can expect to get solutions with a larger number of shorter routes at the beginning of the simulation time. If there are more routes available and they are not built to use the whole working day length, it will be easier to place new customers in a good position in vehicle routes.

On this way, future approaches will integrate new mechanisms that handle these issues and will be able to respond better and react faster to the changing environment. New approaches will provide solutions which are comparable to the solutions obtained in the static case.

Acknowledgments. Authors acknowledge funds from the Associated Teams Program MOMDI of the French National Institute for Research in Computer Science and Control INRIA (`http://www.inria.fr`), the Spanish Ministry of Science and Innovation plus European FEDER under contracts TIN2008-06491-C04-01 (M* Project `http://mstar.lcc.uma.es`), TIN2011-28194 (roadME Project `http://roadme.lcc.uma.es`), and CICE, Junta de Andalucía under contract P07-TIC-03044 (DIRICOM project `http://diricom.lcc.uma.es`). Briseida Sarasola is supported by grant AP2009-1680 from the Spanish government.

References

[1] Alvarenga, G.B., Silva, R.M.A., Mateus, G.R.: A hybrid approach for the dynamic vehicle routing problem with time windows. In: Proceedings of the Fifth International Conference on Hybrid Intelligent Systems, pp. 61–67. IEEE Computer Society, Washington, DC (2005)

[2] Attanasio, A., Cordeau, J.F., Ghiani, G., Laporte, G.: Parallel tabu search heuristics for the dynamic multi-vehicle dial-a-ride problem. Parallel Computing 30(3), 377–387 (2004)

[3] Beaudry, A., Laporte, G., Melo, T., Nickel, S.: Dynamic transportation of patients in hospitals. OR spectrum 32(1), 77–107 (2010)

[4] Bent, R., Van Hentenryck, P.: Dynamic vehicle routing with stochastic requests. In: Gottlob, G., Walsh, T. (eds.) Proceedings of the 18th International Joint Conference on Artificial Intelligence, pp. 1362–1363. Morgan Kaufmann Publishers Inc., San Francisco (2003)

[5] Bent, R., Van Hentenryck, P.: Scenario-based planning for partially dynamic vehicle routing with stochastic customers. Operations Research 52(6), 977–987 (2004)
[6] Bertsimas, D.J., Van Ryzin, G.J.: A stochastic and dynamic vehicle routing problem in the euclidean plane. Operations Research 39(4), 601–615 (1991)
[7] Bertsimas, D.J., Van Ryzin, G.J.: Stochastic and dynamic vehicle routing with general demand and interarrival time distributions. Advanced Applied Probability 25, 947–978 (1993)
[8] Bianchi, L.: Notes on dynamic vehicle routing -the state of the art-. Technical report, Istituto Dalle Molle Di Studi Sull Intelligenza Artificiale (2000)
[9] Blanton Jr., J.L., Wainwright, R.L.: Multiple vehicle routing with time and capacity constraints using genetic algorithms. In: Forrest, S. (ed.) Proceedings of the 5th International Conference on Genetic Algorithms, pp. 452–459. Morgan Kaufmann Publishers Inc., San Francisco (1993)
[10] Bosman, P.A.N., La Poutré, H.: Computationally Intelligent Online Dynamic Vehicle Routing by Explicit Load Prediction in an Evolutionary Algorithm. In: Runarsson, T.P., Beyer, H.-G., Burke, E.K., Merelo-Guervós, J.J., Whitley, L.D., Yao, X. (eds.) PPSN 2006. LNCS, vol. 4193, pp. 312–321. Springer, Heidelberg (2006)
[11] Branchini, R.M., Armentano, V.A., Løkketangen, A.: Adaptive granular local search heuristic for a dynamic vehicle routing problem. Computers & Operations Research 36(11), 2955–2968 (2009)
[12] Branke, J.: Evolutionary optimization in dynamic environments. Kluwer Academic Publishers (2002)
[13] Branke, J., Middendorf, M., Noeth, G., Dessouky, M.: Waiting strategies for dynamic vehicle routing. Transportation Science 39(3), 298–312 (2005)
[14] Chitty, D.M., Hernandez, M.L.: A Hybrid Ant Colony Optimisation Technique for Dynamic Vehicle Routing. In: Deb, K., et al. (eds.) GECCO 2004, Part I. LNCS, vol. 3102, pp. 48–59. Springer, Heidelberg (2004)
[15] Christofides, N., Beasley, J.: The period routing problem. Networks 14(2), 237–256 (1984)
[16] Cordeau, J.F., Laporte, G.: A tabu search heuristic for the static multi-vehicle dial-a-ride problem. Transportation Research Part B: Methodological 37(6), 579–594 (2003)
[17] Dantzig, G.B., Ramser, J.H.: The truck dispatching problem. Operations Research, Management Sciences 6(1), 80–91 (1959)
[18] de Oliveira, S.M., de Souza, S.R., Silva, M.A.L.: A solution of dynamic vehicle routing problem with time window via ant colony system metaheuristic. In: Proceedings of the 2008 10th Brazilian Symposium on Neural Networks, SBRN 2008, pp. 21–26. IEEE Computer Society, Washington, DC (2008)
[19] Fabri, A., Recht, P.: On dynamic pickup and delivery vehicle routing with several time windows and waiting times. Transportation Research Part B: Methodological 40(4), 335–350 (2006)
[20] Fagerholt, K., Foss, B.A., Horgen, O.J.: A decision support model for establishing an air taxi service: a case study. Journal of the Operational Research Society 60(9), 1173–1182 (2009)
[21] Fiegl, C., Pontow, C.: Online scheduling of pick-up and delivery tasks in hospitals. Journal of Biomedical Informatics 42(4), 624–632 (2009)
[22] Fisher, M.: Vehicle routing. In: Monma, C.L., Ball, M.O., Magnanti, T.L., Nemhauser, G.L. (eds.) Network Routing. Handbooks in Operations Research and Management Science, vol. 8, pp. 1–33. Elsevier (1995)

[23] Gambardella, L.M., Rizzoli, A.E., Oliverio, F., Casagrande, N., Donati, A.V., Montemanni, R., Lucibello, E.: Ant Colony Optimization for vehicle routing in advanced logistics systems. In: Proceedings of MAS 2003 - International Workshop on Modeling & Applied Simulation, pp. 3–9 (2003)
[24] Garrido, P., Riff, M.C.: DVRP: a hard dynamic combinatorial optimisation problem tackled by an evolutionary hyper-heuristic. Journal of Heuristics 16, 795–834 (2010)
[25] Gendreau, M., Guertin, F., Potvin, J.Y., Séguin, R.: Neighborhood search heuristics for a dynamic vehicle dispatching problem with pick-ups and deliveries. Transportation Research Part C: Emerging Technologies 14(3), 157–174 (2006)
[26] Gendreau, M., Guertin, F., Potvin, J.Y., Taillard, E.: Parallel tabu search for real-time vehicle routing and dispatching. Transportation Science 33(4), 381–390 (1999)
[27] Gendreau, M., Potvin, J.Y.: Dynamic vehicle routing and dispatching (1998)
[28] Ghiani, G., Guerriero, F., Laporte, G., Musmanno, R.: Real-time vehicle routing: Solution concepts, algorithms and parallel computing strategies. European Journal of Operational Research 151, 1–11 (2003)
[29] Haghani, A., Jung, S.: A dynamic vehicle routing problem with time-dependent travel times. Comput. Oper. Res. 32, 2959–2986 (2005)
[30] Haghani, A., Yang, S.: Real-time emergency response fleet deployment: Concepts, systems, simulation & case studies. In: Dynamic Fleet Management, pp. 133–162 (2007)
[31] Hanshar, F.T., Ombuki-Berman, B.M.: Dynamic vehicle routing using genetic algorithms. Applied Intelligence 27, 89–99 (2007)
[32] Housroum, H., Hsu, T., Dupas, R., Goncalves, G.: A hybrid GA approach for solving the dynamic vehicle routing problem with time windows. In: 2nd International Conference on Information & Communication Technologies: Workshop ICT in Intelligent Transportation Systems, ICTTA 2006, vol. 1, pp. 787–792 (2006)
[33] Hvattum, L.M., Løkketangen, A., Laporte, G.: Solving a dynamic and stochastic vehicle routing problem with a sample scenario hedging heuristic. Transportation Science 40, 421–438 (2006)
[34] Ichoua, S., Gendreau, M., Potvin, J.Y.: Diversion issues in real-time vehicle dispatching. Transportation Science 34, 426–438 (2000)
[35] Ichoua, S., Gendreau, M., Potvin, J.Y.: Vehicle dispatching with time-dependent travel times. European Journal of Operational Research 144, 379–396 (2003)
[36] Jih, W.R., Hsu, J.Y.J.: Dynamic vehicle routing using hybrid genetic algorithms. In: Proceedings of the IEEE International Conference on Robotics and Automation, Detroit, Michigan, vol. 1, pp. 453–458 (1999)
[37] Jun, Q., Wang, J., Zheng, B.: A hybrid multi-objective algorithm for dynamic vehicle routing problems. In: Bubak, M., Albada, G.D., Dongarra, J., Sloot, P.M. (eds.) Proceedings of the 8th International Conference on Computational Science, Part III, ICCS 2008, pp. 674–681. Springer, Heidelberg (2008)
[38] Khouadjia, M.R., Alba, E., Jourdan, L., Talbi, E.-G.: Multi-Swarm Optimization for Dynamic Combinatorial Problems: A Case Study on Dynamic Vehicle Routing Problem. In: Dorigo, M., Birattari, M., Di Caro, G.A., Doursat, R., Engelbrecht, A.P., Floreano, D., Gambardella, L.M., Groß, R., Şahin, E., Sayama, H., Stützle, T. (eds.) ANTS 2010. LNCS, vol. 6234, pp. 227–238. Springer, Heidelberg (2010)
[39] Khouadjia, M.R., Jourdan, L., Talbi, E.G.: Adaptive particle swarm for solving the dynamic vehicle routing problem. In: IEEE/ACS International Conference on Computer Systems and Applications (AICCSA 2010), pp. 1–8. IEEE Computer Society (2010)
[40] Kilby, P., Prosser, P., Shaw, P.: Dynamic VRPs: A study of scenarios. Technical report, University of Strathclyde, U.K. (1998)

[41] Kritzinger, S., Tricoire, F., Doerner, K.F., Hartl, R.F.: Variable Neighborhood Search for the Time-Dependent Vehicle Routing Problem with Soft Time Windows. In: Coello, C.A.C. (ed.) LION 2011. LNCS, vol. 6683, pp. 61–75. Springer, Heidelberg (2011)
[42] Larsen, A.: The Dynamic Vehicle Routing Problem. PhD thesis, Technical University of Denmark (2000)
[43] Larsen, A., Madsen, O.B.G., Solomon, M.M.: Partially dynamic vehicle routing-models and algorithms. Journal of the Operational Research Society 53(6), 637–646 (2002)
[44] Larsen, A., Madsen, O.B.G., Solomon, M.M.: The a priori dynamic traveling salesman problem with time windows. Transportation Science 38(4), 459–472 (2004)
[45] Larsen, A., Madsen, O.B.G., Solomon, M.M.: Recent developments in dynamic vehicle routing systems. In: Golden, B., Raghavan, S., Wasil, E. (eds.) The Vehicle Routing Problem: Latest Advances and New Challenges. Operations Research/Computer Science Interfaces Series, vol. 43, pp. 199–218. Springer, US (2008)
[46] Lund, K., Madsen, O.B.G., Rygaard, J.M.: Vehicle routing problems with varying degrees of dynamism. Technical report, IMM, The Department of Mathematical Modelling, Technical University of Denmark (1996)
[47] De Magalhães, J.M., Pinho De Sousa, J.: Dynamic VRP in pharmaceutical distribution -a case study. Central European Journal of Operations Research 14(2), 177–192 (2006)
[48] Mitrović-Minić, S., Krishnamurti, R., Laporte, G.: Double-horizon based heuristics for the dynamic pickup and delivery problem with time windows. Transportation Research Part B: Methodological 38(8), 669–685 (2004)
[49] Montemanni, R., Gambardella, L.M., Rizzoli, A.E., Donati, A.V.: A new algorithm for a dynamic vehicle routing problem based on ant colony system. Journal of Combinatorial Optimization 10, 327–343 (2005)
[50] Osman, I.H.: Metastrategy simulated annealing and tabu search algorithms for the vehicle routing problem. Annals of Operations Research 41(4), 421–451 (1993)
[51] Pavone, M., Bisnik, N., Frazzoli, E., Isler, V.: A stochastic and dynamic vehicle routing problem with time windows and customer impatience. Mobile Networks and Applications 14, 350–364 (2009)
[52] Potvin, J.Y., Xu, Y., Benyahia, I.: Vehicle routing and scheduling with dynamic travel times. Comput. Oper. Res. 33, 1129–1137 (2006)
[53] Prins, C.: A simple and effective evolutionary algorithm for the vehicle routing problem. Computers & Operations Research 31(12), 1985–2002 (2004)
[54] Psaraftis, H.N.: Dynamic vehicle routing problems. Vehicle Routing: Methods and Studies 16, 223–248 (1988)
[55] Psaraftis, H.N.: Dynamic vehicle routing: status and prospects. Annals of Operations Research 61, 143–164 (1995)
[56] Rego, C.: Node-ejection chains for the vehicle routing problem: Sequential and parallel algorithms. Parallel Computing 27(3), 201–222 (2001)
[57] Rizzoli, A., Montemanni, R., Lucibello, E., Gambardella, L.: Ant colony optimization for real-world vehicle routing problems. Swarm Intelligence 1, 135–151 (2007)
[58] Sarasola, B., Khouadjia, M.R., Alba, E., Jourdan, L., Talbi, E.G.: Flexible variable neighborhood search in dynamic vehicle routing. In: 8th European event on Evolutionary Algorithms in Stochastic and Dynamic Environments (EvoSTOC 2011), April 27-29 (2011)
[59] Savelsbergh, M.W.P., Sol, M.: The general pickup and delivery problem. Transportation Science 29(1), 17–29 (1995)

[60] Schilde, M., Doerner, K.F., Hartl, R.F.: Metaheuristics for the dynamic stochastic dial-a-ride problem with expected return transports. Computers & OR 38(12), 1719–1730 (2011)

[61] Schmid, V., Doerner, K.F.: Ambulance location and relocation problems with time-dependent travel times. European Journal of Operational Research 207(3), 1293–1303 (2010)

[62] Sun, L., Hu, X., Wang, Z., Huang, M.: A knowledge-based model representation and on-line solution method for dynamic vehicle routing problem. In: Shi, Y., van Albada, G.D., Dongarra, J., Sloot, P.M.A. (eds.) ICCS 2007: Proceedings of the 7th International Conference on Computational Science, Part IV. LNCS, pp. 218–226. Springer, Heidelberg (2007)

[63] Taillard, É.: Parallel iterative search methods for vehicle routing problems. Networks 23(8), 661–673 (1993)

[64] Tian, Y., Song, J., Yao, D., Hu, J.: Dynamic vehicle routing problem using hybrid ant system. In: Proceedings of the IEEE Conference on Intelligent Transportation Systems, vol. 2, pp. 970–974 (2003)

[65] van Hemert, J., La Poutré, J.A.H.: Dynamic Routing Problems with Fruitful Regions: Models and Evolutionary Computation. In: Yao, X., Burke, E.K., Lozano, J.A., Smith, J., Merelo-Guervós, J.J., Bullinaria, J.A., Rowe, J.E., Tiňo, P., Kabán, A., Schwefel, H.-P. (eds.) PPSN 2004. LNCS, vol. 3242, pp. 692–701. Springer, Heidelberg (2004)

[66] Wang, J.Q., Tong, X.N., Li, Z.M.: An improved evolutionary algorithm for dynamic vehicle routing problem with time windows. In: ICCS 2007: Proceedings of the 7th International Conference on Computational Science, Part IV, pp. 1147–1154. Springer, Heidelberg (2007)

[67] Weicker, K.: Performance Measures for Dynamic Environments. In: Guervós, J.J.M., Adamidis, P.A., Beyer, H.-G., Fernández-Villacañas, J.-L., Schwefel, H.-P. (eds.) PPSN 2002. LNCS, vol. 2439, pp. 64–76. Springer, Heidelberg (2002)

[68] Xu, J., Goncalves, G., Hsu, T.: Genetic algorithm for the vehicle routing problem with time windows and fuzzy demand. In: 2008 IEEE World Congress on Computational Intelligence, WCCI 2008, pp. 4125–4129 (2008)

[69] Yang, J., Jaillet, P., Mahmassani, H.: Real-time multivehicle truckload pickup and delivery problems. Transportation Science 38, 135–148 (2004)

[70] Zhao, X., Goncalves, G., Dupas, R.: A genetic approach to solving the vehicle routing problem with time dependent travel times. In: 16th Mediterranean Conference on Control and Automation, pp. 413–418 (2008)

Chapter 13
Low-Level Hybridization of Scatter Search and Particle Filter for Dynamic TSP Solving

Juan José Pantrigo and Abraham Duarte

Abstract. This work presents the application of the Scatter Search Particle Filter (SSPF) algorithm to solve the Dynamic Travelling Salesman Problem (DTSP). SSPF combines sequential estimation and combinatorial optimization methods to efficiently address dynamic optimization problems. SSPF obtains high quality solutions at each time step by taking advantage of the best solutions obtained in the previous ones. To demonstrate the performance of the proposed algorithm, we conduct experiments using two different benchmarks. The first one was generated for us and contains instances sized 25, 50, 75 and 100-cities and the second one are dynamic versions of TSPLIB benchmarks. Experimental results have shown that the performance of SSPF for the DTSP is significantly better than other population based metaheuristics, such as Evolutionary Algorithms or Scatter Search. Our proposal appreciably reduces the execution time without affecting the quality of the obtained results.

13.1 Introduction

Dynamic optimization problems are characterized by an initial problem definition and a collection of "events" over the time. These "events" define changes on the data of the problem [26]. Therefore, dynamic optimization methods arise from strategies to adapt for non-stationary conditions. In dynamic optimization problems, a key question is how to use information found in previous time steps to obtain high quality solutions in subsequent ones, without restarting the computation from scratch.

Dynamic optimization problems play an important role in industrial applications. Many real-life problems belong to this category, particularly in transportation, telecommunications and manufacturing areas [26]. Surprisingly, compared to

Juan José Pantrigo · Abraham Duarte
Universidad Rey Juan Carlos,
c/ Tulipán s/n Móstoles Madrid, Spain
e-mail: {juanjose.pantrigo,abraham.duarte}@urjc.es

E. Alba et al. (Eds.): Metaheuristics for Dynamic Optimization, SCI 433, pp. 291–308.
springerlink.com

the amount of research undertaken on stationary optimization problems, relatively little work has been devoted to dynamic problems, despite the potential economic advantages in doing so [8, 25]. Any advance in this field would translate to increased company profits, lower consumer prices and improved services [26].

Unlike stationary problems, dynamic problems often lack well defined optimization functions, standard benchmarks or criteria for comparing solutions [3, 8, 25, 26]. In the last decade, mainly used strategies have been specific heuristics [26] and manual procedures [2, 25]. More recently, has been proposed a metaheuristic approach. Metaheuristics are high-level general strategies for designing heuristics procedures [4]. The relevance of metaheuristics is reflected in their application for solving many different real-world complex problems, mainly combinatorial [4, 11]. Since the initial proposal of Glover about Tabu Search in 1986, many metaheuristics have emerged to design good general heuristics methods for solving different problems. The well known metaheuristics procedures are Genetic Programming, GRASP, Simulated Annealing or Ant Colony Optimization. The reader can find a review of this methods in [4, 11].

In dynamic optimization problems, metaheuristic-based approaches usually consider one of the two following strategies after events: (i) restart the search method from scratch or (ii) start from the best solutions found before the last event. In the first approach, the derived problem is processed as unrelated with respect to its origin problem. Therefore, the useful information obtained in the previous time steps is wasted, increasing the computation time. On the other hand, starting the method from the best solutions found in the previous time step could implicate a loss of diversity in the solution set. Moreover, the best solutions found in previous time steps could be close to local optima in the current one. As a consequence, the search algorithm could get stuck in a local optimum.

The filtering problem concerns about updating the present state of knowledge and predicting with drawing inferences about the future state of the system [6]. Sequential Monte Carlo population-based algorithms (also called particle filters) are a special class of filters in which theoretical distributions on the state space are approximated by simulated random measures (also called particles) [6]. Assuming relative small problem data changes, the new optimum location should be related to the solution of the previous problem definition. Thus, the actual search process could use previous knowledge for a more efficient search. A reasonable trade-off between the analysis of the prior problem solution and actual problem computational effort must be found. Diverse methods consisting of the low-level hybridization of Particle Filters and metaheuristics have been successfully applied to dynamic optimization problems [22–24]. The term low-level hybridization refers to the functional composition of a single optimization method. In this hybrid class of optimization methods, a given function of a metaheuristic is replaced by another metaheuristic [28]. In this work, we apply the Scatter Search Particle Filter (SSPF) for the Dynamic Travelling Salesman Problem. SSPF combines sequential estimation strategies (Particle Filter) [1, 6] and metaheuristics methods (Scatter Search) [4] in two different stages.

In the Particle Filter (PF) stage, a particle set is propagated and updated to obtain new particle sets in every time step. In the Scatter Search (SS) stage, some solutions from the particle set are selected and combined, in order to obtain better solutions. SSPF was firstly presented in [23] and it was applied to multidimensional object tracking (articulated and multiple object tracking). This work considers a dynamic variant of the Travelling Salesman Problem [9] in which distances among cities vary over the time. The problem instances used for this study have 25, 50, 75 and 100 cities placed on the Euclidean space, and a modification of approximately 15% of the previous data graph is generated as successive events for the next problem instance. Also, dynamic versions of TSPLIB benchmark are used in order to compare the performance of different approaches. Experimental results have shown that the proposed algorithm has an acceptable performance when applied to the Dynamic Travelling Salesman Problem (DTSP). Specifically, the CPU for the rest of the derived problem instances significantly decreases with respect to the initial graph problem. This reduction in the search time will not affect the quality of the solution of derived graphs.

13.2 Dynamic Travelling Salesman Problem

The Travelling Salesman Problem (TSP) consists of finding the shortest tour connecting a fixed number of locations (cities), visiting each city exactly once [9]. The instances of this problem can be represented as a graph $G = \{V, E, W\}$, where V is a set of vertexes representing the cities, E is a set of edges which model the paths connecting cities and W is a symmetric matrix of weights that store the distances among cities. We suppose that there is an edge connecting every pair of cities. The TSP can be described as the problem of finding a Hamiltonian circuit with minimum length in the graph G [30].

The TSP has been one of the most considered problems in Combinatorial Optimization and Operations Research. This problem belongs to the NP-hard class, as demonstrated in [17]. A relevant review of different approaches used to solve this problem can be found in [12].

The Dynamic Travelling Salesman Problem (DTSP) is a generalization of the TSP in which G is time-dependent. The DTSP is general enough to be a benchmark problem. Moreover, this problem has got several practical applications such as modeling the traffic in cities along time [9] or fluctuating set of active machines [15].

Two different DTSP varieties in the literature have been described. The first one consists of inserting or deleting cities into a given problem instance [13, 14]. A different approach has been taken in [9]. They keep a constant number of cities but allow distance changing among them. The second model is applied to describe traffic jams and motorways connecting cities. In this context, before the traffic jam has occurred, good old solutions may not be optimal and the salesman needs to be re-routed. In this work, we focus on this second approach.

Ant Systems (AS) and Evolutionary Computation (EC) have been the most common metaheuristics employed for solving the DTSP. In [9, 14, 15, 26] different AS implementations were applied to DTSP. The main reason for using AS to dynamic problems is based on the pheromone concept. Pheromone can be exploited as positive and negative reinforcement and, therefore, as a way for transferring knowledge. When a change is detected in the problem instance, a partial decomposition-reconstruction procedure is performed over old solutions [26]. This process determines which elements of ant's solutions must be discarded in order to satisfy the feasibility of the new conditions.

Evolutionary Computation has been successfully applied to several combinatorial problems, including TSP. There have many algorithms based on EC to solve static TSP that can be directly applied to DTSP [30]. EC is considered to be one of the best algorithms for solving DTSP, taking into account the characteristics such that relative large population, statistical convergence, global optimization and considerable robustness [30, 31]. Generally, these implementations are based on the Inver-Over operator [16], one of the fastest algorithms in solving TSP [16, 30].

13.3 Sequential Estimation Algorithm: Particle Filter

Many interesting problems in science and engineering require estimation of the state of a system that changes over time using a sequence of noisy measurements made on the system [1]. Tracking problems, which consist of the estimation of the position of one or multiple targets moving in a scenario along time [20], are important examples of sequential estimation problems. The state-space modelling of these systems focuses on the state vector, which contains all relevant information required to describe the system under investigation. In tracking problems, for example, this information describes kinematic characteristics of the target as position, orientation, velocity, etc.

A particle filter (PF) is based on a large population of discrete representations (called particles) of the probability density function (*pdf*) which describes the evolution of a given system [6]. Particle Filters are algorithms in which theoretical distributions in the state-space are approximated by simulated random measures (also called particles) [6]. The state-space model consists of two processes: (i) an observation process $p(Z_{1:t}|X_t)$, where X_t denotes the system state vector and Z_t is the observation vector at time t, and (ii) a transition process $p(X_t|X_{t-1})$. Assuming that observations $\{Z_0, Z_1, \ldots, Z_t\}$ are sequentially measured in time, the goal is the estimation of the new system state at each time step. In the framework of Sequential Bayesian Modeling, the posterior *pdf* is estimated in two stages:

(a) Evaluation: the posterior *pdf* $p(X_t|Z_{1:t})$ is computed using the observation vector $Z_{1:t}$:

$$p(X_t|Z_{1:t}) = \frac{p(Z_t|X_t)p(X_t|Z_{1:t-1})}{p(Z_t|Z_{1:t-1})} \tag{13.1}$$

(b) Prediction: the posterior *pdf* $p(X_t|Z_{1:t-1})$ is propagated at time step t using the Chapman-Kolmogorov equation:

$$p(X_t|Z_{1:t-1}) = \int p(X_t|X_{t-1})p(X_{t-1}|Z_{1:t-1})dX_{t-1}. \tag{13.2}$$

The aim of the PF algorithm is to recursively estimate the posterior *pdf* $p(X_t|Z_{1:t})$. This *pdf* is represented by a set of weighted particles $\{(\mathbf{x}_t^0,\pi_t^0),\ldots,(\mathbf{x}_t^N,\pi_t^N)\}$, where the weights $\pi_t^i \propto p(Z_{1:t}|X_t = \mathbf{x}_t^i)$ are normalized.

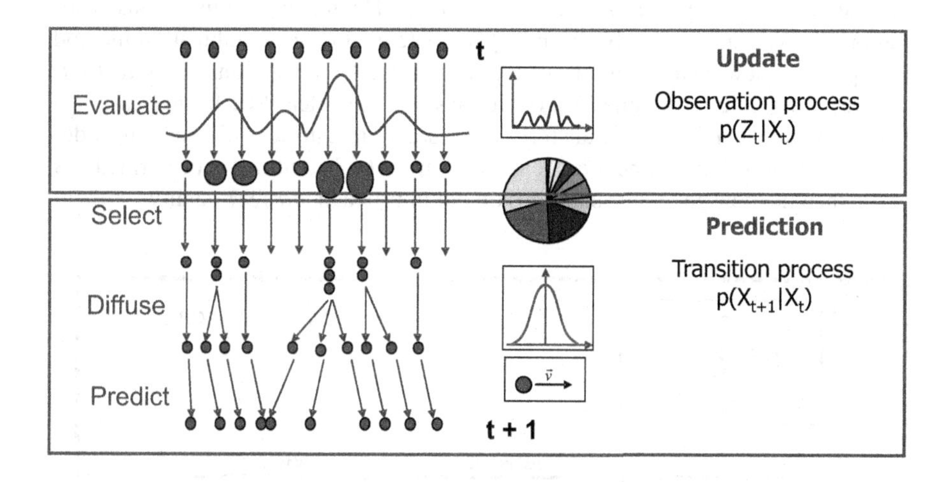

Fig. 13.1 Particle Filter scheme.

Figure 13.1 represents a schema for the PF algorithm. From an algorithmic point of view, PF directs the temporal evolution of a particle set. Particles in PF evolve according to the system model and they are selected or eliminated with a probability which depends on their weight, determined by the *pdf* [6]. In visual tracking problems this *pdf* represents the probability of a target being in a given position in the image. As a consequence, the utility of the particle filter algorithm for dynamic opimization problems lies in the description of the temporal evolution of the system state.

Therefore, Particle Filters can be seen as algorithms handling the particles time evolution. Particles in PF move according to the state model and are multiplied or died according to their weights or fitness values as determined by the likelihood function [6].

13.4 Population Based Metaheuristic: Scatter Search

Scatter Search (SS) [11, 18] is a population-based metaheuristic that provides unifying principles for recombining solutions based on generalized path construction in Euclidean spaces. In other words, SS systematically (never randomly) generates disperse set of points (solutions) from a chosen set of reference points throughout weighted combinations. This concept is introduced as the main mechanism to generate new trial points on lines jointing reference points. SS metaheuristic has been successfully applied to several hard combinatorial problems. A relevant review of this method can be found in [18].

In Figure 13.2 an outline of the SS is shown. SS procedure starts by choosing a subset (called $RefSet$) from a solution set S of $PopSize = |S|$ initial feasible ones. The solutions in $RefSet$ are obtained by choosing the h best solutions and the r most diverse ones in S. Then, new solutions are generated by making combinations of solution subsets (pairs typically) from $RefSet$. The goal of the combination method is to produce new better solutions using information from solutions in the $RefSet$. The resulting solutions, called trial solutions, can be infeasible. In that case, repairing methods are used to transform these solutions into feasible ones. In order to improve the solution fitness, a local search from trial solutions is performed. SS ends when the new generated solutions do not improve the $RefSet$ quality.

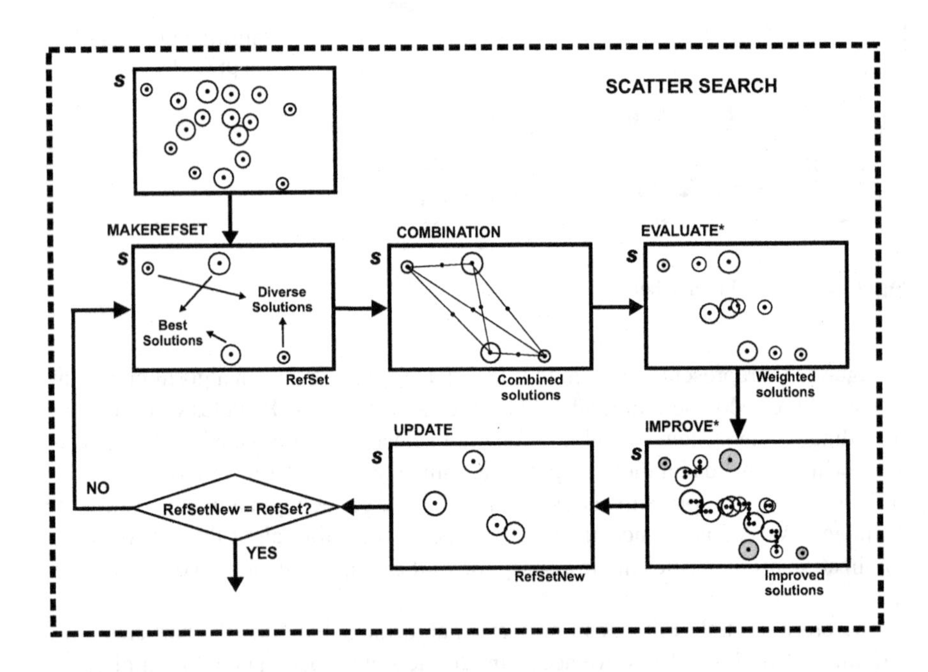

Fig. 13.2 Scatter Search scheme.

13.5 Scatter Search Particle Filter

Dynamic optimization problems deals with optimization techniques, but also with prediction tasks. This assumption is supported by the fact that the optimization method for changing conditions needs from adaptive strategies. Therefore, one key aspect is how to efficiently use important information found in previous events in order to find high quality solutions for new derived problem instances.

Usually in metaheuristics, two approaches can be used depending on the problem change rate. If it is high, each problem is tackled as a different one, so the computation is restarted from scratch. If change rate is low, the last solution (trajectorial metaheuristic) or a set of last solutions (population based metaheuristic) are used as starting point in the new search. For instance, Genetic Algorithms use the previous population as initial set in the next time step. On the other hand, use the previous pheromone deposition in each node as initial pheromone distribution of subsequent steps. The same idea can be extended to other metaheuristics. In Scatter Search, the $RefSet$ obtained in the previous time step can be used as a new $RefSet$ for the next one. In addition, the $RefSet$ could be improved with diverse solutions.

Making a decision of what information is propagated to the next time steps is very important. This is because it is possible that the search algorithm get stuck near local optimum. As a consequence, a reasonable trade-off between both restart from scratch and restart from previous optimum must be found. Therefore, it could not be appropriate to use optimization procedures in the prediction stage. Sequential estimation algorithms, like particle filters, are well-suited in prediction stages, but they are not good enough for solving dynamic optimization problems. Optimization strategies performed with this kind of algorithms are usually very computationally inefficient.

Then, from our viewpoint dynamic optimization problems needs from both optimization and prediction tasks. The key question is how to hybridize these two kinds of algorithms to obtain a new one which combines both techniques. In order to ask this question, a novel hybrid algorithm called *Scatter Search Particle Filter* (SSPF) is proposed to solve the Dynamic TSP.

13.5.1 Scatter Search and Particle Filter Hybridization

SSPF hybridizes both Scatter Search (SS) and Particle Filter (PF) frameworks in two different stages:

- In the *Particle Filter stage*, a particle (solution) set is propagated and updated to obtain a new one. This stage is focused on the evolution in time of the best solutions found in previous time steps. The aim for using PF is to avoid the loss of diversity in the solution set.
- In the *Scatter Search stage*, a fixed number of solutions from the particle set are selected and combined to obtain better ones. This stage is devoted to improve the quality of a reference subset of good solutions in such a way that the final solution is also improved.

Figure 13.3 shows the hybridization of SS and PF algorithms to obtain the SSPF algorithm. As stated in previous sections, PF algorithm can be factorized in *prediction* and *update* stages. SS optimization is performed between these two stages. PF connection with SS is achieved by means of the selection procedure. Solutions found by SS are incorporated into the particle set of PF using the inclusion procedure.

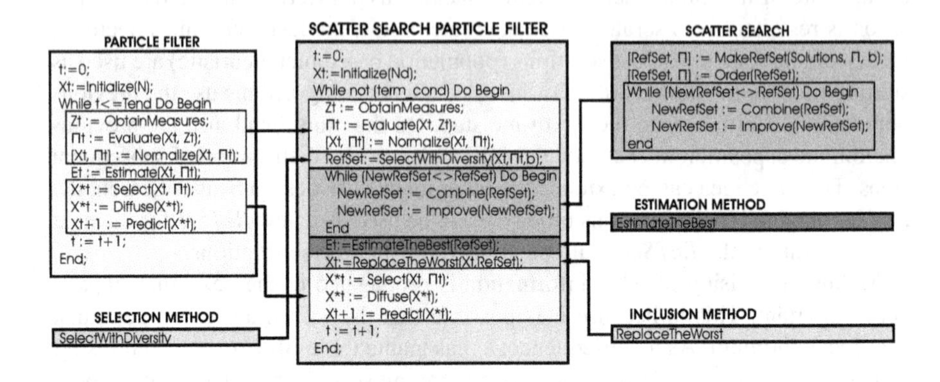

Fig. 13.3 SSPF construction starting from SS and PF.

Figure 13.4 depicts a graphical template of the SSPF algorithm. Dashed lines separate the two main components in the SSPF scheme: PF and SS optimization, respectively. SSPF starts with an initial population of N particles drawn from a known pdf (Figure 13.4: INITIALIZE stage). Each particle represents a possible solution of the problem. Particle weights are computed using a weighting function (Figure 13.4: EVALUATE stage). SS stage is later applied to improve the best obtained solutions of the particle filter stage. A Reference Set (*RefSet*) is created selecting a subset of b ($b << N$) particles from the particle set (Figure 13.4: MAKEREFSET stage). This subset is composed by the $b/2$ best solutions and the $b/2$ most diverse ones of the particle set. New solutions are generated and evaluated, by combining all possible pairs of particles in the *RefSet* (Figure 13.4: COMBINE and EVALUATE stages). In order to improve the solution fitness, a local search from each new solution is performed (Figure 13.4: IMPROVE stage). Worst solutions in the *RefSet* are replaced when there are better ones (Figure 13.4: UPDATEREFSET stage). SS stage ends when new generated solutions *NewRefSet* do not improve the quality of the *RefSet*. Once the SS stage is finished, the "worst" particles in the particle set are replaced with the *NewRefSet* solutions (Figure 13.4: INCLUDE stage). Then, a new population of particles is created by selecting the individuals from particle set with probabilities according to their weights (Figure 13.4: SELECT and DIFFUSE stages). Finally, particles are projected into the next time step by following the update rule (Figure 13.4: PREDICT stage).

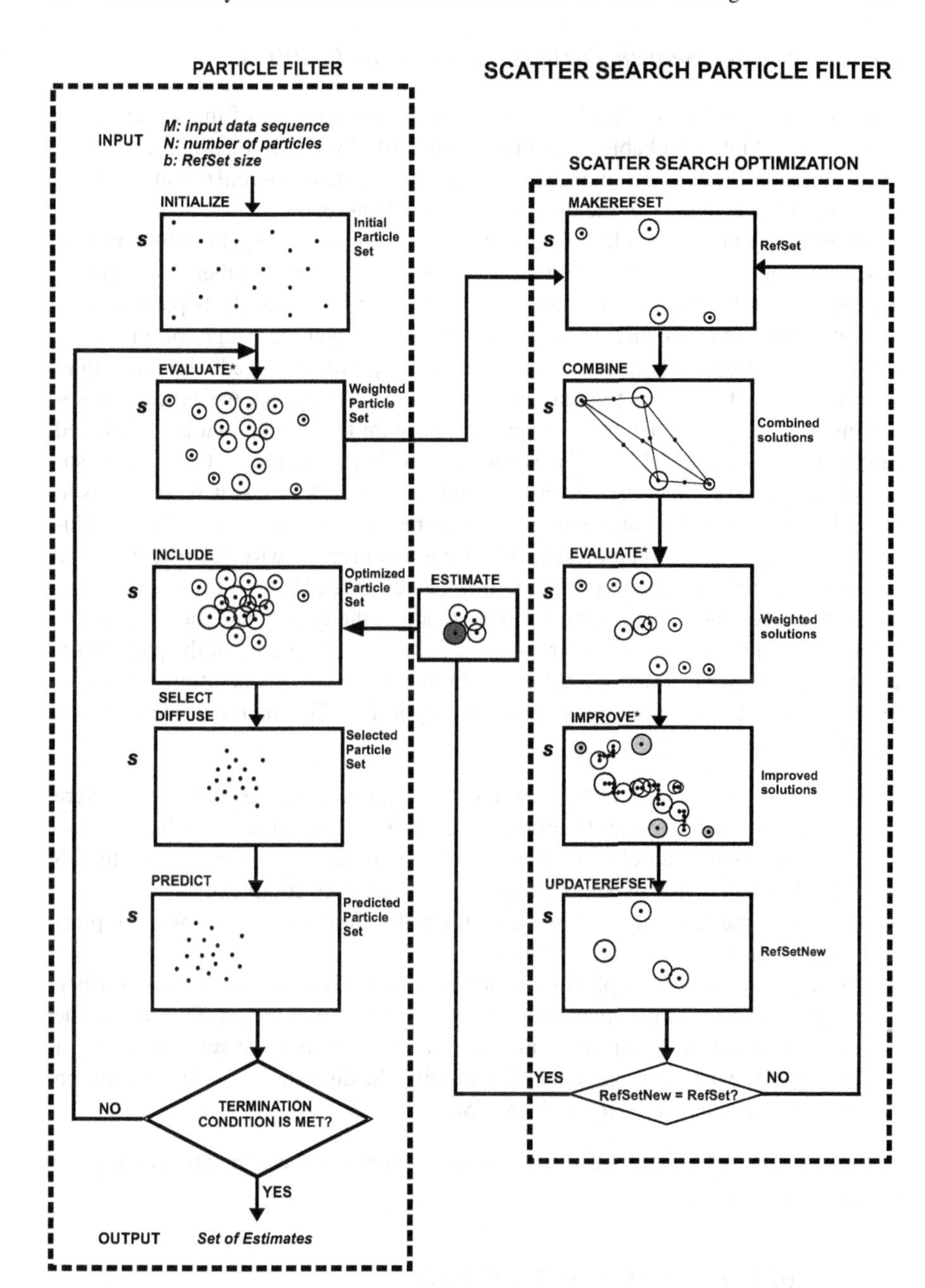

Fig. 13.4 Scatter Search Particle Filter scheme. Weight computation is required during EVALUATE and IMPROVE stages (*).

13.5.2 Scatter Search Particle Filter Main Features

The aim of the SSPF is to lead the search process to a region of the search space in which it is highly probable to find new better solutions than the initial computed ones. PF increases the performance of general SS in dynamic optimization problems by improving the quality of the diverse initial solution set S.

In order to obtain the solution set for the next event $S(t+1)$, PF performs two tasks over the set $S(t)$: select the best solutions in t and predict their most probable location in the event $t+1$. Firstly, the selection procedure selects particles in a weighted random procedure, in such a way that the larger the weight of a particle, the larger the probability to select it. Secondly, PF performs a prediction procedure over the selected solutions to obtain the set $S(t+1)$. As results, we expect that solutions in $S(t+1)$ will be closer to global optimum than other solutions obtained randomly. As a complement, PF performs a diffusion procedure to the selected solutions to preserve the needed diversity in the set $S(t+1)$. In this way, solutions to be included in the *RefSet* in the time $t+1$ will be selected from a set of better solutions than a randomly obtained set. This is the main reason why SSPF reduces the required number of evaluations for the fitness function, and hence the computational load. PF allows parameter tuning in order to adjust the quality and the diversity of the set S, used by SS. On the other hand, SS improves the quality of the particle set allowing the better estimation of the pdf, by including $RefSet$ solutions in the set S. This fact yields to an highly configurable algorithm. The main considered SSPF algorithm parameters are:

- The size of the particle set N is the number of particles in the particle set. There should be enough particles to support a set of diverse solutions, avoiding the loss of diversity in the particle set. Therefore, N influences the performance of the SS stage. The value of N depends on the problem instance complexity.
- The size of the reference set b is the number of solutions in the $RefSet$. A typical b used in the literature is $b = 10$ [5, 18].
- The diffusion stage is applied to avoid the loss of diversity in S. It is performed by applying a random displacement with maximum amplitude A. This amplitude A is a measure of the diversity produced in the new particle set. Therefore, A influences the performance of the SS by tuning the diversity of the initial solution set, and hence, the diversity of the $RefSet$.

In this research field it is usual to perform a preliminary experimentation to achieve the parameter setting.

13.6 Applying SSPF to Solve the DTSP

The main details of the SSPF implementation to solve the dynamic TSP are described in this section. The parameter setting as well as the combination and the improvement methods are detailed.

In our implementation of the SSPF, solutions (particles) are represented as paths over cities. The number of particles N in the particle set S is chosen according to the problem size. Specifically, N varies from 100 for the 25-cities problem instances to 1000 for the 100-cities problem instances. The $RefSet$ is created by selecting the 5 best solutions and the 5 most diverse ones in S, according to the scatter search algorithm.

In order to find the most diverse solutions, the distance metric for *R-permutation problems* was used [18]. DTSP is considered as an R-permutation problem [18]. In these problems, relative positioning of the elements is more important than absolute positioning. As a result, the distance between two permutations p and q for R-permutation problems is defined as:

$$d(p,q) = \textit{number of times } p_{i+1} \textit{ doesn't immediately follow } p_i \text{ in } q \quad \text{for } i = 1,\ldots,n-1.$$

Voting method [18] has been used as the combination procedure over all pairs of solutions in the $RefSet$. In this procedure, each reference solution votes for its first sector not included in the combined solution. The voting determines the element to be assigned to the next free position in the combined solution. An example can be seen in Figure 13.5.

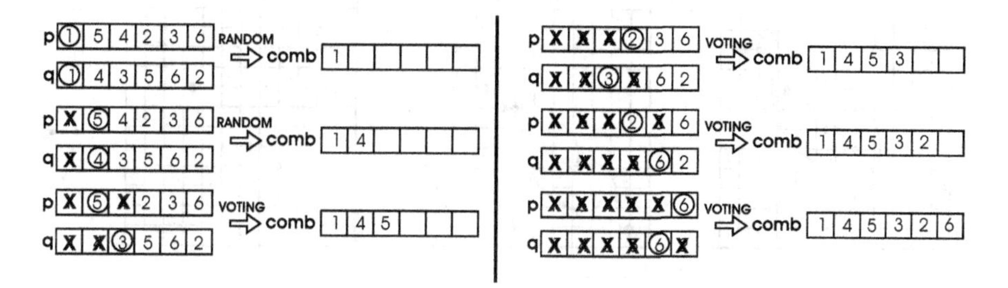

Fig. 13.5 Voting Method.

The *2-opt method* [21, 29] is usually employed as improvement stage in the SS scheme (Figure 13.4). Given a solution, consider all pairs of edges connecting four different cities. Remove two edges from the solution tour. Then there is a unique way of reconnecting the two remaining paths in such a way that a new tour is obtained. If the new tour is shorter, then it replaces the old tour and the procedure is repeated until no improvement is produced.

13.7 Experimental Results

Our experimental design considers the performance comparison of SSPF with respect to the methods Scatter Search and Evolutionary Algorithms. The computational experiments were conducted in an Intel Pentium architecture. All algorithms were coded in MATLAB 6.1, without optimization and by the same programmer to have comparable results. Different implementations are applied to several instances of DTSP and results are compared. The following sections are devoted to describe the considered problem instances, the implemented algorithms and the obtained results.

13.7.1 Problem Instances

Unfortunately, as far as the authors know, there are no benchmarks for the DTSP. Thus, we generated two sets of instances, called as synthetic and standard-based data. Synthetic data are composed by four different graph sequences. They were created, using 25, 50, 75 and 100 cities. Each sequence is composed by 10 different derived graphs. In order to know the value of the optimum, cities are located in the Euclidean space, along the diagonal as shown in figure 13.6. In the first frame, cities are located in lexicographic order along the diagonal. Subsequent frames are generated by performing exchanges of cities in groups of three as shown in figure 13.6. The average probability of node exchange in the graph sequence is $p_{change} = 0.15$.

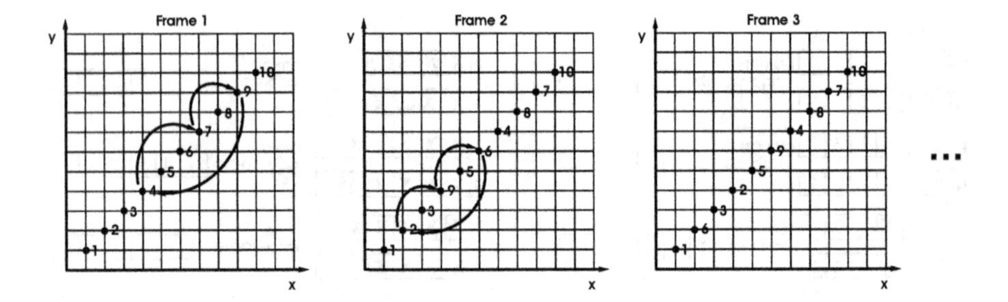

Fig. 13.6 Graph sequence generation process for synthetic instances.

Standard-based data are built as dynamic version of benchmarks from the public-domain library TSPLIB [27]. Specifically, these instances are dynamic modifications of BAYG29, BERLIN51 and ST70. These graphs belong to Euclidean symmetric class. We built each sequence starting from the original graph. Subsequent 4 graphs are obtained from the previous one introducing a perturbation in the actual location of each city according to a Gaussian distribution. Figure 13.7 shows the sequence derived from BERLIN51.

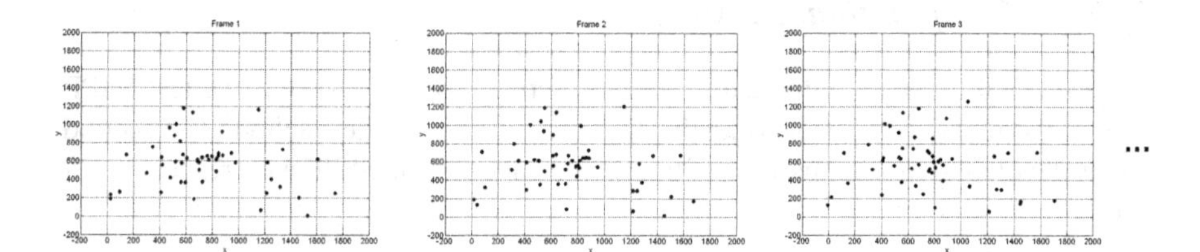

Fig. 13.7 Graph sequence generation process for standard-based instances. The first frame corresponds with the original TSPLIB instance

13.7.2 Implementation Details of the Considered Algorithms

We have implemented different versions of the Scatter Search and Evolutionary Algorithms in order to compare their performance with respect to the SSPF. The main details are described in this section. All procedures use as stopping criteria *10 million objective function evaluations or no improvement in the mean population fitness*. Solutions are coded as tours connecting cities.

13.7.2.1 Scatter Search Implementation

We have developed two different Scatter Search implementations. The first one, called SS1, considers each graph in the sequence is totally decoupled one from the others. Therefore, computation is restarted from scratch after events. In the second implementation (called SS2) the problems are supposed to be quite related, so the $RefSet$ obtained in the previous time step is used as a new $RefSet$ for the next one. SS2 does not use auxiliary methods to add diversity in the $RefSet$.

SS parameters *PopSize* and b for both implementations SS1 and SS2 were set to 100 and 10, respectively, as recommended in [5]. In order to obtain comparable results, we use the same $RefSet$ composition, combination and improvement methods as in SSPF implementation.

13.7.2.2 Evolutionary Algorithm Implementation

The SSPF algorithm is also compared with an Evolutionary Algorithm (EA). EA performs the main stages of a standard genetic algorithm, including an improvement stage. The algorithm uses voting method as crossover operator and 2-opt method as improvement method.

EA parameters were set to $PopSize = 100$, crossover probability $p_c = 0.25$ and mutation probability $p_m = 0.01$ as recommended in [21]. Finally, improvement probability was set to $p_i = 0.25$.

13.7.3 Computational Testing

Experimental results are divided in to two sections. In the first one, we test SSPF over the synthetic instances to justify the convenience of the method. The second one is devoted to the comparison of the performance of the SSPF with respect to SS and EA.

13.7.3.1 SSPF in Synthetic Data

In this section, experimental results obtained by applying our proposal to synthetic data are shown. In table 13.1, mean value of the execution time for the first graph is compared to the mean value of the execution time for the rest of the graph sequence. The proposed strategy, based on particle filter and scatter search hybridization, seems to be more advantageous than the classical SS one, in which an execution from scratch is performed. In this table, the column *Ratio* represents the average time SSPF improvement with respect to the corresponding time of the SS1 solution. As it can be seen, ratio between execution times is always in favor of our algorithm.

Table 13.1 Average execution time values over 10 runs for each graph sequence

Number of Cities	Size of Particle Set	Average Time SS Solution	Average Time SSPF Solution	Ratio
25	100	0.3×10^6	0.2×10^6	0.69
50	100	2.1×10^6	1.1×10^6	0.55
75	500	6.8×10^6	3.0×10^6	0.44
100	1000	14.7×10^6	6.6×10^6	0.44

Figure 13.8 shows the average execution time per frame over 10 runs of the same graph sequence. Each sequence is composed by 10 similar graphs. In this figure, relative execution time is represented for each frame. As it can be observed, execution time for the 2nd to 10th graphs (SSPF improvement) is significantly smaller than the execution time for the first graph (SS approach) in all considered instances.

Results show that SSPF achieves the best solution in all instances. Moreover, it is faster than SS1 implementation without loss of quality.

13.7.3.2 SSPF vs. SS & EA in Standard-Based Data

This section presents a comparison between SSPF and the implementations of SS and EA. Results obtained by these algorithms (SS1, SS2, EA and SSPF) over all standard-based data (BAYG29, BERLIN52 and ST70) are presented in figure 13.9. Because initial conditions and initial procedures performed are the same in SS1,

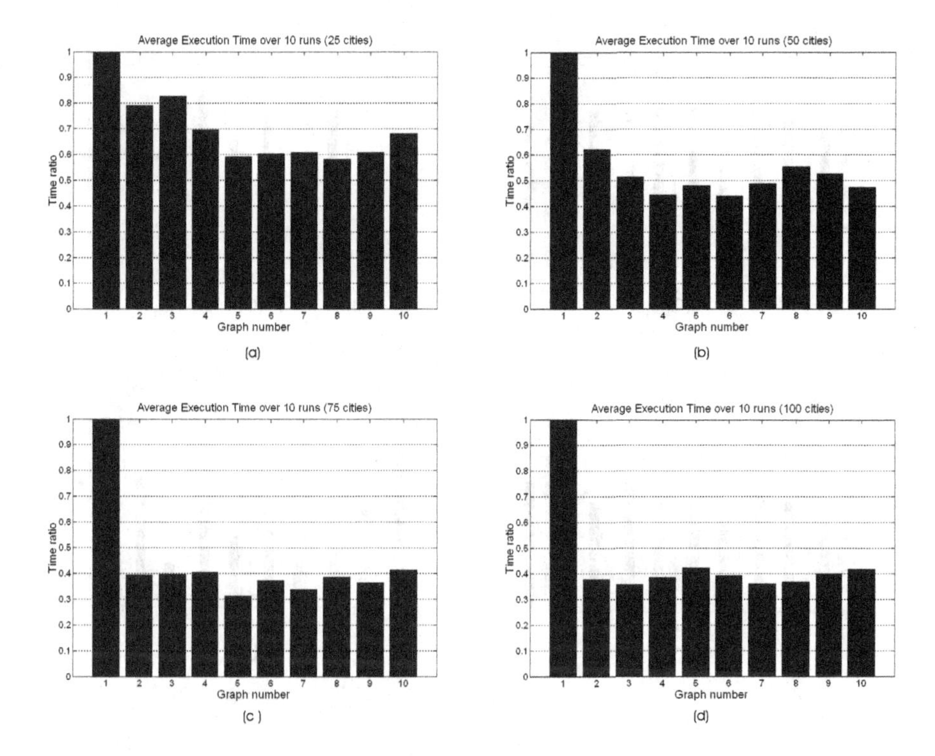

Fig. 13.8 Execution time per graph in (a) 25, (b) 50, (c) 75 and (d) 100-cities problem.

SS2 and SSPF, solutions found in the first graph are exactly the same one. As the EA approach is different to the other ones, the solution and the time required to found this solution in the first graph are also dissimilar.

Quality of estimation performed by SS1 and SSPF is similar in subsequent graphs. However, execution time is significantly lower in the SSPF approach, as explained in previous sections. In the SS2 implementation. the search procedure is trapped in a local optimum (maybe in in the neighbourhood of the optimum found in the previous time step). This yields SS2 that achieves the lowest execution time, but with very poor quality. Finally, EA finds good quality solutions, but the time required to obtain it is larger than using SSPF.

In table 13.2, a resume of main results obtained using different implementations is shown. Average execution time and path lengths values demonstrate the better performance of SSPF.

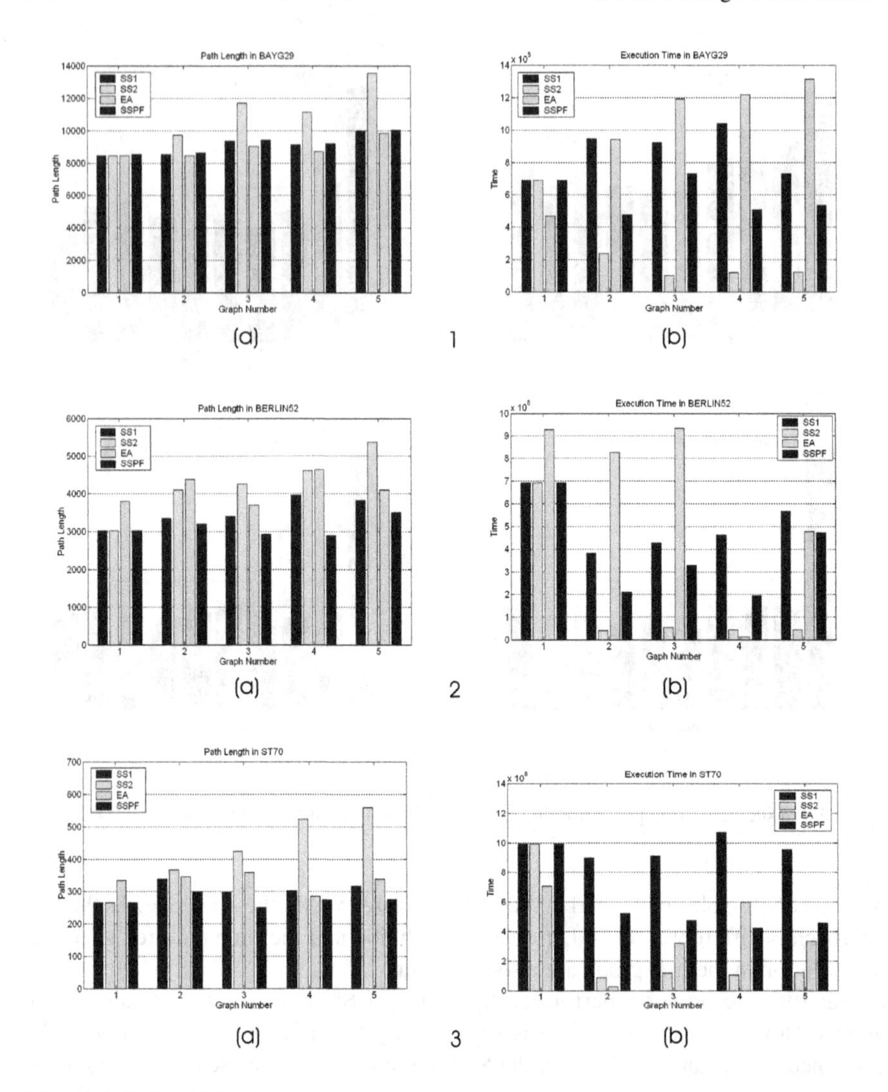

Fig. 13.9 Voting Method.

Table 13.2 Average execution time and path lengths over all instances

	SS1		**SS2**		**EA**		**SSPF**	
	Length	*Time*	*Length*	*Time*	*Length*	*Time*	*Length*	*Time*
BAYG29	0.86×10^6	0.91×10^4	0.25×10^6	1.09×10^4	1.02×10^6	0.89×10^4	0.59×10^6	0.91×10^4
BERLIN52	5.07×10^6	3.51×10^3	1.75×10^6	4.27×10^3	6.37×10^6	4.12×10^3	3.79×10^6	3.11×10^6
ST70	9.65×10^6	302.97	2.84×10^6	427.58	3.9×10^6	331.15	5.72×10^6	272.15

13.8 Conclusions

In this work we successfully applied the Scatter Search Particle Filter (SSPF) algorithm to the Dynamic Travelling Salesman Problem (DTSP). Experimental results show that SSPF appreciably increases the performance of Scatter Search and Evolutionary Algorithm methods in a challenging dynamic optimization problem, without losing quality in the estimation procedure. This improvement becomes even more significant for large instances.

References

[1] Arulampalam, M., et al.: A Tutorial on Particle Filter for Online Nonlinear/Non-Gaussian Bayesian Tracking. IEEE Trans. on Signal Processing 50(2), 174–188 (2002)
[2] Beasley, J., Sonander, J., Havelock, P.: Scheduling Aircraft Landings at London Heathrow using a Population Heuristic. Journal of the Operational Research Society 52, 483–493 (2001)
[3] Beasley, J., Krishnamoorthy, M., Sharaiha, Y., Abramson, D.: The displacement Problem and Dynamically Scheduling Aircraft Landings, Working paper (2002), `http://graph.ms.ic.ac.uk/jeb/displace.pdfm`
[4] Blum, C., Roli, A.: Metaheuristics in Combinatorial Optimization: Overview and Conceptual Comparison. ACM Computing Surveys 35(3), 268–308 (2003)
[5] Campos, V., Laguna, M., Marti, R.: Scatter Search for the Linear Ordering Problem. In: New Ideas in Optimization. McGraw-Hill (1999)
[6] Carpenter, J., Clifford, P., Fearnhead, P.: Building robust simulation based filters for evolving data sets. Tech. Rep., Dept. Statist., Univ. Oxford, Oxford, U.K. (1999)
[7] Dorigo, M., Gambardella, L.: Ant colony system: A cooperative learning approach to the traveling salesman problem. IIEEE Transactions on Evolutionary Computation 1(1), 53–66 (1997)
[8] Dror, M., Powell, W.: Stochastic and Dynamic Models in Transportation. Operations Research 41, 11–14 (1993)
[9] Eyckelhof, C., Snoek, M.: Ant Systems for A Dynamic DSP: Ants Caught in a Traffic Jam. In: Proc. of ANTS 2002 Conference (2002)
[10] Glover, F.: A Template for Scatter Search and Path Relinking. In: Hao, J.-K., Lutton, E., Ronald, E., Schoenauer, M., Snyers, D. (eds.) AE 1997. LNCS, vol. 1363, pp. 13–53. Springer, Heidelberg (1998)
[11] Glover, F., Kochenberger, G.: Handbook of metaheuristics. Kluwer Academic Publishers (2002)
[12] Gutin, G., Punnen, A.: The traveling salesman problem and its variations. Kluwer Academic Publishers (2004)
[13] Guntsh, M., Middendorf, M., Schmeck, H.: An Ant Colony Optimization Approach to Dynamic TSP. In: Proc. GECCO-2001 Conference, pp. 860–867. Morgan Kaufmann Publishers, San Francisco (2000)
[14] Guntsch, M., Middendorf, M.: Pheromone Modification Strategies for Ant Algorithms Applied to Dynamic TSP. In: Boers, E.J.W., Gottlieb, J., Lanzi, P.L., Smith, R.E., Cagnoni, S., Hart, E., Raidl, G.R., Tijink, H. (eds.) EvoIASP 2001, EvoWorkshops 2001, EvoFlight 2001, EvoSTIM 2001, EvoCOP 2001, and EvoLearn 2001. LNCS, vol. 2037, pp. 213–222. Springer, Heidelberg (2001)

[15] Guntsch, M., Middendorf, M.: Applying Population Based ACO to Dynamic Optimization Problems. In: Dorigo, M., Di Caro, G.A., Sampels, M. (eds.) Ant Algorithms 2002. LNCS, vol. 2463, pp. 111–122. Springer, Heidelberg (2002)
[16] Tao, G., Michalewicz, Z.: Inver-over Operator for the TSP. In: Eiben, A.E., Bäck, T., Schoenauer, M., Schwefel, H.-P. (eds.) PPSN 1998. LNCS, vol. 1498, pp. 803–812. Springer, Heidelberg (1998)
[17] Karp, R.: Reducibility among Combinatorial Problems. In: Miller, R., Thatcher, J. (eds.) Complexity of Computer Computations, pp. 85–103. Plenum Press (1972)
[18] Laguna, M., Marti, R.: Scatter Search methodology and implementations in C. Kluwer Academic Publisher (2003)
[19] Larsen, A.: The dynamic vehicle routing problem. PhD Thesis (2000)
[20] MacCormick, J.: Stochastic Algorithm for visual tracking. Springer (2002)
[21] Michalewitz, Z.: Genetic Algorithms + Data Structures = Evolution Programs. Springer (1996)
[22] Pantrigo, J.J., Sánchez, Á., Gianikellis, K., Duarte, A.: Path Relinking Particle Filter for Human Body Pose Estimation. In: Fred, A., Caelli, T.M., Duin, R.P.W., Campilho, A.C., de Ridder, D. (eds.) SSPR&SPR 2004. LNCS, vol. 3138, pp. 653–661. Springer, Heidelberg (2004)
[23] Pantrigo, J.J., Sánchez, A., Montemayor, A.S., Duarte, A.: Multi-Dimensional Visual Tracking Using Scatter Search Particle Filter. Pattern Recognition Letters 29(8), 1160–1174 (2008)
[24] Pantrigo, J.J., Hernández, A., Sánchez, A.: Multiple and Variable Target Visual Tracking for Video Surveillance Applications. Pattern Recognition Letters 31(12), 1577–1590 (2010)
[25] Sadeh, N., Kott, A.: Models and Techniques for Dynamic Demand-Responsive Transportation Planning. Technical Report, CMURI- TR-96-09, Robotics Institute, Carnegie Mellon University (1996)
[26] Randall, M.: Constructive Meta-heuristics for Dynamic Optimization Problems. Technical Report, School of Information Technology, Bond University (2002)
[27] Reinelt, G.: TSPLIB. University of Heidelberg (1996), `http://www.iwr.uni-heidelberg.de/groups/comopt/software/TSPLIB95/`
[28] Talbi, E.-G.: A Taxonomy of Hybrid Metaheuristics. Journal of Heuristics 8(5), 541–564 (2002)
[29] Vizeacoumar, F.T.: Implementation. Project report Combinatorial Optimization CM-PUT - 670 (2003)
[30] Zhang-Can, H., Xiao-Lin, H., Si-Duo, C.: Dynamic traveling salesman problem based on evolutionary computation. In: Proceedings of the 2001 Congress on Evolutionary Computation, vol. 2, pp. 1283–1288 (2001)
[31] Liu, Z., Kang, L.: A Hybrid Algorithm of n-OPT and GA to Solve Dynamic TSP. In: Li, M., Sun, X.-H., Deng, Q.-n., Ni, J. (eds.) GCC 2003, Part II. LNCS, vol. 3033, pp. 1030–1033. Springer, Heidelberg (2004)

Chapter 14
From the TSP to the Dynamic VRP: An Application of Neural Networks in Population Based Metaheuristic

Amir Hajjam, Jean-Charles Créput, and Abderrafiãa Koukam

Abstract. In this paper, we consider the standard dynamic and stochastic vehicle routing problem (dynamic VRP) where new requests are received over time and must be incorporated into an evolving schedule in real time. We identify the key features which make the dynamic problem different from the static problem. The approach presented to address the problem is a hybrid method which manipulates the self-organizing map (SOM) neural network similarly as a local search into a population based memetic algorithm, it is called memetic SOM. The approach illustrates how the concept of intermediate structure provided by the original SOM algorithm can naturally operate in a dynamic and real-time setting of vehicle routing. A set of operators derived from the SOM algorithm structure are customized in order to perform massive and distributed insertions of transport demands located in the plane. The goal is to simultaneously minimize the route lengths and the customer waiting time. The experiments show that the approach outperforms the operations research heuristics that were already applied to the Kilby et al. benchmark of 22 problems with up to 385 customers, which is one of the very few benchmark sets commonly shared on this dynamic problem. Our approach appears to be roughly 100 times faster than the ant colony algorithm MACS-VRPTW, and at least 10 times faster than a genetic algorithm also applied to the dynamic VRP, for a better solution quality.

14.1 Introduction

The vehicle routing problem (VRP) is one of the most widely studied problems in combinatorial optimization. In the standard VRP, a fleet of vehicles must be routed to visit a set of customers at minimum cost, subject to vehicle capacity constraint and route duration constraint. In the static version of the problem, it is assumed that

Amir Hajjam · Jean-Charles Créput · Abderrafiãa Koukam
Laboratoire Systèmes et Transports, U.T.B.M., 90010 Belfort Cedex, France
e-mail: amir.hajjam@utbm.fr

E. Alba et al. (Eds.): Metaheuristics for Dynamic Optimization, SCI 433, pp. 309–339.
springerlink.com

all customers are known in advance to the planning process. However, many real world routing problems include some dynamic elements. The information data often tends to be uncertain or even unknown at the time of the planning. It may be the case that customers, driving times or service times, are unknown before the day of operation has begun, but become available in real-time. Due to the recent advances in information and communication technologies, such as geographic information systems (GIS), global positioning systems (GPS) and mobile phones, companies are now able to manage vehicle routes in real-time. Hence, with the increased access to these services, the need for robust real-time optimization procedures will be of critical importance, for small to big distribution companies, whose logistics are based on a high reactivity to the customer demand.

Depending on the application area, some authors propose to classify problems by their degree of dynamism and the objectives and constraints of the problem. As introduced by Larsen [26], the degree of dynamism may vary between 0 and 1, and has to be computed considering a given period of observation called a working day or planning horizon. The simplest measure is the ratio between the number of dynamic requests, which arrive during the working day, and the total number of requests including the static requests which are known in advance. For instance, if the degree of dynamism is 0.4, then 4 customers out of 10 arrive while the working day has begun. Other measures of dynamism are also introduced by Larsen in order to take into account both dynamic request occurrence times and time windows. In the standard dynamic VRP case, the author argues that a problem is more dynamic if immediate requests occur at the end of the working day and less dynamic as soon as immediate requests are received relatively early during the planning horizon. For example, two problem instances respectively with all arrivals at the beginning of the day or all arrivals at the end of the day have different degrees of dynamism, the latter being highly dynamic whereas the former being static. But we could argue that the crucial point is that customers would have to be served as quickly as possible regardless of the moment of the day they sent their demands. For example, the problem of fire fighting is highly dynamic whenever the disaster appears at the beginning or at the end of the day. Hence, it is necessary to look at the importance given to the real-time objectives and constraints.

Here, an ideal application field would be medical services. An application example would be medical interventions of doctors that require a high to moderate level of dynamicity, which would be constrained by the limited amount of doctors and resources available to perform the service. Hence, the degree of dynamism considered in this paper will be measured by a waiting time of roughly 15% to 30% of the working day. We assume that no information is available about requests locations prior to optimization. Only the overall capacity of the system and hence the total number of requests that the system could serve within a day are supposed to be known. We argue that this is a reasonable assumption since real-time performance mainly depends on the vehicle resources available.

In this paper, we propose a heuristic approach performing empirical evaluations by discrete time simulation on the standard benchmarks of Kilby et al. [24]. We shall focus on a hybrid method which follows metaphors of biologic systems, and hence naturally exhibits intrinsic parallelism and which may be considered intuitive and easy to implement. The approach presented in this paper is based on the concept of the "intermediate structure" pointed out by Glover [20] about applications of neural networks to optimization. The intermediate structure, called the network, represents transport lines that continuously distort and modify their shapes in the plane according to the demand distribution. Such paradigm that we called "adaptive meshing" in earlier papers [9, 13] partially has its origin from the Kohonen selforganizing map (SOM) [25]. Starting from the SOM, we define insertion and deformation operators that are applied to the network, and managed inside a population based metaheuristic similarly as in a memetic algorithm [29], which is an evolutionary algorithm embedding a local search process and mutation operators. The standard SOM is used similarly as a local search process combined with a mapping operator, responsible for the massive insertions of customers to the network, a fitness evaluation, a selection operator, and a specific operator dedicated to customer insertions according to the maximum duration constraint. Such operators perform elementary moves in the plane and a particular point is that almost all operators are based on the nearest point findings implemented on the top a cellular decomposition of the plane by a spiral search algorithm [1].

Successive generations of construction heuristics, improvement heuristics, and metaheuristics were developed by the operations research (OR) community to solve the static traveling salesman problem (TSP) [1] and the VRP [5, 18]. Metaheuristics often encapsulate a construction method followed by the application of one or more improvement heuristics performing local search. From stage to stage, such heuristics were enriched reusing the past enhancements to build new sophisticated neighborhood search structures, which operate on graphs. Then, a question is how the complex data structures of the very powerful OR heuristics for the TSP or VRP, often based on k-d-trees for nearest point search, managing neighborhood lists, "do not look bit" tables or various solution coding schemes, should be reused in a dynamic setting and be implemented in a distributed and parallel way. In our approach, there is no distinction between a construction phase and an improvement phase, as usual in metaheuristics, but rather a distinction between a deployment phases followed by an improvement phase with different intensities. The deployment phase is only responsible to deploy the network from scratch using a high intensity for the network moves. Tour construction and tour improvement are performed simultaneously at any moment, on both phases, based on closest point finding insertions and deformations. Hence, there is no need to introduce new insertion procedures to deal with the on-line arrivals of new demands, or to restart the algorithm, as we would do in order to apply traditional methods into a dynamic setting. As they arrive, new demands are simply inserted on-line in a buffer of requests, in constant time, leading to a very weak impact on the internal data structures and thus to the course of the optimization process. One of the goals of our work is to exploit the natural properties of the on-line SOM algorithm to be applied in a dynamic setting. An important point of using an evolutionary framework is to allow simplicity and flexibility when designing the algorithm. We already applied the approach, which is called memetic SOM, to the static TSP [11], the static VRP [14] and VRPTW [10], and to several other extensions of these problems considering combination of clustering k-median and vehicle routing problems [11, 14]. Also, we link the approach to further possible parallel implementations by considering two strategic levels for the parallel computation: the population based "metaheuristic level" which corresponds to the cooperation of many autonomous local search processes, each one embedding a complete solution, and the "heuristic level" which corresponds to a cellular decomposition of the data, each cell possibly being associated with a processor and a part of the data. Also, the Euclidean nature of the problem is directly reflected into the Euclidean nature of the algorithm, thus allowing application to large instances, as this was done for the TSP [11] on instances with up to 85900 cities.

The following Section II states the dynamic VRP considered in this paper with its constraints and objectives and considered as a straightforward extension of the static VRP. Section III illustrates the philosophy of the intermediate structure concept on previous applications that were presented in earlier papers. We will find the details of the memetic SOM approach in Section IV. In Section V we shall present the real-time simulator able to gauge the efficiency of the method. Then, Section VI reports experiments carried out on the Kilby *et al.* benchmark and the comparisons made with a state-of-the-art ant colony approach and a genetic algorithm already studied on these benchmarks. It also presents a summary of the memetic SOM performances against classical heuristics on the TSP, VRP, and the dynamic VRP. The last section is devoted to the conclusion and further research.

14.2 Dynamic Euclidean Vehicle Routing Problem

As for static vehicle routing problems, a lot of versions of the dynamic problem exist depending on the application areas. For an overview and classification of the numerous versions of realftime routing and dispatching problems, we refer the reader to the general surveys and classifications given in [18, 19, 26, 31, 32]. One of the simplest versions is the standard dynamic VRP with capacity and time duration constraints [24], called "dynamic VRP" in this paper, which is a straightforward extension of the classical static VRP [3]. In this problem, the customers are the only elements who have a dependence on time. Customers are not known in advance but arrive as the day progresses, and the system has to incorporate them into the already designed routes in real time. Problems fitting this model appear frequently in industry. Most parcel delivery services, replenishment of stocks in a manufacturing context, waste collection, and dispatch of emergency services can be modeled in this way. Geographically dispersed failures to be serviced by a mobile repairman also fit this model.

In the different versions of the dynamic VRP presented, there are a very few dynamic routing problems except the dynamic VRPTW or dynamic PDPTW that are recognized as standard problems well suited to allow comparative evaluations of heuristics and metaheuristics on a common set of benchmarks. For example, we only found two papers on the dynamic VRP that shared detailed results on a common test set. They are first an adaptation of the ant colony approach MACS-VRPTW [17] by Montemanni et al. [28], and second a genetic algorithm by Goncalves et al. [21]. They shared results on the Kilby et al. [24] test set with 22 problems of sizes from 50 with up to 385 customers. This paper tries to go one step further in that direction considering the dynamic VRP as a standard dynamic problem, and yielding a comparative study with these two methods on the Kilby et al. test set. Then, we restrict the scope of our work to the dynamic VRP, with capacity and time duration constraints.

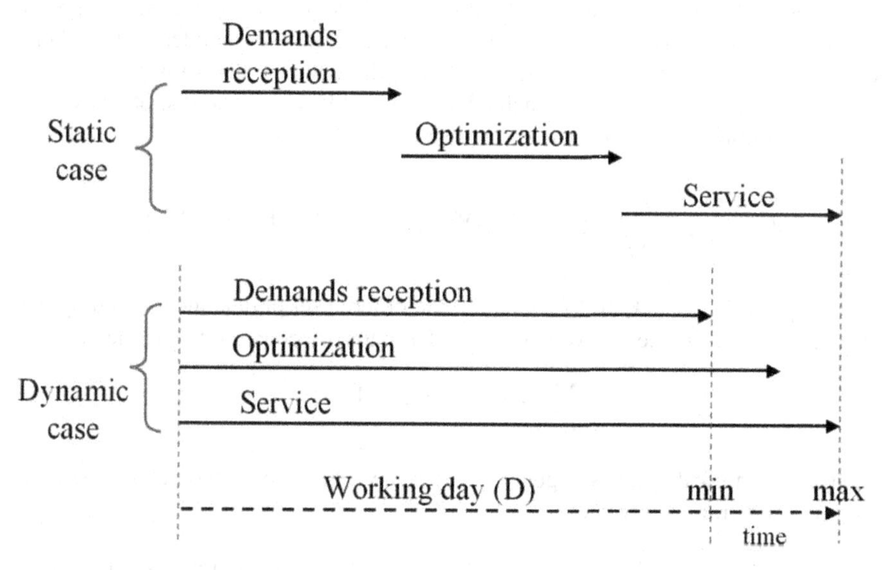

Fig. 14.1 Static vs dynamic VRP.

In the static VRP, vehicles must be routed to visit a set of customers at minimum cost, assuming that all orders for all customers are known in advance. In the dynamic VRP however, new tasks enter the system and must be incorporated into the vehicle schedules and served as the day progresses. In real-time distribution systems, demands arrive randomly in time and the dispatching of vehicles is a continuous process of collecting demands, forming and optimizing tours, and dispatching requests to vehicles in order to process requests at the required geographic locations. In the case of the static VRP, the three phases of demands reception, routes optimization, and vehicles traveling are clearly separated and sequentially performed, the output of a given phase being the input of the subsequent one. At the opposite, as illustrated in Figure 14.1, we can see the dynamic VRP as an extension of the static VRP where these three time-dependent processes are merged into an approximately same period of time. This period of time is called the working day or planning horizon of length D. Here, we precisely define the working day length D as the length of the collecting period, knowing that the optimization period and the vehicle traveling period would have to be of approximately the same length.

The static VRP is defined on a set $V = v_0, v_1, ..., v_N$ of vertices, where vertex $v - 0$ is a depot at which are based m identical vehicles of capacity Q, while the remaining N vertices represent customers, also called requests, orders, or demands. A non-negative cost, or travel time, is defined for each edge $(v_i, v_j) \in V \times V$. Each customer has a non-negative load $q(v_i)$ and a non-negative service time $s(v_i)$. A vehicle route is a circuit on vertices. The VRP consists of designing a set of m vehicle routes of least total cost, each starting and ending at the depot, such that each customer is visited exactly once by a vehicle, the total demand of any route does not exceed Q, and the total duration of any route does not exceed a preset bound T. As it is mostly done in practice [5], we address the Euclidean VRP where each vertex vi has a location in the plane, and where the travel cost is given by the Euclidean distance $d(v_i, v_j)$ for each edge $(v_i, v_j) \in V \times V$. Then, the objective for the static problem is the total route length (Length) defined by

$$Length = \sum_{i=1,..,m} \left(\sum_{j=1,...,k_i-1} d\left(v_j^i, v_{j+1}^i\right) + d\left(v_0, v_1^i\right) + d\left(v_{k_i}^i, v_0\right) \right), \tag{14.1}$$

where $v_i^j \in V, 0 \leq j \leq k_i, 0 \leq ki \leq N$, are the ordered set of demands served by the vehicle i, $1 \leq i \leq m$, i.e. the vehicle route. The capacity constraint is defined by:

$$\sum_{j=1,...,k_i} q\left(v_j^i\right) \leq Q, i \in 1, ..., m \tag{14.2}$$

then, assuming without loss of generality that the vehicle speed has value 1 the time duration constraint is given by:

$$\sum_{j=1,...,k_i} s\left(v_j^i\right) + \sum_{j=1,...,k_i-1} d\left(v_j^i, v_{j+1}^i\right) + d\left(v_0, v_1^i\right) + d\left(v_{k_i}^i, v_0\right) \leq T, i \in 1, ..., m \tag{14.3}$$

The problem is NP-hard. Then, for large instances, using heuristics is encouraged in that they have statistical or empirical guaranty to find good solutions possibly for large scale problems with several hundreds of customers.

It is argued in the literature that in a real-life situation the objective function often consists of a mixture of customers, waiting time costs, or system response time, and travel (or routing) costs. The analysis made by several authors [2, 19] confirms that there is a trade-off between travel costs and system time in a dynamic routing system and that the travel costs can be reduced in return for an increase in the system time. In weakly dynamic systems the focus is on minimizing routing costs. On the other hand, when operating a strongly dynamic system, minimizing the expected system response time is the main objective. In dynamic settings, the waiting time is often more important than the travel cost. We will then define the dynamic VRP as a multiobjective problem, since the "interesting" or "good" solutions are always a compromise between these two criteria. For example, the static problem case can be seen as a particular case of the dynamic problem where the waiting time is completely discarded, the optimization process starting whenever all requests to be served are known. We define the dynamic VRP as a bi-objective problem by adding to the classical objective and constraints of the standard VRP a supplementary objective which consists of minimizing the average customer waiting time. The customer waiting time is the delay between the occurrence time of a demand and the instant the service of the demand begins. It is often called "response time" or "system time" (Bertsimas and Simchi-Levi 1996). Hence, in addition to the classical objective and constraints defined above, we add a supplementary criterion to be considered when evaluating solutions. This criterion is the average customer waiting time (WT):

$$WT = \sum_{i \in \{1,\ldots,N\}} W_i \Big/ N \qquad (14.4)$$

where W_i is the waiting time of demand i, defined by $W_i = sti - ti$ where $t_i \in [0,\ D]$ is the demand occurrence time, and sti is the time when the service starts for that demand.

In order to evaluate the customer waiting time we need to, not only consider travel distances and service times, but also consider the "real time" at which the service is really performed. Real time includes the possible extra times during which the vehicle may be waiting or driving back to the depot before some new requests are dispatched to it. Only the evaluation of (14.4) depends on a real-time realization. The evaluation of (14.1)-(14.3) only depends on the ordering of demands in a route, the same way as for the static VRP. We consider the waiting time as the essential criterion to gauge the dynamicity of the system. Hence, it is an important criterion to evaluate the effectiveness of algorithms on this problem.

14.3 Method Principle

In this section, we illustrate the "philosophy" of the neural network based approach proposed to address the Dynamic VRP. We present the main characteristics of the approach and explain why it may naturally deal with the dynamic and stochastic version of the VRP, without changing quite nothing in its implementation when passing from a static to a dynamic context. There is no need to introduce new insertion procedures or to design new mechanisms to deal with the on-line arrivals of new demands along the working day since the approach is already based on massive insertions to an "intermediate" independent structure representing routes. While the very powerful OR heuristics to the static VRP are often based on internal data structures difficult to implement and to modify dynamically, the advantage of our approach would be on the simplicity of updating the evolving internal data structures. Furthermore, we argue that a promising characteristic concerns its potential for a parallel and distributed implantation on multi-processors or grid systems. We can distinguish two levels for the parallel computation that are, the heuristic level which deals with the problem dependent operations distributed in the plane and the metaheuristic level which deals with a population of independent local search processes.

14.3.1 Biologic Metaphor and the Intermediate Structure Paradigm

One way to explain the "philosophy" of the approach may be by referring the reader to some well known concepts in the Artificial Intelligence domain like emergent computation, bio-inspired methods, and soft-computing concepts including neural network, evolutionary algorithms, or hybrid systems. The approach can be seen as following a biologic metaphor where customers constitute external stimuli to which a "biologic organism", the network of transport lines, may respond dynamically adapting its shape continuously to absorb, neutralize, or satisfy the external stimuli. More generally, we can exploit this metaphor to address a large class of spatially distributed problems of terrestrial transportation and telecommunications, such as facility location problems, vehicle routing problems, or dimensioning mobile communication networks [12, 13]. These problems involve the distribution of a set of entities over an area (the demand) and a set of physical systems (the suppliers) which have to respond optimally relatively to the demand. This optimal response constitutes the solution to the optimization problem. Thus, a distributed bioinspired heuristic to address such problems is a simulation process of such spatially distributed entities (vehicles, antenna, customers) interacting in an environment which produces the "emergence" of a solution by the many local and distributed interactions. The approach presented involves an "intermediate structure", which is a network or a graph in the plane, representing the transport lines that continuously distort and modify their shapes according to a demand distribution. The important point is that tour construction and tour improvement operations are all based on massive and distributed insertions and line deformations. Customers are chosen randomly in the plane and are repeatedly presented online and many times to a simple insertion procedure based on nearest point search. The closest point search in the network structure is performed on the top of a cellular decomposition of the plane by a spiral search algorithm that is known to perform in constant time for bounded distributions [1]. This paradigm of "intermediate structure" and quite "instantaneous adaptation", that we called "adaptive meshing" in previous applications, has from a part its origin and inspiration from the Kohonen self-organizing map (SOM) neural network, which was applied to the TSP since a long time and which can address large size problems with up to 85900 cities [11]. The SOM algorithm is a neural network approach dealing, when applied in the plane, with visual patterns moving and adapting to distributed data. Its main "emergent" property is to allow adaptation by density and topology preservation of a planar graph (the transport network) to an underlying data distribution (demand set). It can also be seen as a center based clustering algorithm with topological relationships between cluster centers. Here, we generalize the SOM algorithm giving rise to a class of "closest point findings" based operators that are embedded into a population based metaheuristic framework. The structure of the metaheuristic is similar to the memetic algorithm, which is an evolutionary algorithm incorporating a local search [29]. The SOM is a (long) stochastic gradient descent performed during the many generations allowed, and used as a "local search" similarly as in a classical memetic algorithm. This is why the approach has been called memetic SOM in previous work and we will maintain the name in this paper. The approach follows two types of metaphors. It follows a self-organization metaphor at the level of the interacting problem components, or heuristic level, and an evolution based metaphor at the population based metaheuristic level. Since demands are conceptually separated from the routes representation, which is an independent network or graph in the plane which continuously adjusts itself to the data, this leads to a straightforward application from a static to a dynamic setting. As they arrive, new demands are simply inserted on-line in a buffer of demands, in constant time, leading to a very weak impact on the course of the optimization process.

Figure 14.2 (a-c) illustrates the application from a static to a dynamic setting. Figure 14.2(a) presents a bus transportation system where vehicle routes, modeled as paths with a common arrival point and depicted by lines in the figure, are adapted to a given distribution of customers and represented by dots in the figure. The application concerned a set of 780 employees of an enterprise located over a geographic area of $73km \times 51km$ around the towns of Belfort and Montbeliard in the East of France [8]. Vehicle routes and customers are shown juxtaposed in the figure over the underlying road network, represented by thin lines in the figure. The problem tackled, called VRP-Cluster, is a combination of the Euclidean k-median problem with a classical VRP. It consists of positioning bus stops, or cluster centers, according to customer locations (k-median problem) and simultaneously generating vehicle routes among bus-stops (VRP). Bus-stops define clusters where customers are grouped and to which they have to walk to take the bus. As illustrated in Figure 14.2 (b-c), application to a dynamic context mainly results from considering an evolving static VRP where the starting locations of the vehicles (the filled circles in the figure) evolve step by step as the vehicles move in the plane and perform their services along the working day. Hence, the system must monitor and update the vehicle locations, their capacities, and the vehicle travel durations on a rate defined by a decomposition of the day within many short time-slices. Figure 14.2 (b) presents a version of the dynamic case where vehicles perform the service as soon as possible, whereas Figure 14.2(c) presents a case where the vehicle starting times are slightly delayed giving rise to a longer horizon for route optimization, hence to longer vehicle paths in the figure. In this paper, we will experiment different degrees of dynamism and gauge different trade-offs between waiting time and length minimization by simply delaying the departure of vehicles.

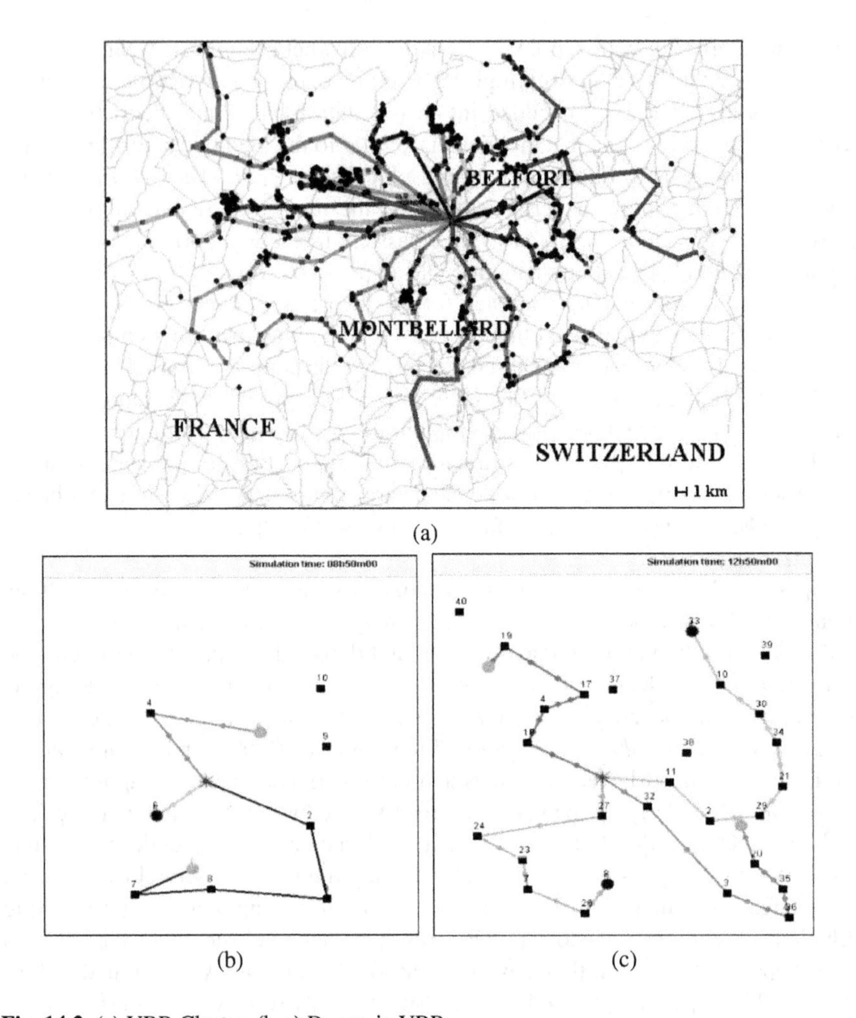

Fig. 14.2 (a) VRP-Cluster, (b-c) Dynamic VRP.

14.4 The Metaheuristic Embedding Framework

When introduced into a population based metaheuristic frame-work, the SOM is a long run process applied to a population of solutions. This process is interrupted at each cycle, called a generation, by the application of problem dependent operators. A generation occurs in such a way that at most $O(N)$ basic iterations are performed at each generation, N being the problem size. The main components of the method which are intended for driving the search are:

- a self-organizing map procedure based on closest point findings and route deformations as a low level stochastic process,
- problem-oriented insertion operators interleaving with SOM,

- a random perturbation with a decreasing intensity,
- a mapping operator which massively assigns requests to their closest vehicle route points,
- a fitness function incorporating constraints evaluation,
- a population based metaheuristic strategy with selection operators,
- a search performed within two phases (deployment, improvement).

The optimization process is divided within two phases, that are, a deployment phase followed by an improvement phase. It is worth noting that there is no distinction between a "construction" phase and an "improvement" phase, as usually done in OR metaheuristics, where the construction phase builds admissible solutions whereas improvement phase only improves the already constructed solution by swaps or exchanges of customers. Here, the distinction is only based on the intensity of the moves, since tour insertions and improvement operations operate simultaneously at any moment during the deployment and improvement phases. Hence, there is no need to add new specific insertion operators into the algorithm when passing from a static to a dynamic setting. The approach can be said "naturally on-line" and "intrinsically" customized for an application into a dynamic setting.

We claim that an interesting property of the approach is its intrinsic potential for parallel and distributed implementations in multi-processors, multi-core, grids, or P2P networks. The approach exhibits two strategic levels of parallel execution. On the one hand, we can exploit the "metaheuristic level" that corresponds to the cooperation of many autonomous local search processes, each one embedding a complete solution. It is worth noting that since the communication times at the level of the selection operators are relatively small, the long running times of independent local search processes favor parallel execution of the method. Also, this is why the optimization simulator has been structured as presented in Figure 14.4 (b). A population of agent-solvers embeds local search processes applied to the encapsulated solutions, whereas a meta-solver plays the role of a scheduler of the agent-solvers and applies selection operators to the population of agents. On the other hand, we could also exploit the "heuristic level" of the approach, which is problem dependent, and based on a cellular decomposition of the data. Each cell is associated with a part of the problem data and hence could be allocated to a given processor. Furthermore, this should favor the application to very large size problems for which the actual memory size of personal computers or workstations is notably insufficient.

14.5 The Evolutionary Algorithm Embedding Self-Organizing Maps

14.5.1 Memetic SOM

The approach is similar to a memetic algorithm [29], that is, a hybrid evolutionary algorithm embedding a local search. It is a simplified version of the approach presented in [14] which was applied to the static VRP. As illustrated previously by Figure 14.3(b), a population of Pop independent processes, called agentsolvers, perform local search and other operations such as fitness evaluation and request insertions, each on a single encapsulated solution. A main loop, called metasolver loop, manages the scheduling of the local search processes and applies selection operators to the population of agent-solvers, i.e. of solutions. Each agent-solver exactly encapsulates one solution, which is defined by a graph of routes, called the network, where each route is represented by an independent path with $max(5N'/m,5)$ vertices, m being the total number of vehicles available in the system, and N' the current number of available demands at time t, t being the current real-time. Each route starts at the vehicle location computed for the time $t+Topt$ and ends at the depot, $Topt$ being the optimization time-slice duration. Each route ends at the depot letting the vehicles go back to the depot each time they have no demand to be served in their schedule. The number of vertices by route corresponds to the maximum number of customers a vehicle can handle during a given optimization time-slice. It has been adjusted empirically to allow a good trade-off between the number of customers visited, equilibration of route lengths, and computation speed. The main loop manages two optimization phases, that are, a deployment phase followed by an improvement phase, each one executing a fixed number of iterations (called generations) which is set to the problem size N, N being the total number of demands received within a day. It is worth noting that the knowledge of the problem size N is considered in this paper as a reasonable assumption in order to adequately dimension the memory and computational resources. The meta-solver and agent-solver behaviors can be stated in pseudo-code as follows:

The deployment phase starts its execution with solutions having randomly generated vertex coordinates into a rectangle area containing the demands. The improvement phase follows the deployment phase. Then, once the improvement phase has finished, the algorithm restarts at the beginning. The main difference between the deployment and improvement phases is that the former is responsible for creating an initial ordering from random initialization. It follows that SOM processes embedded in the deployment loop have a larger initial neighborhood, proportional to N', N' being the number of demands that are currently in the system at the moment of parameters initialization. However, the improvement loop is intended for simply performing local improvements using SOM processes with smaller neighborhoods and applying fewer iterations. The parameter values of the SOM operators are set exactly as in [14], except that the tmax value and radius of neighborhoods α final and σ_{init} depend on the instantaneous number of available demands N' at the time of parameters initialization, rather than on the total number of demands N.

An important operator is the SOM algorithm. At each generation, a predefined number ($niter$) of basic SOM iterations, proportional to the current problem size N', are performed letting the long SOM decreasing run being interrupted and combined with the application of other operators. Such operators can be also a specialization of the SOM operator in order to perform request insertions, or to introduce perturbations, a mapping/assignment operator for generating admissible solutions, a fitness evaluation, and the selections at the population level. Below is a detailed description of the operators:

1. Self-organizing map operator. It is the standard SOM applied to the graph network.

   It is denoted by its name and its parameters, as $SOM(\alpha_{init}, \alpha_{final}, \sigma_{init}, \sigma_{final}, t_{max})$.

   A SOM operator is executed performing niter basic iterations by solution, at each generation. Parameter t_{max} is the number of iterations defining a long decreasing run ideally performed within N generations and applied to a given solution. When parameters initialization take place, it is stated as $t_{max} = N \times$ *nite* r, with *niter* adjusted depending on the number of available demands as given in the pseudo-code above. Other parameters define the initial and final intensity and neighborhood for the learning law. The operator is used to deploy the network toward customers from scratch in deployment phase, or to introduce punctual moves to exit from local minima during the improvement phase.

Algorithm 14.1. Meta-solver main loop

```
1: Initialize population with Pop agent-solvers with routes randomly generated.
2: Initialize agent-solvers and their SOM parameters for the deployment phase.
3: Gen = 0
4: while not(a stop order is received from the company) do
5:     Look at the received messages from the company and update vehicles and request
       set according to the optimization protocol (see section 0).
6:     if a "request" order is received then
7:         add the new received demands to the end of the demand buffer
8:     end if
9:     if an "optimizer" order is received then
10:        update the vehicle locations, their capacities, travel duration, and route sizes,
           at the future time t + Topt, at the same time refresh the current request set
           according to the future time t + Topt
11:    end if
12:    Activate each agent-solver in turn, each one executing a single agent-solver
       generation
13:    Save the best solution encountered, and send it back to the company.
14:    Apply selection operator SELECT to the agent-solver population
15:    Apply elitist selection operator SELECT_ELIT.
16:    Gen = Gen + 1
17:    if Gen = N then
18:        swap agent-solvers and their SOM parameters to the improvement phase
19:    end if
20:    if Gen = 2N then
21:        Gen=0
22:        randomize population
23:        reset the best solution
24:        then swap agent-solvers and their SOM parameters to the deployment phase
25:    end if
26: end while
```

Algorithm 14.2. Agent-solver generation

1: In deployment mode only, apply a standard SOM operator, with parameters $(\alpha_{init}, \alpha_{final}, \sigma_{init}, \sigma_{final}, t_{max}) = (0.5, 0.5, max(2 \times N'/m, 5), 4, N \times niter)$, to the network, performing $niter = max(N'/4, 5)$ iterations.
2: In improvement mode only, apply the derived SOM operator, denoted SOMVRP, with parameters $(\alpha_{init}, \alpha_{final}, \sigma_{init}, \sigma_{final}, t_{max}) = (0.5, 0.5, 10, 4, N \times niter)$, to the network, performing $niter = max(N'/m, 5)$ iterations.
3: In improvement mode only, apply the derived SOM operator, denoted SOMVRP, with parameters
$(\alpha_{init}, \alpha_{final}, \sigma_{init}, \sigma_{final}, t_{max}) =$
$(0.9, 0.5, max(2 \times N'/m, 5), max(2 \times N'/m, 5)/2, N \times niter)$, to the network, performing $niter = max(N'/m, 5)$ iterations.
4: Apply mapping operator MAPPING to the solution network to assign each demand to its nearest vertex and move vertices to the demand locations.
5: Apply fitness evaluation operator FITNESS to the solution.
6: 5. Apply derived operator SOMDVRP, to perform greedy insertions of the residual demands according to the time duration constraint.

2. SOM derived operators. Two operators are derived from the SOM algorithm for dealing with the VRP. The first operator, denoted SOMVRP, is a standard SOM restricted to be applied to a single randomly chosen vehicle/route at each generation, using customers already inserted into the route. It helps to eliminate the remaining crossing edges in routes. While capacity constraint is greedily tackled by the mapping/assignment operator below, the second operator, denoted SOMDVRP, deals specifically with the time duration constraint. It performs few greedy insertion moves at each generation. Given a randomly chosen customer that is not yet already assigned to a vehicle, the competitive step selects to be the winner the vehicle vertex for which the route time increase is minimum, the route time duration constraint for that vehicle being satisfied. The evaluation of the route time increase is done moving the vertex to the customer location and including the customer into the route.
3. Mapping/assignment operator. This operator, denoted MAPPING, generates a VRP solution by inserting customers into routes and modifying the shape of the network accordingly, at each generation. The operator first greedily maps customers to their nearest vertex for which the corresponding vehicle capacity constraint is satisfied, and to which no customer has been yet assigned. The capacity constraint is then greedily tackled by the customer assignment. Then, the operator moves the route vertices to the location of their assigned customer (if exist) and regularly dispatches (by translation) other vertices along edges formed by two consecutive customers in a route. The result is a vehicle route where assigned vertices alternate with the many more not assigned vertices. At this stage, few customers may not be inserted because of capacity constraint violation.

4. Fitness operator denoted $FITNESS$. Once the assignment of customers to routes has been performed, this operator evaluates a scalar fitness value for a given solution. This value has to be maximized, it is used by the selection operator at the population level. Taking care of time duration constraint, the fitness value is sequentially computed following routes one by one and removing a customer from the route if it leads to a violation of the time duration constraint. The value returned is $fitness = sat - 10 - 5 \times Length$, where sat is the number of customers that are successfully assigned to routes, and Length is the length of the routes defined by the ordering of such customers. The value sat is then considered as a first objective and admissible solutions are such that $sat = N'$, N' being the current number of customers in the system at a given optimization time-slice.
5. Selection operators. Based on fitness maximization, at each generation the operator denoted $SELECT$ replaces $Pop/5$ worst solutions, which have the lowest fitness values in the population, by the same number of best solutions, which have the highest fitness values in the population. An elitist version $SELECT_{ELIT}$ replaces $Pop/10$ worst individuals by the single best individual encountered during the run.

14.5.2 Spiral Search Algorithm

By the evolutionary dynamics, the goal is to make the closest point assignment coincide to the right assignment, which minimizes objectives and satisfies constraints. The algorithm can be seen as a massive and parallel insertion method to the nearest points. To perform N closest point findings in expected $O(N)$ time for uniform distributions, we have implemented the spiral search algorithm of Bentley, Weide and, Yao [19] based on a cell partitioning of the area. It performs an optimal nearest point search with expected $O(1)$ time complexity for uniform or bounded distributions, with $O(N)$ space complexity. Hence, a cell based decomposition of the area within $O(N1/2 \times N1/2)$ cells is performed during the initialization phase of the memetic algorithm. Each cell has a (non null) memory capacity proportional to an estimation of the number of demands at that location. The memory is allocated once. The contents of the memory cells are updated each time a given operator (SOM or mapping) has to be applied. Vertices of the network are introduced into the cells and the subsequent (at most) $O(N)$ closest point findings will be based on their content. The cell contents are not updated after each move. This may introduce a relaxation on the requirement of finding the true nearest neighbor. But this drawback is balanced by the limited number of iterations performed and by the fact that vertex coordinates are modified after each move.

The choice of a spiral search algorithm based on a geometric partitioning of the area, rather than a standard k-d tree search method or a Delaunay-Voronoïmethod [30] was drawn from "heuristic" arguments. We did not perform evaluations to yield a firm conclusion about the superiority of a method over another in the context of the SOM. Here, the closest point findings concern the network vertices, rather than customers as usual for swapping operations in standard local search approaches. The insertions of the network vertices into the cells are to be done many times, more precisely, before each application of a given operator to a solution. The insertions take $O(N)$ computation time, for at most $O(N)$ closest point findings subsequently performed. Using a k-d tree would require $O(N \times log(N))$ computation time to build the balanced tree at each time. Thus, it is not clear whether the amortized computation cost would be inferior in that case. Also, we link the spiral search to further possible parallel implementations of the approach, in order to deal with very large instances for example. The k-d tree constitutes a hierarchical structure which is adequate in a context of shared memory. On the contrary, geometric partitioning according to a given topology may have some advantages when dealing with multiprocessor implantations. Here, a given cell would only have to "communicate" with its fixed 8 neighboring cells, each one being associated to a given part of the data and/or the network.

14.5.3 Algorithm Complexity

In our experiments, a given working day is divided into $O(N)$ time slices with constant computation time each, N being the problem size. Hence, the computation time allowed in experiments is $O(N)$. The number of generations performed, as the problem size grows, will depend on the complexity of the closest point findings based operators. With a constant population size, a SOM neighborhood proportional to N, and N basic iterations performed by generation, the time complexity to execute a given generation is $O(N^2)$ in the worst case. However, we claim that the spiral search mechanism considerably improves the nearest search. This point was confirmed empirically in [11], a linear time was achieved for constant neighborhood size operators applied on some unstructured TSPLIB test cases. The memetic SOM space complexity is $O(N)$, as usual for SOM. It is worth noting that this space complexity allows dealing with large size instances of several thousands of customers, on standard computer workstations.

14.6 Real-Time Simulation and Optimizer

This section presents the real-time simulator developed in Java which allows the dynamic solving of a dynamic VRP. To make things concrete, we assume that a transport company centralizes the optimization procedure, receives the orders from the environment, monitors the vehicle locations, and dispatches the continuously generated and optimized routes to the vehicles. Hence, we assume the existence of a communication system between the company, the customers and the drivers, and that communication times are negligible relatively to the rest of the real-time activities. In this section, we detail the simulator structure and the main parameters that will allow controlling the dynamic optimization process. It is worth noting that the simulator can be described independently of any optimization al-gorithm or policy that could be applied.

14.6.1 Simulator Architecture

The simulator consists of two main components implemented within two Java threads which communicate through an asynchronous protocol. The first thread plays the role of a real-time scheduler which decomposes the working day into many short time-slices based on a timer clock, in order to simulate the vehicles activities implemented as simple state machines. The second thread plays the role of a background task which encapsulates the optimization process which continuously optimizes routes using the remaining CPU resources. During each time-slice, the optimization process solves a continuously evolving static VRP, with evolving vehicle capacities and starting locations, and with an evolving set of currently available requests. The idea has been discussed many times in the literature [21, 24, 26] and is clearly different from the rough strategy which consists in restarting the optimizer each time a new event occurs.

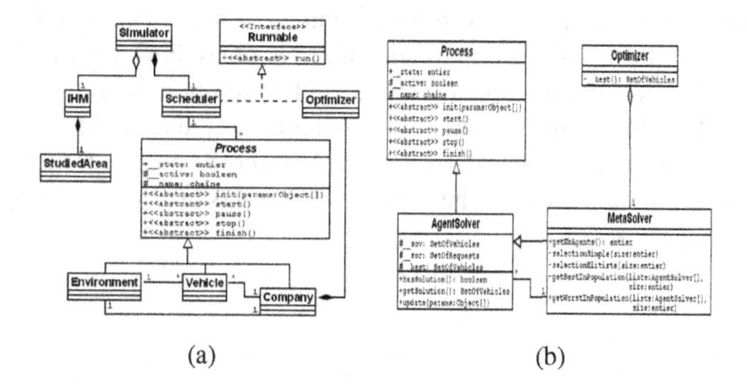

Fig. 14.3 Real-time simulator and asynchronous optimizer.

The architecture of the simulator is presented in Figure 14.3 in the UML class diagram style. Figure 14.3(a) presents the real-time scheduler while Figure 14.3(b) presents the structure of the optimizer. Three types of real-time processes are implemented and scheduled based on the timer clock. They are the Environment, Company, and Vehicle objects in Figure 14.3(a). The company is the center entity which receives demands from the environment, centralizes the de-mands, controls the optimization task, and dispatches orders to the vehicles. Figure 14.3(b) shows the structure of the asynchronous optimization process which manages a population of agents, called AgentSolver objects, that are responsible for generating solutions and constitute the population of the metaheuristic approach. Each agent solver then encapsulates a single evolving solution. A particular agent, called MetaSolver, plays the role of a scheduler of the agent solvers activities; it performs a selection between the solutions in a similar way of an evolutionary algorithm. Since this section is mainly devoted to the real-time simulation and not to the optimization algorithm, we shall only focus here to the real-time aspects of the system and to the communication protocols. The solving policy and heuristics will be detailed further in the paper.

14.6.2 Asynchronous Protocol

On the one hand, the company receives new orders from the environment and communicates with the vehicles in a synchronous way, as both processes share the same real-time clock. On the other hand, the communication between the company and the optimizer is asynchronous, using mailboxes to exchange information. Here, the asynchronous execution mode of the optimizer is intended to allow the consumption of all the remaining available CPU resources. Figure 14.1 shows the structured information shared by the two asynchronous processes. The company controls the optimization process following a master/slave scheme. We can distinguish simple orders and structured orders sent by the company. Simple orders are the start, stop commands performed respectively at the beginning or at the end of a working day, and the dispatch of the new customer demands as soon as they arrive in the system. Structured orders concern the transfer of a complete solution, in the *SetOfVehicles* object, as well as, from time to time, the transfer of all the available requests once removing the ones already served, in the *SetOfRequests* object. Hence, the mailboxes have a size proportional to the number of requests in $O(N)$, N being the total number of demands arriving within a day.

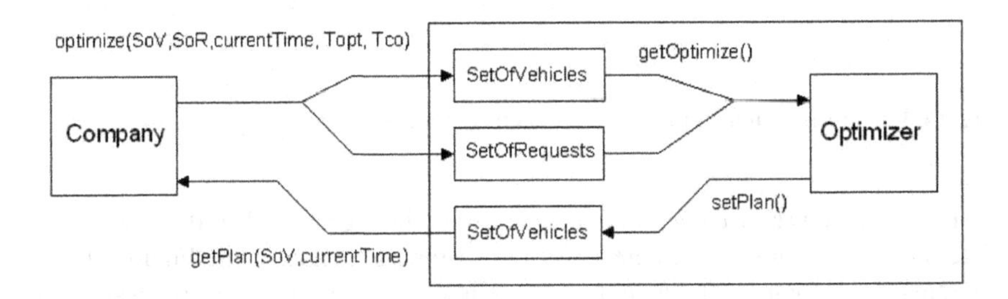

Fig. 14.4 Asynchronous data transfers using mailboxes.

It should be noted that the exchanges, when considering the direction from the company to the optimizer, mainly concern the updates of the vehicle locations, together with the arrival of the new requests. On the contrary, the exchanges from the optimizer to the company concern transfers of the built vehicle routes. But since it can occur in real life situations that autonomous vehicles could by their own modify their routes or themselves participate to the optimization process and modify their plans, we choose to implement a complete bi-directional exchange of vehicle routes. Then, the optimizer systematically chooses as a result the best solution between the received solution by the company and its best generated solution. However in this paper, the vehicles are supposed to strictly follow the optimized routes provided by the optimizer, and not try to optimize the routes by themselves. Hence, the simulator has a general structure allowing possible distributed computations as done in some multi-agent approaches like [30], where the company and the vehicles simulate a market based protocol to construct the routes in a distributed way. Bidirectional exchange of vehicle routes is intended to allow further developments where a competitive solving could take place between the company and vehicles on the one hand, and the optimizer on the other hand.

In order to control the computation time allowed to simulate a working day, we decompose it into many time slices. Two parameters, denoted To and ToR next in the paper, define the amount of computation time allowed to simulate a basic time slice. They respectively control the compression of the working day into a small period of computation time, and the real-time precision of the system. Parameters To and ToR are respectively expressed in computation time units and real-time units. Since the arrivals of new requests have to be tackled as soon as they arrive, ToR can be seen as the basic unit of the real-time clock. It discretized the arrival of the re-quests along the working day. Here, this value is chosen to be in $O(D/N)$. Whereas, the To value corresponds to the few milliseconds of computation time allowed to simulate a period of ToR units of real-time. In the experiments presented in this paper, the ToR value will be adjusted to an integer value compatible with the benchmark test cases unit of time, taking ToR as the smallest integer greater than $0.1 \times D/N$. The To parameter will be adjusted to evaluate the performance of the system from large to very short computation time allowed, hence choosing To from $To = 200ms$ to $To = 20ms$.

The unit of real-time being defined, we now introduce the main parameters governing the frequency of the route updates between the company and the optimizer. They are the optimization time-slice Topt and the commitment horizon Tco, which are both expressed in real-time units. The optimization time-slice Topt defines the time between two consecutive route updates occurring between the company and the optimizer. The commitment horizon Tco is a period of time which defines the requests in routes that cannot be reallocated to other vehicle routes. The period starts from the current time, it is a commitment to the drivers that cannot be changed. We necessarily have $Topt \leq Tco$, and $ToR \leq Topt$. Each $Topt$ units of time, the company respectively gets back the new routes generated by the optimization process, and sends the current vehicle routes containing the actual vehicle positions. Hence, each time a route update is sent by the company at time t using the '"optimize()" procedure, the optimizer anticipates the vehicle positions at their future positions at time $t + Topt$. This is done within the "getOptimize()" procedure. Then, the optimization is performed with the anticipated solution at time $t + Topt$, when the solution will get back by the company using the "getPlan()" procedure. As well, the optimizer anticipates the vehicle positions at time $t + Tco$ since only the part of the routes behind this point can evolve through the optimization process. The optimizer regularly updates the mailbox with the new solution following its own internal rate Tg by calling the "setPlan()" procedure. The new requests are sent to the optimizer whenever they appear, using the "request()" procedure.

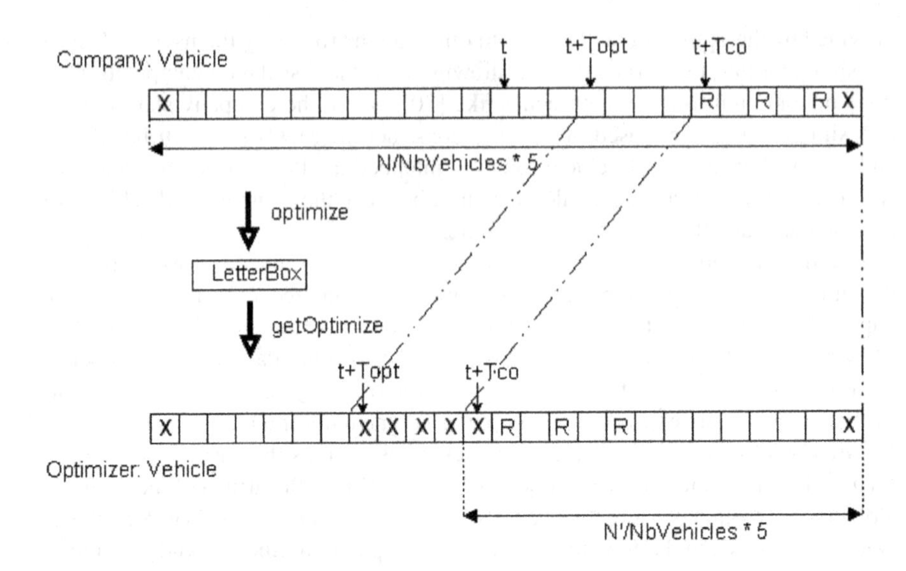

Fig. 14.5 Data structure for transfer from the company to the optimizer.

Figure 14.5 illustrates how the data structure representing a given vehicle route is affected by the transfer from the company to the optimizer. Since a route is defined as an ordered set of requests, the lookahead consists in finding the vehicle position at time $t + Topt$, and then fixing the requests from that date to the date $t + Tco$ since the committed requests will not be affected by the optimization procedure. It should be noted also in Figure 14.5 that the vehicle buffer size is adjusted each Topt to the value $5 \times N'/NbVehicles$, with N' the number of currently available requests in the system at time $t + Tco$, in order to be sufficiently large to insert new requests as they arrive during the next time-slice. The sizes of the Topt and Tco windows can be fixed independently from each other. But in a real time setting, and as showed in the Kilby *et al.* paper [24], the reactivity of the system drastically diminishes with the augmentation of the commitment horizon Tco. In this case, further requests will not be inserted into the committed portion of a route and then will be served later. In order to ensure a maximum of reactivity and dynamism of the system, we set $Tco = Topt$ in all the experiments presented in the paper, with Topt as small as possible. The Topt value is set to $Topt = 10 \times ToR$, thus taking an optimization time-slice in $O(D/N)$, allowing a single request occurrence on average for a single optimization time-slice.

14.7 Experimental Results

14.7.1 Experiments Overview

In this paper, we define the dynamic VRP as a straightforward extension of the static VRP. The length objective (1), the constraints of capacity (2) and time duration (3) are defined exactly the same way as for the static case problem. This allows compatible evaluations according to static optimal values and allows a standard formulation of the problem. To take into account the degree of dynamism of the optimization process, a second objective is defined related to the real-time execution. It is the customer waiting time WT defined in (4). However, in order to give a supplementary insight into the real-time execution, an auxiliary criterion that we think useful to consider is the maximum vehicle finishing time MT defined by (5), i.e. the date of arrival at the depot of the last vehicle once all demands have been served. We think that this auxiliary criterion will help to gauge the excess part of the vehicle services performed behind the working day, once all the demands have been already received, and indirectly to evaluate the part of the instance that is solved as a static problem due to system congestion.

The proposed memetic SOM was programmed in Java and has been ran on a AMD Athlon 2 GHz computer. All the tests performed with the memetic SOM are done on a basis of 10 runs per instance.

For each test case is evaluated the percentage deviation, denoted "%Length", to the best known route length, of the mean solution value obtained, i.e.

$$\%Length = (meanLength - Length*) \times 100/Length \tag{14.5}$$

where $Length*$ is the best known value taken from the VRP Web, and "mean Length" is the sample mean based on 10 runs. The average computation times are also reported based on 10 runs. The average customer waiting time (4) and the maximum vehicle finishing time (5) are expressed as a fraction of the working day in order to compare data with different working days. The waiting time is expressed as a percentage of the working day length D by

$$\%WT = meanWT \times 100/D, \tag{14.6}$$

whereas the maximum finishing time is expressed as an excess deviation to the working day by:

$$\%MT = (meanMT - D) \times 100/D \tag{14.7}$$

This setting also guarantees that it is possible to serve all the demands for the problems considered. Finally, to make things concrete and realistic, the vehicle speed defined in the benchmarks of 1 distance-unit by 1 time-unit can be seen as a vehicle speed of 1 km/mn, or equivalently of 60 km/h. In order to be concrete, we will express the real-time in minutes and the distances in km when reported by their absolute values in some graphics. The working days are roughly between 4 hours and 17 hours, with an exception of a single test case having a 195 hours working day. It is worth noting that the parameter N and the total load of the demands are known before optimization in order to adequately dimension the system. Hence, the working day D can be decomposed into the many required time-slices. We assume that such values are necessarily known in advance in order to model a concrete real-life situation where a limited number of vehicles are intended to serve a maximum amount of demands, and to reasonably dimension the real-time simulator memory and the optimization system.

14.7.2 Influence of the Main Simulation Parameters

In this chapter, we apply to the dynamic VRP a simplified version of the memetic SOM algorithm that was studied in [14] for the static VRP. Since an analysis of the role of the operators and of the algorithm internal parameters was previously performed in the above mentioned paper, we will restrict the focus to an analysis of the few parameters that have an important impact to the dynamic and real-time implementation of the approach. In this section, we study the influence of three parameters and their impact to the length objective and waiting time trade-offs. The parameters studied are the population size Pop of the metaheuristic, the computation time allowed by the choice of the basic temporization To, or timer-clock, of the real-time simulator. The degree of dynamism is adjusted by simply delaying the starts of vehicles, from immediate starts (maximum dynamism) to an half of the day starts (medium dynamism). To respectively implement a high or medium degree of dynamism, delay start is expressed with two values: immediate start at time 0, or delay start at time $D/2$. A delay start at D or $2D$ is used to simulate a static VRP solver in a second set of experiments. The computation times are fast or long depending on the choices $To = 30ms$ or $To = 200ms$ to simulate a given time slice of a working day. Three population sizes are considered: $Pop = 1$, $Pop = 10$, and $Pop = 50$. The experiments were done with the 22 dynamic instances of Kil-by *et al.* performing 10 runs by instance and reporting the average lengths and average waiting times. These experiments are for a part reported with details in Table 1 and Table 2 next in the chapter.

14.7.3 Trace Analysis

We analyze the trace execution of a typical simulation run in order to illustrate how real-time services are adequately simulated and performed along a working day, and verify that the CPU computation time is uniformly distributed along the day. The execution traces presented in Fig.6(a-b) illustrate how the route length and waiting time are evolving for the two cases of dynamism, that is, a maximum and a medium degree of dynamism, respectively, simulated by an immediate start of vehicles or a delay start of D/2. The problem considered is the c50 test case of the benchmark set, having a working day of $D = 351mn$, simulated with the timer-clock $To = 30ms$, and using a metaheuristic population size of $Pop = 10$. In Figure 14.6(a) is shown the length evolution, whereas in Figure 14.6(b) is shown the waiting time evolution for both cases of dynamism. The tradeoff between these two criteria suggests that one would have to choose between two incompatible scenarios: choose to drastically minimize the customer waiting time or choose to minimize length and drivers working time at an expense of the customer waiting time.

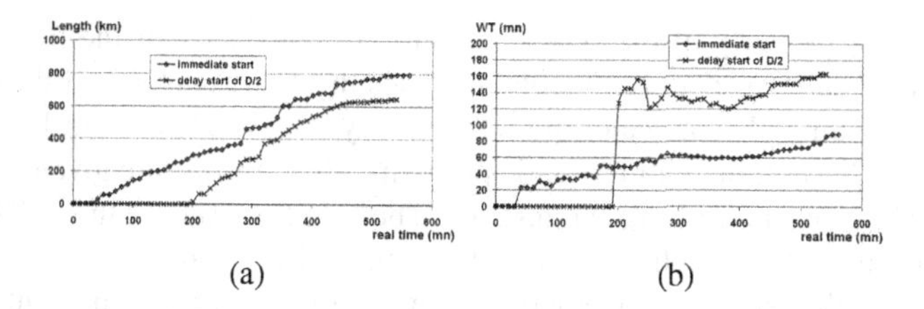

Fig. 14.6 Trace analysis.

As well, in order to verify that the CPU computation time is uniformly distributed along the simulated working day, we report in Figure 14.7(a) the simulated real-time as a function of the processor time, measured by Java system calls every period of Topt units of real-time. The test case is the same as above with immediate vehicle starts and clock $To = 30ms$. The straight line obtained in the figure clearly shows that the computation time is uniformly distributed along the day. We also report in 14.7(b) the number of generations performed by the memetic SOM algorithm as a function of the measured computation time. Again, the number of generations performed at each time step looks roughly proportional to the computation time allowed, with smooth variations on the curve probably due to the varying problem size along the day, or the varying CPU consumption of the vehicles activities implemented as simple state machines.

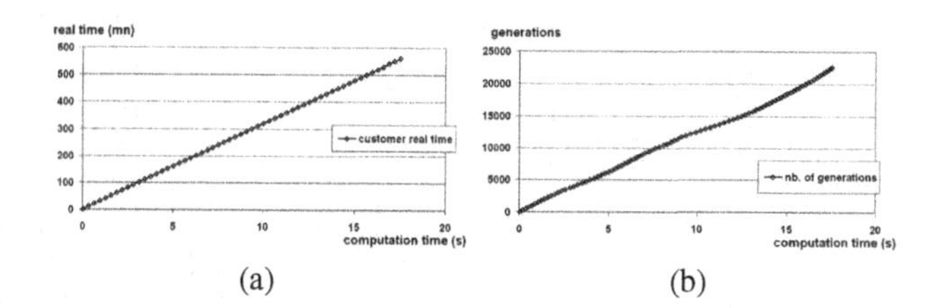

Fig. 14.7 Repartition of the CPU computation time.

14.7.4 Comparative Evaluation

We report detailed results of the experiments performed on the Kilby et al. [24] benchmarks in Table I and Table II. Here, such results are mainly given in order to allow further comparisons with heuristic algorithms for the dynamic VRP. In Table 14.1 are given the results when considering a high degree of dynamism implemented by vehicle immediate starts. Results in Table I are only given for further comparisons. In Table 14.2 , results are presented against the two other approaches found in the literature [21, 28] that have used the benchmark set with a medium degree of dynamism, considering that half of the demands were known in advance. It is worth noting that we simulate the same degree of dynamism by a vehicle delay start time at D/2. As we argued along this paper, we consider the degree of dynamism as a property of the system rather than a property of the instance. The first column "Name-size" of each table indicates the name and size of the instance. The second column "D" indicates the working day length, and the third column the best known value obtained for the static problem. Then, results are given within five columns for a given algorithm configuration. The columns "%Length", "%WT", and "%MT" are respectively defined by equations 14.5, 14.6, and 14.7, as the percentage routes length, percentage average customer waiting time, and percentage maximum finishing time. The column "±%CI" is the 95% confidence interval for the routes length. Finally, the column "Sec" reports the computation times in seconds. Two algorithm configurations are considered respectively with fast ($To = 30ms$) and long ($To = 200ms$) computation times. The metaheuristic population size was set to $Pop = 10$.

Table 14.1 Evaluation on the 22 instances of Kilby et al (1998) with maximum degree of dynamism.

			memetic SOM (fast, immediate start, Pop=10)					memetic SOM (long, immediate start, Pop=10)				
Name-size-veh	D	Best	%Length	±%CI	Sec[a]	%WT	%MT	%Length	±%CI	Sec[a]	%WT	%MT
c50	351	524.61	44.77	3.04	17.56	26.15	59.86	48.03	3.24	111.57	26.78	56.24
c75	346	835.26	38.19	2.73	16.60	21.74	53.21	35.62	2.77	109.94	22.35	56.18
c100	399	826.14	44.21	2.79	18.05	22.11	44.51	37.63	4.17	118.37	24.42	45.84
c100b	468	819.56	33.06	5.21	18.89	16.73	29.00	24.74	4.18	124.04	17.35	30.30
c120	794	1042.11	23.41	1.71	29.21	13.25	17.61	20.70	1.89	190.21	12.92	17.83
c150	399	1028.42	62.88	2.50	17.85	19.57	42.88	58.29	1.47	119.71	20.44	47.49
c199	399	1291.29	60.70	3.17	17.05	18.34	36.49	55.30	2.04	111.65	18.37	37.54
f71	211	237	45.15	3.54	9.04	22.68	36.73	42.11	4.38	58.56	23.51	36.21
f134	11741	11620	75.03	7.59	47.22	3.41	15.75	62.65	4.24	316.19	3.81	19.27
tai75a	769	1618.36	24.72	2.17	16.47	17.72	36.80	20.13	2.91	107.14	18.25	36.96
tai75b	905	1344.62	28.35	3.69	18.87	10.98	33.26	26.55	7.73	123.77	11.23	34.48
tai75c	782	1291.01	31.13	2.68	16.69	15.17	36.34	27.08	1.67	108.72	15.60	36.68
tai75d	789	1365.42	25.06	3.21	16.26	15.84	31.63	20.50	3.76	106.81	16.44	33.08
tai100a	897	2041.34	28.13	2.92	39.59	16.23	41.15	27.23	2.75	255.73	16.26	40.27
tai100b	799	1940.61	30.29	4.59	36.02	18.77	44.17	25.41	2.12	234.20	19.73	44.21
tai100c	905	1406.2	45.29	4.24	34.33	10.16	21.29	35.12	4.10	232.15	10.88	26.20
tai100d	782	1581.25	43.13	3.59	32.62	14.14	33.35	40.86	3.99	213.99	14.72	34.62
tai150a	1062	3055.23	29.01	3.31	44.95	15.06	35.35	27.31	1.85	291.58	15.87	35.09
tai150b	988	2656.47	40.56	3.21	37.39	14.35	21.00	35.23	2.70	249.37	15.07	24.18
tai150c	1081	2341.84	52.64	3.97	41.45	10.55	22.63	46.53	2.45	275.21	11.28	25.26
tai150d	1025	2645.39	40.19	2.37	38.60	14.24	20.41	35.44	5.15	253.74	14.72	21.80
tai385	4816	24431.44	55.40	3.76	100.50	10.56	33.51	46.75	2.06	650.69	10.40	33.01
Average without tai385			40.28	0.70	26.89	16.06	33.97	35.83	0.69	176.79	16.67	35.12
Average all			40.97	0.68	30.24	15.81	33.95	36.33	0.66	198.33	16.38	35.22

[a] Time per run in AMD Athlon (2 GHz) seconds, Java program.

When looking at the results of Table 14.1 and Table 14.2 , one should observe the different tradeoffs between route lengths (%Length) and waiting times (%WT), and note that the maximum finishing times (%MT) have similar values for both degrees of dynamism, corroborating the trace analysis made above. Then, a medium degree of dynamism will favor the drivers working period to be smaller, but at an expense of the customer waiting time. In Table 14.2, the approach is compared with an ant colony approach, that is, an adaptation of the well known MACS-VRPTW approach of Gambardella *et al.* [17] that is considered as one of the best performing approaches to the static VRP. The application to the dynamic VRP is due to Montemanni et al. [28]. Also, it is compared with the genetic algorithm of Goncalves *et al.* [21]. In Table II, the memetic SOM is compared with the ant colony approach of Montemanni *et al.* [28] and to the genetic algorithm of Goncalves *et al.* [21]. It is

Table 14.2 Comparative evaluation on the 22 instances of Kilby et al. (1998) with medium dynamism.

			memetic SOM (fast, start time D/2, Pop=10)					memetic SOM (long, start time D/2, Pop=10)					Montemanni et al. (2005)		Goncalves et al. (2007)	
Name-size-veh	D	Best	%Length	±%CI	Sec[a]	%WT	%MT	%Length	±%CI	Sec[a]	%WT	%MT	%Length	Sec[b]	%Length	Sec[c]
c50	351	524.61	18.58	2.48	16.52	45.49	50.37	17.82	2.14	110.82	46.45	55.19	30.00		13.99	
c75	346	835.26	24.59	2.88	17.00	39.59	56.99	22.65	2.20	106.01	39.53	50.58	24.75		15.89	
c100	399	826.14	14.33	2.68	18.44	42.14	47.69	13.39	2.06	118.60	42.13	46.12	29.03		24.07	
c100b	468	819.56	9.76	4.19	19.49	34.82	33.06	12.88	2.02	127.54	34.92	33.97	24.95		13.60	
c120	794	1042.11	9.47	2.22	29.33	29.26	18.11	7.95	2.62	189.07	29.32	17.12	46.34		37.22	
c150	399	1028.42	32.05	2.20	17.89	36.98	43.23	32.14	1.54	114.52	36.50	41.08	41.58		30.59	
c199	399	1291.29	33.23	1.72	17.22	35.71	37.89	30.29	2.18	113.69	35.35	40.05	42.88		30.18	
f71	211	237	30.89	3.99	9.73	48.76	47.20	27.03	2.94	63.19	49.07	47.01	47.26		19.41	
f134	11741	11620	53.14	8.37	48.38	18.12	18.60	46.98	5.34	318.72	18.59	20.21	38.42		35.29	
tai75a	769	1618.36	19.99	3.02	17.10	31.41	42.08	15.40	2.07	108.87	31.21	39.17	20.18		15.18	
tai75b	905	1344.62	23.93	4.01	18.76	30.75	32.49	22.54	3.69	123.42	31.07	34.08	26.73		13.86	
tai75c	782	1291.01	20.67	2.93	16.72	34.55	36.57	21.93	2.68	108.63	34.70	36.55	28.12		25.64	
tai75d	789	1365.42	17.26	2.94	17.13	33.10	38.71	14.89	3.28	112.53	33.54	40.20	11.98		10.22	
tai100a	897	2041.34	15.98	3.67	40.12	36.97	43.04	13.40	2.21	262.21	37.35	43.81	18.94		18.65	
tai100b	799	1940.61	17.13	2.78	36.35	39.01	45.48	15.19	2.13	239.28	39.05	47.32	20.99		15.48	
tai100c	905	1406.2	23.39	2.99	34.92	25.74	23.38	24.58	3.11	229.39	26.44	24.69	17.76		24.02	
tai100d	782	1581.25	28.21	2.73	33.90	31.78	38.61	24.71	3.68	221.12	31.75	39.09	30.34		20.66	
tai150a	1062	3055.23	19.75	2.52	44.73	34.89	34.71	20.72	3.05	291.58	34.70	35.08	25.69		20.51	
tai150b	988	2656.47	21.36	3.22	38.83	32.79	25.69	20.89	2.38	254.67	32.69	26.81	25.24		24.49	
tai150c	1081	2341.84	21.30	1.44	42.40	27.87	25.40	18.61	2.01	277.16	27.53	26.14	28.79		24.43	
tai150d	1025	2645.39	20.16	3.06	40.44	35.30	26.15	16.22	1.87	269.60	35.20	29.40	21.12		20.55	
tai385	4816	24431.44	38.03	2.21	95.76	29.33	27.03	38.86	1.65	623.00	28.64	27.33	-		-	
Average without tai385			**22.63**	0.63	**27.40**	34.52	36.45	**19.82**	0.53	**179.08**	34.62	36.84	28.62	1500	21.62	1500
Average all			23.33	0.66	30.51	34.29	36.02	20.58	0.51	199.26	34.35	36.41				

[a] Time per run in AMD Athlon (2 GHz) seconds, Java program.
[b] Time per run in Pentium IV (1.5 GHz) seconds, C program.
[c] Time per run in Pentium IV (2.4 GHz) seconds, Java program.

worth noting that the authors do not report the customer waiting time. Nevertheless, we tried to follow the same experimental setting. The authors have used the same benchmark set without the largest test case named *tai385*, and using a medium degree of dynamism. As explained by the authors, a medium degree of dynamism is achieved when half of the demands are considered as known in advance. Here, a medium degree of dynamism is achieved by delaying the vehicle starts to the half of the working day. Since the time distribution is uniform, half of the demands are then expected to be known beforehand. As shown in Table 14.2, the memetic SOM outperforms both the ant colony approach and the genetic algorithm. It improves the solution quality using lesser computation time. Computation time can be roughly an hundred times lesser.

Finally, we report in Figure 14.8(a) synthetic presentation of the evaluations previously performed with the memetic SOM in [11, 14], as well as of the ones of this paper, against state-of-the-art operations research heuristics considering the trade-offs between objective minimization and computation time. Starting to look at the figures from 14.8(d) and in reverse order to figure 14.8(a), the results are given from standard static routing problems to the already studied more complex dynamic VRP. The problems are respectively, the static TSP, the static VRP with capacity constraint only, the static VRP with time duration constraint, and the dynamic VRP, all problems being Euclidean problems. The aim is to suggest how a massive and distributed insertion method, originally based on the neural network self-organizing map algorithm, can be adequately applied in a dynamic setting and to complex problems in a way competitive with the very sophisticated operations research heuristics specifically dedicated to deal with a given problem at hand.

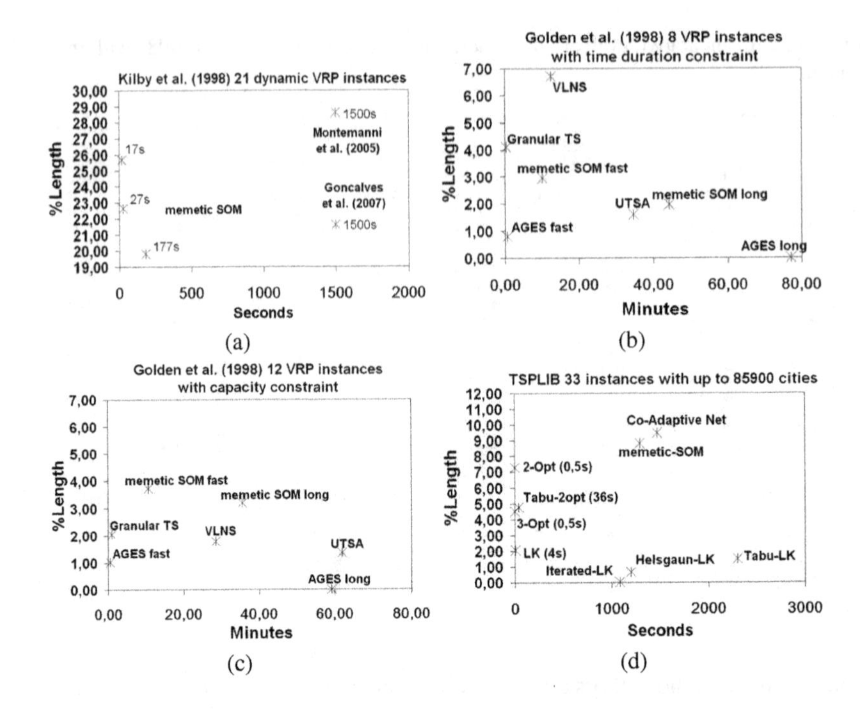

Fig. 14.8 Evaluation of the memetic SOM against state-of-the-art heuristics and metaheuristics. Dynamic VRP (a), static VRP with time duration constraint (b), static VRP with capacity constraint only (c), and static TSP (d).

Results in Figure 14.8(d) illustrate the performance of the memetic SOM on the 33 TSPLIB problems [33] of size larger than 1000 cities, with up to 85900 cities. It is worth noting that these problems were used in the last DIMACS challenge from which are reported the results, the computation times being normalized here to our AMD Athlon (2GHz) computer using Dongarra's factors [15]. Certainly, and as illustrated in Figure 14.8(d), neural networks based approaches do not compete with state-of-the-art OR heuristics for the TSP, such as the 2-Opt, 3-Opt and Lin and Kernighan local search heuristics, which are from a long time the best performing approaches to the TSP according to both length minimization and computation time spent. For example, referring to the Johnson and McGeoch paper [23], the new implementation of the Lin and Kernighan heuristic by Helsgaun [22] is clearly one of the most effective approaches for the TSP.

Nevertheless, and as it is the case also for the SOM based approach called Co-Adaptive Net algorithm [4], our approach was evaluated on many more test problems than previously considered in the literature for neural networks and more importantly on larger TSP's. To be competitive, solution quality produced by neural networks, as well as computation time, would have to be improved both by at least a factor of ten. Other OR powerful heuristics for the TSP are, in the most often cases, a reuse of local search techniques embedded in a restarting or metaheuristic framework, as, for example, the Iterated-LK, Tabu-2-opt, or Tabu-LK, as given in the DIMACS challenge and reported in the Figure 14.8(d). However, the design and implementation of classical OR heuristics is not trivial since there are many implementation and design decisions to be made that have a great influence on performances. Here, on the contrary, we have focused on a heuristic which follows a metaphor in biologic systems, hence which exhibits a high degree of intrinsic parallelism and which may be considered intuitive and easy to implement.

In Figure 14.8(b-c) we turn to the static VRP, reporting results for the large size 20 test problems of Golden *et al.* [7] with sizes from 240 with up to 480 customers. The benchmark contains twelve problems with the capacity constraint only and height problems with the supplementary constraint of time duration. Comparison is presented against some of the recent heuristics presented in the survey paper [5] that cover the global range of metaheuristic performances for the static VRP. We used the numerical results reported in [5] with computation times normalized to our computer. The selected approaches are the Active Guided Evolution Strategy (AGES) [27], the Granular Tabu Search (GTS) [34], the Unified Tabu Search Algorithm (UTSA) [6], and the Very Large Neighborhood Search (VLNS) [16]. Two configurations of the memetic SOM "fast" and "long" are reported for respectively short and long computation times. From what we know, the AGES approach is, at the date of writing, the overall winner considering both solution quality and computation time for the static VRP. AGES is however considered complicated. On the contrary, UTSA is recognized to be simple (easy to understand and implement) and flexible (easy to extend) but more time consuming. Considering the instances with the capacity constraint in Figure 14.8(c), the memetic SOM is less efficient on accuracy than the other approaches, computation times being comparable or lesser than the ones of UTSA and VLNS. Considering the instances with the time-duration constraint in Figure 14.8 (b), the length value becomes closest to the one of UTSA for slightly spending more computation time. For such instances, GTS yields worst quality results but computes very quickly. The memetic SOM performs better than VLNS considering both quality solution and computation time. Hence, the more complex the problem becomes with new constraints added, the more competitive the memetic SOM becomes with OR powerful heuristics.

Finally, Figure 14.8(a) presents a summary of the experiments done on the dynamic VRP. The memetic SOM looks to be very faster and more efficient than the few approaches that were applied to the Kilby *et al.* benchmarks. The results are reported for different computation times allowed, set by a timer clock at respectively $To = 20, 30$, and 200ms, thus illustrating the "anytime" nature of the algorithm that is able to yield competitive results even for very short computation times.

As we explained in the introduction, the results corroborate the idea that the approach presented in this paper was from the beginning designed to be applied in a dynamic and Euclidean setting. Hence, the approach looks more simple and flexible than traditional OR approaches based on neighborhood swapping operators that need complex implementation tricks to yield their effective power [23] when applied in the Euclidean plane. Here, the performances can be explained by the many nearest point searches performed in the plane in a distributed way by spiral search. Furthermore, there is no distinction between a construction and an improvement procedure as usual, but rather a distinction between a deployment and improvement phase where solution construction and solution improvement are a unified process with a decreasing intensity operating on an intermediate structure that continuously adapts and distorts itself in the plane to an underlying distribution of the demands. No new insertion procedure needs to be added to a previously designed improvement procedure in order to dynamically add the new demands in real-time, as usually done with improvement heuristics based on swapping operators that are subsequently applied to dynamic versions of a vehicle routing problem [24, 31].

Furthermore, the internal data structures of the memetic SOM are weakly impacted by the arrivals of new demands, an insertion into the memory being done in constant time $O(1)$. Subsequent insertions are then done naturally by the repeated massive and distributed nearest point searches. Similar implementation mechanisms look not to be considered in the ant colony and genetic algorithm when compared to our approach. For example, an ant would have to perform a complete tour in order to introduce a new demand in a route, thus theoretically performing $O(N^2)$ operations to do so. As well, adding a new demand in the memory theoretically needs modifying the graph structure by adding as many edges as necessary. And such conclusion arises also for the genetic algorithm where a complete examination of a solution-chromosome structure is needed to perform an insertion. The conclusion is that the ant colony algorithm and the genetic algorithm look far from being the best candidates for an application into a dynamic setting without saying anything about how their structures are adequately updated in the dynamic case.

14.8 Conclusions

We have presented the dynamic VRP as a straightforward extension of the classic and standard VRP, and a hybrid heuristic approach to address the problem using a neural network procedure as a search process embedded into a population based evolutionary algorithm, called memetic SOM. Based at the origin on the standard self-organizing map algorithm, the memetic SOM reuses the concept of an intermediate structure representing routes that adapt to an underlying distribution of demands by the many route distortions performed in the plane. By extension, these mechanisms become opera-tors in the population based metaheuristic. They lead to route improvements performed at the same time of customer insertions. Massive insertions are performed based on a spiral search algorithm for the nearest point search implemented on the top of a cell decomposition of the plane. That is why we think that the approach naturally deals with dynamic and real-time arrivals of demands distributed in the plane with a weak impact on the evolving structures.

This paper concludes a set of studies where the memetic SOM was successively applied to many routing problems. It was applied to the static TSP with problem sizes with up to 85900 cities, to the static VRP and the VRPTW, and to different combined bus-stop positioning and routing problems. While the approach looks far from being competitive on the TSP when compared to the very powerful operations research heuristics, it becomes more competitive when considering more complex problems such as the static VRP with duration constraint, and specifically competitive when applied to the dynamic VRP. The results look encouraging in that the approach clearly outperforms the few heuristic approaches already applied to the dynamic VRP and evaluated in an empirical way on a common benchmark set. We claim that the memetic SOM is simple to understand and implement, as well as flexible in that it can be applied from a static to a dynamic setting with slight modifications. Also, we think that the memetic SOM is a good candidate for parallel and distributed implementations at different levels, at the level of the population based metaheuristic and at the level of the cellular partition of the plane.

Further research should focus on a better evaluation of the method against simple policies, or heuristics and metaheuristics approaches that were applied to more complex dynamic vehicle routing problems, such as the dynamic VRP with time-windows or the dynamic pick-up and delivery problem with time-windows. These approaches would be easily customized to the standard and simplest dynamic VRP presented in this paper. Hence, this paper has reported evaluations to allow further comparisons on the basis of a standard formulation of the dynamic VRP and a standard test set. It would be of interest to better study and normalize the dynamic and real-time benchmarks in a similar way that is done for the static problems, in order to favor future empirical evaluations of algorithms on dynamic unstructured large size problems.

References

[1] Bentley, J.-L., Weide, B.W., Yao, A.C.: Optimal expected-time algorithms for closest point problems. ACM Trans. Math. Softw. 6(4), 563–580 (1980)
[2] Bertsimas, D.J., Levi, S.D.: A New Generation of Vehicle Routing Research: Robust Algorithms, Addressing Uncertainty. Operations Research 44(2), 286–304 (1996)
[3] Christofides, N., Mingozzi, A., Toth, P.: The vehicle routing problem, pp. 315–338. Wiley (1979)
[4] Cochrane, E.M., Beasley, J.E.: The co-adaptive neural network approach to the euclidean travelling salesman problem. Neural Network 16(10), 1499–1525 (2003)
[5] Cordeau, J.-F., Gendreau, M., Hertz, A., Laporte, G.T., Sormany, J.-S.: New heuristics for the vehicle routing problem. In: Langevin, A., Riopel, D. (eds.) Logistics Systems: Design and Optimization, pp. 279–297. Springer, US (2005)
[6] Cordeau, J.-F., Laporte, G., Mercier, A.: A unified tabu search heuristic for vehicle routing problems with time windows. The Journal of the Operational Research Society 52(8), 928–936 (2001)
[7] Metaheuristics in Vehicle Routing. In: Crainic, T.G., Laporte, G. (eds.) Fleet Management and Logistics, pp. 33–56. Kluwer, Boston (1999)
[8] Creput, J.-C., Koukam, A.: Clustering and routing as a visual meshing process. Journal of Information and optimization sciences 28(4), 573–601 (2007)
[9] Creput, J.-C., Koukam, A.: Interactive meshing for the design and optimization of bus transportation networks. Journal of Transportation Engineering 133(9), 529–538 (2007)

[10] Creput, J.-C., Koukam, A.: Self-organization in evolution for the solving of distributed terrestrial transportation problems. In: Prasad, B. (ed.) Soft Computing Applications in Industry. STUDFUZZ, vol. 226, pp. 189–205. Springer, Heidelberg (2008)
[11] Creput, J.-C., Koukam, A.: A memetic neural network for the euclidean traveling salesman problem. Neurocomputing 72, 1250–1264 (2009)
[12] Creput, J.-C., Koukam, A., Hajjam, A.: Self-organizing maps in evolutionary approach for the vehicle routing problem with time windows. International Journal of Computer Science and Network Security 7(1), 103–110 (2007)
[13] Creput, J.-C., Koukam, A., Lissajoux, T., Caminada, A.: Automatic mesh generation for mobile network dimensioning using evolutionary approach. IEEE Trans. Evolutionary Computation 9(1), 18–30 (2005)
[14] Creput, J.-C., Koukam, A.: The memetic self-organizing map approach to the vehicle routing problem. Soft Computing - A Fusion of Foundations, Methodologies and Applications 12, 1125–1141 (2008)
[15] Dongarra, J.: Performance of various computers using standard linear equations software. Technical Report CS-89-85, Department of Computer Science, University of Tennesse, US (2006)
[16] Ergun, O., Orlin, J.B., Steele-Feldman, A.: Creating very large scale neighborhoods out of smaller ones by compounding moves: A study on the vehicle routing problem. MIT Sloan Working Paper No. 4393-02 (October 2002)
[17] Gambardella, L.M., Taillard, É., Agazzi, G.: Macs-vrptw: A multiple colony system for vehicle routing problems with time windows. In: New Ideas in Optimization, pp. 63–76. McGraw-Hill (1999)
[18] Gendreau, M., Laporte, G., Potvin, J.-Y.: Metaheuristics for the capacitated VRP, pp. 129–154. Society for Industrial and Applied Mathematics, Philadelphia (2001)
[19] Ghiani, G., Guerriero, F., Laporte, G., Musmanno, R.: Real-time vehicle routing: Solution concepts, algorithms and parallel computing strategies. European Journal of Operational Research 151 (2003)
[20] Glover, F.: Optimization by ghost image processes in neural networks. Computers and Operations Research 21(8), 801–822 (1994); Heuristic, Genetic and Tabu Search
[21] Gonçalves, G., Hsu, T., Dupas, R., Housroum, H.: Une plate-forme de simulation pour la gestion dynamique de tournées de véhicules. Journal Européen des Systèmes Automatisés 41(5), 515–539 (2007)
[22] Helsgaun, K.: An effective implementation of the lin-kernighan traveling salesman heuristic. European Journal of Operational Research 126(1), 106–130 (2000)
[23] Johnson, D., McGeoch, L.: Experimental analysis of heuristics for the stsp. In: Du, D.-Z., Pardalos, P.M., Gutin, G., Punnen, A. (eds.) The Traveling Salesman Problem and Its Variations of Combinatorial Optimization, vol. 12, pp. 369–443. Springer, US (2004)
[24] Kilby, P., Prosser, P., Shaw, P.: Dynamic vrps: a study of scenarios. Technical Report APES-06-1998, University of Strathclyde, UK (1998)
[25] Kohonen, T.: Self-organization and associative memory, 3rd edn. Springer, New York (1989)
[26] Larsen, A., Madsen, O.B.G., Solomon, M.M.: Recent developments in dynamic vehicle routing systems. In: Sharda, R., Voß, S., Golden, B., Raghavan, S., Wasil, E. (eds.) The Vehicle Routing Problem: Latest Advances and New Challenges. Operations Research/-Computer Science Interfaces Series, vol. 43, pp. 199–218. Springer, US (2008)
[27] Mester, D., Braysy, O.: Active-guided evolution strategies for large-scale capacitated vehicle routing problems. Computers and Operations Research 34(10), 2964–2975 (2007)

[28] Montemanni, R., Gambardella, L., Rizzoli, A., Donati, A.: Ant colony system for a dynamic vehicle routing problem. Journal of Combinatorial Optimization 10, 327–343 (2005)

[29] Moscato, P.: A gentle introduction to memetic algorithms. In: Handbook of Metaheuristics, pp. 105–144. Kluwer Academic Publishers (2003)

[30] Preparata, F.P., Shamos, M.I.: Computational geometry: an Introduction. Springer, New York (1985)

[31] Psaraftis, H.N.: Dynamic vehicle routing: Status and prospects. Annals of Operations Research 61, 143–164 (1995)

[32] Psaraftis, H.N.: Dynamic vehicle routing problems, pp. 223–248. Elsevier Science Ltd. (1998)

[33] Reinelt, G.: Tsplib - a traveling salesman problem library. ORSA Journal on Computing 3(4), 376–384 (1991)

[34] Toth, P., Vigo, D.: The granular tabu search and its application to the vehicle-routing problem. INFORMS Journal on Computing 15(4), 333–346 (2003)

Chapter 15
Insect Swarm Algorithms for Dynamic MAX-SAT Problems

Pedro C. Pinto, Thomas A. Runkler, and João M.C. Sousa

Abstract. The satisfiability (SAT) problem and the maximum satisfiability problem (MAX-SAT) were among the first problems proven to be $\mathcal{NP}$-complete. While only a limited number of theoretical and real-world problems come as instances of SAT or MAX-SAT, many combinatorial problems can be encoded into them. This puts the study of MAX-SAT and the development of adequate algorithms to address it in an important position in the field of computer science. Among the most frequently used optimization methods for the MAX-SAT problem are variations of the greedy hill climbing algorithm. This chapter studies the application to dynamic MAX-SAT (i.e. MAX-SAT problems with structures that change over time) of the swarm based metaheuristics ant colony optimization and wasp swarm optimization algorithms, which are based in the real life behavior of ants and wasps, respectively. The algorithms are applied to several sets of static and dynamic MAX-SAT instances and are shown to outperform the greedy hill climbing and simulated annealing algorithms used as benchmarks.

Pedro C. Pinto
Bayern Chemie GmbH, MBDA Deutschland, Department T3R,
Liebigstr. 15-17 D-84544 Aschau am Inn, Germany
e-mail: pedro.caldas-pinto@mbda-systems.de

Thomas A. Runkler
Siemens AG, Corporate Technology, Intelligent Systems and Control,
CT T IAT ISC, Otto-Hahn-Ring 6, 81730 Munich, Germany
e-mail: thomas.runkler@siemens.com

João M. C. Sousa
Technical University of Lisbon, Instituto Superior Técnico, Dep. of Mechanical Engineering,
IDMEC-LAETA, Avenida Rovisco Pais, 1049-001 Lisbon, Portugal
e-mail: jmsousa@ist.utl.pt

E. Alba et al. (Eds.): Metaheuristics for Dynamic Optimization, SCI 433, pp. 341–369.
springerlink.com

15.1 Introduction

This chapter introduces an application of ant colony optimization (ACO) and wasp swarm optimization (WSO) to solve static and dynamic instances of the maximum satisfiability problem (MAX-SAT). The satisfiability problem (SAT) is a type of constraint satisfaction problem (CSP) that is central in the theory of computation. In SAT, the binary variables that define a problem can only take two values, 1 or 0, corresponding to "true" of "false", respectively. Logic formulas are constructed with these binary variables and the problem is defined as finding an assignment to the variables that result in "true" for each of the logic formulas. MAX-SAT is the related optimization problem extension of the SAT problem. In MAX-SAT, the goal is to find a solution that minimizes the number of formulas violated (or, possibly, the sum of their weights).

In a dynamic MAX-SAT problem, clauses can be added to or removed from an instance over time. This implies a model of a system which is subject to different constraints at different points in time. These constraints could reflect the state of the environment or of a subsystem, or the input by a user who controls the system interactively. Another way of defining the dynamic problem is to keep the same fixed set of clauses but allow certain variables to be set to true or false at different points in time, and thus adding or removing extra constraints to/from the problem. These extra constraints could represent, for example, sensor information or user input, in which a variable has a physical meaning and may be forced to assume a value at a certain time, so that the other variables of the problem have to adapt [18].

For MAX-SAT optimization, exact methods and approximate methods algorithms have been studied.

Several approximate optimization algorithms have been introduced in the literature in the last two decades for the optimization of combinatorial problems. Evolutionary algorithms have been successfully applied to constraint problems [21] and dynamic problems [3], [40]. One of the algorithms most often used as a basis for MAX-SAT optimization is greedy hill climbing (GHC) which, despite its simplicity, often produces excellent results [42]. Simulated annealing (SA) was one of the first metaheuristics to be introduced in [20], with tremendous success, and is perhaps the best well known algorithm of the class. Another increasingly well known metaheuristic is ant colony optimization, which despite its relatively young age has already been applied with success to a wide range of problems [11], [2], [37], including MAX-SAT. Wasp swarm optimization is another algorithm of the metaheuristics class, which in the last years has been increasingly applied to diverse optimization problems [38],[9].

While only a limited number of theoretical and real-world problems come as instances of SAT or MAX-SAT, many of these are combinatorial problems with quite natural CSP-like formulations, which can be easily encoded into SAT. This puts the study of SAT and the optimization variant MAX-SAT to an important position in the field of computer science, and there is a lot of work done in developing algorithms that encode problems into SAT and MAX-SAT as SAT problems due to recent advances in SAT solvers[4],[25]. The new solvers are capable of solving significantly large and hard real-world problem instances, which more traditional SAT solvers are incapable of, and it is hoped that techniques capable of encoding those problems into SAT and MAX-SAT can be applicable to them. The practical results vary in quality, as the encoding of some problems may easily lead to instances with billions of clauses, rendering the problem more intractable than the original. Better encoding methods continue being developed, so to an extensive number of problems practically applying MAX-SAT solvers remains an important, open question [25].

In [28] and [29], we have shown results of the application of ACO and WSO to optimize dynamic MAX-SAT problems. The goal of this article is to expand on those first results in the following way: firstly, a direct comparison between the algorithms is enabled. Secondly, the benchmark problems chosen for this study are significantly more challenging. In third place, and just as important, the ACO and WSO results are given together here with the results obtained with the GHC algorithm. The comparison with GHC allows for an immediate evaluation of the quality of the solutions, as the GHC is particulary well suited for MAX-SAT and thus used as a benchmark in many MAX-SAT related publications. It also allows for an easy reconstruction and verification of the results published here, since its algorithmic formulation, given here, is extensively known and thus straightforward to implement.

15.2 MAX-SAT Optimization Problem

This section provides a brief summary of solving SAT and MAX-SAT problems. For further information, the reader is referred to [34],[19],[24],[6], which provide an excellent overview of the subject.

The satisfiability problem and the associated maximum satisfiability problem were the first problems proven to be $\mathcal{NP}$-complete [13],[15]. In a satisfiability problem, or SAT, the variables $\{x_1, ...x_n\}$ of the problem can take two values, 1 and 0, which correspond to $x_i =$"true" and $x_i =$"false", respectively. Logic formulas are constructed with binary variables and the operators $\wedge$ (*and*), $\vee$ (*or*), and $\neg$ (*not*). $x_1 \vee x_2$ means that either x_1 or x_2 or both need to be true. $x_1 \wedge x_2$ means that both x_1 and x_2 need to be true. $\neg x_1$ means that x_1 needs to be false. A literal l_i is the propositional variable x_i or its negation $\neg x_i$. In the first case the literal is called positive, and negative in the later. The literal l_i is interpreted as "true" if it is positive and the variable is assigned the value 1 or if it is negative and the variable is assigned the value 0, and "false" otherwise. The disjunction of the literals $l_1, ..., l_n$ is:

$$l_1 \vee l_2 \vee ... \vee l_n. \tag{15.1}$$

The constraints are given as a set of clauses C_i, where each clause is a disjunction of literals. The length of a clause is the number of different literals in that clause. A conjunction of clauses $C_1, C_2, ..., C_m$ is:

$$C_1 \wedge C_2 \wedge ... \wedge C_m. \tag{15.2}$$

A CNF-*formula* (Conjunctive Normal Form), or instance, is a conjunction of clauses. The number of variables in a CNF-formula is denoted by n, and the number of clauses denoted by m.

An instance is satisfiable if a satisfying truth assignment exists of the involved binary variables and unsatisfiable otherwise. k-SAT is defined as the class of problems where the instance consists only of clauses with at most k literals. $\{k\}$-SAT is defined as the class of problems where the instance has only clauses with exactly k literals. Cook [10] proved that the class of k-SAT problems, for $k \geqslant 3$, is NP-complete.

A solution of an instance consists of an instantiation of all the variables which does not violate any of the constraints, i.e., a consistent labeling of each variable with a value from its domain.

MAX-SAT is the related optimization problem extension of the SAT problem. In MAX-SAT, the goal is to find a solution that minimizes the number of constraints violated, or their weight, instead of finding a solution that satisfies all of them like in SAT.

In case of weighted MAX-SAT, each clause has an associated weight, which means that it is more important to satisfy certain clauses than others.

The 3-satisfiability optimization problem, or MAX-3SAT, is a special case of k-SAT, where each clause contains at most $k = 3$ literals. It is known that 3-SAT experiences dramatic transitions from easy to difficult and then from difficult back to easy when the ratio of the number of clauses to the number of variables increases and becomes therefore simple to prove if no solution exists [43].

15.2.1 Phase Transitions

It is known that the class of random 3-SAT instances introduced above is subject to threshold phenomena with respect to the density parameter, which for SAT and MAX-SAT is the number of constraints to number of variables ratio [6],[24]. Low-density problems are satisfiable with high probability and easily shown to be satisfiable by basic techniques. High density problems are unsatisfiable with high probability. Instances with a low density tend to have many solutions, which makes it relatively easy to find one. Instances with a high density offer many different possibilities to prove a contradiction. Instances near the density threshold that are satisfiable have few solutions, and those that are not have few different ways to prove unsatisfiability, which makes them hard to solve (i.e. long computational times are necessary).

The critical value of this order parameter for 3-SAT is around 4.13. A 3-SAT is almost always satisfiable when the clause/variable ratio is below this critical value and is almost always unsatisfiable beyond it, making a sharp transition from satisfiability to unsatisfiability. The computational cost is also low when the probability of satisfiability is close to one or zero, being the highest around the 4.13 ratio.

MAX-3SAT follows the same pattern of 3-SAT to enter the computationally difficult region, but it follows an easy-hard pattern as the clause/variable ratio increases, instead of an easy-hard-easy pattern (see Fig. 15.1).

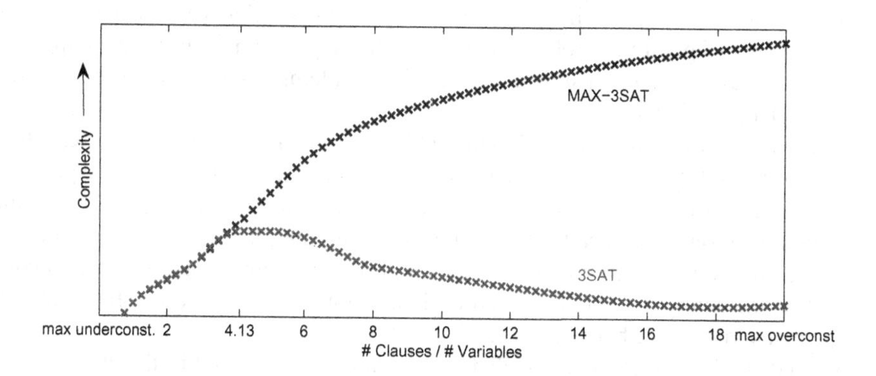

Fig. 15.1 Typical phase transition of 3SAT and MAX-3SAT [43].

The discrepancy between the different patterns of the complexity phase transitions of 3-SAT and MAX 3-SAT indicates that optimization is more difficult than decision. The optimal solution to a MAX 3-SAT can be used to determine if the corresponding 3-SAT is satisfiable or not, but the reverse is obviously not true. Thus, a MAX 3-SAT is at least as hard as its corresponding 3-SAT. When a problem is over-constrained, a small subset of the problem is very likely to be over-constrained as well, so that the problem can be declared unsatisfiable when the subproblem is detected to be unsatisfiable. In this way, the more constrained the satisfaction problem is, the quicker it can be found that no solution exists. However, in an over-constrained MAX-SAT problem, the task of finding an optimal solution to minimize the total number of violated constraints is typically hard, since every possible variable assignment has to be considered.

15.3 The Dynamic MAX-SAT Optimization Problem

The formulation of the dynamic MAX-SAT problem introduced in this section follows the formulation given in [18]. Within MAX-SAT there are basically two ways of turning a problem dynamic. The first possibility is to start from a given instance and allows clauses to be dynamically added to or retracted from this instance over time. Using this definition a system which is subject to different constraints at different points in time (e.g. environment change, new user input) can be modeled. Another way of defining the dynamic problem is to keep the same fixed set of clauses but allow certain binary variables to be set to true or false at different points in time, and thus generating or removing extra constraints to the problem. This way of making the problem dynamic has its basis on the representation of sensor information or user input, in which one variable has a physical meaning and may be forced to assume a value at times, so that the other variables of the problem have to adapt. The factors influencing the choice and performance of an algorithm applied to dynamic MAX-SAT (and dynamic problems in general) are described in [3] and include:

- Severity of change: in first place, some exploitable similarity between the current and next stage of the problem must exist. If the problem is changed completely, it can be regarded as a simple sequence of problems that have to be solved from scratch.
- Frequency of change: how often the environment changes is another critical characteristic of the dynamic problem. If the rate of changes is slow, then the problem can be considered static or quasi-static, in which case restarting the algorithmic computations periodically may be the most computationally efficient method. On the other hand, if the rate of changes is faster, algorithms that are able to cope with the changes without a restart of the computations are likely to be more efficient and produce better results.
- Predictability of change: whether there is a pattern or trend in the changes that can be used to predict the occurrence of the next change, and/or its severity. If available, the information can be used to adjust the algorithms.

The notion of a dynamic SAT problem can be captured by the following definition:

An **instance of the dynamic SAT problem** (DynSAT) over a set V of variables is given by a function $\Omega : N \mapsto CNF(V)$, where N is the set of nonnegative integers, and $CNF(V)$ is the set of all propositional formula in conjunctive normal form which only use the variables in V. For practical purposes, Ω is restricted to a finite number of clauses.

If a DynSAT instance does not change after a finite number of time steps, i.e., if $\exists_n : \forall_m > n : \Omega(m) = \Omega(n)$, this instance is called an n-stage DynSAT instance. A DynSAT instance is cyclic with period Δ if $\forall_n : \Omega(n+\Delta) = \Omega(n)$.

The second way of defining DynSAT is to use a fixed set of clauses but to allow certain propositional variables to be set to true or false at different points in time. This second definition can be formalized in the following way:

An instance of DynSAT over a set V of proportional variables is given by a CNF formula F over V and a second-order function $\Psi : N \mapsto CNF(V \mapsto \{true, false, free\})$, where $free$ means that the variable is unassigned.

For each time n, $\Psi(n)$ determines for each variable appearing in F whether it is fixed to true, fixed to false, or not fixed. The notion of $n-stage$ and *cyclic DynSAT* remains the same.

The two definitions of dynamic MAX-SAT given above are equivalent in the sense that each DynSAT instance according to the first definition can always be transformed into an equivalent DynSAT instance according to the second definition and vice versa, and thus results obtained for one case are valid for both. The proof of this proposition is based on the following two observations: Given DynSAT instance (F, Ψ), for each variable v which is fixed at time n we add the unit clause $\neg v$ to F if $\Psi(n)(v) = false$ (in case $\Psi(n)(v) = free$, then v is not fixed at time n and no unit constraints need to be added). This gives a sequence Ω where $\Omega(n)$ consists of F with necessary unit clauses added. Clearly, Ω is satisfiable if (F,Ψ) is satisfiable.

In the same manner, given a DynSAT instance Ω, we let F be the CNF formula consisting of all the clauses mentioned by Ω. The set of propositional variables is then extended by adding an indicator variable v_i for each clause c_i in F. Now, another CNF formula F' is obtained by replacing each clause c_i in F by $c_i \vee \neg v_i$. Finally, $\Psi(n)$ is defined such that for the indicator variables $\Psi(n)(v_i)$ is $true$ if c_i appears in $\Omega(n)$ and $false$ otherwise. For all original problem variables v, $\Psi(n)(v) = free$. (F',Ψ) is satisfiable exactly if Ω is satisfiable.

The first way of defining the problem is conceptually simpler and a slightly more obvious generalization from conventional SAT from a theoretical point of view. This makes it slightly more adequate for theoretical considerations. The second definition, on the other hand, reflects actual dynamic systems in a more direct way, and it is easier to apply directly to existent regular MAX-SAT problems without changing their characteristics by adding out of place clauses. For both reasons, we focus on dynamic problems formalized as according to the second definition. Like in the regular MAX-SAT problem the optimization objective is the minimization of the number of unsatisfied clauses, or cost, at each stage of the problem.

15.3.1 Dynamic MAX-SAT: Practical Example

A simple, dynamic MAX-SAT problem is formulated next to serve as an example. Let us have a regular MAX-SAT problem defined by a set of clauses with a fixed number of variables being fixed to true or false in every stage, which means that for the MAX-SAT problem the variable set X is defined in instance $\Phi(i)$ as $x \in \{0,1,free\}$. In this example, the problem is composed of an instance Φ with 3 clauses and 5 variables:

$$\Phi = (x_1 \vee \neg x_2 \vee x_4) \wedge (x_3 \vee x_4 \vee \neg x_5) \wedge (x_1 \vee x_3 \vee \neg x_4) \tag{15.3}$$

Ψ is the function that determines which variables of X are fixed for each instance and their values, from the set of existing variables: $X_1 = \Psi(X_1)$. In this example, it fixes two variables randomly at each stage:

$$Stage\ 1 : X = [0, free, 1, free, free], \tag{15.4}$$

In order to assure the variability of the problem, at least one of the variables fixed at stage i cannot be fixed in stage $i+1$.

$$\begin{aligned} Stage\ 2 &: X = [0, free, free, free, 1] \\ Stage\ 3 &: X = [free, 1, free, free, 1] \\ &\vdots \end{aligned} \tag{15.5}$$

Stage 1 can then be formulated as a modified set of constraints and variables. The new constraints are $x_1 = 0$ and $x_3 = 1$, and the variables are x_2,x_4 and x_5. The assignment $x_2 = 1$, $x_4 = 1$ and $x_5 = 0$ solves the problem, since

$$(0 \vee 0 \vee 1) \wedge (1 \vee 1 \vee 1) \wedge (0 \vee 1 \vee 0) = 1. \tag{15.6}$$

Therefore, the solution of the first stage of the dynamic problem is $S = \{0,1,1,1,0\}$. In this case, the cost of the solution is 0 (no clauses are left unsatisfied) and the solution is not unique, which is not a concern since the problem here consists only in finding one solution at all.

15.4 ACO Applied to MAX-SAT

Ant colony optimization [12] is a bio-inspired multi-agent algorithm that belongs to a special group of metaheuristics which attempt to emulate behavior characteristics of social insects, in this case ant colonies. In ACO, the behavior of each agent in the optimization mimics the behavior of real life ants and how they interact with each other in order to find food sources and carry resources to the colony efficiently. Collection of food by a group of ants is a multiple stage process.

In the beginning, each ant follows more or less a random walk, leaving a *pheromone* trail behind. Eventually, a few ants will come upon a food source and start the process of moving the food to the nest. Other ants converge on the source and, by means of *stigmergy* [16], soon the shortest path between the food source and the nest is being used by the vast majority of ants in the group (the evaporation over time of the pheromones makes less used and longer paths less and less attractive to the ants). The ants still keep some exploring capability, enabling the discovery of other food sources. ACO has been used to optimize a wide range of problems, such as the satisfiability problem [31],[26],[28], the traveling salesman problem [36], supply-chain logistics [33], [32], routing problems [7], and sorting problems [17].

One of the first applications of ant colony optimization (ACO) to the regular MAX-SAT problem was in [31] by Roli, Blum, and Dorigo with the ant colony system (ACS) algorithm, and studies on the viability of ACO applied to the problems have been conducted in [26] and [35].

Before explaining the mechanism of the algorithm, it is necessary to introduce the fundamentals required to apply ACO to the MAX-SAT optimization problem. These fundamental concepts are the problem representation, the formulation of the pheromone track and the heuristics. There are several ways of mathematically representing the problem to optimize. After choosing the representation, the formulation of the pheromone track, τ and the heuristic information η can be determined. The formulation of the probabilistic rule p, which is used to decide on which action to perform, is determined by the information stored in the pheromones and the heuristics.

Representation of the problem. The structure to learn is an assignment $X = \{x_1, x_2, ..., x_n\}$ composed of n variables x_i, $i = \{1, ..., n\}$. Each variable x_i can assume two different states v_j, $j \in \{0, 1\}$, corresponding to "false" and "true", respectively. The function f to minimize is a simple count of the number of unsatisfied clauses, or the weight of those unsatisfied clauses for weighted MAX-SAT problems, where a different weight is assigned to each clause. No restriction on the candidate lists was implemented.

Pheromone update. To direct the ants, artificial pheromones are placed onto every candidate assignment $< x_i, v_i >$ of the graph. Each one of the two variable states is differentiated within the $n \times 2$ pheromone matrix τ, which is represented in full by $\tau_{<x_i,v_i>}$. This means that each assignment $< x_i, v_i >$ has its own entry in the pheromone matrix, which stores information on the desirability of including the assignment in a learned solution.

The pheromones are updated at the end of each iteration in order to preserve the information about high quality solutions X found in previous iterations. The update has the following steps:

- application of the evaporation coefficient ρ to each entry of τ

$$\tau \leftarrow \tau \cdot (1 - \rho) \tag{15.7}$$

- update each entry $\tau_{<x_i,v_i>}$ in the pheromone matrix τ of each $< x_i, v_i >$ appearing in the current learned solution T:

$$\tau_{<x_i,v_i>} \leftarrow \tau_{<x_i,v_i>} + 1/f(T), \tag{15.8}$$

  where $f(T)$ is the current cost, or the cost before any new assignment of the variable.
- update each entry $< x_i, v_i >$ in the pheromone matrix τ for the current learned solution T_{best}:

$$\tau_{<x_i,v_i>} \leftarrow \tau_{<x_i,v_i>} + 1/f(T_{best}), \tag{15.9}$$

Heuristics. In addition to the pheromones, heuristic information is used to guide the ants to a desirable assignment $< x_i, v_i >$. The heuristics η are represented in a similar way as the pheromones τ. Each assignment $< x_i, v_i >$ has its own heuristics $\eta_{<x_i,v_i>}$.

Let $f_{ij}(< x_i, v_i >)$ be the cost $f(T \cup < x_i, v_i >)$, for a given solution T, with variable x_i being assigned to value v_i.

The heuristic information is then defined as:

$$\eta_{<x_i,v_i>} = f(T \cup < x_i, v_i >) - f(T). \tag{15.10}$$

Probabilistic rule. In the ACO algorithm, each ant can change one variable state (or leave it unchanged) in the solution T in a single iteration. Thus, one ant can choose an assignment v_i for variable x_i. The decision about the state is based on the pheromones ρ and the heuristics η in a probabilistic way.

The probability of variable x_i being assigned the value v_1 is given by

$$p_{<x_i,v_1>} = \frac{\tau^{\alpha}_{<x_i,v_1>} \cdot \eta^{\beta}_{<x_i,v_1>}}{\sum_{k=0}^{1} \tau^{\alpha}_{<x_i,v_k>} \cdot \eta^{\beta}_{<x_i,v_k>}}, \tag{15.11}$$

where α and β are parameters that balance the importance of pheromone and heuristics in p, respectively. The ranges of α and β are problem dependent. A study about the appropriate variable ranges for a wide selection of problems is given in [12].

Algorithm 15.1. ACO Algorithm for MAX-SAT

Input: Set of unassigned variables $x_1,..,x_n$
Output: T with variables $x_1,..,x_n$ assigned

`// Initialization`
pheromones τ: initialize each entry of τ with τ_0;
define N_{max} as max number of iterations;
T_{best} = empty solution construct;
$N_{iter} = 0$;
`// Optimization`
repeat
  T = empty solution construct;
  `// variable assignment`
  while *Convergence is not reached* **do**
    `// go over all m ants`
    for $k = 1$ *to* m **do**
      choose variable x_i from X randomly;
      choose variable assignment v_i according to (15.11);
      compute cost benefit for the assignment $< x_i, v_i >$;
    end
    find ant *BestAnt* with highest cost benefit;
    assign variable value e_i chosen by ant *BestAnt* to T;
  end
  `// Pheromone update`
  if $f(T) < f(T_{best})$ **then**
    T_{best} = copy of T;
  end
  Update τ according to (15.7);
  Update τ according to (15.8) using T;
  Update τ according to (15.9) using T_{best};
  N_{iter}++;
until N_{iter}=N_{max};
return T_{best}

The algorithm proposed here is presented in Algorithm 15.1. Starting from an empty assignment, the ants collaboratively build a complete solution T at each iteration. Within one iteration, every ant picks randomly an edge and assigns a state to that edge based on the pheromones and heuristics. In more detail, each ant performs the following two steps:

1. Random selection of the next variable to be evaluated from the set of candidate variables.
2. Assignment of a variable value, from $v_i \in \{0,1\}$ of x_i. This assignment is made probabilistically using (15.11), in a balance between the pheromone information for the edge contained in $\tau_{<x_i,v_i>}$ and the locally computed heuristic information provided by (15.10).

The ant that found the assignment with the highest cost improvement changes the variable. The current variable assignment T and the best solution found so far T_{best} are used to update the pheromone information using (15.7), (15.8), (15.9) in order to guide the ants to higher quality solutions in next iterations.

In our paper we iterate this $N_{max} = 200$ times, which is a value high enough to allow the pheromone matrix to saturate and the solutions to converge in all the analyzed problems. The output of the optimization process is the variable assignment T_{best}.

The methods used to update the pheromones are the basis of the different ACO algorithms. Three pheromone update approaches were followed, named static ACO, restart ACO (ACOR), and dynamic ACO (ACOD). Static ACO and Restart ACO are applications of the ACO algorithm introduced for static MAX-SAT. Static ACO has no adaptation to the dynamic problem, therefore it should produce solutions of decreasing quality as the information contained in the pheromone matrix saturates. Restart ACO depends on knowing the existence of a new stage, and the pheromone matrix is reset when this is detected. In dynamic MAX-SAT the pheromone information is adjusted in order to prevent its saturation by defining the evaporation coefficient, ρ, as a function of the diversity, div_i, of the last solution computed:

$$\rho_i = \begin{cases} \rho_0 & \text{if } \text{div}_{i-1} > \rho_0 \\ K \cdot \text{div}_{i-1} & \text{otherwise} \end{cases} \tag{15.12}$$

The diversity div_i is defined as the lowest Hamming distance (number of bits which differ between two binary strings) between solution i and any solution computed previously, divided by the number of variables. The concept was used in [1] by R. Battiti and M. Protasi to adapt a reactive local search algorithm with good results.

15.5 WSO Applied to MAX-SAT

Group living within species of social animals often results in occasional conflict among the elements of the group, due to the finite number of resources generally available, may they be food, shelter, or desirable mates. In many species of animals, when several unacquainted individuals are placed together in a group, they engage each other in contests for dominance. Dominance behavior has been described in hens, cows, ponies, lizards, rats, primates, and social insects, and most notoriously wasps [41]. Some of the contests are violent fights, some are fights that do not lead to any serious injury, and some are limited to the passive recognition of a dominant and a subordinate. In the initial period after placement, contests will be extremely frequent, becoming less and less frequent with time and being replaced by stable dominance-subordination relations among all group members [5]. Once an hierarchical order is obtained it lasts for long periods of times, only with minor modifications caused by the occasional successful attempt of a subordinate to take over. The formation of this hierarchy organizes the group in such a way that internal conflicts do not supersede the advantages of group living.

Theraulaz introduced in [38] a model for the organization characteristic of a wasp colony. The model describes the nature of interactions between an individual wasp and its local environment with respect to task allocation. The colony's self-organized allocation of tasks is modeled using what is known as response thresholds. An individual wasp has a response threshold for each zone of the nest. Based on a wasp's threshold for a given zone and the amount of stimulus from brood located in that zone, a wasp has a certain probability to become engage in the task of foraging for that zone. A lower response threshold for a given zone amounts to a higher likelihood of engaging in activity given a stimulus.

Just as before, it is necessary to introduce the fundamentals required to apply wasp swarm optimization (WSO) to the MAX-SAT optimization problem. These fundamental concepts are the problem representation and the formulation of the stimulus of each wasp. The formulation of the probabilistic rule p, which is used to decide on which action to perform, is determined stochastically by the information stored in the stimulus of each wasp.

Representation of the problem. The representation of the problem is the same as in the ACO formulation, with $X = \{x_1, x_2, ..., x_n\}$ being the assignment of variables to learn, where each variable x_i, $i = \{1, ..., n\}$ can have two different assignments v_i, $v_i \in \{0, 1\}$ and the function f to minimize is the number of unsatisfied clauses, or the weight of those unsatisfied clauses for weighted MAX-SAT problems, where a different weight is assigned to each clause. In the formulation followed here, each wasp w_i, $i = 1, ..., n$, represents a candidate assignment, with $F_{<x_i,v_i>}$ the force of wasp representing assignment $< x_i, v_i >$. An active wasp is represented by $< w_i, v_1 >$ and an inactive wasp by $< w_i, v_0 >$. Algorithm 15.2 shows the basic mechanism of the optimization.

Stimulus of each wasp. The probability of wasp w_i bidding to change the state of x_i from unassigned (v_0) to assigned (v_1) is given by:

$$P_{<x_i,v_1>\leftarrow<x_i,v_0>} = \frac{\eta^{\alpha}_{<x_i,v_1>}}{\eta^{\alpha}_{<x_i,v_1>} + \Theta^{\alpha}_{<x_i,v_1>}} \quad (15.13)$$

where η is the problem heuristics as defined in (15.10). In the first iteration, each variable is forcibly assigned to a state, since in MAX-SAT no variables are allowed to be left unassigned. In the following iterations, both wasps are allowed not to bid. The parameter α is in our paper set to 2.

Each wasp has a certain force $F_{<x_i,v_1>}$ that influences its chance of winning the tournament. In the approach followed here, the update of $F_{<x_i,v_1>}$ is influenced by the heuristics formulated in (15.10) for ACO, and by wasp w_i winning or losing the current contest:

$$F_{<x_i,v_1>} = F_{<x_i,v_1>} \cdot \Delta + \eta_{<x_i,v_1>} \tag{15.14}$$

where Δ is given the value $1+k_w$ if the wasp wins and $1-k_w$ if the wasp loses. The parameter k_w was roughly set by experimentation to 0.1 here.

Probabilistic rule. Each wasp can change one variable state (or leave it unchanged) in the solution X in a single iteration. Thus, one wasp can choose an assignment $< x_i, v_i >$ for variable x_i. The decision about the state is based on the force $< x_i, v_i >$. The probability of variable x_i being assigned the value 1 is (similarly to the ACO probabilistic rule) given by:

$$p_{<x_i,v_1>} = \frac{F^{\beta}_{<x_i,v_1>}}{\sum_{k=0}^{1} F^{\beta}_{<x_i,v_k>}}, \tag{15.15}$$

where we set $\beta = 2$.

15.6 SA and GHC Applied to MAX-SAT

In order to evaluate the quality of the introduced algorithms, the quality of the achieved solutions is compared with the results for other metaheuristics, namely greedy hill climbing (GHC) and simulated annealing (SA). Tabu GHC is a combination of tabu search (TS) and GHC. Once a potential solution has been determined, it is marked as "taboo", which means that the algorithm is forbidden to consider that possibility again. Tabu GHC is a simple but highly effective method introduced by F. Glover in [14] as a method to enhance the performance of a local search method by using memory structures to guide the search process. The most widely applied feature of tabu search is the use of a short term memory to escape from local minima. This version is denoted as simple tabu search. However, TS is often used combined with another algorithm that helps to guide the solution building process into the right direction. This algorithm can be GHC. However, TS can also be used with other algorithms, namely SA, ACO, and WSO. In order to prevent the local search from an immediate return to a previously visited solution and to avoid cycles, in TS moves to recently visited solutions are forbidden. This can be implemented by explicitly memorizing previously visited solutions and forbidding moving to those. Usually, the move is forbidden for a number tl of iterations. The parameter tl is

Algorithm 15.2. WSO Algorithm for MAX-SAT

```
Input: Set of unassigned variables x_1,..,x_n
Output: T with variables x_1,..,x_n assigned
// Initialization
define N_max as max number of iterations;
T_best = empty solution construct;
N_iter = 0;
// Optimization
repeat
    T = empty solution construct;
    // For N_iter iterations
    choose variable x_i from T randomly;
    update F_<x_i,v_i> according to (15.14);
    choose if variable is candidate for assignment according to (15.13);
    choose variable assignment e_i according to (15.15);
    Assign chosen variable value < x_i, v_i > to T;
    if cost T > cost T_best then
        T_best = copy of T;
    end
    N_iter++;
until N_iter=N_max;
return T_best
```

Algorithm 15.3. GHC Algorithm for MAX-SAT

```
Input: Set of unassigned variables x_1,..,x_n
Output: T with variables x_1,..,x_n assigned
T = empty solution construct;
T_best = empty solution construct;
TL : tabu list (FIFO) with last 100 tested variables;
repeat
    choose variable x_i not in the tabu list;
    apply best action to x_i;
    add x_i to TL;
    if cost T > cost T_best then
        T_best = copy of T;
    end
until T_best has not changed last 20 times;
return T_best
```

called the *tabu tenure* and if it is set to a value too high for the problem in question, the algorithm becomes too restrictive and may be unable to find the best solutions. If the number is too low, the memory effect may not work and the algorithm may fall into a local minima more easily. Henceforth, every time GHC is mentioned it is implicit that it is with a tabu implementation.

Algorithm 15.4. SA Algorithm for MAX-SAT

Input: Set of unassigned variables $x_1,..,x_n$
Output: T with variables $x_1,..,x_n$ assigned

```
T_start = 30; // start temperature
T_end = 0.000005; // end temperature
γ_SA = 0.99; // temperature decrease factor
β_SA = 20; // iteration factor

T = empty solution construct;
t = t_start;
repeat
    for β_SA · |X| times do
        choose variable x_i randomly ;
        choose best local action for variable (possible actions: 0 and 1);
        calculate cost benefit for best action: ΔS;
        if (ΔS > 0) or (exp(ΔS/t) > rand(0,1)) then
            apply best action to T;
        end
    end
    t = t · γ_SA;
until T > T_end;
return T
```

GHC is used with a tabu list through which the algorithm keeps a list of the last 100 structures and allows only changes that lead to a structure not contained in the list. The local change can lead to an increased cost in order to escape local extrema. After 20 changes without any performance increase, the algorithm returns the overall highest quality solution (see Algorithm 15.3).

The SA algorithm randomly chooses a variable and then chooses the best value for the variable according to the cost and applies the best action given the net benefit is positive. Otherwise, with a negative cost benefit, the action is chosen with a certain probability, depending on the temperature t (see Algorithm 15.4). After a predefined number of iterations, the temperature is decreased. These steps are repeated until the temperature reaches a minimum value.

15.7 Experiments

The dynamic benchmark instances were generated from a selection of static instances from the MAX-SAT competition of 2007 and from the SATLIB benchmark set. The MAX-SAT competition of 2007 benchmarks can be downloaded on the official MAX-SAT 2007 website (www.maxsat07.udl.es/) and the SATLIB benchmark sets can be found on the website www.cs.ubc.ca /hoos/SATLIB. The chosen benchmarks, UUF and RAMSEY, and their characteristics are shown in Table 15.1. The original, static instance is in this study the first stage of the dynamic problem, the second stage is created from it.

Table 15.1 MAX-SAT benchmarks:

Instance family	# instances	# Variables	# Clauses
UUF50	10	50	218
UUF100	10	100	430
UUF175	10	175	753
UUF250	10	250	1025
Ramsey 3k	42	6-190	5-5985
Ramsey 4k	42	10-190	10-9690
Ramsey 5k	42	15-190	21-20349

Each time step before the instance changes and the new stage commences is considered to be an iteration of each algorithm, i.e., there is a cost function evaluation at each time step. For simulation purposes, each stage has a known duration so restart optimization strategies can be applied for comparison with the non-restart algorithms. The duration of each stage is an important parameter of the system, since if all stages last long the speed of the optimization is not critical. In a real-world problem, it is possible (even likely) that the stages have an unknown or variable duration.

From the benchmarks UUF 250 and RAMSEY 5k, instances with 20 and 100 stages were drawn. The effect of the duration of each stage in solution quality and runtime for each algorithm is evaluated for a series of instances with 20 stages and increasing duration. The two 100-stage instances were evaluated for fixed durations chosen based on the 20-stage problem evaluation, in order to provide a proper representation of the problem and to evaluate the behavior of the algorithms for long running problems. The stages of the dynamic UUF250 instance have a duration of 100 iterations, such as the RAMSEY stages. The number of fixed variables per stage, ψ, was set to $\psi = 6$. A basic evaluation of the effect of several values for ψ in the optimization with the algorithms is included in section 8.

The fixed parameters used in the algorithms were set to:

- GHC: the algorithm keeps a list of the last $TL = 100$ structures. After 20 changes without any decrease in cost, the algorithm returns the overall highest quality solution.
- SA: the start temperature was set to $T_0 = 30$ and the end temperature to $T_{end} = 0.000005$. The temperature decrease factor is $\gamma_{SA} = 0.99$ and the iteration factor $\beta_{SA} = 20$.
- ACO: $\alpha = 1$, $\beta = 3$, $\rho = 0.05$. These parameters were fitted by basic experimentation, and generally recommended in [12]. The number of ants m is set to 30. The effect of different number of ants is roughly evaluated below in the results section. The maximum number of iterations was set to $N_{max} = 200$.
- WSO: The parameters were set to $k_w = 0.1$, $\Theta_t = 1$, $\alpha = 2$, $\beta = 2$, and $N_{max} = 200$. A value of β equal to approximately 2 has been found to give the best results in applications to several problems [27],[8],[9]. In our algorithmic application, all wasps are part of the bidding in every iteration, therefore Θ_t is set to 1.

In order to evaluate the performance of the algorithms, two metrics are reported:

- Cost $f(T)$: cost of the achieved solution, T, which is defined as either the number of unsatisfied clauses or the sum of their weight. The cost determines how close the solutions are to an unknown optimal solution. The lower the cost, the closer the solution is to the optimum.
- Runtime: CPU time needed for the algorithm to find the reported best solution of the problem (in seconds) in a Windows XP PC with a 1.83 GHz processor and 1.00 GB RAM. The algorithm implementation was done in Matlab 7.1.

The benchmarks used in this study are used as the only benchmark set for a large number of approximate algorithms [30],[23],[34]. Nevertheless, one should mention that these instances do not pose anymore a reasonable challenge to state-of-the-art exact algorithms for MAX-SAT. However, judging from the computational results reported in the literature, exact algorithms seem to be limited (with the notable exception being SAT-encoded Steiner tree problems) to instances with a few hundred variables which indicates that when high quality solutions for large MAX-SAT instances are required, metaheuristic approaches are currently the most efficient solution techniques [39]. Moreover, the computational results reported in the literature so far do not give a comprehensive picture on which algorithm should be preferred, because in most researches, algorithm designers applied their algorithms only to a specific benchmark set. All of this indicates that the study of new metaheuristics and local search methodologies in general remains relevant. As it was reasoned in Section 3.1, a threshold can be expected for the number of variables/clauses after which an exact optimization strategy will generally be less efficient than approximate optimization, even if that threshold is progressively pushed further. Moreover, metaheuristics may help to flatten the landscape, with algorithms that can cope with a wide range of benchmarks.

15.8 Analysis and Results

The analysis of dynamic MAX-SAT instances is computationally demanding. For this reason, only one representative problem of each benchmark set was chosen as a template to create dynamic MAX-SAT instances. With the objective of giving a more comprehensive overview of the quality of the algorithms, it was opted to report the optimization results for all instances of the benchmark sets UUF and RAMSEY sets for the static problem (i.e., the first stage of the dynamic problems). The static problem results are also used to fine tune the number of ants m in the ACO algorithm. The SA algorithm is also included in the study for comparison purposes.

15.8.1 Results for Static MAX-SAT

The nondeterministic algorithms (ACO, WSO, and SA) were each run 10 times for all the instances of each benchmark set. To note that, from Table 1, there are 10 instances for each benchmark of the UUF family and 42 for each of the RAMSEY family. This means that the algorithms were executed 100 times for each benchmark of the UUF family, and 420 times for each benchmark of the RAMSEY family, ensuring that the study is statistically significant since the complexity of the instances within the same set of a benchmark family is identical. For each algorithm the average and standard deviation of the cost and runtime needed to reach the final solution are reported. The results for the benchmarks are shown in Figs. 15.2 and 15.3.

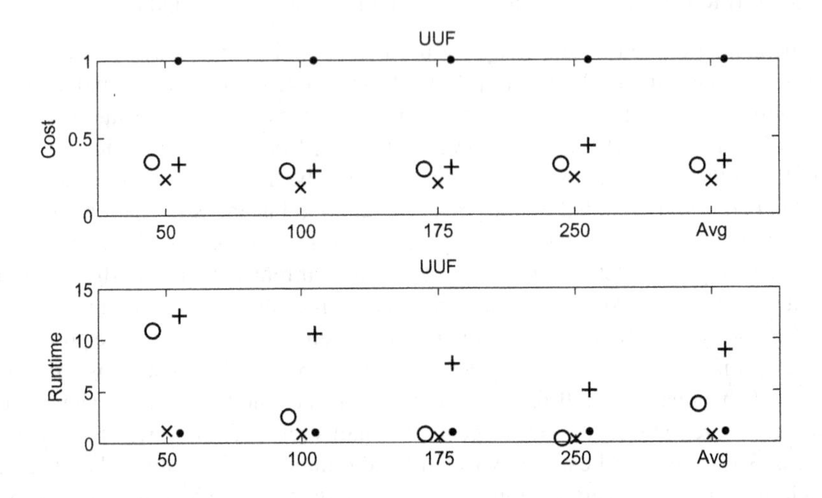

Fig. 15.2 Learned solution cost and runtime (normalized for GHC): +: ACO, ×: WSO, ∘: SA, •: GHC.

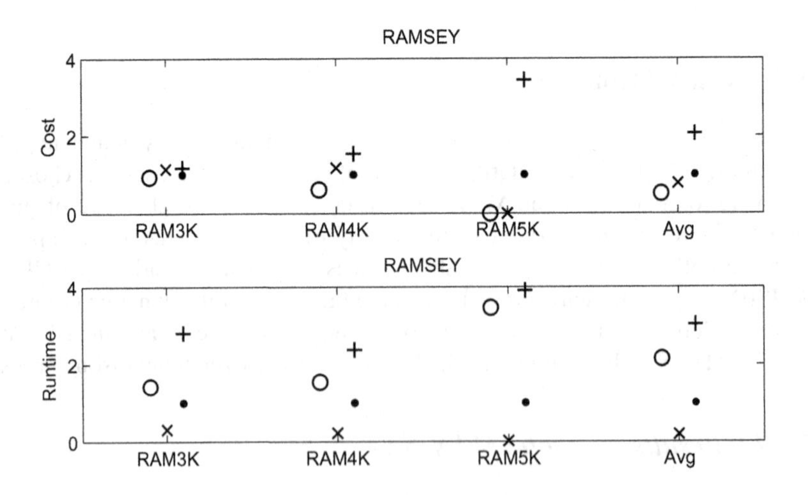

Fig. 15.3 Learned solution cost and runtime (normalized for GHC): +: ACO, ×: WSO, ∘: SA, •: GHC.

Each figure corresponds to a benchmark and contains two subfigures. The first shows the cost of the benchmark instance and the second shows the runtime. Each subfigure contains the average results of the four algorithms: ACO, WSO, GHC, and SA. The results are shown for the different family of instances of each benchmark and the average over all instances. All the measures are normalized by the corresponding measures of GHC. As mentioned earlier in the chapter, the cost of the learned solutions is the number of unsatisfied clauses (or weight of the unsatisfied clauses), therefore having a normalized cost smaller than one means that ACO, WSO, or SA performs better than GHC, and viceversa. Since the normalization gives the same weight in the average to each benchmark instance irrespective of its size, both in terms of number of clauses and variables. Without it, the average would be dominated by the largest or hardest to learn instances.

Tables 15.2 and 15.3 contain the results for the average over all benchmark instances, separated for each instance class, with the objective of giving a quick but comprehensive overview of the results. The tables show the measures for the normalized cost (Table 15.2) and the normalized runtime (Table 15.3). In addition to the information shown in the figures, the tables also report the standard deviation for every sample size as an indication for statistical significance.

Table 15.2 Average normalized cost results

Algorithm	UUF	RAMSEY	Avg
GHC	1.000 ± 0.000	1.000 ± 0.000	1
ACO	0.341 ± 0.072	2.056 ± 1.221	1.064
WSO	0.211 ± 0.028	0.774 ± 0.670	0.585
SA	0.312 ± 0.029	0.515 ± 0.474	0.589

Table 15.3 Average normalized runtime results

Algorithm	UUF	RAMSEY	Avg
GHC	1.000 ± 0.000	1.000 ± 0.000	1
ACO	8.890 ± 3.224	3.031 ± 0.784	15.257
WSO	0.717 ± 0.399	0.186 ± 0.151	1.294
SA	3.652 ± 4.936	2.147 ± 1.140	7.874

Figures 15.2 and 15.3 show the performance of WSO (cross) and ACO (plus) in comparison to GHC (dot) and SA (circle). It can be seen clearly that ACO, SA, and WSO perform always better than GHC for the UUF benchmarks. The RAMSEY benchmark has GHC performing better than ACO, but on average worse than SA and WSO.

Figures 15.4 shows the influence of the number of ants m for one representative instance of the UUF benchmark and for 5 optimizations with the algorithm for each

m (results normalized by the corresponding results obtained with GHC). The left hand side plot shows the influence on the cost of an increasing number of ants, and the right hand side plot shows the runtime with an increasing number of ants. The runtime increases proportionally with the number of ants, and the cost decreases with the number of ants until a certain point, from which on having more ants does not bring a benefit. The average curve is shown for the cost and runtime graphs, and the minimum of each run curve is shown for the cost graph. Based on the cost evolution, m was set to 30 in the algorithm as a value that achieves a good balance between optimization speed and quality.

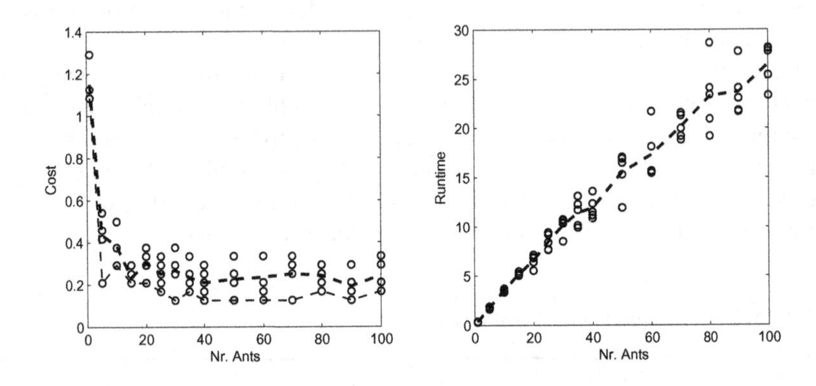

Fig. 15.4 Cost and runtime (in seconds) for an increasing number of ants in ACO for the UUF benchmarks.

15.8.2 Influence of Runtime on Optimization Cost with ACO, WSO, and SA

The runtime becomes increasingly important as problems become harder. Figs. 15.5 and 15.6 show the quality of the learned solutions for an increasing $clauses/variables$ ratio for the benchmark families RAMSEY-4K and RAMSEY-5k. It should be noted that the hardest to learn of these instances have around 20,000 clauses and 200 variables, so that the ratio is around 100, and that the benchmark family RAMSEY-5K is entirely satisfiable. Of all the algorithms, only WSO and SA managed to find the optimum solution for each case, and WSO took the least time. For RAMSEY-4K on the other hand, SA achieved on average the best solutions.

15.8.3 Results for Dynamic MAX-SAT

A comparison of ACO and WSO algorithms with GHC was carried out for the set of dynamic MAX-SAT instances. The restart algorithms are GHC and ACOR, and the continuous optimization (non-restart) algorithms are ACOS, ACOD, and WSO. Each 20-stage and 100-stage problem was run for 5 samples for each algorithm.

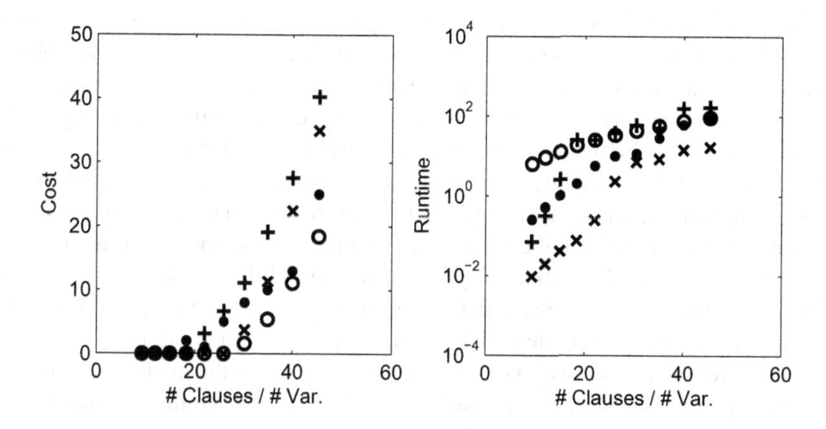

Fig. 15.5 Cost and runtime (in seconds) for increasingly hard RAMSEY 4K instances. +: ACO, ×: WSO, ∘: SA, •: GHC.

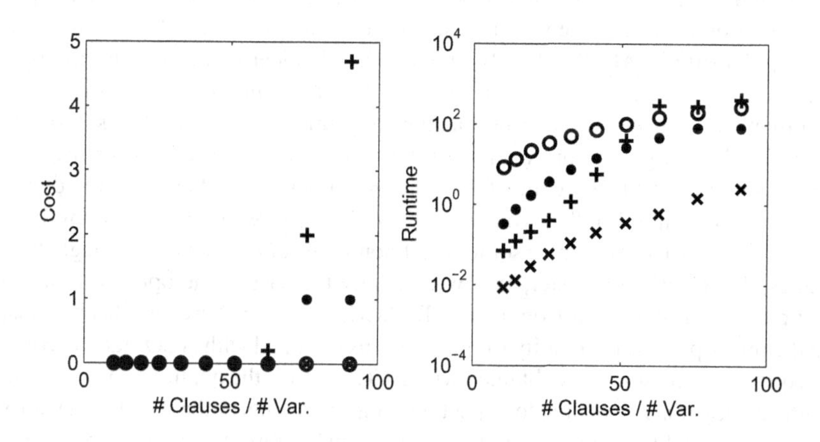

Fig. 15.6 Cost and runtime (in seconds) for increasingly hard RAMSEY 5K instances. +: ACO, ×: WSO, ∘: SA, •: GHC.

The solutions obtained for each stage are evaluated in terms of the cost of the best solution achieved in the stage, and the runtime necessary to compute the best solution.

Figures 15.7 and 15.8 show the optimization results for the incoming instance stages in terms of cost. Each stage can be separated into two basic parts, the *initial* phase and the *stable* phase. There are no clear start and end points that define each phase, but looking at the graph the meaning is clear. The initial phase concerns the period between the first iteration of stage 1 (beginning of the process) and the point where after some stages the algorithms have entered a phase of stable behavior. This means that the mean quality of the solutions is not improving or deteriorating (with the exception of static ACO) from stage to stage.

Figures 15.9 and 15.10 show the average stage results normalized with the results of GHC, in terms of cost and runtime. In the same manner as for the static MAX-SAT learning, a normalized cost smaller than 1 means that the solution learned with ACO is of better quality than GHC, and the closest to 0 the better the quality is.

Finally, Figs. 15.11 and 15.12 show a pair-wise statistical testing of all the algorithms. Each plot describes the significance of the algorithm stated above in comparison with the algorithms stated below. Each figure contains a boxplot corresponding to an algorithm with lines at the lower quartile, median, and upper quartile data values. The whiskers are lines extending from each end of the box to show the extent of the rest of the data. Outliers are data with values beyond the ends of the whiskers. If there are no data outside the whisker, a dot is placed at the bottom whisker. The notches represent a robust estimate of the uncertainty about the medians for box-to-box comparison. Boxes whose notches do not overlap indicate that the medians of the two groups differ at the 5% significance level. A detailed explanation of boxplots is given in [22].

It is important to distinguish the results obtained with the restart and non-restart algorithms, due to the fact that, in a real-world problem, it may not be simple to restart the optimization even in case a change of state is detected. Analyzing the non-restart algorithms, ACOD, ACOS, and WSO, it is clear that ACOS is inadequate. After the initial phases, the quality of the solutions achieved decreases drastically. In effect, and considering the runtime, the algorithm ceases to try new solutions, an indicator that the pheromone matrix is saturated. The difference between the quality of the solutions achieved with ACOD and WSO is not significant for the dynamic RAMSEY instance. For the dynamic UUF, WSO achieves a better score. Both algorithms are shown to guarantee solutions of non-deteriorating quality through the 100 stages, though WSO is faster, as it was already the case in the optimization of the static problem. In comparison to ACOR, ACOD and WSO are significantly faster. In the initial phases, the quality of the solutions obtained with ACO static is equal to the quality of the solutions obtained with ACOD. Since the quality remains constant with ACOD, it is reasonable to expect that improvements of the ACO metaheuristic available in the literature increase the solution quality throughout the whole dynamic instance.

Overall, these results suggest that continuous optimization is generally more efficient and provide significantly more stable solutions than random restart when solving dynamic instances of MAX-SAT problems. The quality of the best solution in each stage is more or less constant through the iterations. The continuous optimization is faster than restart in all cases.

15.8.4 Influence of the Number of Iterations per Stage on the Optimization with ACO and WSO

Figures 15.13 and 15.14 show the influence of the stage duration in the optimization with ACO and WSO for the UUF250 and RAMSEY dynamic instances, respectively.

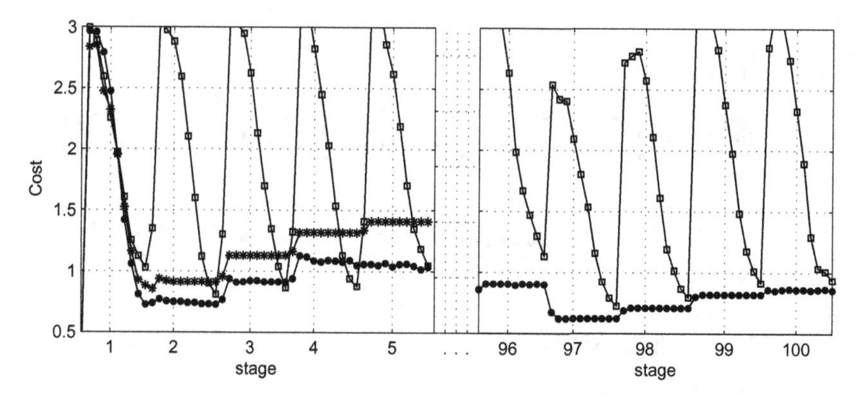

Fig. 15.7 Algorithms (•: ACOD, □: ACOR, ⋆: ACOS) applied to a dynamic UUF instance, all iterations, first 5 stages (initial phase) and last 5 stages (stable phase).

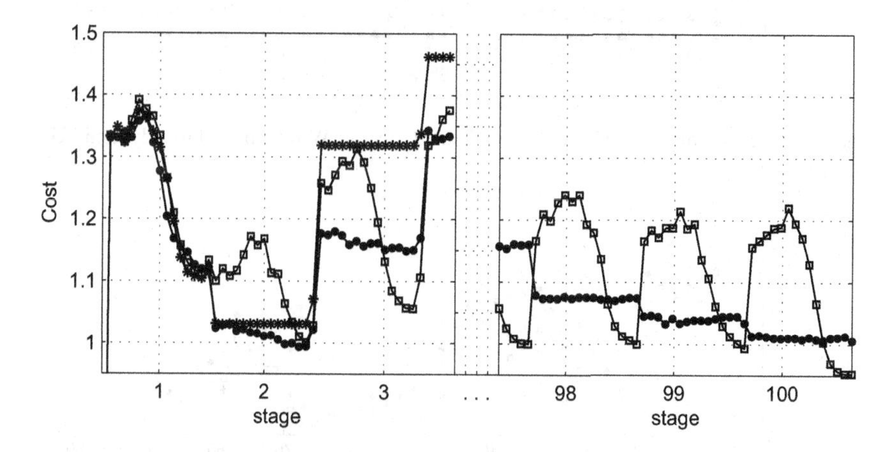

Fig. 15.8 Algorithms (•: ACOD, □: ACOR, ⋆: ACOS) applied to a dynamic RAMSEY instance, all iterations, first 3 stages (initial phase) and last 3 stages (stable phase).

For the UUF250 instance, it can be seen that until a stage duration of around 60 iterations, the average solution quality achieved with ACOD surpasses the quality achieved with ACOR. For 60 iterations, ACOR is approximately 3 times slower than ACOD or WSO in computing the solution. For stage durations higher than 60, the quality achieved with ACOR is higher than the quality achieved with ACOD, but while the runtime of ACOD remains constant, for a stage duration of 100 iterations the runtime of ACOD is already approximately 10 times higher than the runtime of ACOD. For the dynamic UUF250 instance, WSO always achieves better quality solutions than the ACO algorithms for all stage durations. The optimization speed is directly comparable to the optimization speed of ACOD.

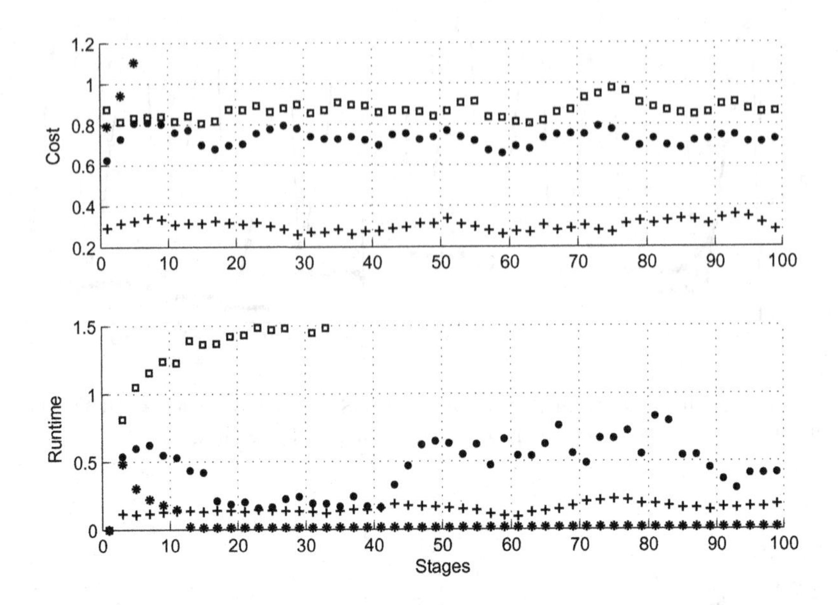

Fig. 15.9 Algorithms (•: ACOD, □: ACOR, ⋆: ACOS, +: WSO) applied to a dynamic UUF instance.

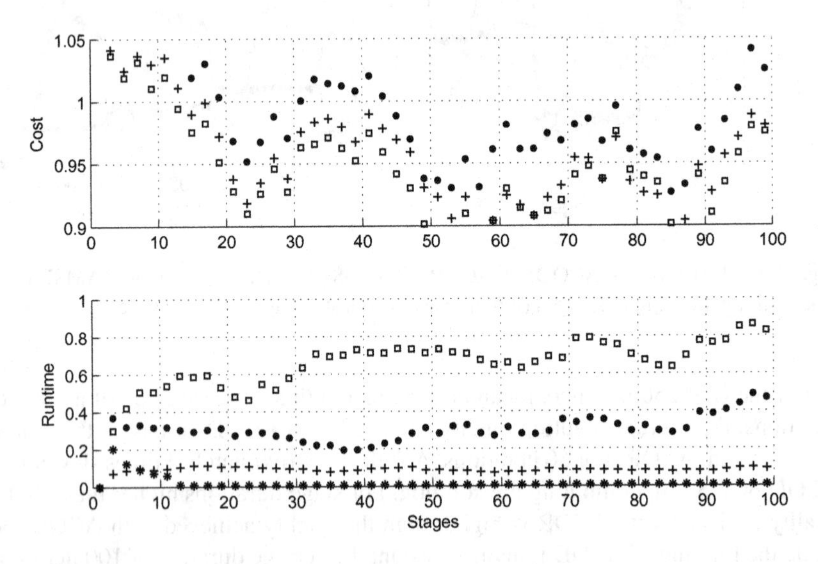

Fig. 15.10 Algorithms (•: ACOD, □: ACOR, ⋆: ACOS, +: WSO) applied to a dynamic RAMSEY instance.

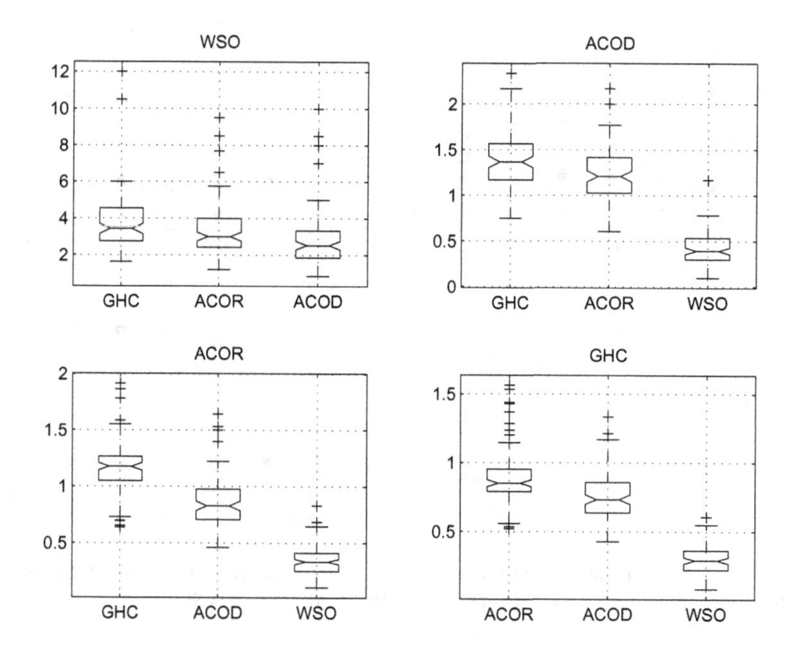

Fig. 15.11 Significance plots of the algorithms applied to a dynamic UUF instance.

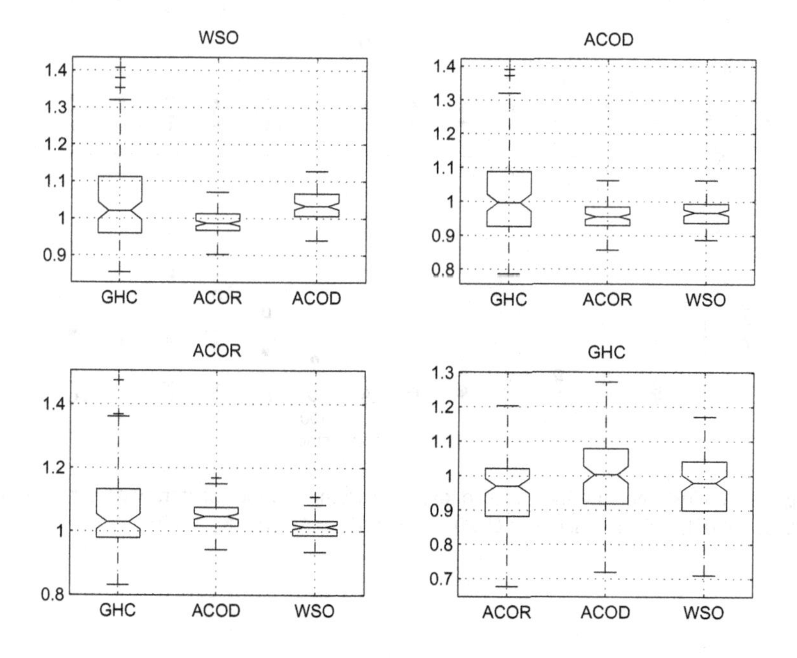

Fig. 15.12 Significance plots of the algorithms applied to dynamic RAMSEY instance.

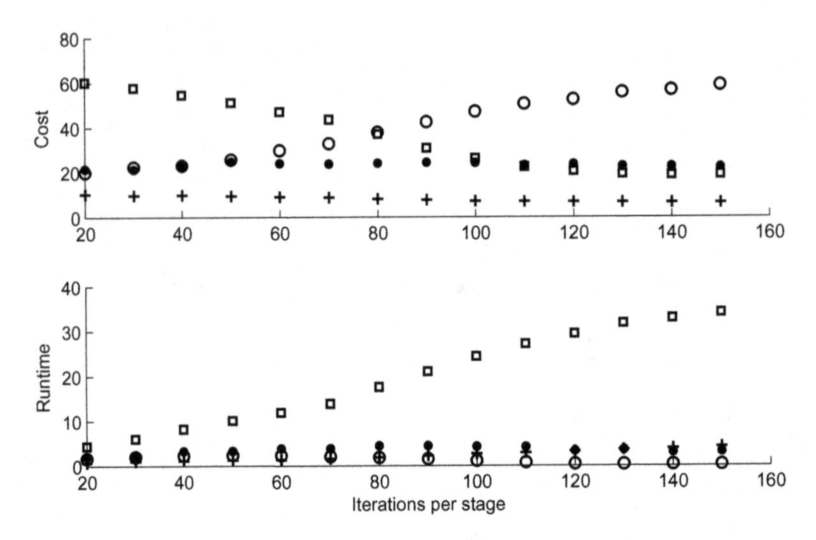

Fig. 15.13 Cost and runtime (in seconds) for varying number of iterations per stage for a UUF250 dynamic instance, •: ACOD, □: ACOR, ○: ACOS, +: WSO.

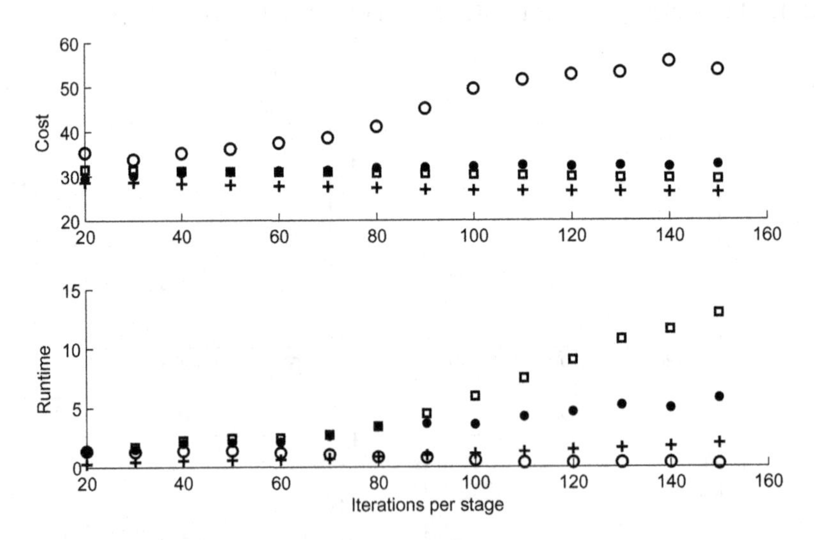

Fig. 15.14 Cost and runtime (in seconds) for varying number of iterations per stage for a RAMSEY dynamic instance, •: ACOD, □: ACOR, ○: ACOS, +: WSO.

15.9 Conclusions

The analysis of the results achieved for the static MAX-SAT problem shows that ACO and WSO are very competitive even when solving relatively large instances. In its first application to MAX-SAT, WSO clearly beats greedy hill search and simulated annealing and is practically equivalent to ACO in solution quality. The performance of WSO and ACO is best for the UUF benchmark family. It is likely that by adjusting the algorithm parameters for each individual problem, the results would be even better. However, it is important to remember that a big advantage of metaheuristic algorithms is to provide a quick framework to get quality solutions to problems and thus spending a disproportionate amount of time determining which parameters are best for each individual problem detracts from that goal. The analysis of the results obtained in dynamic MAX-SAT optimization shows that the static ACO (ACOS) algorithm quickly saturates and does not provide useable results. The introduced dynamic ACO (ACOD) shows a significantly better performance, overall comparable with WSO. Both algorithms are shown to guarantee solutions of non-deteriorating quality through all the stages of the dynamic problems, though WSO is faster, as it was also the case in the optimization of the static problems. In comparison to restart ACO (ACOR), ACOD and WSO are significantly faster. Overall, the results suggest that continuous optimization is generally more efficient and provides significantly more stable solutions than random restart, when solving dynamic instances of MAX-SAT problems.

Further studies will focus on the optimization effect of considering a large range of instance families. There are also several different implementations of ACO algorithms and GHC based algorithms that have not be explored and can provide different results. We consider that the implemented study shows the capacity of WSO and ACO to handle large, dynamic problems and their advantages in optimizing these problems.

Acknowledgements. This work was supported by the Portuguese foundation for Science and Technology, through IDMEC under LAETA.

References

[1] Battiti, R., Protasi, M.: Reactive search, a history-based heuristic for MAX-SAT. ACM Journal of Experimental Algorithmics 2 (1997)
[2] Blum, C., Dorigo, M.: The hyper-cube framework for ant colony optimization. IEEE Transactions on Systems, Man, and Cybernetics-Part B 34(2), 1161–1172 (2004)
[3] Branke, J., Schmeck, H.: Designing evolutionary algorithms for dynamic optimization problems. In: Advances in Evolutionary Computing: Theory and Applications, pp. 239–262. Springer-Verlag New York, Inc., New York (2003)
[4] Cadoli, M., Schaerf, A.: Compiling problem specifications into SAT. In: Programming Languages and Systems, pp. 387–401 (2001)
[5] Chase, I.D.: Models of hierarchy formation in animal societies. Behavioral Sciences 19, 374–382 (1974)

[6] Cheeseman, P., Kanefsky, B., Taylor, W.M.: Where the really hard problems are. In: Proceedings of the 12th International Joint Conference on Artificial Intelligence, Sidney, Australia, pp. 331–337 (1991)
[7] Cicirello, V.A., Smith, S.F.: Ant colony for autonomous decentralized shop floor routing. In: Proceedings of the 5th International Symposium on Autonomous Decentralized Systems, pp. 383–390 (2001)
[8] Cicirello, V.A., Smith, S.F.: Wasp nests for self-configurable factories. In: Agents 2001, Proceedings of the 5th International Conference on Autonomous Agents, pp. 473–480. ACM Press (2001)
[9] Cicirello, V.A., Smith, S.F.: Wasp-like agents for distributed factory coordination. Autonomous Agents and Multi-agent systems 8, 237–266 (2004)
[10] Cook, S.A.: The complexity of theorem-proving procedures. In: Proceedings of the Third Annual ACM Symposium on Theory of Computing, pp. 151–158. ACM, New York (1971)
[11] Dorigo, M., Maniezzo, V., Colorni, A.: Ant system: Optimization by a colony of cooperating agents. IEEE Transactions on Systems, Man, and Cybernetics-Part B 26(1), 29–41 (1996)
[12] Dorigo, M., Stützle, T.: Ant Colony Optimization. MIT Press (2004)
[13] Garey, M.R., Johnson, D.S.: Computers and Intractability: A Guide to the Theory of $\mathcal{NP}$-Completeness. WH Freeman Publishers (1979)
[14] Glover, F., Laguna, M.: Tabu search. John Wiley & Sons, Insc., New York (1993)
[15] Goodrich, M.T., Tamassia, R.: Algorithm Design - Foundations, Analysis, and Internet Examples. John Wiley & Sons, Inc. (2001)
[16] Grasse, P.: La reconstruction du nid et les coordinations inter-individuelles chez bellicositermes natalensis et cubitermes sp. la theorie de la stigmergie: Essai d'interpretation du comportement des termites constructeurs. Insectes Sociaux 6, 41–81 (1959)
[17] Hartmann, S.A., Runkler, T.A.: Online optimization of a color sorting assembly buffer using ant colony optimization. In: Proceedings of the Operations Research Conference, pp. 415–420 (2007)
[18] Hoos, H.H., O'Neill, K.: Stochastic local search methods for dynamic SAT - an initial investigation. In: Leveraging Probability and Uncertainty in Computation, Austin, Texas, pp. 22–26. AAAI Press (2000)
[19] Hoos, H.H., Stützle, T.: Local search algorithms for SAT: an empirical evaluation. In: Journal of Automated Reasoning, special Issue " SAT 2000", pp. 421–481 (1999)
[20] Kirkpatrick, S., Gelatt, C.D., Vecchi, M.P.: Optimization by simulated annealing. Science 220(4598), 671–680 (1983)
[21] Liu, J., Zhong, W., Jiao, L.: A multiagent evolutionary algorithm for constraint satisfaction problems. IEEE Transactions on Systems, Man, and Cybernetics-Part B 36(1), 54–73 (2006)
[22] McGill, R., Tukey, J.W., Larsen, W.A.: Variations of Boxplots. In: The American Statistician, pp. 12–16. American Statistical Association (1978)
[23] Mills, P., Tsang, E.: Guided local search for solving SAT and weighted MAX-SAT problems. Journal Automated Reasoning 24(1-2), 205–223 (2000)
[24] Mitchell, D., Selman, B., Levesque, H.: Hard and easy distributions of SAT problems. In: 10th National Conference on Artificial Intelligence, San Jose, CA, pp. 459–465 (1992)
[25] Tamura, N., Taga, A., Kitagawa, S., Banbara, M.: Compiling Finite Linear CSP into SAT. In: Benhamou, F. (ed.) CP 2006. LNCS, vol. 4204, pp. 590–603. Springer, Heidelberg (2006)

[26] Pimont, S., Solnon, C.: A generic ant algorithm for solving constraint satisfaction problems. In: 2th International Workshop on Ant Algorithms, Brussels, Belgium, pp. 100–108 (2000)

[27] Pinto, P., Runkler, T.A., Sousa, J.M.C.: Wasp swarm optimization of logistic systems. In: Ribeiro, et al. (eds.) Adaptive and Natural Computing Algorithms, 7th International Conference on Adaptive and Natural Computing Algorithms, Coimbra, Portugal, pp. 264–267. Springer, NewYork (2005)

[28] Pinto, P., Runkler, T.A., Sousa, J.M.C.: Ant colony optimization and its application to regular and dynamic MAX-SAT problems. In: Advances in Biologically Inspired Information Systems: Models, Methods, and Tools, pp. 283–302 (2007)

[29] Pinto, P.C., Runkler, T.A., Sousa, J.M.C.: Wasp Swarm Algorithm for Dynamic MAX-SAT Problems. In: Beliczynski, B., Dzielinski, A., Iwanowski, M., Ribeiro, B. (eds.) ICANNGA 2007. LNCS, vol. 4431, pp. 350–357. Springer, Heidelberg (2007)

[30] Resende, M.G.C., Pitsoulis, L.S., Pardalos, P.M.: Approximate solution of weighted MAX-SAT problems using GRASP. In: Satisfiability Problem: Theory and Applications. DIMACS Series on Discrete Mathematics and Theoretical Computer Science, vol. 35, pp. 393–405. American Mathematical Society (1997)

[31] Roli, A., Blum, C., Dorigo, M.: ACO for maximal constraint satisfaction problems. In: Metaheuristics International Conference, pp. 187–192 (2001)

[32] Silva, C.A., Runkler, T.A., Sousa, J.M.C., Sá da Costa, J.M.G.: Distributed supply chain management using ant colony optimization. European Journal of Operational Research 199(2), 349–358 (2009)

[33] Silva, C.A., Sousa, J.M.C., Runkler, T.A., Sá da Costa, J.: Distributed optimization of logistic systems and its suppliers using ant colony optimization. International Journal of Systems Science 37(8), 503–512 (2006)

[34] Smyth, K., Hoos, H., Stützle, T.: Iterated Robust Tabu Search for MAX-SAT. In: Xiang, Y., Chaib-draa, B. (eds.) Canadian AI 2003. LNCS (LNAI), vol. 2671, pp. 129–144. Springer, Heidelberg (2003)

[35] Solnon, C.: Ants can solve constraint satisfaction problems. IEEE Transactions on Evolutionary Computation 6, 347–357 (2002)

[36] Stützle, T., Hoos, H.H.: The MAX-MIN ant system and local search for the traveling salesman problem. In: Proceedings of the 4th International Conference on Evolutionary Computation, vol. 8, pp. 308–313. IEEE Press (1997)

[37] Stützle, T., López-Ibánez, M., Dorigo, M.: A Concise Overview of Applications of Ant Colony Optimization. In: Wiley Encyclopedia of Operations Research and Management Science. John Wiley & Sons (2011)

[38] Theraulaz, G., Goss, S., Gervet, J., Deneubourg, J.L.: Task differentiation in polistes wasps colonies: A model for self-organizing groups of robots. In: From Animals to Animats: Proceedings of the 1st International Conference on Simulation of Adaptive Behavior, pp. 346–355. MIT Press (1991)

[39] Stützle, T., Hoos, H., Roli, A.: A review of the literature on local search algorithms for MAX-SAT. Technical report aida-01-02. Technical report, Technische Universität Darmstadt (2006)

[40] Wang, H., Yang, S., Ip, W., Wang, D.: IEEE Transactions on Systems, Man, and Cybernetics-Part B 39(6), 1348–1361 (2009)

[41] Wilson, E.O.: The insect societies. Harvard University Press (1971)

[42] Winston, W., Goldberg, J.: Operations Research: Applications and Algorithms. Cengage Learning, 4th edn. (2003)

[43] Zhang, W.: Phase Transitions and Backbones of 3-SAT and Maximum 3-SAT. In: Walsh, T. (ed.) CP 2001. LNCS, vol. 2239, pp. 153–167. Springer, Heidelberg (2001)

Chapter 16
Dynamic Time-Linkage Evolutionary Optimization: Definitions and Potential Solutions

Trung Thanh Nguyen and Xin Yao

Abstract. Dynamic time-linkage optimization problems (DTPs) are special dynamic optimization problems (DOPs) where the current solutions chosen by the solver can influence how the problems might change in the future. Although DTPs are very common in real-world applications (e.g. online scheduling, online vehicle routing, and online optimal control problems), they have received very little attention from the evolutionary dynamic optimization (EDO) research community. Due to this lack of research there are still many characteristics that we do not fully know about DTPs. For example, how should we define and classify DTPs in detail; are there any characteristics of DTPs that we do not know; with these characteristics are DTPs still solvable; and what is the appropriate strategy to solve them. In this chapter these issues will be partially addressed. First, we will propose a detailed definition framework to help characterising DOPs and DTPs. Second, we will identify a new and challenging class of DTPs where it might not be possible to solve the problems using traditional methods. Third, an approach to solve this class of problems under certain circumstances will be suggested and experiments to verify the hypothesis will be carried out. Two test problems will be proposed to simulate the property of this new class of DTPs, and discussions of real-world applications will be introduced.

Trung Thanh Nguyen
School of Engineering, Technology and Maritime Operations, Liverpool John Moores Univ.
e-mail: T.T.Nguyen@ljmu.ac.uk

Xin Yao
CERCIA, School of Computer Science, Univ. of Birmingham
e-mail: X.Yao@cs.bham.ac.uk

E. Alba et al. (Eds.): Metaheuristics for Dynamic Optimization, SCI 433, pp. 371–395.
springerlink.com

16.1 Dynamic Time-Linkage Problems - From Academic Research to Real-World Applications

The class of dynamic time-linkage problems (DTPs) is a common and special type of dynamic optimization problems (DOPs). They are defined as problems where "... there exists at least one time $0 \leq t \leq t^{end}$ for which the dynamic optimization value at time t is dependent on at least one earlier solution..." [7]. In other words, a time-linkage problem is an online control problem where the algorithm is the actual controller to control the future behaviour of the system and any decision made at the current moment by the algorithm might consequently influence how the future problem would be. For example, in a car navigation system, if at the current present we found an optimal routing decision and apply this optimal decision to every cars on the same road, all cars might follow the same "optimal" route and consequently make the route no longer optimal, or even congested in the future.

DTPs are very common in real-world applications. In [19, chap. 3], we have made a survey on a large set of recent real-world dynamic optimization applications and the survey results show that a large number of the surveyed problems were mentioned by the authors as having the time-linkage properties (45% in the combinatorial domain, and 81% in the continuous domain, out of 56 surveyed applications. - see Figure 16.1 for a summary). In the continuous domain, a majority of time-linkage problems are control problems where the current value of the control variables would determine how the systems behave in the future. In the combinatorial domain, the time-linkage properties exist in a wide range of problems, for example scheduling, routing, allocation/layout/assignment, planning and path finding. Detailed list of type of problems that have the time-linkage properties can be found in [19].

Despite the popularity of time-linkage problems, this type of problem still has not attracted much attention from the EDO academic research community. Only very few recent studies proposed using EAs to solve time-linkage problems [6, 8, 9, 20]. There are also only few test problems with the time-linkage property in EDO research [6, 8, 9, 20, 28].

From the practical side, the time-linkage property has also not been studied sufficiently in evolutionary research. In the survey in [19] we found that although some references using EAs/meta-heuristics do mention the time-linkage property when solving real-world problems online [2, 4, 13, 14, 16–18, 21, 24, 29, 30], none of them equip their EAs with the ability to handle this property online. In these references, the time-linkage property is either ignored, e.g. in [13, 14, 17, 18], or handled using a separate component/heuristics, e.g. in [2, 4], or handled in an offline way, e.g. in [24, 30].

The lack of EDO studies in DTPs despite their popularity creates an important gap, which should be addressed if we want to apply EAs effectively to solving real-world DTPs. To make this important class of problems more accessible to the community, it is necessary to (i) clearly define/characterise DTPs and distinguish DTPs from DOPs and other types of time-dependent problems; (ii) study the characteristics of DTPs, the suitability of EAs in solving DTPs; and (iii) based on what we

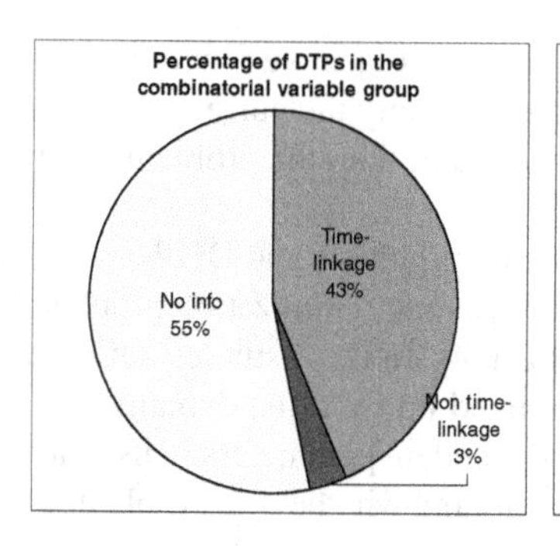

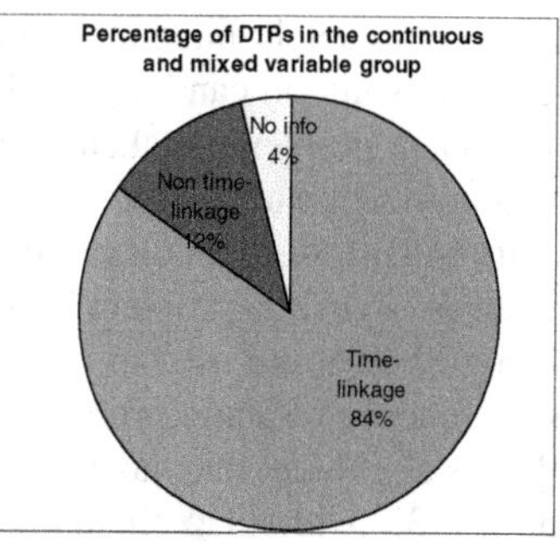

Fig. 16.1 Percentage of problems with the time-linkage properties among the 56 surveyed applications in [19]

learn from the characteristics of DTPs, develop new EA techniques to solve DTPs more effectively.

In the next three sections of this chapter, we will outline our attempts to accomplish the three tasks above to help closing the gaps between academic research and real-world DTPs.

16.2 A Framework for Defining DOPs and DTPs

This section describes a formal definition framework with a view to defining/characterising, and distinguishing DOPs and DTPs. We will firstly discuss the current research gaps in defining/characterising DOPs and DTPs, and then propose a definition framework to help closing the gaps.

16.2.1 Research Gaps and Motivations

This subsection discusses two gaps in defining DOPs and DTPs: (i) the lack of a formal definition to distinguish DOPs/DTPs and other time-dependent problems; and (ii) the lack of a formal definition to fully represent DTPs.

16.2.1.1 Dynamic Problems and Dynamic Optimization Problems

It is necessary to distinguish between *dynamic problems* (also called dynamic environments or time-dependent problems), which are problems that change over time, and *dynamic optimization problems* (DOPs), which belong to a special class of dynamic problems that are *solved online by an optimization algorithm as time goes by*. Of these two types of problems, only DOPs are of interest to evolutionary dynamic optimization (EDO). This is because, no matter how the problem changes, from the perspective of an EA or an optimization algorithm in general, a time-dependent problem is only different from a static problem if it is solved in a dynamic way, i.e. new solutions are produced to react to changes as time goes by. Otherwise, if future changes can be completely integrated into a static objective function, or if a single robust-to-changes solution can be provided, or if only the current static instance of the time-dependent problem is taken into account, then the problem can be solved using static optimization techniques.

In spite of this difference between dynamic problems and DOPs, in many EDO studies the terms "dynamic problem" and "dynamic optimization problem" are not clearly distinguished or are used interchangeably. In these studies, DOPs are either defined as a sequence of static problems linked up by some dynamic rules [3, 22, 23, 31, 32] or as a problem that have time-dependent parameters in its mathematical expression [5, 7, 33] without explicitly mentioning whether the problems are solved online or not.

In definitions like those cited above, although the authors may assume that the problems are solved online by the algorithm as time goes by (as mentioned by the authors elsewhere or as shown by the way their algorithms solve the problems), this assumption was not captured explicitly in the definitions. As a result, although these definitions can be used to effectively represent time-dependent problems, they do not clearly show whether a time-dependent problem is different from a static problem from the perspective of an optimization algorithm and hence might not be able to clearly distinguish a DOP from the general time-dependent problems.

It might be better to explicitly state that, from the perspective of an optimization algorithm, a time-dependent problem is a DOP only if it is solved online by the algorithm when time goes by in the definitions of DOPs, i.e. a time-dependent problem is a DOP only if it is solved using dynamic optimization techniques. Such explicit descriptions, however, are lacking in most current EDO formal definitions. In this chapter a formal definition framework will be introduced in an attempt to close this gap. This definition is an extended version of a previous study [20].

16.2.1.2 Dynamic Time-Linkage Problems

A DTP is firstly a DOP, hence it also has all the characteristics of a regular DOP. The additional feature of a DTP, which makes it different from normal DOPs, is that the dynamic of a parameter may depend not only on the time variable, but also on earlier decisions made by the algorithm. It means that at the current time t^{now} the value of the parameter $\gamma(t^{now})$ of a function f may depend on the value of the variable $x(t), 0 \leq t \leq t^{now}$ found by the algorithm at at least one point before t^{now}.

Due to the lack of EDO studies on DTPs, so far there has been only one formal definition for DOPs that mentioned DTPs [6] [7]. Equation 16.1 shows this definition for DOPs (including DTPs) with the time variable $t \in \mathbb{T} = \left[0, t^{end}\right], t^{end} > 0$.

$$\begin{aligned}
&\max\left\{F_\gamma(x(t))\right\} \text{ subject to } C_\gamma(x(t)) = feasible \text{ with} \\
&F_\gamma(x(t)) = \int_0^{t^{end}} f_{\gamma(t)}(x(t))\,dt \\
&C_\gamma(x(t)) = \begin{cases} feasible & \text{if } \forall t \in \left[0, t^{end}\right] : C_{\gamma(t)}(x(t)) = feasible \\ infeasible & \text{otherwise} \end{cases}
\end{aligned} \tag{16.1}$$

where γ are the time-dependent parameters of f and C is the constraint.

The definition above might not be detailed enough if we want to characterise one important property of DTPs: *algorithm-dependent*. We consider DTPs algorithm-dependent because the structure of a DTP in the future may depend on the current value of $x(t)$, which in turn depends on the algorithm used to solve the problem. At a particular change step t, different algorithms might provide different solutions $x(t)$, hence changing the future problem in different ways. Because of this property, we believe that in order to define a DTP in an unambiguous way, the algorithm used to solve a problem instance should be considered a part of the problem instance itself. The original definition in Eq. 16.1 does not fully encapsulate this property.

Another reason for us to formulate an extended definition is that in (Eq. 16.1) the time-linkage feature is not explicitly expressed. Instead, this feature is encapsulated in the expression of $f_{\gamma(t)}$. It would be better if the time-linkage property can be captured explicitly in the definition. This has been partially done in [7] and here we will extend that concept further by including previous solutions that affect future function values in the definition.

The definition in Eq. 16.1 also does not explicitly distinguish DOPs from general time-dependent problems. It would be better to provide a new definition that is able to distinguish DTPs from DOPs, and distinguish DOPs from other time-dependent problems.

16.2.2 A Definition Framework

To contribute in closing the gaps above, in this chapter we will propose a new definition framework which describes DTPs and DOPs in a more detailed level. It is hoped that the framework will help in defining and characterising DTPs and DOPs better and can be used as a basis for future theoretical works. The definition framework can also help in generating benchmark problems that are able to capture the characteristics of DTPs and DOPs, as will be described in the next section. Within this chapter we will focus on the single-objective case only. Details of the definition framework will be described below.

Definition 16.1 (Full-description form).
Given a finite set of functions $F = \{f_1(x), ..., f_n(x)\}$; a *full-description form* of F is a tuple

$$\left\langle \widehat{f}_\gamma(x), \{\mathbf{c}_1, ..., \mathbf{c}_n\} \right\rangle$$

where $\widehat{f}_\gamma(x)$ is a mathematical expression with its set of parameters $\gamma \in \mathbb{R}^m$, and $\{\mathbf{c}_1, ..., \mathbf{c}_n\}, \mathbf{c}_i \in \mathbb{R}^m$ is a set of vectors; so that:

$$\begin{aligned} &\widehat{f}_\gamma(x) \overset{\gamma=\mathbf{c}_1}{\rightarrow} f_1(x) \qquad (16.2) \\ &\cdots \\ &\widehat{f}_\gamma(x) \overset{\gamma=\mathbf{c}_n}{\rightarrow} f_n(x) \end{aligned}$$

Each function $f_i(x), i = 1 : n \in N^+$ is called an *instance of the full-description form* at $\gamma = \mathbf{c}_i$. From now on we will refer to the full-description form $\left\langle \widehat{f_\gamma}(x), \{\mathbf{c}_1, ..., \mathbf{c}_n\} \right\rangle$ as $\widehat{f}$.

Example 16.1. The combination of the expression $\widehat{f_\gamma} = ax + b$ with $\gamma = \{a, b\}$ and the following set of parameter values:
$\{\{a = 1, b = 0\}, \{a = 0, b = 1\}, \{a = 1, b = 1\}\}$ is the full-description form of the following set of functions: $\{f_1 = x; f_2 = 1; f_3 = x + 1\}$ because

$$\begin{aligned} \widehat{f_\gamma} &= ax + b \overset{a=1,b=0}{\rightarrow} f_1 = x \\ \widehat{f_\gamma} &= ax + b \overset{a=0,b=1}{\rightarrow} f_2 = 1 \\ \widehat{f_\gamma} &= ax + b \overset{a=1,b=1}{\rightarrow} f_3 = x + 1 \end{aligned} \tag{16.3}$$

The implication of a full-description form is that it can be used to represent different functions at different times by changing the parameters. It should be noted that, however, a full-description form is not unique: one set of functions can be represented by multiple full-description forms and one full-description form can be used to represent multiple set of functions. What is unique is a combination of (a) a full-description form $\widehat{f}$; (b) a given set of functions $\{f_1(x), ..., f_n(x)\}$ represented by $\widehat{f}$; and (c) the way the parameters of $\widehat{f}$ can be changed to transform f_i to f_j $\forall i, j = 1 : n$.

In real-world problems, changes in the parameters are usually controlled by some specific time-dependent rules or functions. For example, in dynamical systems changes of parameters can be represented by a linear, chaotic or other nonlinear equations of the time variable t. The dynamic rules, which govern how the parameters of a full-description form will change, can be defined mathematically as follows.

Definition 16.2 (Dynamic driver). Given a tuple $\left\langle \widehat{f}, \gamma_t, t \right\rangle$ where t is a time variable, $\widehat{f}$ is a full-description form of the set of functions $F = \{f_1(x), ..., f_n(x)\}$ with respect to the set of m-element vectors $\{\mathbf{c}_1, ..., \mathbf{c}_n\}, \mathbf{c}_i \in \mathbb{R}^m$, and $\gamma_t \in \mathbb{R}^m$ is an m-element vector containing all m parameters of $\widehat{f}$ at the time t;
we call a mapping $D(\gamma_t, t) : \mathbb{R}^m \times \mathbb{N}^+ \longrightarrow \mathbb{R}^m$ a *dynamic driver* of $\widehat{f}$ if

$$\gamma_{t+1} = D(\gamma_t, t) \in \{\mathbf{c}_1, ..., \mathbf{c}_n\} \forall t \in \mathbb{N}^+ \tag{16.4}$$

and

$$\gamma_{t+1} \text{ is used as the set of parameters of } \widehat{f} \text{ at the time } t + 1.$$

Definition 16.3 (Time-dependent problem). Given a tuple $\left\langle \widehat{f}, D(\gamma_t, t) \right\rangle$ where t is a time variable, $\widehat{f}$ is a full-description form of the set of functions $F = \{f_1(x), ..., f_n(x)\}$ with respect to the set of m-element vectors $\{\mathbf{c}_1, ..., \mathbf{c}_n\}, \mathbf{c}_i \in \mathbb{R}^m$, $\gamma_t \in \mathbb{R}^m$ is the parameter-vector of $\widehat{f}$ at the time t, and $D(\gamma_t, t)$ is a dynamic driver of $\widehat{f}$;

we call $\widehat{f}_{D(\gamma_t)} = \left\langle \widehat{f}, D(\gamma_t, t) \right\rangle$ a *time-dependent problem* with respect to the time variable t. In this problem changes can be represented as changes in the parameters and are controlled by $D(\gamma_t, t)$.

The inclusion of dynamic drivers and full-description form helps in distinguishing time-dependent problems. As discussed earlier, many existing definitions represent a time-dependent problem as a sequence of multiple static problems. These representations might be ambiguous because there might be multiple ways to transform one static problem to another and hence it is not clear what type of dynamic the considered time-dependent problem has. The dynamic driver in Definition 16.3 represents the actual dynamic of the problem and hence it helps distinguish one time-dependent problem from another.

In some existing definitions [7, 22], it has already been implied that changes in time-dependent problems can be represented as changes in the parameter space. In some other studies[10, 27], changes in time-dependent problems were considered as a "time function", which can be separated from the "structure function", i.e. the structure of the objective function. In this chapter these concepts will be formulated in a more detailed level and will be explicitly defined: *most common types of changes in time-dependent problems can be represented as changes in the parameter space if we can formulate the problem in a general enough full-description form.* This is true even in extreme cases where there is no correlation between the functions before and after a change. For example, a function-switching change from $f(x)$ at $t = 0$ to $g(x)$ at $t \geq 1$, $t \in \mathbb{N}^+$ can be expressed as $\widehat{f}(x) = a(t) f(x) + b(t) g(x)$ where $a(t)$ and $b(t)$ are two time-dependent parameters given by

$$\begin{cases} a(t) = 1 \text{ and } b(t) = 0 \text{ if } t = 0 \\ a(t) = 0 \text{ and } b(t) = 1 \text{ otherwise.} \end{cases}$$

Dimensional changes, as found in some real-world systems, can also be represented as changes in the parameter, given that the maximum number of variables is taken into account in the full-description form. For example, the function $\sum_{i=1}^{n} x_i^2$ with dimension n varies from 1 to 2 can be represented as the full-description form $\sum_{i=1}^{2} b_i(t) x_i^2$ with $b_i(t) \in \{0, 1\}$ depending on t.

Definition 16.4 (Time unit). When a time-dependent problem is being solved, a *time unit*, or a unit for measuring time periods in the problem, represents the time durations needed to complete *one function evaluation* of that problem.[1] The number of evaluations (or time units) that have been evaluated so far since we started solving the problem is measured by the variable $\tau \in N^+$.

Definition 16.5 (Change step and frequency of change). When a time-dependent problem is being solved, a *change step* represents the moment when the problem changes. The number of change steps that have occurred so far in a time-dependent

[1] As mentioned in [5] and [22], from the perspective of optimization algorithms time is discrete and the smallest time unit is one function evaluation.

problem is measured by the variable $t \in N^+$. Obviously t is a time-dependent function of τ- the number of evaluations made so far since we started solving the problem; $t(\tau) : \mathbb{N}^+ \longrightarrow \mathbb{N}^+$. Its dynamic is controlled by a problem-specific *time-based dynamic driver*:

$$t(\tau+1) = D_T(t(\tau), \tau), \tag{16.5}$$

where $D_T(t(\tau), \tau)$ is the problem-specific time-based dynamic driver. It decides the *frequency of change* of the problem and can be described as follows:

$$\begin{cases} D_T(t(\tau), \tau) = t(\tau) + 1 & \text{when a change occurs} \\ D_T(t(\tau), \tau) = t(\tau) & \text{otherwise} . \end{cases} \tag{16.6}$$

$D_T(t(\tau), \tau)$ is responsible for mapping the actual wall-clock duration of a change step to the corresponding number of function evaluations.

Definition 16.6 (optimization algorithms and dynamic solutions).
Given a time-dependent problem $\widehat{f}_{D(\gamma_t)} = \left\langle \widehat{f}, D(\gamma_t, t) \right\rangle$ at the change step t (see Definition 16.3) and a set P_t of k_t solutions $\mathbf{x}_1, ..., \mathbf{x}_{k_t} \in S_t$ where $S_t \subseteq \mathbb{R}^d$ is the search space[2],
an *optimization algorithm* G to solve $\widehat{f}_{D(\gamma_t)}$ can be seen as a mapping $G_t : \mathbb{R}^{d \times k_t} \to \mathbb{R}^{d \times k_{t+1}}$ capable of producing *a solution set* P_{t+1} of k_{t+1} optimized solutions $\mathbf{x}_1^G, ..., \mathbf{x}_{k_{t+1}}^G$ at the next change step $t+1$:

$$P_{t+1} = G_t(P_t) . \tag{16.7}$$

Generally, at a change step $t^e \in \mathbb{N}^+$ the set of *dynamic solutions* $X_{f_t}^{G_{[t^b, t^e]}}$ that we get by applying an algorithm G to solve $\widehat{f}_{D(\gamma_t)}$ with a given initial population $P_{t^b - 1}$ during the period $[t^b, t^e]$, $t^b \geq 1$ is given by:

$$X_{f_t}^{G_{[t^b, t^e]}} = \bigcup_{t=t^b}^{t^e} P_t = \bigcup_{t=t^b}^{t^e} G_t(P_{t-1}) . \tag{16.8}$$

In real-world time-dependent problems, some time-dependent rules that change the problems' parameters may have the *time-linkage* feature, i.e. they take solutions found by the algorithm up to the current time step as their parameters. In such cases, the time-linkage dynamic rules can be defined mathematically as follows.

Definition 16.7 (Time-linkage dynamic driver). Given a tuple $\left\langle \widehat{f}, \gamma_t, t, X_{\widehat{f}}^{G_{[1,t]}} \right\rangle$ where t is a time variable, $\widehat{f}$ is a full-description form of the set of functions $F = \{f_1(x), ..., f_n(x)\}$ with respect to the set of m-element vectors $\{\mathbf{c}_1, ..., \mathbf{c}_n\}, \mathbf{c}_i \in \mathbb{R}^m$, $\gamma_t \in \mathbb{R}^m$ is an m-element vector containing all m parameters of $\widehat{f}$ at the time t, ; and

[2] Here we are considering search spaces $\subseteq \mathbb{R}^d$. However, the definition can be generalized for other non-numerical encoding algorithms by replacing $\mathbb{R}^d$ with the appropriate encoding space.

$X_{\widehat{f}}^{G[1,t]}$ is a set of k d-dimensional solutions achieved by applying an algorithm G to solve $\widehat{f}$ during the period $[1,t]$;
we call a mapping $D\left(\gamma_t, X_{\widehat{f}}^{G[1,t]}, t\right) : \mathbb{R}^m \times \mathbb{R}^{d\times k} \times \mathbb{N}^+ \longrightarrow \mathbb{R}^m$ a *time-linkage dynamic driver* of $\widehat{f}$ if

$$\gamma_{t+1} = D\left(\gamma_t, X_{\widehat{f}}^{G[1,t]}, t\right) \in \{\mathbf{c}_1, ..., \mathbf{c}_n\} \forall t \in \mathbb{N}^+ \tag{16.9}$$

and γ_{t+1} is used as the set of parameters of $\widehat{f}$ at the time $t+1$.

When $X_{\widehat{f}}^{G[1,t]}$ does not have any influence on the future of $\widehat{f}$, $D\left(\gamma_t, X_{\widehat{f}}^{G[1,t]}, t\right)$ becomes a regular dynamic driver with no time-linkage feature.

Definition 16.8 (Dynamic optimization problem and dynamic time-linkage problem). Given a tuple $\left\langle \widehat{f}, \widehat{C}, D_P, D_D, D_T, G\right\rangle$;
a *dynamic optimization problem* in the period $\left[1, \tau^{end}\right]$ function evaluations, $\tau^{end} \in \mathbb{N}^+$ can be defined as

$$\text{optimise} \left\{ \sum_{\tau=1}^{\tau^{end}} \widehat{f}_{\gamma\left(t_\tau, X_{\widehat{f}}^{G[1,t]}\right)}(\mathbf{x}_t) \right\} \tag{16.10}$$

subject to $\widehat{C}^{i=1:k\in N^+}_{\gamma\left(t_\tau, X_{\widehat{f}}^{G[1,t]}\right)}(\mathbf{x}_t, t_\tau) \le 0$; and $\mathbf{l}\left(t_\tau, X_{\widehat{f}}^{G[1,t]}\right) \le \mathbf{x} \le \mathbf{u}\left(t_\tau, X_{\widehat{f}}^{G[1,t]}\right)$ where

$\widehat{f}$: full-description form of the objective function
$\widehat{C}^1 ... \widehat{C}^k$: full-description forms of k dynamic constraints[3]
D_P : dynamic driver for objective and constraint parameters (see below)
D_D : dynamic driver for domain constraints (see below)
D_T : dynamic driver for times and frequency of changes (Eq. 16.6)
G : algorithm used to solve the problem
$\tau \in \left[1, \tau^{end}\right] \cap \mathbb{N}$: number of function evaluations done so far
t_τ, or $t(\tau) \in \mathbb{N}^+$: current change step; it is controlled by D_T (Eq. 16.6)
$X_{\widehat{f}}^{G[1,t]}$: set of solutions achieved by applying G to solve $\widehat{f}$ during $[1,t]$
$\gamma_{t_\tau} \in \mathbb{R}^p$: time-dependent parameters of $\widehat{f}$ and $\widehat{C}^i$; $\gamma_{t_\tau+1} = D_P\left(\gamma_{t_\tau}, X_{\widehat{f}}^{G[1,t]}, t\right)$
$\mathbf{l}(t_\tau), \mathbf{u}(t_\tau) \in \mathbb{R}^n$ are domain constraints; $\begin{cases} \mathbf{l}(t_\tau+1) = D_D\left(\mathbf{l}(t_\tau), X_{\widehat{f}}^{G[1,t]}, t_\tau\right) \\ \mathbf{u}(t_\tau+1) = D_D\left(\mathbf{u}(t_\tau), X_{\widehat{f}}^{G[1,t]}, t_\tau\right) \end{cases}$

If $X_{\widehat{f}}^{G[1,t]}$ has any influence on the future behaviour of this dynamic optimization problem, or in other words if any of the dynamic drivers D_P, D_D, D_T is a time-linkage dynamic driver, then the problem becomes a *dynamic time-linkage problem*. □

The new definition brings us some advantages. First, with the introduction of the *change step*, the *optimization algorithm* and the *dynamic solutions* produced by the algorithm at each change step, the definition clearly defines DOPs/DTPs as *time-dependent problems that are solved online in a dynamic way*, and hence distinguishes DOPs/DTPs from other time-dependent problems. Second, we can now classify DOPs based on three distinguished components: the *full-description forms*, the *dynamic drivers*, and the *algorithm*. This separation facilitates us in characterising DOPs and evaluating the impact of each component on the difficulty of the problems. Third, the definition supports an important feature of DTPs that has not been fully considered before: *algorithm-dependent*. Fourth, the definition encapsulates different aspects of DOPs/DTPs such as dynamic rules, change frequencies, changes in constraints, changes in domain range, changes in objective functions in a more detailed level. Fifth, the definition framework also facilitates generating benchmark DTPs, as will be shown in the next section.

16.3 The Prediction-Deceptive Problem in DTPs

In this section we will study some interesting characteristics of DTPs and the possibility to solve this class of problems. Specifically, although it is believed that DTPs can be solved to optimality with a perfect prediction method to predict future function values [6, 8], in this section we will discuss a new class of DTPs where it might not be possible to solve the time-linkage problems to optimality because there is not always the possibility to perfectly predict the future. In addition, in this type of DTPs if we try to predict the future based on information from the past, we may even get worse results than not using a predictor at all. We will then carry out some experiments to verify the finding and will also discuss under which situation can we solve this particular type of DTPs.

16.3.1 Time-Deceptive and the Anticipation Approach

According to [7], a dynamic problem is said to be time-deceptive toward an optimizer if the problem is time-linkage and the optimiser cannot efficiently take into account this time-linkage feature during its optimization process.

Bosman[6] illustrates this property by proposing the following test problem:

$$\text{given } n = 1; h(x) = e^x - 1; \ \max_{x(t)} \left\{ \int_0^{t^{end}} f\left(x\left(t\right), t\right) dt \right\} \tag{16.11}$$

$$f(x_t, t) = \begin{cases} -\sum_{i=1}^{n} \left(x\left(t\right)_i - t\right)^2 & \text{if } 0 \le t < 1 \\ -\sum_{i=1}^{n} \left[\left(x\left(t\right)_i - t\right)^2 + h\left(\left|x\left(t-1\right)_i\right|\right)\right] & \text{otherwise,} \end{cases}$$

The test problem above is a DTP because for any $t \geq 1$, the current value of $f\left(x, t\right)$ depends on $x\left(t-1\right)$ found at the previous time step.

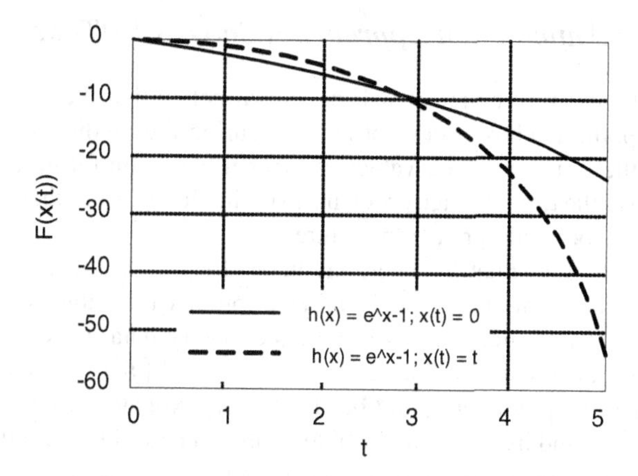

Fig. 16.2 This figure (reproduced from [6]) illustrates the time-deceptive property. We can see that the trajectory of $f(x_t)$ when we optimize the present (dash line, with optimum solution $x(t) = t$) is actually worse than the trajectory of $f(x_t)$ with a simple solution $x(t) = 0$ (the solid line). To solve this problem to optimality, we need to use a predictor to predict the trajectory of function values given different outcomes of current solutions, then choose the one that gives us the maximum profit in the future.

An interesting property is revealed when we try to optimise the above problem using the traditional approach: optimising the present. That property is: *the trajectory formed by optimum solutions at each time step might not be the optimal trajectory.* For example, in Figure 16.2 we can see that the trajectory of $f(x^*,t)$ when we optimise the present (with optimum solution $\mathbf{x}^*(t) = t$ at the time step t) is actually worse than the trajectory of a $f(x^0,t)$ with a simple solution $\mathbf{x}^0 = 0 \; \forall t$. It means that the problem is deceptive because an optimiser following the traditional approach is not able to take into account the time-linkage feature.

Bosman [6, 7] suggested that DTPs can be solved to optimality by estimating the values of the function for future times given a trajectory $\cup_{t=0}^{t^{now}} \{f_t, t\}$ of history data and other previously evaluated solutions. From that estimation, we can choose a future trajectory with optimal future function values. In other words, it is suggested that time-linkage problems can be "*solved to optimality*" by prediction methods and the result could be "*arbitrarily good*" if we have a "*perfect predictor*"[6–8][4]. The authors also made some experiments on the test problem mentioned in Eq. 16.11 and on the dynamic pickup problem, showing that under certain circumstances prediction methods do help to improve the performance of the tested algorithms.

[4] A predictor, as defined in [7, line 8-12, pg 139], is "*a learning algorithm that approximates either the optimization function directly or several of its parameters... When called upon, the predictor returns either the predicted function value directly or predicted values for parameters*". Hence, perfect predictors should be ones that can predict values exactly as the targets.

16.3.2 Can Anticipation Approaches Solve All DTPs?

Contrary to the existing belief, it will be shown below that there might be cases where the hypothesis above does not hold: if during the predicted time span, the trajectory of the future function values changes its function form, it might not be possible to solve the time-linkage problems to optimality because there is not always the possibility to perfectly predict the future.

Let us consider the situation where predictors help in achieving optimal results first. At the current time $t^{now} \geq 1$, in order to predict the values of $f(x(t))$ at a future time t^{pred}, a predictor needs to take the history data, for example the previous trajectory of function values $Z^{[0,t^{now}-1]} = \bigcup_{t=0}^{t^{now}-1} \{f_t, t\}$, as its input. Given that input, a perfect predictor would be able to approximate correctly the function form of $Z^{[0,t^{now}-1]}$ and hence would be able to predict precisely the future trajectory $Z^{[t^{now},t^{pred}]}$ if it has the same function form as $Z^{[0,t^{now}-1]}$. One example where predictors work is the problem in Eq. 16.11. In this problem, for each trajectory of $x(t)$ the trajectory of $f(x(t))$ always remains the same. For example with $x(t) = t$, the trajectory is always $1 - e^{t-1}$ or with $x(t) = 0$ the trajectory is always $-t^2$ (see Figure 16.2). As a result, that problem is predictable.

Now let us consider a different situation. If at any particular time step $t^s \in [t^{now}, t^{pred}]$, the function form of $Z^{[t^{now},t^{pred}]}$ changes, the predicted trajectory made at t^{now} to predict $f(x(t))$ at t^{pred} is no longer correct. This is because before t^s there is no information about how the the function form of $Z^{[t^{now},t^{pred}]}$ would change. Without such information, it is impossible to predict the optimal trajectory of function values after the switch, regardless of how good the predictor is. It means that the problem cannot be solved to optimality because it is not possible to perfectly predict the future.

To illustrate this situation, let us consider the following simple problem where the trajectory of function values changes over time (illustrated in Figure 16.3).

$$\widehat{F}(x_t) = a_t f(x_t) + b_t g(x_t) + c_t h(x_t) \quad 0 \leq x_t \leq 1, \tag{16.12}$$

where $\widehat{F}(x)$ is the full-description form[5] of a dynamic function; $f(x_t) = x_t$; $g(x_t) = x_t + (d-2)$; $h(x_t) = x_t + d$; a_t, b_t and c_t are the time-dependent parameters of $\widehat{F}(x_t)$. Their dynamic drivers are set out as follows:

$$\begin{cases} a_t = 1; b_t = c_t = 0 \quad \text{if } (t < t^s) \\ a_t = 0; b_t = 1; c_t = 0 \text{ if } (t \geq t^s) \text{ and } \left(\widehat{F}_{t^s-1}\left(x^G_{t^s-1}\right) \geq 1\right), \\ a_t = 0; b_t = 0; c_t = 1 \text{ if } (t \geq t^s) \text{ and } \left(\widehat{F}_{t^s-1}\left(x^G_{t^s-1}\right) < 1\right) \end{cases} \tag{16.13}$$

where $t^s > 1$ is a pre-defined time step, $d \in \mathbb{R}$ is a pre-defined constant, and $x^G_{t^s-1}$ is a single solution produced at $t^s - 1$ by an algorithm G. Eq. 16.13 means that with

[5] The concepts like *full-description forms*, *time-dependent parameters* and *dynamic drivers* have been defined and described in Section 16.2.

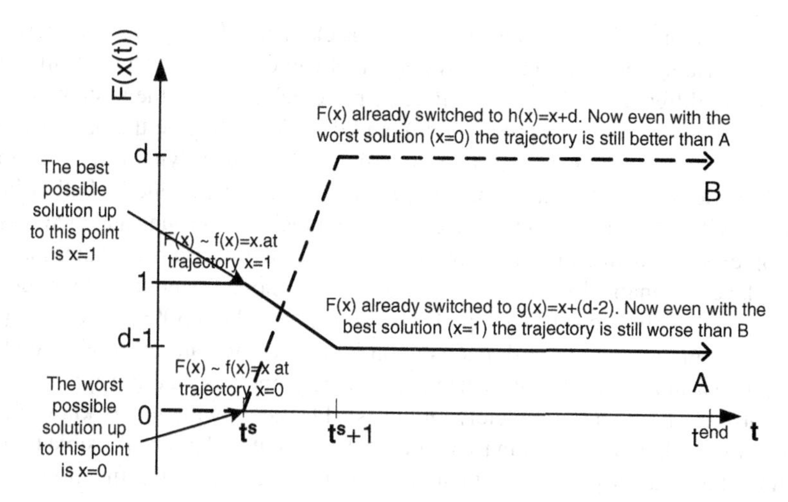

Fig. 16.3 This figure illustrates a situation where even the best predictor + the best algorithm (A) still perform worse than the worst predictor + the worst algorithm (B) due to the prediction-deceptive property of the problem in eq.16.12. Assume that we want to predict the trajectory of $F(x)$ from $[0, t^{end}]$. In case A, the best predictor allows us to predict $F(x) \sim f(x) = x$ in just only one time step $[0, 1]$. With that perfect prediction the algorithm is able to find the best solution $x = 1$, which is valid until t^s. Now at t^s although the history data tells the predictor that the trajectory must still be $F(x) \sim f(x) = x$, according to eq.16.13 the actual $F(x)$ does switch to $g(x) = x + (d-2)$, which is the worst trajectory. In other words, the best predictor chose the worst trajectory to follow. On the contrary, in the case B the worst predictor+worst algorithm actually get benefit from the switch: the terrible solution $(x = 0)$ they found during $[0, t^s]$ does help them to switch to $F(x) \sim h(x) = d + x$, whose trajectory after t^s is always better than A regardless of the value of x.

$t < t^s$, the form of $\widehat{F}(x_t)$ is always equal to $f(x_t)$; with $t \geq t^s$, depending on the solution of $x^G_{t^s-1}$ the form of $\widehat{F}(x_t)$ would switch to either $g(x_t)$ or $h(x_t)$.

In the above problem, because at any $t \geq t^s$ the values of a_t, b_t and c_t (and consequently the value of the function $\widehat{F}$) depend on the solution found by G at $t^s - 1$, according to the definition in [7] the problem is considered time-linkage.

This problem has a special property: at any $t < t^s$ one can only predict the value of $\widehat{F}$ up to $t^s - 1$. Before t^s, history data does not reveal any clue about the switching rule in Eq. 16.13, hence it is impossible to predict (i) whether the function will switch at t^s; (ii) which value $x^G_{t^s-1}$ should get to switch $\widehat{F}(x_t)$ to $g(x_t)$ / $h(x_t)$ and (iii) which form, g or h, would provide better future trajectory.

Even worse, even a predictor that can perfectly learn the current function form of the system might still be deceived to provide worse result than not using any predictor while solving this time-linkage problem! Figure 16.3 illustrates the situations where the best predictor could provide the worst result while the worst predictor could provide better results after t^s!

Problems like this example, i.e. time-linkage problems with function forms switching from one to another, are very common in real-world systems. One common class of problems with this property is the class of hybrid systems. According to [26], hybrid systems are real-life systems that can evolve according to different dynamics at different times. At each time step the behaviour of the system is controlled by only one dynamics (one mode), and then depending on the behaviour of the system, at some point the system may switch from one dynamics to another (switch mode). Examples vary from simple systems like the bouncing ball (where the state switches from falling to bouncing when it meets the ground and vice versa) to complex systems like the autopilot in programmes commercial airplanes (where the airplane automatically switches from one flying mode to another). Our survey of real-world applications [19, chap. 3] also shows that about 30% of the surveyed applications in the continuous domain or continuous+combinatorial domain are hybrid systems, and all of them have the time-linkage properties. In these applications, if we solve the problems completely *online* as a black-box without any knowledge, we will not be able to solve them to optimality because it will not be possible to predict how the systems will switch their function forms and how the function forms after the switch will be.

Summarising, the example problem proposed in this section illustrates a common class of DTPs (but has not been studied by the EC community yet) where it is not guaranteed to get optimal results because it is impossible to find a perfect predictor to predict the function values using history data. We call this class *time-linkage problems with unpredictable optimal function trajectories*. The example illustrates a special case where any predictor that relies on past data can be deceived and hence provide the worse results than not using predictor at certain time steps. We call these types of problems the *prediction-deceptive time-linkage problems*.

In Section 16.3.4, some experiments will be carried out to demonstrate a prediction-deceptive time-linkage problem and its effect on the performance of an algorithm that predicts the future function values based on history data.

16.3.3 Solving Prediction-Deceptive Time-Linkage Problems

Prediction-deceptive DTPs are challenging and only under limited circumstances can we solve them to optimality. The answer of whether we can solve them to optimality or at least to avoid being deceived would depend on whether we have to solve them totally online or partially online, and whether do we have to solve the problem as a complete black box or can we get any problem-specific information.

If we have to solve the problem online as a black box, there is not much thing that we can do. Knowing that the problem is prediction-deceptive, we might try not to use anticipation approaches to avoid being deceived. However, there is no guarantee that other approaches would work better.

In real-world applications, however, it might be possible to solve the problem in a partially online way and also there might be some problem-specific information available so that the problem can be solved as a partial black-box. Our survey of real-world applications in [19] shows that in most of the surveyed hybrid systems, the problems are not totally black box because the mathematical function forms of the possible switch-modes and the switching rules have already been calculated offline based on observation data from real systems or from simulation, e.g. see [1, 11, 12, 15]. However, because there are modelling errors or disturbances, these mathematical function forms might not exactly reflect the current status of the actual systems. Because of that, the problems still need to be solved online. During the online phase the actual function form of the system will be learned/predicted based on history data to "correct" any mis-modelling due to errors/disturbances.

In time-linkage problems with function forms switching from one to another and with the knowledge about switching rules like these, it might still be possible to solve them using prediction method while avoid being deceived. In order to do that, the solver needs to take into account not only the current function value and the future consequent values of the current function forms, but also the consequent function switches and the future values of the new function after a switch has been made.

Specifically, given a time-linkage problem with switching function forms and the knowledge of the switching rules, in order to solve the problem to optimality during the period $\left[t^{now}, t^{end}\right]$, at the current moment t^{now} an algorithm needs to find the solution $x\left(t^{now}\right)$ and a set of switching times $\{T_1, ..., T_{n-1}, T_n\}$ where $T_n = t^{end}$ to optimise the future trajectory and future switches:

$$\text{optimise} \tag{16.14}$$

$$\left\{ f\left(x\left(t^{now}\right)\right) + \sum_{t=t^{now}+1}^{T_1} f_{pred}\left(x(t)\right) + \sum_{T_i=T_1}^{T_{(n-1)}} \sum_{t=T_i+1}^{T_{(i+1)}} f_{switch}\left(x(t), x(T_i)\right) \right\}$$

where f_{pred} is the estimated function form of the current dynamic model of the system and f_{switch} is the expected function form of the dynamic model that the system will switch into under the estimated value of $x(T_i)$.

In summary, for time-linkage problems with switching function forms where the knowledge of the switching rules is available, it is possible to solve the problem more effectively if during the optimization process we take into account not only the current function value and the future consequent values of the current function forms, but also the consequent function switches and the future values of the new function after a switch has been made. In other words, it is possible to solve the problem more effectively if the algorithm optimises the problem using the objective function described in Equation 16.14.

16.3.4 Experimental Studies

In this section some experiments will be carried out to verify: (1) The impact of the time-deceptive property in time-linkage problems on optimization algorithms that follow the optimising-the-present approach; (2) The efficiency of the learning-the-current-function-form approach in solving time-deceptive time-linkage problems; (3) The impact of prediction-deceptive property in time-linkage problems on optimization algorithms that follow the learning-the-current-function-form approach; and (4) The efficiency of our proposed approach in solving prediction-deceptive time-linkage problems when information about switching rules is available

Points (1) and (2) have already been illustrated in [6, 7], but here the verification will be re-done again because these results will be needed for verifying points (3) and (4).

16.3.4.1 Test Problems

Problem DTP1

In [6], a test time-linkage problem with the time-deceptive property has been proposed. This problem will be used in this section to verify the points (1) and (2) above. The test problem has been described in Equation 16.11, page 380). In this subsection the problem (with $n = 1; h(x) = x^2$) is presented in a slightly different way to make it conform to our definition framework in Section 16.2 and make the change severity level adjustable:

$$\max_{x(t)} \left\{ \sum_{0}^{t_{end}} \widehat{F}(x_t) \right\} \tag{16.15}$$

where

$$\widehat{F}(x_t) = f^1 = \begin{cases} -\Sigma_{i=1}^{n} (x(t)_i - s.t)^2 & \text{if } 0 \le t < \lfloor 1/s \rfloor \\ -\Sigma_{i=1}^{n} \left[(x(t)_i - s.t)^2 + [x(t - \lfloor 1/s \rfloor)_i]^2 \right] & \text{otherwise} \end{cases}$$

and $s \in R$ is the change severity, $0 < s \le 1$.

The problem is named DTP1. Experiments on this problem will be presented in Subsection 16.3.4.3.

Problem DTP2

To verify points (3) and (4), we need to create a problem with the prediction-deceptive property. To maintain continuity and to re-use the results we got from the process of verifying points (1) and (2), the original Bosman problem in Equation 16.15 is modified to make it a prediction-deceptive problem. Particularly, up to the change step t^{switch} the problem is similar to DTP1, but at t^{switch} the problem switches its function form depending on the function value found by the algorithm at t^{switch}. If the found function value is high, the problem switches to a low-value trajectory. Vice versa, if the value found at t^{switch}, the problem switches to a high-value trajectory. Details of the problem are as follows:

$$\max_{x(t)} \left\{ \sum_{0}^{t_{end}} \widehat{F}(x_t) \right\}, \widehat{F}(x_t) = a_t f^1(x_t) + b_t f^2(x_t) + c_t f^3(x_t) + d_t f^4(x_t) \tag{16.16}$$

where $\widehat{F}(x)$ is the full-description form[6] of the mathematical descriptions f^1, f^2, f^3, f^4 (given in Equation 16.17); a_t, b_t, c_t, d_t are the time-dependent parameters of $\widehat{F}(x_t)$ (their dynamic drivers are given in Equation 16.18).

Below are the descriptions of f^1, f^2, f^3, and f^4 :

$$\begin{cases} f^1(x_t,t) = \begin{cases} -\sum_{i=1}^{n} (x(t)_i - s.t)^2 & \text{if } 0 \leq t < \lfloor 1/s \rfloor \\ -\sum_{i=1}^{n} \left[(x(t)_i - s.t)^2 + [x(t - \lfloor 1/s \rfloor)_i]^2 \right] & \text{otherwise} \end{cases} \\ f^2(x_t,t) = -60 \\ f^3(x_t,t) = -40 \\ f^4(x_t,t) = -10 \end{cases} \quad (16.17)$$

where $s \in R$ is the change severity, $0 < s \leq 1$.

Below are the descriptions of the dynamic drivers of the time-dependent parameters a_t, b_t, c_t, d_t:

$$\begin{cases} a_t = 1; b_t = c_t = d_t = 0 \quad \text{if } \left(t \leq t^{switch}\right) \\ a_t = 0; b_t = 1; c_t = d_t = 0 \text{ if } \left(t > t^{switch}\right) \text{ and } \left(-36 \leq \widehat{F}(x_{t^{switch}})\right) \\ a_t = b_t = 0; c_t = 1; d_t = 0 \text{ if } \left(t > t^{switch}\right) \text{ and } \left(-50 \leq \widehat{F}(x_{t^{switch}}) < -36\right) \\ a_t = b_t = 0; c_t = 0; d_t = 1 \text{ if } \left(t > t^{switch}\right) \text{ and } \left(\widehat{F}(x_{t^{switch}}) < -50\right) \end{cases} \quad (16.18)$$

where $t^{switch} > 1$ is a pre-defined change step, and $x_{t^{switch}}$ is a single solution produced at t^{switch} by the solver. Equation 16.18 means that with $t \leq t^{switch}$, the form of $\widehat{F}(x_t)$ is always equal to $f^1(x_t)$; with $t \geq t^{switch}$, depending on the solution of $x_{t^{switch}}$ the form of $\widehat{F}(x_t)$ would switch to either $f^2(x_t), f^3(x_t)$ or $f^4(x_t)$. In other words, Equation 16.18 defines the switching rule of the problem.

Equation 16.16 is a prediction-deceptive problem because it will deceive any good predictor to choose a high-value trajectory during the period $\left[0, t^{switch}\right]$. After the change step t^{switch}, however, such a high-value trajectory may lead the solver to the worst possible trajectory of $f^2(x_t,t) = -60$, which may eventually affect the total score of the solver and make a solver with predictor to perform worse than a solver without a predictor!

The problem is named DTP2. Experiments and discussions on this prediction-deceptive problem will be presented in Subsection 16.3.4.3.

16.3.4.2 Test Algorithms

To carry out the experiments, three different versions of GA are developed to represent the three different approaches in solving time-linkage problems: first, a standard GA (Algorithm 16.1, page 388) to represent the tradition *optimise-the-present* approach; second, a combination of GA + predictor (linear least-square regression) to represent the predict-the-future-based-on-history-data approach proposed in [7] (Al-

[6] The concepts like *full-description forms*, *time-dependent parameters* and *dynamic drivers* have been defined and described in Section 16.2.

gorithm 16.2, page 389); and third, a combination of GA + predictor + knowledge (about the switching rules) to represent the approach proposed in Section 16.3.3 (Algorithm 16.3, page 390). It should be noted that, for the purpose of simplicity Algorithm 16.3 was designed to solve only the cases where the switching rules are known and the switching time is also known (as found in the real-world applications in [24, 25]). In addition, we assume that the function to be estimated has a quadratic form. Of course in reality this assumption is not always true and it might be necessary to estimate the form of the function as well. In such case, powerful function approximation models like neural networks can be used to represent the function to be predicted. The simple assumption that the function form is quadratic was used because our purpose is not to propose a state-of-the-art or an efficient algorithm but just to show a proof of principle for the four points mentioned at the beginning of this section.

Algorithm 16.1. Standard GA

1. *initialization*
2. *Search*: for each generation
   a. Standard GA's crossover
   b. Standard GA's mutation
   c. Evaluation: For each individual $x(t_{now})$, evaluate $f(x(t_{now}))$
   d. Standard GA's selection

To create a fair testing environment, all three algorithms use the same set of parameters. Table 16.1 shows the detailed parameters of the algorithms and all other settings for the experiment.

To evaluate the performance of the algorithms, two measures are used. The first one is *performance plot* - the plot of the trajectory of the best function values that the algorithms achieved at each change step. The trajectory of the variable $\mathbf{x}$ as time goes by is also plotted to study the behaviours of the algorithms. The second measure is the total function values, which is calculated as the summation of the best function values taken after each 1/s change steps (1,000 change steps): $\text{totalVal} = \sum_{i=1}^{10} f\left(x\left(t^{begin} + \lfloor i/s \rfloor\right)\right)$. The first measure is a DOP standard metrics. The second measure is not a DOP standard metrics and is needed to evaluate how the time-linkage property affects the performance of the tested algorithms. Detailed experimental results are given in the next subsection.

16.3.4.3 Experimental Results

GA vs GA+Predictor in time-deceptive problems (DTP1)

Here we verify the suggestion of Bosman [7] that in time-deceptive DTPs, learning from the past to predict the future can be useful. Figure 16.4a, where the mean

Algorithm 16.2. GA + Predictor

List of parameters:
Pred: A linear least-square regression to approximate quadratic functions with 2 variables
s Change severity
h^{len} The length of the predicted future horizon

1. *initialization*
2. *Prediction*: After m generations, use the predictor *Pred* to estimate the current function form based on history data
   - Input:
     a. Solutions achieved in previous $1/s$ change steps: $\forall x(t), (t^{now} - \lfloor 1/s \rfloor) \le t \le t^{now}$.
     b. The previous $1/s$ change steps $t, (t^{now} - \lfloor 1/s \rfloor) \le t \le t^{now}$
     c. All corresponding function values $f(x(t))$.
   - Output: the estimated function form f_{pred}
3. *Search*: for each generation
   a. Standard GA's crossover
   b. Standard GA's mutation
   c. Evaluation: For each individual $x(t_{now})$, evaluate
   $$\text{Fitness}(x(t_{now})) = \left\{ f(x(t^{now})) + \sum_{t=t^{now}+1}^{t^{now}+h^{len}} f_{pred}(x(t)) \right\}$$
   d. Standard GA's selection

and standard deviation of function values of GA and GA+Predictor in the problem DTP1 are shown, confirms the advantage of this approach. The figure shows that although GA+Predictor has worse function values in the first few change stages, in the longer run it performs much better (has higher total values) than the traditional GA, which only focuses on optimising the present. The results confirm the advantage of maximising future values over just optimising the present in this particular problem.

GA vs GA+Predictor in prediction-deceptive problems (DTP2)

Predicting the future using data from the past, however, is not always beneficial in solving DTPs. In problems like the DTP2 where a high function value might switch the system to a low-value trajectory and vice versa, predicting future using data from the past might make the algorithm perform worse than not using a predictor. This behaviour is confirmed in the experiment. Figure 16.5a shows that GA+Predictor actually has lower total values than the GA without a predictor. This is due to that, since the eighth changing stage, the high-value trajectory that GA+Predictor predicted during the period $[0, t^{switch}]$ leads the algorithm to a worse trajectory than what GA achieves.

Algorithm 16.3. GA + Predictor + Knowledge about the switching rules

List of parameters:

Pred:	A linear least-square regression to approximate quadratic functions
s	Change severity
h^{len}	The length of the predicted future horizon
f_{switch}	Expected full-description form of the switching rules
$\{T_1,...,T_{n-1},T_n\}$	Set of switching times within current horizon $t^{now} < T_i \leq t^{now} + h^{len}$

1. *initialization*
2. *Prediction*: Same as step 2 in Algorithm 16.2.
3. *Search*: for each generation
   a. Standard GA's crossover
   b. Standard GA's mutation
   c. Evaluation: For each individual $x(t_{now})$,
      i. *Calculate current function value*: $A = f(x(t^{now}))$
      ii. *Calculate the expected future function/variable values until the first switching time*:
      $$B = \sum_{t=t^{now}+1}^{T_1} f_{pred}(x(t))$$
      iii. Estimate the variable $x(T_1)$ given the estimated outcome of f_{pred} during the period $[t^{now}+1, T_1]$
      iv. *Calculate the expected future values after the first switching time*:
      $$C = \sum_{T_i=T_1}^{T_{(n-1)}} \sum_{t=T_i+1}^{T_{(i+1)}} f_{switch}(x(t), x(T_i))$$
      v. *Calculate the fitness value of* $x(t_{now})$: Fitness $(x(t_{now})) = A + B + C$
      vi. *Update*: update the set of switching times for the next future horizon
   d. Standard GA's selection

Table 16.1 Test settings for GA, GA+Predictor and GA+Predictor+Knowledge.

Algorithm parameters	Pop size	25
	Elitism	No
	Selection method	Non-linear ranking
	Mutation method	Uniform, $P = 0.15$
	Crossover method	Arithmetic, $P = 0.8$
	Prediction method	Least-square regression for quadratic function
Test problem settings	Number of runs	50
	Change frequency	25 function evaluations (one generation)
	Change severity s	0.001
	Learning frequency	Every 10 generations
	Number of change steps	$11/s$ (11,000 change steps, $t^{end} = 11{,}000$)
	Predicted future horizon h^{len}	$5/s$
	Switching time	$8/s$

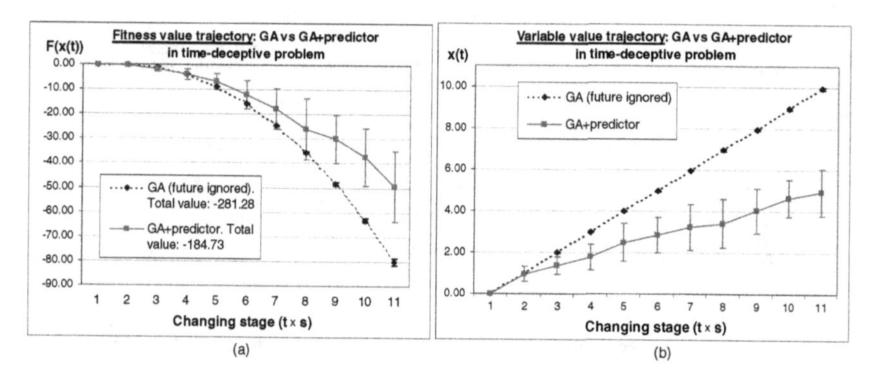

Fig. 16.4 Plots of the mean (and standard deviation) of highest function values over 50 runs: GA without predictor vs GA with predictor in a time-deceptive problem (DTP1). (a) Fitness values, (b) Variable values.

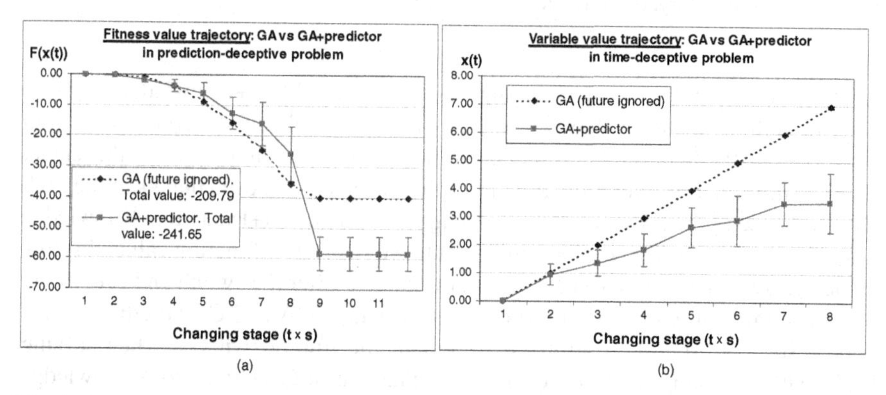

Fig. 16.5 Plots of the mean (and standard deviation) of highest function values over 50 runs: GA without predictor vs GA with predictor in the prediction-deceptive problem (DTP2). The switching time is at the 8th changing stage. (a) Fitness values, (b) Variable values.

GA vs GA+Predictor vs GA+Predictor+Knowledge in prediction-deceptive problems (DTP2)

In this subsection we verify the efficiency of our proposed approach described in Section 16.3.3, which suggests that the knowledge of the switching rules, if available, should be taken into account when anticipating the future. Figure 16.6a shows that the new approach does help improve the performance of the algorithm (GA+Predictor+Knowledge) and avoid being deceived into the wrong trajectories. As can be seen in Figure 16.6a, during the first six changing stages GA+Predictor+Knowledge follows exactly the same trajectory as GA+Predictor to maximise the function value trajectory in the period when the system has not switched to the other mode yet. However, from the sixth changing stage, GA+Predictor+Knowledge follows a different route from that of the original

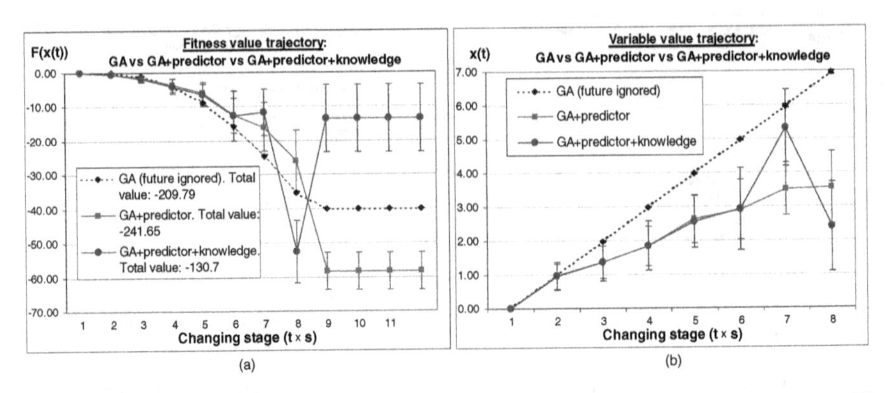

Fig. 16.6 Plots of the mean (and standard deviation) of the highest function values over 50 runs: GA without predictor vs GA+predictor vs GA+predictor+switching_knowledge in the prediction-deceptive problem DTP2. The switching time is at the 8th changing stage. (a) Fitness values trajectory, (b) Variable values trajectory.

GA+Predictor. At the sixth changing stage, GA+Predictor+Knowledge chose a slightly higher function value, which leads it to a completely different route from those of GA+Predictor and normal GA at the seventh changing stage. At this stage, the algorithms chose a very low function value, which is achieved thanks to the high value it chose in the previous changing stage. Although GA+Predictor+Knowledge has to sacrifice its current performance to achieve such a low function value, this low value helps the algorithm to reach a better trajectory after the switch and eventually it has a significantly higher total function values than GA and GA+Predictor. This good result confirms the usefulness of anticipating future switches when solving DTPs with switching function-forms. The behaviour of GA+Predictor+Knowledge in choosing the variables to achieve a high total function value is also shown in Figure 16.6b.

Another note is that when taking into account the future, the problem becomes more complex and it is getting more difficult to get high-precision results, as can be seen by looking at the standard deviations of the results in Figures 16.4, 16.5, 16.6. We can see that the traditional GA (future ignored) achieves very consistent results (standard deviations of the mean best values are almost zero) over 50 runs. However, when the algorithm has to predict the current function-form (GA+Predictor) and hence has to optimise not only the present but also the future, the problem becomes more complex and the standard deviations of the mean best values over 50 runs become higher. When the algorithm has to predict the current function-form *and* also has to anticipate any possible future mode-switching, the problem becomes even more complex and hence the level of inconsistency (standard deviation) increases even higher. This phenomenon shows the trade-off in taking into account the future when solving DTPs.

16.4 Conclusions

In this chapter we have firstly introduced a detailed definition framework to help characterising DTPs and DOPs, and to distinguish these two classes of problems from other types of time-dependent problems.

Then we have identified a challenging class of time-linkage problems where existing prediction approach might fail to find the optimal results. We named this class *prediction-deceptive time-linkage problems*.

An approach to solve this class of problems under certain circumstances has been suggested. Some test algorithms have been developed to implement this approach. Experiments were also made to verify the advantage and disadvantage of the anticipation approach in solving DTPs, to illustrate the impact of the prediction-deception property on algorithm performance, and to evaluate the efficiency of our proposed approach in solving prediction-deceptive time-linkage problems. To test problems were also proposed in this chapter to simulate the new characteristics of DTPs.

Although the experiments (and the algorithms + test problems) in this chapter are over-simplified, and the advantages of a predictor/ predictor+knowledge are expected, such simplifications are necessary to prove the principle and to show the potentiality of EAs because this research is just a beginning step and is the first EDO study in this topic. To the best of our knowledge, previously this class of problems has not been taken into account in existing academic EDO research despite their popularity in real-world scenarios.

For future works we plan to do more experiments on more realistic scenarios with a more powerful predictor integrated with state-of-the-art EAs. Especially, more research will be carried out to investigate the situation where the algorithm needs to determine multiple switching times during the optimization process. The possibility of combining time-linkage handling techniques with normal environmental dynamic handling techniques will also be investigated and comparisons between the new predictive algorithms and existing DO algorithms will also be carried out.

Acknowledgements. The authors are particularly grateful to P. Rohlfshagen, J. Rowe, P. Bosman and Y. Jin for their helpful comments. We would also like to thank S. Yang, T. Ray, C. Li, and L. Xing for their fruitful discussions. This work was partially supported by an EPSRC grant (No. EP/E058884/1) on "Evolutionary Algorithms for Dynamic optimization Problems: Design, Analysis & Applications", an ORS Award and a School of Computer Science PhD Studentship.

References

[1] Ahmad, A.Z., Liu, K.-Z.: A new model predictive control approach to dc-dc converters based on combinatory optimization. In: Proceedings - 34th Annual Conference of the IEEE Industrial Electronics Society, IECON 2008, Orlando, FL, United states, pp. 460–465 (2008)

[2] Akanle, O.M., Zhang, D.Z.: Agent-based model for optimising supply-chain configurations. International Journal of Production Economics 115(2), 444–460 (2008)

[3] Aragon, V.S., Esquivel, S.C.: An evolutionary algorithm to track changes of optimum value locations in dynamic environments. Journal of Computer Science and Technology 4(3), 127–134 (2004)
[4] Jason, A.D., Atkin, E.K., Burke, J.S.: Greenwood, and Dale Reeson. On-line decision support for take-off runway scheduling with uncertain taxi times at london heathrow airport. Journal of Scheduling 11(5), 323–346 (2008)
[5] Bäck, T.: On the behavior of evolutionary algorithms in dynamic environments. In: IEEE International Conference on Evolutionary Computation, pp. 446–451. IEEE (1998)
[6] Bosman, P.A.N.: Learning, anticipation and time-deception in evolutionary online dynamic optimization. In: Yang, S., Branke, J. (eds.) GECCO Workshop on Evolutionary Algorithms for Dynamic Optimization (2005)
[7] Bosman, P.A.N.: Learning and anticipation in online dynamic optimization. In: Yang, S., Ong, Y.-S., Jin, Y. (eds.) Evolutionary Computation in Dynamic and Uncertain Environments. SCI, vol. 51, pp. 129–152. Springer (2007)
[8] Bosman, P.A.N., Poutré, H.L.: Learning and anticipation in online dynamic optimization with evolutionary algorithms: the stochastic case. In: GECCO 2007: Proceedings of the 9th Annual Conference on Genetic and Evolutionary Computation, pp. 1165–1172. ACM, New York (2007)
[9] Branke, J., Mattfeld, D.: Anticipation and flexibility in dynamic scheduling. International Journal of Production Research 43(15), 3103–3129 (2005)
[10] Dreo, J., Siarry, P.: An ant colony algorithm aimed at dynamic continuous optimization. Applied Mathematics and Computation 181(1), 457–467 (2006)
[11] Fiacchini, M., Alamo, T., Alvarado, I., Camacho, E.F.: Safety verification and adaptive model predictive control of the hybrid dynamics of a fuel cell system. International Journal of Adaptive Control and Signal Processing 22(3), 142–160 (2008)
[12] Houwing, M., Negenborn, R.R., Heijnen, P.W., De Schutter, B., Hellendoorn, H.: Least-cost model predictive control of residential energy resources when applying μCHP. In: Proceedings of the Power Tech 2007 Conference, Lausanne, Switzerland, Paper 291 (July 2007)
[13] Jin, N., Termansen, M., Hubacek, K., Holden, J., Kirkby, M.: Adaptive farming strategies for dynamic economic environment. In: Proceedings of the IEEE Congress on Evolutionary Computation CEC 2007, pp. 1213–1220 (2007)
[14] Kanoh, H.: Dynamic route planning for car navigation systems using virus genetic algorithms. International Journal of Knowledge-based and Intelligent Engineering Systems 11(1), 65–78 (2007)
[15] Long, C.E., Polisetty, P.K., Gatzke, E.P.: Deterministic global optimization for nonlinear model predictive control of hybrid dynamic systems. International Journal of Robust and Nonlinear Control 17(13), 1232–1250 (2007)
[16] Morimoto, T., Ouchi, Y., Shimizu, M., Baloch, M.S.: Dynamic optimization of watering satsuma mandarin using neural networks and genetic algorithms. Agricultural Water Management 93(1-2), 1–10 (2007)
[17] Moser, I., Hendtlass, T.: Solving dynamic single-runway aircraft landing problems with extremal optimisation. In: IEEE Symposium on Computational Intelligence in Scheduling (2007)
[18] Ngo, S.H., Jiang, X., Le, V.T., Horiguchi, S.: Ant-based survivable routing in dynamic wdm networks with shared backup paths. The Journal of Supercomputing 36(3), 297–307 (2006)

[19] Nguyen, T.T.: Continuous Dynamic Optimisation Using Evolutionary Algorithms. PhD thesis, School of Computer Science, University of Birmingham (2011), http://etheses.bham.ac.uk/1296, http://www.cs.bham.ac.uk/txn/theses/phd_thesis_nguyen.pdf
[20] Nguyen, T.T., Yao, X.: Dynamic Time-Linkage Problems Revisited. In: Giacobini, M., Brabazon, A., Cagnoni, S., Di Caro, G.A., Ekárt, A., Esparcia-Alcázar, A.I., Farooq, M., Fink, A., Machado, P. (eds.) EvoWorkshops 2009. LNCS, vol. 5484, pp. 735–744. Springer, Heidelberg (2009)
[21] Rocha, M., Neves, J., Veloso, A.: Evolutionary algorithms for static and dynamic optimization of fed-batch fermentation processes. In: Ribeiro, B., et al. (eds.) Adaptive and Natural Computing Algorithms, pp. 288–291. Springer (2005)
[22] Rohlfshagen, P., Yao, X.: Attributes of Dynamic Combinatorial Optimisation. In: Li, X., Kirley, M., Zhang, M., Green, D., Ciesielski, V., Abbass, H.A., Michalewicz, Z., Hendtlass, T., Deb, K., Tan, K.C., Branke, J., Shi, Y. (eds.) SEAL 2008. LNCS, vol. 5361, pp. 442–451. Springer, Heidelberg (2008)
[23] Rohlfshagen, P., Yao, X.: On the role of modularity in evolutionary dynamic optimisation. In: Proceedings of the 2010 IEEE Wolrd Congress on Computational Intelligence, WCCI 2010, Spain, pp. 3539–3546 (2010)
[24] Sonntag, C., Su, W., Stursberg, O., Engell, S.: Optimized start-up control of an industrial-scale evaporation system with hybrid dynamics. Control Engineering Practice 16(8), 976–990 (2008)
[25] Summers, S., Bewley, T.R.: Mpdopt: A versatile toolbox for adjoint-based model predictive control of smooth and switched nonlinear dynamic systems. In: Proceedings of the 46th IEEE Conference on Decision and Control 2007, pp. 4785–4790 (2007)
[26] Tafazoli, S., Sun, X.: Hybrid system state tracking and fault detection using particle filters. IEEE Transactions on Control Systems Technology 14(6), 1078–1087 (2006)
[27] Tfaili, W., Dréo, J., Siarry, P.: Fitting of an ant colony approach to dynamic optimization through a new set of test functions. International Journal of Computational Intelligence Research 3, 205–218 (2007)
[28] Ursem, R.K., Krink, T., Jensen, M.T., Michalewicz, Z.: Analysis and modeling of control tasks in dynamic systems. IEEE Transactions on Evolutionary Computation 6(4), 378–389 (2002)
[29] Wang, J., Tao, X., Cho, H.: Microassembly of micro peg and hole using an optimal visual proportional differential controller. Proceedings of the Institution of Mechanical Engineers, Part B (Journal of Engineering Manufacture) 222(B9), 1171–1180 (2008)
[30] Wang, N., Ho, K.-H., Pavlou, G.: Adaptive Multi-topology IGP Based Traffic Engineering with Near-Optimal Network Performance. In: Das, A., Pung, H.K., Lee, F.B.S., Wong, L.W.C. (eds.) NETWORKING 2008. LNCS, vol. 4982, pp. 654–666. Springer, Heidelberg (2008)
[31] Weicker, K.: An Analysis of Dynamic Severity and Population Size. In: Deb, K., et al. (eds.) PPSN 2000. LNCS, vol. 1917, pp. 159–168. Springer, Heidelberg (2000)
[32] Weicker, K.: Evolutionary algorithms and dynamic optimization problems. Der Andere Verlag (2003)
[33] Woldesenbet, Y.G., Yen, G.G.: Dynamic evolutionary algorithm with variable relocation. IEEE Transactions on Evolutionary Computation 13(3), 500–513 (2009)

Index